Accountability, Social Responsibility and Sustainability

Accounting for Society and the Environment

PEARSON

At Pearson, we have a simple mission: to help people make
more of their lives through learning

We combine innovative
learning technology with trusted content and
educational expertise to provide engaging and effective
learning experiences that serve people wherever and
whenever they are learning.

From classroom to boardroom, our curriculum materials,
digital learning tools and testing programmes
help to educate millions of
people worldwide – more than any other private enterprise.

Every day our work helps learning flourish,
and wherever learning flourishes, so do people.

To learn more please visit us at www.pearson.com/uk

Accountability, Social Responsibility and Sustainability

Accounting for Society and the Environment

Rob Gray, Carol A. Adams and Dave Owen

Harlow, England • London • New York • Boston • San Francisco • Toronto • Sydney • Auckland • Singapore • Hong Kong
Tokyo • Seoul • Taipei • New Delhi • Cape Town • São Paulo • Mexico City • Madrid • Amsterdam • Munich • Paris • Milan

Pearson Education Limited
Edinburgh Gate
Harlow CM20 2JE
United Kingdom
Tel: +44 (0)1279 623623
Web: www.pearson.com/uk

First published 2014 (print)

ISBN: 978-0-273-68138-0 (print)
       978-0-273-77798-4 (PDF)
       978-0-273-77797-7 (eText)

British Library Cataloguing-in-Publication Data
A catalogue record for the print edition is available from the British Library

Library of Congress Cataloging-in-Publication Data
A catalog record for the print edition is available from the Library of Congress
Gray, Rob.
  Accountability, social responsibility, and sustainability : accounting for society and the environment / Rob Gray, Carol A. Adams, and Dave Owen.
    pages cm
  ISBN 978-0-273-68138-0
  1. Social accounting.   2. Environmental auditing.   3. Social responsibility of business.   I. Title.
  HD60.G71 2014
  658.4'08—dc23

                                                                                        2013040064

10 9 8 7 6 5 4 3 2 1
18 17 16 15 14

Print edition typeset in 10/12 pt Ehrhardt MT Std by 75
Print edition printed and bound by Ashford Colour Press Ltd, Gosport

NOTE THAT ANY PAGE CROSS REFERENCES REFER TO THE PRINT EDITION

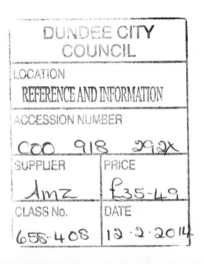

Brief contents

Contents

Preface

Accountability, Social Responsibility and Sustainability has been a long time in preparation. The present text represents a complete rewrite and development from *Accounting and Accountability* which we published in 1996.[1] In *Accounting and Accountability* we (somewhat ambitiously) sought to articulate the whole field of social (and environmental and sustainability) accounting (and auditing and reporting) as we then understood it. We hoped that such a text might help teachers teach social accounting, that it might help students study social accounting and that it might provide a helpful platform for new researchers in this emerging field. To some degree at least we probably succeeded in these ambitions and it is with some (albeit qualified) pride that we note that the text became (as far we can tell) the most widely cited source in the field. In many regards, the text has stood up well to the test of time, but the last two decades have seen so many changes that even its fond parents have had to recognise that the book was becoming really rather long in the tooth. It is not just that there has been a range of theoretical and empirical developments in social accounting and related fields, nor that it has become increasingly obvious that there are important parts of the field that we either missed or skated over but, perhaps most importantly, the political, social and economic contours of the world look to have changed beyond recognition – taking the worlds of education and scholarship with them. Oh, how we wish that this really was the case!

On the face of it, the world has made enormous strides towards a recognition of the crucial interactions of social, environmental and sustainability concerns with the worlds of business, finance and accounting. Accountants, businesses, financial markets, politicians and universities all apparently embrace sustainability with zeal. Recycling is perhaps now a fact of life and climate change appears to be largely taken for granted. There is widespread recognition that economics and wealth are not the sole determinants of happiness or well-being and there have been truly startling advances in the efficiencies with which manufacturing and services employ environmental resources. Waste reduction is no longer thought of as contentious whilst global initiatives for matters as diverse as corporate reporting, the literacy of peoples, drought and biodiversity are everywhere. One might be forgiven for thinking that social and environmental accounting and management are now so much part of the mainstream that recognition of something identifiably 'social accounting' is rapidly becoming something of an anachronism.

It is a great deal more complex than that – and a great deal more complex than we formerly recognised when we wrote *Accounting and Accountability*. On the one hand, there are these astonishing strides forward that we need to recognise and integrate into our

---

[1]Which itself was a total rewrite and development of *Corporate Social Reporting* published in 1987. Many of the comments in this preface are made from the perspective of having worked in this field for 30+ years.

understandings of social, environmental and sustainability accounting. There is, genuinely, an enormous amount of good news concerning social and environmental initiatives that we can celebrate and study. This good news, to varying degrees, has either been enthusiastically embraced by conventional businesses and other organisations or has clear application to them.

But there is also a really extremely disturbing dark side to all of this. Alongside all this good news, the environmental state of the planet, the levels of inequality between peoples, the numbers of people in poverty or children dying through drought continue to get seriously worse. Despite the exceptional steps forward made by environmental management, environmental accounting and voluntary reporting, the accountability of organisations is no better and perhaps, under the veil of all the good news, is actually getting worse.

It is this recognition of the centrality of *conflict*: between good news and bad news; between the haves and have nots; between cleaner rivers and the loss of biodiversity; between cleaner technology and increased pollution; between increasing awareness of sustainability and declining life-support systems; that represents the core *motif* informing our comprehensive rewriting of the book.

And there is one further motivation which underlies a lot of what follows. Our principal audience has been, and remains, teachers, students and researchers. These are the people with the time and capacity to consider newer and more challenging ideas, to look at things in new and unconventional ways and to come up with new solutions to increasingly urgent problems. The growth in the teaching of corporate social responsibility (CSR), sustainability, environmental management and social accounting has been significant over the last 20 years. The growth in the research community committed to these issues globally has also been astonishing. But this apparently encouraging trend has occurred simultaneously with an increased commodification of both students and universities as well as a deeply pernicious constraining and narrowing of what it means to be an academic. So rather than an increasing cohort of informed intelligent and able people with a desire for change, we fear that society is encouraging the formation of an increasingly informed cohort of intelligent people who see little further than the next grade mark, the next job or the next journal article. This may be an overly pessimistic view and perhaps we mis-read the causes of (what we see as) the most educated members of society becoming less radicalised and less politically and socially active. However, if social and environmental accounting and sustainability require pretty drastic insights, ideas and initiatives (as we believe they do) our fear is that such initiatives look less and less likely to emerge through education and research. That is a very gloomy conclusion indeed – and we can only hope that it is incorrect. This text is written as part of our attempt to re-open the challenging, even scary, implications of considering the possibility of a fairer society with truly sustainable sensibilities: a society and a process that would be supported by an accounting, management and reporting system that is authentically sensitive to humanity and nature. Whether we succeed at all in this is quite another matter of course.

The text of *Accountability, Social Responsibility and Sustainability* differs from its predecessor in a number of observable ways. At a general level, we have made a number of changes of orientation in addition to the changed emphasis arising from our comments above. The text, whilst still predominantly an accounting-based text, has been written from a wider management and organisational perspective. This will be apparent in a range of places but especially where we try both to give a context to different issues we address and to recognise both managers' and society's views in our discussions. In addition, both CSR and sustainability are given more attention and (hopefully) are treated in more nuanced ways. The final broad change probably lies in the recognition that the field of social, environmental and sustainability accounting, reporting and management now possesses a quite enormous literature. We have done our best to digest much of this and to make wide reference to the literature for those wanting to follow issues further. Equally, though, where other easily accessible sources do the work for us, we have not sought to duplicate that effort. There are lots of wheels which no longer need inventing.

The structure of the text is very loosely similar to *Accounting and Accountability* in that we start with theoretical reflections, then move onto areas of practice before looking forward to possibilities for the future. The present text takes four chapters to lay down some of theoretical bases of social accounting and draws its palate very widely. There is less emphasis on history (which *Accounting and Accountability* covered in some depth) and somewhat more on reflection and analysis. The initial empirical chapters are organised, as might be expected, into chapters on community and society, employees and unions and environmental issues. And, as might be expected, there is a thorough exploration of the 'external social audits'. However, there are new chapters which explore: finance and financial markets; the whole controversy of the 'triple bottom line' and sustainability; the crucial emerging challenges of governance and attestation; and one chapter which tries to open up the sorely under-examined areas of social accounting for non-profit and other types of organisation. The final sections offer our own hostages to fortune and show how far innovative research and practice *have* managed to come by outlining how an organisation which *really* wanted to account for social, environmental and sustainability issues might go about it. Needless to say, no organisation anywhere in the world (as far as we know) comes close to this ideal.

We close this preface with a suggestion – actually probably more like a warning. This suggestion relates to how we understand the broad intellectual field of social accounting as one which is practicable but often ignored by practice; as one which is sufficiently theoretically coherent to offer a challenge to piecemeal pragmatism but is sufficiently practically orientated to draw telling (if abstract) critiques from the more penetrating theorists of academe. This sounds a bit obscure; what do we mean? The academic field of social accounting – or at least that field as we have represented it here – includes a wide diversity of issues and approaches from the explicitly practical (e.g. costing of energy) through the innovative and radical (accounting for the un-sustainability of large business and financial markets) through to some deeply challenging questioning of humankind's fundamental interaction with its own species, with other species and with the planet. Whilst there are important ways in which these differing approaches can be complementary there are also major – and *very important* – tensions and conflicts between these different strands. The considerable range of initiatives from business are, of course, of a predominantly practical nature but, *importantly,* are rarely (if ever) theoretically coherent or designed to challenge the status quo or develop real accountability. If sustainability requires major change, it is thus very unlikely that business-led initiatives (at least alone) will be effective. Equally, whilst the theoretical challenges of social accounting – whether they be from traditional critical theory or from the perspective of post-modernity – are often neither obviously practical nor practicable, this does not mean that such critiques are not justified nor that they do not deserve the very serious attention of anyone with a real concern for the future of people and the planet. For us, social accounting is constantly challenged by the need to navigate between these extremes: offering theoretically coherent solutions of a practicable nature and resisting the twin sirens of exquisite theory or immediate practicality. In a sense, this becomes some kind of a commitment to *pragmatism,* in which theory alone will not solve the problems but, equally, recognising that allowing current orthodoxy and business practice to determine what is 'practical' is a certain recipe for disaster. These tensions ensure that the study of and research into social accounting can never be a comfortable or straightforward endeavour.

Acknowledgements

After 30+ years in this emerging and challenging field there are so many people to whom we owe a debt of one sort or another. We could not possibly mention them all, so to all our families, close friends and colleagues, to all our students and to all those who have supported, challenged and developed this 'social accounting project' may we simply say, thank you. Working with these issues is both harrowing and frustrating – it is the company of dedicated companions of good heart that make it all worthwhile.

Blurb + Short Bios

The late 20th and early 21st centuries can perhaps be typified by lurches from crisis to crisis – economic crises, social crises, environmental crisis and political crises. As the world becomes more populated and apparently more wealthy it is also becoming more unequal, possibly more unstable and certainly more destructive of its natural environment. Making any sense of this complexity and the life-threatening effects of un-sustainability is perhaps the single biggest challenge for all of society. But crucial to any such understanding is a realistic appreciation of the central role(s) played by organisations, businesses, managers, finance, financial markets and, inevitably, accounting and accountability in how humanity manages its relationships between its members and navigates its relationships with the planet and with other species. *Accountability, Social Responsibility and Sustainability* is one attempt to address the broad and complicated interactions between organisational life, civil society, markets, inequality and environmental degradation through the lens(es) of accounting, accountability, responsibility and sustainability. Placing the way in which organisations are controlled and the metrics by which they are run at the heart of the analysis, the text explores how current ways of managing organisations and measuring their success is antithetical to the very concerns of societal well-being and environmental stewardship that are the *sine qua non* of any civilised society. Alternative ways of measuring and managing are explored and the key motifs of conflict and accountability are offered as essential components of a more civilised economic realm.

The text starts from the point that it is increasingly urgent for all organisations to face – honestly – what environmental management, CSR and sustainability can do for (and to) organisations and, most importantly, what they cannot do. Simply talking about CSR and sustainability is not enough and only when the overwhelming waves of rhetoric that clutter up the whole CSR and environmental debates around business and finance are grounded in sensible and realistic systems of representation and accountability will humanity start to make any serious progress on any alternative to its current headlong flight towards gross un-sustainability.

Accountability, Social Responsibility and Sustainability is a very substantial revision and redevelopment of the earlier seminal texts *Corporate Social Reporting* (published in 1987) and *Accounting and Accountability* (published in 1996). This text offers a deeper and more nuanced guidance on theory and recognises the crucial role played by the very act of framing how we as scholars and practitioners approach the central tensions between the economic, the social and the environmental. The theory is extensively supported by review and analysis of developments in practice as well as a critical assessment of the extensive range of realistic and important possibilities to which politics and practice continues to be resistant.

Accountability, Social Responsibility and Sustainability is written for all scholars, students, teachers, practitioners, researchers and policy-makers who recognise the central role accounting, finance, accountability, CSR and sustainability play in the future of society and the planet.

Rob Gray is Professor of Social and Environmental Accounting at the University of St Andrews. He was Director of the Centre for Social and Environmental Accounting Research (CSEAR) from its inception in 1991 until 2012.

Carol A. Adams is Professor at the Monash Sustainability Institute, Monash University and a member of the Global Reporting Initiative Stakeholder Council. She is founding editor of the *Sustainability Accounting, Management and Policy Journal*.

Dave Owen is an Emeritus Professor at the International Centre for Corporate Social Responsibility (Nottingham University Business School). He is also an Honorary Fellow of the Centre for Social and Environmental Accounting Research (University of St Andrews).

About the authors

Rob Gray is a qualified chartered and chartered certified accountant and is the author/co-author of over 300 articles, chapters, monographs and books – mainly on social and environmental accounting, sustainability, social responsibility and education. He founded CSEAR (The Centre for Social and Environmental Accounting Research) in 1991 and for 21 years was its Director. He serves on the editorial boards of 15 learned journals and travels extensively overseas in response to academic and professional invitations. He has worked with a wide range of international and local commercial and non-commercial organisations including collaborations with the United Nations from time to time. In 2001, he was elected the British Accounting Association Distinguished Academic Fellow and in 2004 became one of the 14 founding members of the British Accounting Association Hall of Fame. He was awarded an MBE in the Queen's 2009 Birthday Honours List and was elected to the Academy of Social Science in 2012.

Carol A. Adams is a Chartered Accountant (ICAS), Professor at the Monash Sustainability Institute, Monash University and Visiting Professor at the Adam Smith Business School, University of Glasgow. Her research in corporate financial and sustainability reporting and change management to integrate sustainability has been cited over 4,000 times. She has played an active role in global corporate reporting initiatives including the development of the AA1000 Standards and UN Global Compact guidelines for universities and is currently a member of the Stakeholder Council of the Global Reporting Initiative and a member of the Capitals Technical Collaboration Group for the International Integrated Reporting Council (IIRC). She provides advice to international organisations on mainstreaming sustainability, has led the development of international award winning sustainability reports, management and governance processes and writes on topical issues on www.drcaroladams.net. She is founding editor of the *Sustainability Accounting, Management and Policy Journal*.

Dave Owen retired as Professor of Social and Environmental Accounting at the International Centre for Corporate Social Responsibility (Nottingham University Business School) in September 2010. He is now an Emeritus Professor at the Centre and continues to pursue his research interests in the field of social and environmental accounting, auditing and reporting. Dave has published extensively in a wide range of professional and academic journals on topics such as social and environmental accounting education, social investment, corporate social audit and corporate social and environmental reporting and assurance practice. Prior to his retirement, Dave served on the editorial boards of *Accounting, Auditing and Accountability Journal*, *Accounting and Business Research*, *Accounting Forum*, *Accounting, Organizations and Society*, *British Accounting Review*, *European Accounting Review* and *Business Strategy and the Environment*.

Publisher's Acknowledgments

We are grateful to the following for permission to reproduce copyright material:

Figures

Figure 1.1 from The Living Planet Report 2008, Godalming: World Wide Fund for Nature; Figure 4.3 from Internal organisation factors influencing corporate social and ethical reporting, *Accounting, Auditing and Accountability Journal*, 15(2), pp. 223–50 (Adams, C. 2002), Bingley: Emerald Group Publishing; Figure 5.5 from Giving USA, A report of the American Association of Fundraising Counsels (2010), Wellfleet, MA: US National Parks Service; Figures 6.4, 6.5 from The changing portrayal of the employment of women in British banks' and retail companies' corporate annual reports, *Accounting Organizations and Society*, 23, pp. 781–812 (Adams, C.A. and Harte, G. 1998), Oxford: Elsevier Ltd; Figure 8.1 from Financial Systems in Europe, the USA, and Asia, *Oxford Review of Economic Policy*, 20, pp. 490–508 (Allen, F., Chui, M. and Maddaloni, A. 2004), Oxford: Oxford University Press; Figure 12.2 from http://my.clevelandclinic.org/Documents/About-Cleveland-Clinic/overview/CC_UNreport_2012.pdf, Lyndhurst, OH: The Cleveland Clinic Foundation; Figures 12.5, 12.6 from http://www.latrobe.edu.au/sustainability/documents/4906_Creating_Futures_Web.pdf, Melbourne: La Trobe University; Figure 12.7 from Christian Aid Annual Report 2011-12, http://www.christianaid.org.uk/images/2011-2012-annual-report.pdf, London: Christian Aid; Figure 12.8 from http://www.socialauditnetwork.org.uk/files/7613/4633/2646/Stop_Violence_NZ_Social_Accounts_Summary_-_2004.pdf, Nelson, New Zealand: Stopping Violence Services Nelson; Figure 13.1 from Carbon accounting for sustainability and management: Status quo and challenges, *Journal of Cleaner Production*, 36(1), pp. 1–16 (Schaltegger, S. and Csutora, M. 2012), Oxford: Elsevier Ltd. Reproduced with permission.

Text

p. 20 quotation from Sir Winston Churchill 1947, London: Curtis Brown; p. 77 quotation from George Santayana, *Life of Reason, Reason in Common Sense*. Scribner's, 1905, p. 284 Critical Edition, MTT Press, 2011, p. 172; p. 185 quotation from *Financing Change: The financial community, eco-efficiency and sustainable development*, (Schmidheiny, S. and Zorraquin, F.J. 1996) p. xxi, Cambridge, MA: MIT Press; p. 287 quotation from

Multilaterism and building stronger international institutions (Woods, N.) in *Global accountabilities: Participation, pluralism and public ethics* (Woods, N./Ebrahim, A. and Weisband, E. (eds) 2007) pp. 27-44, Cambridge: Cambridge University Press. Reproduced with permission.

In some instances we have been unable to trace the owners of copyright material, and we would appreciate any information that would enable us to do so.

CHAPTER 1

Introduction, issues and context

1.1 Introduction

Planet Earth in the 21st century is a bewildering, complex place. Human beings, or at least the more reflective members of that species, have long been bewildered by – and tried to make sense of – their world. Sense-making and dealing with such bewilderment comes in many forms. Ignoring the issues – whether by keeping such a narrow focus on the world that big issues are excluded from view or by hoping that they might just go away – is probably the most common strategy. However, sense-making of a more constructive kind seems to draw on combinations of religion, reason and mythology coupled with an appealing tendency to impose order where none actually exists. Despite this apparent theme in human history, it seems unlikely that bewilderment was ever so all-embracing. Whilst some of us live in a near-paradise[1] – and our place in paradise is rarely the direct result of our own efforts and achievements – at least 25%, by most estimates, of fellow members of the species live in hell.[2] For some countries of the world, shopping for branded luxuries is, quite bizarrely, considered to be the most sought after of pastimes, an activity representing the very height of personal achievement. In some countries, having enough water to drink is the epitome of paradise whilst in other countries, time spent with family or sharing a meal is the lynchpin of what it is to be alive. The material well-being of a planetary elite has probably never been so high; the inequality of access to material goods and material well-being across the globe has probably never been as great; the trading and business system has never promised, and indeed delivered, so much (not always of the same things); opposition to this nirvana has probably never been so widespread. It is difficult to know for sure, but it is probable that never have so many people died every hour from a lack of water and basic food and amenities. Oh, and by the way, as far as we can tell on the best available evidence, humanity is probably killing the planet and causing irreversible decline in its sustainability. This is almost certainly a 'first'.

This is all part of a seeming barrage of both 'good news' and 'bad news' about the conditions of human existence that we seem to receive from governments, newspapers,

---

[1]A personal statement here might be appropriate in order to recognise the immense good fortune many of us experience in having water to drink, fresh air to breathe, enough food and clothing, largely a freedom from personal violence and, for many of us, quite fabulously beautiful places to live, work, walk and meet friends and family. Life may well not be perfect – we are human after all – but compared to the millions of the less fortunate, it behoves us to recognise our largely undeserved privilege.

[2]Poverty, drought and violence are all experiences nobody would wish. Poverty is notoriously difficult to define but, whilst the number of people living on less that $1 per day has fallen drastically in recent decades and halved towards the end of the 20th century, there are still a quarter of people in this state and maybe as much as 40% still living on less than $2 per day. More detail can be found through discussions in and around the UN's Human Development Index and the UN's Millennium Development Goals.

researchers, businesses, films, etc. To make any sense of it all, it is likely that we must see the 'good news' and the 'bad news' as, to a degree at least, two sides of the same coin. Catastrophic oil spills, destruction of habitat, famine and abject poverty, involuntary unemployment, destruction of the ozone layer, industrial conflict, stock market collapse, major fraud and insider trading, stress-related illness, violence, acid rain and exploitation are all major nega-tive shocks to individuals, communities, nations and whole species of life. But rather than being isolated and unrelated phenomena they are, to a large extent, closely connected. They are the increasingly high price that the world pays for its 'good news'. The medical break-throughs and the level of health care, the rising material standards of living and the increased life expectancy of a proportion of mankind, rising gross national product (GNP) and profit levels, the technological advances, the increased travel opportunities, the rising quality of privilege and perhaps even freedom and stability experienced by many in the West are not unrelated or costless successes. Each economic or social 'advance' is won by an exceptionally successful business and economic system – but at a price. That 'price' is what economists refer to as the **externalities** – the consequences of economic activity which are not reflected in the costs borne by the individual or organisation enjoying the benefits of the activity.

And yet, it is perhaps surprising how rarely the 'good news' and the 'bad news' are actively connected up. The business press celebrates the growing profits measured by con-ventional accounting; financial markets celebrate increasing share prices and returns to investors; business advertising conjures visions of new and better worlds through increasing consumption; governments continue to listen to the blandishments of business about 'unnecessary' regulation (or red tape as it is typically pejoratively called); and, as we shall see in some detail, leading organisations – especially multi-national companies – go to tremen-dous lengths to show us the positive and almost exclusively benign impacts of their leading edge management and careful stewardship. At the same time, elements of the media, non-governmental organisations (NGOs) and civil society organisations parade catastrophes before us – the perfidy of big business; the desperation of Africa; the plight of the oppressed and the homeless; the ruthlessness of mineral extraction; the desecration of virgin wilder-ness and the collapse of eco-systems.

What we want to see is this all 'joined up'. What we believe that society needs to under-stand is the implacable connection between the good and the bad news: the extent to which this year's reported profit was bought at the cost of increased environmental footprint; the extent to which it was exploitation of child labour that allowed me to buy my trainers so cheaply; the extent to which my pension fund is dependent upon sales of weapons to oppres-sive regimes; the extent to which I contribute to climate change and pollution through my preferences for private car transport and air-conditioning . . . and so on.

Now, whilst it is far from immediately obvious why it should be something we are going to call 'social accounting' – or indeed anything connected with accounting at all – that will help us tease out, examine and perhaps ameliorate the negative aspects of modern day life, stay with us. As we start to demonstrate the links between successful business performance and sustainability, as we explore corporate social responsibility and as we try to show you the centrality of accountability to any future civilised society, the role of accounting and the potential of *social* accounting should become apparent. At its very best, *social accounting* can reveal the conflicts, difficulties, inextricable externalities and potential solutions that advanced 21st century international financial capitalism must face up to. It is these sorts of issues and connections that this book will try to justify, explain, examine and then demonstrate.

This chapter is principally concerned with providing the beginnings of the theoretical basis which sets the scene for the rest of the book. In the following sections, we first provide an examination of what is meant by **'social accounting'** (and its myriad synonyms) and then explain why the subject is of crucial importance. We then outline a (necessarily brief) introduction to some of the key elements: sustainability, the state of the world, the nature of the state and civil society and so on. The chapter concludes with an explanation of how this new text is structured.

1.2 What is social accounting?

Social accounting is simultaneously three things: (i) a fairly straightforward manifestation of corporate efforts to legitimate, explain and justify their activities; (ii) an ethically desirable component of any well-functioning democracy and, (iii) just possibly, one of the few available mechanisms to address sustainability that does not involve fascism and/or extinction of the species. This might seem like an unusual introduction to a subject. That is because the subject is unusual.

First of all, 'social accounting' gets called all sorts of names.[3] As it is not enshrined in law, the terminology remains fluid. One will see it labelled as: social accounting; social disclosure; social reporting; social and/or environmental and/or sustainability accounting; social responsibility disclosure; social, environmental and ethical reporting; and any number of combinations of these terms plus other synonyms. To a large extent we shall use the terms interchangeably throughout with 'social accounting' being the most generic term. However, there will be times when we specifically wish to talk about accounting (like management accounting) rather than reporting and when we want to discuss the natural 'environment' specifically. This should be obvious from the context.[4]

Regardless of what we call it, we are concerned with the production of 'accounts' (i.e. 'stories' if you prefer) concerning (typically) organisations' interactions with society and the natural environment.

Gray et al. (1987) defined corporate social reporting as:

> . . . the process of communicating the social and environmental effects of organisations' economic actions to particular interest groups within society and to society at large. As such, it involves extending the accountability of organisations (particularly companies), beyond the traditional role of providing a financial account to the owners of capital, in particular, shareholders. Such an extension is predicated upon the assumption that companies do have wider responsibilities than simply to make money for their shareholders.
>
> (Gray et al., 1987: ix)

Like all definitions, this needs more work. There are, for example, important aspects of social accounting which remain internal to organisations as they seek ways in which they might better understand the social, environmental and, indeed, sustainability impacts of their activities. However, the definition will serve as a starting point. For comparison a related, and slightly later definition might be:

> . . . the preparation and publication of an account about an organisation's social, environmental, employee, community, customer and other stakeholder interactions and

---

[3]Note that 'social accounting' is a term also used by economists to refer to national income accounting – i.e. the way in which gross domestic and gross national product are calculated. This sense of the term is quite different for the organisational accounting we are concerned with here.

[4]Elsewhere, you will also see reference to social audit and/or non-financial reporting. These are much more problematic terms and to be avoided unless they are referring respectively and explicitly to (what we shall call) 'external social audits' and all reporting other than traditional financial accounting.

activities and, where, possible, the consequences of those interactions and activities. The social account may contain financial information but is more likely to be a combination of quantified non-financial information and descriptive, non-quantified information. The social account may serve a number of purposes but discharge of the organisation's accountability to its stakeholders must be the clearly dominant of those reasons and the basis upon which the social account is judged.

(Gray, 2000: 250)

Social accounting can take a potentially infinite range of forms. It can be designed to fulfil any one or more of a wide range of objectives. It can cover a myriad of different subjects, and social accounts can be constructed around almost any type of information or with almost any sort of focus. Social accounting is not a systematic, regulated or well-established activity and so what is covered in the following chapters is limited (in description) only by practice and (in prescription) only by our imaginations. Many of the principal examples *from* practice and the better known suggestions *for* practice are reviewed in this book, although there are many sources through which you can gain familiarity with practice.[5]

Some idea of the relationship between conventional and social accounting and of the extent and potential limits of social accounting may be useful to begin with. This is principally because a great deal of social accounting thinking, research and practice derives from conventional accounting itself. Indeed, it is possible to say that social accounting might be thought of as concerned with:

- the social and environmental (including sustainability) impacts and effects arising from conventional accounting practice;
- ameliorating the social and environmental impacts arising from conventional accounting pratice (including seeking ways to reduce the negative impacts and looking for ways to encourage positive social and environmental effects);
- deriving and developing new methods of accounting that might be implicated in more benign social and environmental effects and which, typically, would advance the case of accountability.

At its most basic, there are four necessary, although not sufficient, characteristics which define conventional western accounting practice (see Bebbington *et al.*, 2001). These four characteristics delineate the world which accountants perceive and lead to conventional accounts being restricted to:

1 the financial description;
2 of specified (priced) economic events;
3 related to defined organisations or accounting entities;
4 to provide information for specified users of that information.

The conventional accounting system effectively creates and then reinforces this profoundly narrow image of all possible interactions between the 'world' and the organisation. In doing this, conventional accounting thus stands as a political and social process in that it makes choices about the world; emphasises certain things and privileges or ignores others, thereby creating, to all intents and purposes, its own social reality (Gambling, 1977; Cooper and Sherer, 1984; Hines, 1988, 1989, 1991).

---

[5]Chapter 4 will, in fact, formally encourage you to actively garner and consult reports. Also, should you wish to look at reports, consider spending time on websites such as the Global Reporting Initiative (www.globalreporting.org/) or Corporate Register (www.corporateregister.com/). Consultation of the CSEAR website at www.st-andrews. ac.uk/~csearweb/ should also be helpful.

In broad terms, social accounting is, at a minimum, an *addendum* to the world created by conventional accounting or, more typically, it offers the prospect of a significantly different (and therefore challenging) view of the world. Social accounting research approaches this challenge by seeking to contest the propriety of the four characteristics of conventional accounting.[6] More specifically, social accounting is about some combination of:

- accounting for different things (i.e. not accounting only for economic events);
- accounting in different media (i.e. not only accounting in strictly financial terms);
- accounting to different individuals or groups (i.e. not only accounting to the providers of finance); and,
- accounting for different purposes (i.e. accounting for a range of purposes and not only to enable the making of decisions whose success would be judged in financial or even only cash flow terms).

Thus we might consider traditional financial accounting as a significantly and artificially constrained set of all accountings. Traditional (financial) accounting is only one particularly narrow form of the whole universe of 'accounting', only one possible version of a whole range of broader, richer 'social accounting'. In effect, social accounting is what you get when the artificial restrictions of conventional accounting are removed. However, whilst we might wish to encompass all possible 'accountings' (which would include everything from descriptions of one's time at university to novels, from journalism to advertising, from prayer to excuses), this might prove to be somewhat impracticable (but see, for example, Lehman, 2006). As a result, the primary focus of social accounting tends to be upon:

- '*Formal*' (as opposed to 'informal') accounts: The primary concern in social accounting tends to be with the larger organisations such as multi-national companies (MNCs, see, for example, Rahman, 1998; Unerman, 2003; Kolk and Levy, 2004)[7] and the focus tends to be upon the visible, external accounts rather than the potentially equally important, but much less visible, internally produced accounts. In MNCs, there is typically a considerable 'distance' (spatial, financial, cultural, etc.) between the reporting entity and those affecting and affected by it, i.e. its stakeholders. In small communities, accounts are given and received informally (between you and your parents, you and your friends, you and your teacher, etc.) because of what Rawls (1972) calls 'closeness'. The greater this absence of closeness, the greater the need for formality in giving and receiving accounts (see, for example, Gray *et al.*, 2006; Unerman and O'Dwyer, 2006).[8]

- Accounts typically prepared *by* organisations or which are (less commonly) prepared and disclosed by others (the 'external social audits'): Most of our attention will be upon the reports that organisations produce about themselves (in the same way as organisations produce their own financial statements) and which, just like financial accounting, are

---

[6]In general, the 'defined organisation' or 'accounting entity' characteristic has been retained in social accounting as this remains the focus of some process of accounting (i.e. one needs 'something' for which to account). There are problems with retaining the entity definition (see, for example, Tinker, 1985; Hines, 1988; and especially Cooper *et al.*, 2005), and attempts have been made to soften, if not remove, the characteristic.

[7]There is, however, a considerable and important interest in both NGOs and social enterprise accountability and associated social accounting. The social accounting in such organisations tends to raise somewhat different issues (see, for example, Ball and Osborne, 2011).

[8]There is a wider and more general point here that the giving and receiving of accounts is a ubiquitous human activity and one which seems to reflect a deep human need. Which needs the accounts satisfy and the form they take is a measure of the circumstances in which the accounts are given and received. For more detail, see Arrington and Puxty (1991) and Arrington and Francis (1993).

visible to us as people external to the organisation. However, it is essential to realise that only a small proportion of such activity is regulated. That is, most of the social, environmental and sustainability reporting we will examine is produced voluntarily – with all the benefits and problems that this brings with it. This topic is explored in more depth in Chapter 10 where we examine the phenomenon known as the 'external social audits' – the practice of external bodies, for example NGOs or researchers, independent of the accountable entity producing reports whether or not the entity wishes it.

● Reports are prepared *about* certain areas of activities: Whereas we tend to assume that we know what a financial report should be about, the contents of a social report can be less obvious. However, most commentators assume that a report will normally cover: the natural environment; employees; and wider 'ethical' issues which typically concentrate upon: consumers and products; and local and international communities.

As we shall see, especially when considering the 'social audits', this can be a very narrow range of concerns. Other issues, such as ethical stances and action on race and gender issues are clearly also important elements of an organisation's social activity. An indication of the potential range of issues that social accounting might need to address is given in Figure 1.1, taken from the *Ethical Consumer* criteria for evaluation of products and companies. Social accounting generally tends to concentrate on the four principal areas we identified above. However, the reader is reminded that this is an artificial limitation of the issues. Some of the effects of this limitation will be re-examined as the book develops.

● Accounts, not just for shareholders and other owners and finance providers but primarily *for* 'stakeholders': What makes social accounting of interest to us is the potential for holding organisations to account – i.e. 'accountability' (see Chapter 3). For this to happen, 'stakeholders' must be informed. Stakeholders are normally understood as 'any group or individual who can affect or is affected by the achievement of the organization's objectives' (Freeman, 1984; see also Friedman and Miles, 2006). At its simplest, we tend to assume that stakeholders comprise the other internal and external participants in the organisation and these are normally assumed to include: members of local communities; employees and trade unions; consumers; suppliers; society-at-large. Of course, this is also a limiting and potentially dangerous assumption which we will examine below.

These then are the basic elements of the social accounting framework – the basic, but often implicit, assumptions that the social and environmental accounting (SEA) literature adopts. They are summarised in Figure 1.2 and are developed further in Chapter 4.

These basic characteristics are, however, *underspecified* in that they do not tell us, for example, why an organisation might self-report, or why it might, or should, report on

Figure 1.1 *The ethical consumer* **criteria for evaluation of products and companies (Ethiscore)**

Environment: Environmental Reporting, Nuclear Power, Climate Change, Pollution & Toxics, Habitats & Resources;

People: Human Rights, Workers' Rights, Supply Chain Policy, Irresponsible Marketing, Armaments;

Animals: Animal Testing, Factory Farming, Other Animal Rights;

Politics: Political Activity, Boycott Call, Genetic Engineering, Anti-Social Finance, Company Ethos;

Product Sustainability: Organic, Fairtrade, Positive Environmental Features, Other Sustainability.

Source: Taken from http://www.ethiscore.org/ June 2006.

Figure 1.2 The basic elements of the conventional corporate social accounting model

- a formal account;
- prepared and communicated by and/or about an 'organisation';
- about social and environmental aspects of the organisation's activities;
- communicated to the internal and external 'participants' of the organisation.

particular aspects and to particular groups of individuals. Clearly international companies do not, for example, communicate to everybody the detail of their environmental impacts, their impacts on communities in lesser-developed countries or their attempts to persuade governments not to pass legislation that might restrict their commercial activities. So why do organisations report at all and, more importantly, why do they *not* report and why *should* they report?

These are questions which raise ethical, social and political – as well as economic – issues. In fact all of business – and, as a result, all of social accounting – implicitly begs a whole range of fundamental questions about the structure of, and power in, society, the role of economic as opposed to social and political considerations, the proper ethical response to issues and so on. Sadly, these matters are rarely made explicit in business education and training and so, as a result, we tend to be ill-equipped to consider them. Therefore, the rest of this chapter will introduce some of these issues – albeit in a simple manner.[9]

1.3 Is social accounting important? Why?

We have already mentioned that social accounting can be undertaken for a wide range of reasons, and one can undertake its study for a similarly wide range of reasons. Social accounting might, for example, simply be interesting because it is new and different, it might attract our support because it talks about 'nice' things (as opposed to 'nasty' economic things) or is concerned, perhaps, with doing 'good things', whatever that means. More substantially, social accounting has the potential to offer alternative, 'other', accounts of the primary economic organs on the planet (typically MNCs, that is), it can allow alternative voices to be heard; and social accounting can potentially expose the conflict between the pursuit of economic objectives and the pursuit of social and environmental ambitions. Certainly, social accounting is important at some levels to companies and other organisations – a conclusion derived from the simple fact that they undertake this costly activity voluntarily. There are many explanations as to why organisations do this[10] but it is certainly seen by them as a means of legitimating activity, managing stakeholders, forestalling legislation, putting the organisation's side of the story and keeping up employee morale as well as keeping up with competitors, creating competitive advantage and signalling the successful management of risk.

So, there is a range of reasons for social accounting. We, however, are going to derive what we see as the crucial importance of social accounting from two critical principles: those of accountability and sustainability. The key principle underlying this text is that of **accountability**. At its simplest, *accountability is a duty to provide information to those who have a right to it*. It is linked closely with notions of (social) responsibility and is an essential

---

[9]See Chapters 2 and 3.
[10]See Chapter 4.

component in a democracy. The greater the power an individual, or an organisation, has over people, resources, communities, etc., the greater the responsibility to provide a full account of stewardship of those people, resources or communities. If our world is to be democratic, then those with the greatest power, large companies and governments, owe the greatest accountability. That accountability is discharged through social, environmental and sustainability accounts (see Chapter 3).

Now, accountability is a principle based on a really important notion – namely democracy. That would be impetus enough to make social accounting important, but there is a much more pressing reason to consider social accounting as a truly urgent matter, a matter of life and death, and that is the notion of **sustainability**. Our contention is that one of the major ways in which we need to be able to hold large organisations accountable is over their contribution to – or detraction from – individual societies' and, ultimately the planet's, capacity to maintain itself, its eco-systems and life itself. It is this capacity to maintain itself that we know as sustainability. We will need to briefly review the evidence (and this we do below in Section 1.5), but there is considerable and chilling evidence that many aspects of planetary sustainability are under the most serious threats.[11] Such threats are likely to arise from a combination of populations (about which social accounting at this level has little to say) and economic activity, organisation and performance (about which social accounting has much to say).[12] If, as we shall seek to show, corporate pursuit of profit, driven by increasingly demanding capital markets, is amongst the principal causes of this exponential growth in un-sustainable activities, then society as a whole has a serious need to know about it. Good, thorough, social accounting should be able to provide appropriate information.

That is, society can only infer the detailed effects that (large) organisations have on society and the planet, and our principal means of doing so is information intermittently provided by the organisations themselves on a voluntary basis. Only with the sort of complete and penetrating data that a good social account should provide are we likely to be able to encourage the urgent debate about the power and activities of financial markets, the power and activities of companies and the power and activities of governments. Thus social accounting, at its ultimate, is motivated by the relationships between international financial capitalism, corporate activity, the role of the state, civil society and planetary systems. Social accounting, therefore, has the potential to play a crucially important part of civilised intercourse on a planetary scale. It is difficult to think of anything much more 'important' than that!

1.4 Crisis? What crisis? Sustainability and the state of the world

The importance of social accounting – both as a study and as a practice – derives from a number of sources. The most important of these is the context within which social and economic intercourse is conducted – that is, the departure point for social accounting is not a set of legal requirements, as with conventional accounting, or a particularly ubiquitous or exemplary practice, as might be the case with say finance or marketing, but rather a series of compelling concerns that all is not well with our world. Whether those concerns are the imbalance between power and responsibility; a concern for a democratic deficit; appalling inequality;

---

[11]For a brief introduction and review of these issues and this evidence see Porritt (2005); Gray (2006a,b); Milne *et al.* (2006).

[12]More rigorously, it is generally considered that planetary threat can be modelled through the **IPAT** equation first formulated by Ehrlich and Holdren (1971) and Commoner (1972). IPAT is the suggestion that Impact = (Population) × (Affluence) × (Technology) and is sometimes shown as IPCT where the 'C' relates to consumption not affluence. For more detail, see Dresner (2002); Meadows *et al.* (2005).

Figure 1.3 Crisis? What crisis?

Inequality	Deforestation
Climate change	Third world debt
Species extinction	Waste disposal
Habitat destruction	Energy usage
Drought	Starvation
Poverty	Population
Desertification = desert	Water depletion
Acid rain	Toxic chemicals
Soil erosion	Nuclear waste
Air pollution	Displacement of ethnic peoples
Water pollution	Child labour
Land pollution	Racism + genocide
Noise pollution	Excess consumption
Resource scarcity	Social alienation
Urban violence	Drug addiction

poverty and drought in the face of plenty; waste and excess; the inevitable exigencies of international financial capitalism; or planetary desecration; there are a range of issues that we, as privileged scholars, students and professionals owe a duty to address. Social accounting is one of the ways in which we might seek to address, redress and re-orientate our relationship with some of the less positive consequences of human existence (see Figure 1.3).

Let us start at the beginning. A significant majority of us in the West are profoundly fortunate – at least in certain material ways. Most of us (and stress this is only most of us) have never known hunger, drought, life-threatening poverty or been directly threatened by war. Such astonishing material well-being, however, comes at a price: whether that price be the exploitation of children, the repression of others, the destruction of the natural environment or whatever (see Figure 1.4). That is, as Milton Friedman is so frequently quoted as saying: 'there is no such thing as a free lunch'. Our well-being comes at a price and that price has, for many, long been morally unacceptable, and it is increasingly looking as though it may prove to be physically un-sustainable.

A range of reports produced by responsible, independent and presumably fairly reliable sources has provided a chilling picture of the planet's capacity to support our levels of extraction, usage, waste and pollution. The United Nations Millennium Ecosystem Assessment; the United Nations Global Environmental Outlook; Kofi Annan's Millennium Development Goals; WWF and the *Limits to Growth* project (e.g. Meadows *et al.*, 2005) all tell us, in fairly incontrovertible terms, that the current population with our current ways of economic organisation and activity are using more than the planet can produce – we are eating into planetary capital.[13] (For an excellent summary see Porritt, 2005.)

---

[13]United Nations Environment Programme (UNEP) (2008) *UNEP Year Book (formerly GEO): An Overview of Our Changing Environment*; United Nations Millennium Ecosystem Assessment (2005) *Living Beyond Our Means: Natural Assets and Human Well Being*; WWF (2008), *Living Planet Report 2008*; United Nations Millennium Development Goals: www.un.org/millenniumgoals/.

Figure 1.4

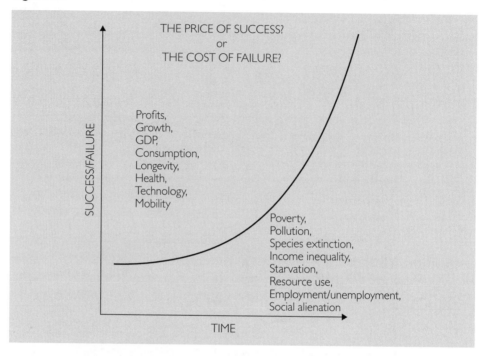

This is expressed graphically by reference to the notion of the *ecological footprint* (Wackernagel and Rees, 1996; Meadows *et al.*, 2005). Ecological footprint is an estimate of the amount of physical space a person, a society – or a species – is currently using to support its way of life. The footprint of the average individual in North America might be about 10 times greater than that for the average person living in Africa for example – regardless of how much space each actually has at their disposal. One representation of the footprint for the human species is shown in Figure 1.5.

Figure 1.5 (and see other estimations such as that from Meadows *et al.*, 2005) suggest that, within living memory, humanity has stopped living off 'planetary income' (as any prudent species would do) and has started to eat into 'planetary capital' – living beyond our means in effect. Such an activity is clearly un-sustainable. Furthermore, when the peoples of India and China for example start to have the sorts of levels of consumption that are associated with average European levels, we find that we need up to three planet Earths to support our ways of living.[14] Humanity's footprint will be three times the available planetary space. This is clearly absurd. It is quite clear that something must be done to change current levels, *inter alia,* of consumption, production, waste, pollution and habitat destruction, otherwise, no species, including humanity, will be able to survive.

As if this were not enough, evidence suggests that the rich, although getting richer, are getting no happier (Layard, 2005); the gap between the rich and the poor, although open to debate in places, would certainly appear not to be getting any smaller (see, for example, Sutcliffe, 2004) and, more disturbingly, in some regards, the situation of the very poorest is getting worse – some of the Millennium Development Goals, notably environmental sustainability, are actually in decline. Thus the undoubted increases in material prosperity: are not

---

[14]These calculations are made on the basis of very positive assumptions about technological change. Technology has made astonishing strides in the ability to make more from less, but there must always be eventual limits to this.

big steps

exceeding the limit

Figure 1.5 Humanity's ecological footprint, planetary carrying capacity and <u>overshoot</u>

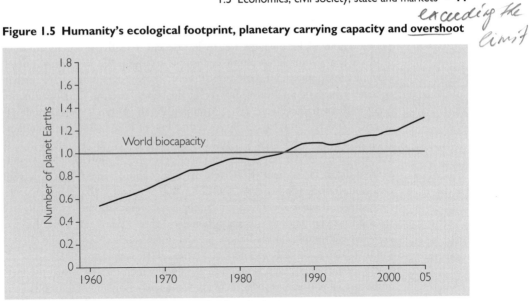

Source: Taken from WWF, 2008: 12.

making the recipients of that prosperity happier; are not reaching peoples equally; are probably contributing to increasing inequality; and are occurring at a time when the un-sustainability of the planet appears to be approaching crisis (see Jackson, 2009).

A backdrop like this certainly gains our attention but – and it is a potentially significant 'but' – what does this have to do with social accounting? Well, it would seem extremely unlikely that there is no connection between the engine of material prosperity – i.e. capitalism – and the apparent consequences of that prosperity. Corporations are, to a very considerable degree, the engines of capitalism and the prosperity that it brings. So, corporations, and other large organisations and institutions, are one of the places to which we might turn our attention if we wish to seek some different balance between the positive and negative consequences of economic growth. Social accounting, with its organisational/institutional focus, seeks to address one, though fundamental, element of our modern world: what are organisations doing (purportedly) on our behalf and can we, if we wished to, control them?

These issues we examine briefly in the next section.

1.5 Economics, civil society, state and markets

If we are to make any systematic sense of these issues, we are going to need to try and avoid too simple or too refined an approach. On the one hand, we need to try and avoid the temptation of the easy answers ('it is all the fault of government'), the easy targets ('it is all the fault of capitalism') or the trite solution ('I recycle my paper, so it is all OK'). On the other hand, there is no obvious advantage in either despairing or producing such complex analyses of the issues that we all have little understanding of what is actually going on. That is, we believe that we need to try and be rigorous, even-handed and, probably, more than a little pragmatic in what follows. There are both heroes *and* villains. The problems are legion, but we must be able to try and do something. Our understanding will influence our choices of actions but, in doing so, will not exclude or close down alternative perspectives and solutions.

The difficulty is obvious. Each of us is bombarded with images of success, of desire, of need, of opportunity, and so on, on a minute by minute basis. The evidence of the success of the system in which we live is clear for all to see. Equally, it is not at all obvious that the people who work for business are any less decent, sensitive or intelligent than anybody else.[15] And yet the volume of corporate critique is startling. We have already seen the issues concerning environmental degradation (Porritt, 2005). Corporations are further accused of significantly and systematically subverting the state – even in developed nations like the UK, USA and Australia (Hertz, 2001; Monbiot, 2000). There is considerable disquiet over corporate standards and, especially, over the role of brands and the dominance of marketing (see, for example, Klein, 2001; and the Adbusters website and magazine). The debate over the impact of MNCs, especially on lesser developed countries (LDCs) and whether or not foreign direct investment (FDI) actually brings more benefit than disadvantage is a serious and complex one (Bailey *et al.*, 1994a,b, 2000; French, 1997; Rahman, 1998; Annisette and Neu, 2004). It seems clear that controlling large organisations – especially in an era of uncontrolled financial markets – is exceptionally difficult (Hatfield, 1998; Levy and Kolk, 2002) and that one simply cannot believe all that organisations say that they do or say that they do not do (Adams, 2002; Kolk and van Tulder, 2004; Leipziger, 2003).

So, even though there is much that is good which may be laid at the feet of corporations and other large organisations, there are some fairly fundamental problems as well. The situation, whatever else we may discover, is unlikely to ever be black and white. Beware of simple explanations.

There are a number of themes that will pervade what follows and amongst them three are worthy of mention at this stage. These are that: (a) there is something inherently problematic in the nature of the corporation itself (see, for example, Bakan, 2004); (b) there is something inherently problematic with international financial capitalism as we currently experience it (see, for example, Kovel, 2002; Porritt, 2005); and (c) there is something inherently problematic with the nature of economics and markets (Thielemann, 2000). In essence, as Thielemann argues, it is the nature of economics to seek to drive out anything which is not-economic: to drive out, what he calls, 'market-alien values'. To be efficient in economic terms means seeking out more and more economic opportunity and this is achieved at the expense of other areas of life which become colonised by the economic. The pupil–teacher relationship, the nurse–patient relationship, the husband–wife relationship and so on are all increasingly dominated by economic concerns, whereas they are more properly thought of as relationships whose essential nature is professional, social or human. The international financial markets are the extreme example of this in which all matter (nature becomes natural resources; poverty becomes relative advantage on costs and so on) and all relationships (between owners or organisations and the employees of that organisation, for example) are reduced to income and expenditure; dividends and costs. The current environmental disasters can relatively easily be seen as an inevitable manifestation of this development (Gray, 1990; Kovel, 2002; Porritt, 2005). Corporations, especially those subject to the strictures of international financial markets, must therefore, as Bakan argues, behave psychotically and any notion of humanity and responsibility must be expunged from them.

So, if such critique has any substance, then we are confronted by a really difficult set of choices. First amongst these is: 'can corporations deliver responsibility and sustainability?' The answer, increasingly (and, as we shall see in Chapter 3) is that they cannot deliver responsibility and sustainability if we continue to rely entirely upon voluntary initiative and the absence of regulation and full accountability for their delivery. Equally, however, it seems that 'consumers', whilst they cannot always be expected to act responsibly, *can* act

---

[15]We guess that most of you reading this book will, at some time, work for a large organisation – thus making this point for us. However, one should be aware of the impact that organisational rationality and structure can have on individual behaviour. See, for example, Jackall (1988), Estes (1996), Schwartz and Gibb (1999).

successfully on occasions (there are, for example, successful movements like fair trade and the periodic embargos). Similarly, the state – both the bureaucracy of state organs and the politicians themselves – whilst seeming generally incapable of serious sustained moral and social leadership, can occasionally be seen to take a stand on an important issue.

So where is the responsibility? And where will change come from? There is no simple – or even unequivocal – answer to either question. It seems likely that responsibility accords with power and, to the extent that we have power, we also have responsibility. This is the theme that is developed in Chapter 3 and which pervades the book. 'How will change occur?' is, however, something of a mystery. Change *does* occur, and sometimes even for the good. If change is to come about in a civilised manner, it seems to us that *all stakeholders* – investors, management, customers, employees, etc. and especially civil society – must be empowered and must find ways to act in line with their power. The development of social accounting must achieve this in one way or another (see, for example, Lehman, 2001) and that is the political *motif* of this text. We hope you find this stimulating and can embrace these notions.

1.6 Summary and structure of the book

This chapter has sought to introduce you to some of the basic elements of social accounting and accountability and to lift the lid on a range of more complex and confusing notions which provide the motivation and context for social accounting. We live in an unusually complex world and at an unusually complex time in human history. Social accounting offers both a means of trying to make sense of some of that complexity whilst, simultaneously, offering a potential means to develop a systematic response to the range of negative effects that are the price for the well-being we are currently enjoying. So, whilst social accounting is often a fairly pragmatic activity, it has within it a very strong thread of responsibility, idealism and morality: which threads of which responsibility, which idealism and which morality are questions which you, individually and collectively, will have to resolve for yourself.

This book tries to take these basic themes and develop them in some detail: both theoretically and practically. The text is broadly organised into three sections. The opening section comprises four chapters which seek to offer a theoretical and reflective framework for our study. Chapter 2 will introduce ways of seeing the world and offer systems thinking as a useful means for doing this. Chapter 3 develops this 'way of seeing' theme to explore, in quite some detail, what is meant by 'social responsibility'. This chapter offers accountability as a central theme for our studies as well as a means by which we might begin to resolve many of the tensions we are now experiencing. Chapter 4 rounds off this excursion by providing a brief history of social accounting before examining the range of theoretical interpretations that are offered in the subject.

The second section of the book comprises eight chapters and represents the core of the book. Chapters 5, 6, 7 and 8, respectively, address the four broad groups of stakeholders that tend to occupy social accounting: society and the community, employees and unions, the natural environment and the financial community. Chapter 9 looks at the emergence of the triple bottom line and (stressing that these are *not* the same thing at all) how accounts for sustainability are being and might be developed. Chapter 10 explores that most important area of accountability, the actions of organisations outside the organisation of interest – the so-called external social audits – before Chapter 11 critically examines issues of governance and assurance before we turn, in Chapter 12, to an exploration of a number of the principal issues for social accounting in (broadly) non-commercial organisations.

We round the book off with two chapters with which to conclude our study and try to peer into the future and offer a series of potential mechanisms through which a more positive future may be achievable. The possibilities of regulation are explored in Chapter 12 whilst Chapter 13 offers exemplars of leading practice, both actual and potential: this is how to do social accountability! The final chapter reviews our journey and offers a few suggestions and conclusions.

References

Adams, C. A. (2002) Internal organisational factors influencing corporate social and ethical reporting, *Accounting, Auditing and Accountability Journal,* **15**(2): 223–50.

Annisette, M. and Neu, D. (2004) Accounting and empire: An introduction, *Critical Perspectives on Accounting,* **15**(1): 1–4.

Arrington, C. E. and Francis, J. R. (1993) Giving economic accounts: Accounting as cultural practice, *Accounting, Organizations and Society,* **18**(2/3): 107–24.

Arrington, C. E. and Puxty, A. G. (1991) Accounting, interests and rationality: A communicative relation, *Critical Perspectives on Accounting,* **2**(1): 31–58.

Bailey, D., Harte, G. and Sugden, R. (1994a) *Making Transnationals Accountable: A significant step for Britain.* London: Routledge.

Bailey, D., Harte, G. and Sugden, R. (1994b) *Transnationals and Governments: Recent policies in Japan, France, Germany, the United States and Britain.* London: Routledge.

Bailey, D., Harte, G. and Sugden, R. (2000) Corporate disclosure and the deregulation of international investment, *Accounting, Auditing and Accountability Journal,* **13**(2): 197–218.

Bakan, J. (2004) *The Corporation: the pathological pursuit of profit and power.* London: Constable and Robinson.

Ball, A. and Osborne, S. P. (2011) *Social Accounting and Public Management: Accountability and the common good.* Abingdon: Routledge.

Bebbington, K. J., Gray, R. H. and Laughlin, R. (2001) *Financial Accounting: Practice and Principles,* 3rd Edition. London: Thomson Learning.

Commoner, B. (1972) The social use and misuse of technology, in Benthall, J. (ed.), *Ecology: the shaping enquiry,* pp. 335–62. London: Longman.

Cooper, D. J. and Sherer, M. J. (1984) The value of corporate accounting reports: Arguments for a political economy of accounting, *Accounting, Organizations and Society,* **9**(3/4): 207–32.

Cooper, C., Taylor, P., Smith, N. and Catchpowle, L. (2005) A discussion of the political potential of social accounting, *Critical Perspectives on Accounting,* **16**(7): 951–74.

Dresner, S. (2002) *The Principles of Sustainability.* London: Earthscan.

Ehrlich, P. R. and Holdren, J. P. (1971) Impact of population growth, *Science,* **171**: 1212–17.

Estes, R. W. (1996) *Corporate Social Accounting.* New York: Wiley.

Freeman, R. E. (1984) *Strategic Management: A stakeholder approach.* Boston, MA: Pitman.

French, H. F. (1997) When foreign investors pay for development, *World-Watch,* May/June: 8–17.

Friedman, A. L. and Miles, S. (2006) *Stakeholder: Theory and practice.* Oxford: Oxford University Press.

Gambling, T. (1977) Magic, accounting and morale, *Accounting, Organizations and Society,* **2**(2): 141–51.

Gray, R. H. (1990) *The Greening of Accountancy: The profession after Pearce.* London: ACCA.

Gray, R. H. (2000) Current developments and trends in social and environmental auditing, reporting and attestation: A review and comment, *International Journal of Auditing,* November: 247–68.

Gray, R. (2006a) Social, Environmental, and Sustainability Reporting and Organisational Value Creation? Whose Value? Whose Creation?, *Accounting, Auditing and Accountability Journal,* **19**(3): 319–48.

Gray, R. H. (2006b) Does sustainability reporting improve corporate behaviour? Wrong question? Right time?, *Accounting and Business Research* (International Policy Forum), **36**: 65–88.

Gray, R., Bebbington, J. and Collison, D. J. (2006) NGOs, civil society and accountability: Making the people accountable to capital, *Accounting, Auditing and Accountability Journal,* **19**(3): 319–48.

Gray, R. H., Owen, D. L. and Maunders, K. T. (1987) *Corporate Social Reporting: Accounting and accountability.* Hemel Hempstead: Prentice Hall.

Hatfield, J. (1998) At the mercy of monsters, *CA Magazine,* September: 10–19.

Hertz, N. (2001) *The Silent Takeover: Global capitalism and the death of democracy*. London: Heinemann.

Hines, R. D. (1988) Financial accounting: In communicating reality, we construct reality, *Accounting, Organizations and Society*, **13**(3): 251–61.

Hines, R. D. (1989) The sociopolitical paradigm in financial accounting research, *Accounting, Auditing and Accountability Journal*, **2**(1): 52–76.

Hines, R. D. (1991) Accounting for nature, *Accounting, Auditing and Accountability Journal*, **4**(3): 27–9.

Jackall, R. (1988) *Moral Mazes: The world of corporate managers*. New York: Oxford University Press.

Jackson, T. (2009) *Prosperity Without Growth? The transition to a sustainable economy*. London: Sustainable Development Commission.

Klein, N. (2001) *No Logo: No Space, No Choice, No Jobs*. London: Flamingo/Harper Collins.

Kolk, A. and Levy, D. (2004) Multinationals and global climate change: issue for the automative and oil industries, *Research in Global Strategic Management*, **9**: 171–93.

Kolk, A. and van Tulder, R. (2004) Internationalization and environmental reporting: The green face of the world's multinationals, *Research in Global Strategic Management*, **9**: 95–117.

Kovel, J. (2002) *The Enemy Of Nature: The end of capitalism or the end of the world?* London: Zed Books.

Layard, R. (2005) *Happiness: lessons from a new science*. New York: Penguin.

Lehman, G. (2001) Reclaiming the public sphere: problems and prospects for corporate social and environmental accounting, *Critical Perspectives on Accounting*, **12**(6): 713–33.

Lehman, G. (2006) Perspectives on language, accountability and critical accounting: An interpretative perspective, *Critical Perspectives on Accounting*, **17**(6): 755–79.

Leipziger, D. (2003) *The Corporate Responsibility Code Book*. Sheffield: Greenleaf.

Levy, D. L. and Kolk, A. (2002) Strategic responses to global climate change: conflicting pressures on multinationals in the oil industry, *Business and Politics*, **4**(3): 275–300.

Meadows, D. H., Randers, J. and Meadows, D. L. (2005) *The Limits to Growth: The 30-year Update*. London: Earthscan.

Milne, M. J., Kearins, K. N. and Walton, S. (2006) Creating adventures in wonderland? The journey metaphor and environmental sustainability, *Organization*, **13**(6): 801–39.

Monbiot, G. (2000) *The Captive State: The corporate takeover of Britain*. London: Macmillan.

Porritt, J. (2005) *Capitalism: as if the world matters*. London: Earthscan.

Rahman, S. F. (1998) International accounting regulation by the United Nations: A power perspective, *Accounting, Auditing and Accountability Journal*, **11**(5): 593–623.

Rawls, J. (1972) *A Theory of Justice*. Oxford: Oxford University Press.

Schwartz, P. and Gibb, B. (1999) *When Good Companies Do Bad Things: Responsibility And Risk In An Age Of Globalization*. New York: Wiley.

Sutcliffe, B. (2004) World inequality and globalisation, *Oxford Review of Economic Policy*, **20**(1): 15–37.

Thielemann, U. (2000) A brief theory of the market – ethically focused, *International Journal of Social Economics*, **27**(1): 6–31.

Tinker, A. M. (1985) *Paper Prophets: a social critique of accounting*. Eastbourne: Holt Saunders.

Unerman, J. (2003) Enhancing organizational global hegemony with narrative accounting disclosures: An early example, *Accounting Forum*, **27**(4): 425–48.

Unerman, J. and O'Dwyer, B. (2006) Theorising accountability for NGO advocacy, *Accounting, Auditing and Accountability Journal*, **19**(3): 349–76.

United Nations Environment Programme (2008) *UNEP Year Book [formerly GEO]: An Overview of Our Changing Environment*. UNEP: Nairos (www.unep.org/yearbook/2008).

United Nations Millennium Development Goals: (www.un.org/millenniumgoals/).

United Nations Millennium Ecosystem Assessment (2005) *Living Beyond Our Means: Natural Assets and Human Well-Being: Statement from the board* (http://www.millenniumassessment.org/en/Products.BoardStatement).

Wackernagel, M. and Rees, W. (1996) *Our Ecological Footprint: Reducing Human Impact on the Earth*. Gabriola Island, BC: New Society Publishers.

WWF (2008) *Living Planet Report 2008*. Gland: WWF – World Wide Fund for Nature.

CHAPTER 2

Ways of seeing and thinking about the world: systems thinking and world views

2.1 Introduction

Once we have some sense of what this thing called 'social accounting' might be and why we might be concerned with it, we can turn to look in detail at the history, practice and potential of social accounting. Before we do that, however, it is essential to introduce some framework for thinking about our world. If our examination of social accounting, the problems that it seeks to address and the potential that it offers is to be in any way systematic, it needs to be framed. That is, we need to 'theorise' our world in some systematic way that allows us to begin to see some of the explanations of why, for example: the human species is in the mess that it is; social accounting has never quite made the mainstream despite its apparent self-evident desirability; and so many intelligent and thoughtful, as well as influential, people still oppose social accounting developments. The only way to do this is to employ some theoretical 'spectacles' in order to give us a perspective on organisational, social, political and economic activity that will allow us to look beyond the superficial and conventional explanations of organisational rationality and behaviour. We want to try to see more clearly the social, environmental and political assumptions and implications of our ways of organisational life. Without access to theory, we will find ourselves quite unable to offer any systematic analysis of what organisational practice actually is, can be or should be. Theory gives us a basis from which to evaluate both current and other potential forms of activity such as social and environmental accounting. Only by careful evaluation will we be able to judge whether, for example, conventional business models have serious or trivial limitations, whether social and environmental accounting is a waste of time or not, and whether the whole issue of the social and environmental effects of economic activity are, indeed, anything to do with commerce (see also Gray *et al.*, 2010).

This chapter, and the two which follow it, will seek to provide us with the conceptual apparatus with which to see the world afresh. In so doing it will, hopefully, allow you to re-arrange a lot of what you already 'know' about the world in which you live. If all observation *is* theory-laden as so many philosophers reckon, we hope to change your perception by changing your theories.

The chapter is organised as follows. In the next section, we introduce something called 'systems thinking'. This is a well-established intellectual position that is especially important (as we shall see in Section 2.3) when trying to understand connections between things and the connectivity of such things as capitalism, multi-national corporations and/or environmental crisis. From there we will move to examine some of the more common political and economic assumptions about the world and, in particular, explore liberal economic democracy in Section 2.4. Section 2.5 will briefly examine some of the failings of liberal economic democracy – both as an ideal and as a description of the world as it is – before we explore more carefully in Section 2.6 what we mean by this thing called 'capitalism' and how it affects the activities and behaviours of corporations. Section 2.7 briefly introduces a range of 'world views' as a way of poking a stick into the basic issue of whether gradual change to deal with the very real problems we face is realistic in any sense. Section 2.8 offers a potential middle road here by introducing a neo-pluralist view of the world, before we return to democracy and the role that information plays in the pursuit of such an ideal. The chapter, therefore, fleshes out the basic outline of social accounting and prepares the ground for later, and rather deeper, examination of responsibility, world views and accountability.

2.2 Systems thinking and general systems theory

The principal advantage of a 'systems perspective' is that it provides an explicit contrast to the 'reductionism' which more typically characterises thought in business, economics, accounting and finance. Furthermore, the systems perspective also provides a wide range of additional insights into the world in which we live and, especially, is something of a *sine qua non* for serious environmental thought (Meadows, 2009).

The genesis of systems thinking is normally attributed to the pioneering work of Ludwig von Bertalanffy (see, for example, von Bertalanffy, 1956) which derived from his concern over the way in which the natural sciences were developing. Von Bertalanffy's conception of what he called **general systems theory (GST)** was an attempt to break down the barriers between knowledge systems and to reverse – or at least slow down – the tendency in scientific thought towards reductionist reasoning (Feyerabend, 2011). The essence of the concern was that:

- the attempt to study a part (of anything) without understanding the whole from which the part comes (reductionism) was bound to lead to misunderstandings – the part can only be understood in its context;
- understanding tends to be directed by and limited to one's own discipline, and natural phenomena are complex and cannot be successfully studied by artificially bounded modes of thought – in Ackoff's famous dictum: . . . *we must stop acting as though nature was organised in the same way as university departments are* (Ackoff, 1960).

This led to an expansion of systems thinking and it was soon recognised that the conception of systems was not restricted to natural science. In fact, most phenomena with which the human species interacts could be usefully considered in a systems way. GST could provide a framework for thought throughout the natural and the social sciences *and* attempt to capture the interactions between the two (but see Bryer, 1979; Hopper and Powell, 1985).

Kast and Rosenweig 1974 define a system as:

An organised unitary whole, composed of two or more components or subsystems and delineated by identifiable boundaries from its environmental suprasystem . . .

(Kast and Rosenweig, 1974 101)

That is, a system is a conception of a part of the world that recognises explicitly that the part is (a) one element of a larger whole with which it interacts (i.e. influences and is itself influenced by); and (b) also contains other parts which are intrinsic to it. Ultimately, therefore, nothing can be understood without a complete understanding of everything else. Although possibly desirable, this is clearly impossible. However, we are counselled by GST to be aware that each attempt to focus on a manageable chunk of experience – one system or subsystem – *must* risk misunderstanding through loss of the interactions between the system and other systems.

With such a wide definition, it is apparent that *everything* can be seen as a system. Systems thinking has been successfully applied to social science and, in particular, to organisations and their internal and external interactions, to systems of thought and, especially relevant, to the interaction between human and other systems of natural ecology. Systems thinking has also had an obvious influence on thinking about organisational systems, management and control systems and management information systems, for example. Unfortunately, attempts have also been made to apply systems thinking in a specifically 'scientistic' and functionalist manner to the articulation and solution of specific problems. This has highlighted an essential tension in the use of GST. That is, GST is especially helpful as a way of thinking – as a mental framework with which to stand back from issues and see them in their broader context. It is not especially good at helping solve specific, closely-defined problems. If one seeks specific and precise solutions, one is in real danger of 'reducing' the problem artificially to produce that solution. In such circumstances, one is using a *constrained* systems thinking approach – which always runs the risk of missing the point of GST by excluding from the problem the complex and irreducable elements of the system under consideration.[1]

Business organisations are often considered in a relatively constrained systems perspective. But business is a complex system, comprising many components and operating within and interacting with a complex array of social and natural systems. Neither is business just part of some system which we might describe as 'economic'. Business, accounting, finance, marketing all also interact with systems which we might call 'social', 'political' and 'ethical' and so on. More substantially, business, accounting, finance, marketing systems are directly related to interactions within and between organisational systems *and* between those organisational systems and individuals, groups, communities, societies, nations and the non-human elements of the planetary natural environment.[2] Conventional economic analysis – and conventional accounting and finance in particular – too often ignores these interactions and so it falls to social accounting to attempt to account for some of these missing elements.

2.3 Using the GST framework

Imagine yourself sat on the moon, comfortably equipped, and looking down on the planet Earth with an all-seeing telescope (Boulding, 1966). We are familiar with the concept of the solar 'system' of which the moon and planet Earth are a part. However, with the dominant exception of the sun (to which we return later), we are not especially concerned here with this 'level' of system.[3]

So we increase the level of resolution of (we focus in) our telescope to look at the Earth. Looking at the planet Earth, we might observe climate, weather, oceans and physical features. These we might observe within the mental constructs of climatology, meteorology, oceanography, geology, geography, etc. These are clearly human-constructed categories and

---

[1] For more detail about systems thinking in general see, for example, von Bertalanffy (1956, 1971, 1972); Ackoff (1960); Emery (1969); Beishon and Peters (1972); Kast and Rosenweig (1974); Checkland (1981); Carter *et al.* (1984); Meadows (2009); and in business and accounting see, for example, Lowe and McInnes (1971); Lowe (1972); Laughlin and Gray (1988); Gray (1992). For a useful summary of limitations see Hopper and Powell (1985).

[2] See, for example, Lowe and McInnes (1971); Lowe (1972); Laughlin and Gray (1988); Gray (1990).

[3] However, before increasing the level of resolution of our telescope to focus on the Earth, we should perhaps ask readers to decrease the level of resolution of their mental telescope, to expand their vision to the metaphysical level – to the (scientifically) unknowable system within which all systems – planetary, solar, galaxy, etc. – may purportedly exist. That is, we must ask readers to answer a personal question as to whether or not deity or deities exist (and if so which) as the creator, controller and/or framework for all systems. The answer to this will logically influence reactions and conclusions later on.

our understanding and familiarity with these categories is automatically influencing what we see and how we see it. This will become more obvious as we move into social systems, but the general rule that 'observation is theory-laden' – that how we think affects what we see – is an important insight of GST. A second insight from GST is that there is a wholeness to the planetary systems that is lost as soon as we begin to break down categories of experience. We do this (isolating experience into categories – reductionism) in order to increase the depth of our knowledge but, in so doing, we immediately lose sense of the completeness (the holism) of the thing experienced. *And* in so doing we risk failure of understanding at the boundaries of our 'oligies' and 'ographies'. Thus, for example, attempts to explain a particular weather phenomenon without reference to the oceans, mountains, seasons and other physical characteristics are obviously doomed to failure.

Increasing the level of resolution of our telescope further will perhaps bring biological systems, 'life' as we humanly understand it, to our attention. We might see vast numbers of species (categories) with one – humanity – seemingly the most ubiquitous and intrusive. Focusing more closely still may bring to sight those systems which are clearly of human construction (i.e. things which do not exist if humans are not present) – for example, nations, states, regions, organisations, households, groups, political parties. That is, we might choose to 'see' the human system as consisting of organisational subsystems. We might equally, however, choose to 'see' systems of activity that we might call 'economic systems', typically distinguished in modern conventional economics by the presence of priced transactions, 'political systems', in which power is exercised and imposed, 'social systems', in which humans organise their activities and support systems, 'ethical' systems, 'metaphysical' systems, etc., etc.

This might all seem self-evident, but we rarely make it explicit. Whilst we may wish to believe that the means by which to assess commercial and financial success (accounting) are no more than a complex set of socially neutral techniques and skills, that economics is a 'science' abstracted from ethics, values, human emotions, exploitation, quality of life and the state of the physical environment, and that 'business' is somehow separate from normal human values, such beliefs are untenable at best and destructive, dishonest and immoral at worst. If we think that economics, business and accounting have anything at all to do with human and non-human systems, then it is the worst sort of reductionism to draw our systems boundaries around those bits we might choose to ignore. Societies, organisations, economics, accounting, ecology are all systems and they interact. Simply assuming that an activity is unrelated to societal or environmental desecration *does not make it so*!

But just to identify economic, social, political, etc. systems is not enough. We need to be able to say something about the *nature* of these systems. That is, there are different ways of conceiving of the systems we think of as 'economic' or 'social'; there are different ways of interpreting the *same* social or economic systems, and there are different ways of conceiving of the interactions between these systems. This is far from trivial. If we wish to have any understanding of conventional business and financial organisation, consider the role(s) it does, can and should play and then extend that to social accounting, we must have some conception of the world in which that activity can, does and should take place.

The most common way of articulating this conception in the western developed nations is by way of **liberal economic democracy**. This is not only the most common conception in the rhetoric of western, and especially North American, British and Australasian, politicians and business leaders, but it is the root metaphor and an essential *but implicit* factor in the way in which business and conventional accounting are taught and conceived. By making those assumptions explicit we can identify many of the systems interactions, identify the limitations and myths of the conception and thereby identify the contestable intellectual foundations of much of conventional business and management. However, the issue to beware of is that these criticisms may also be extended to social accounting and may undermine the claims to which social accounting might aspire.

2.4 Liberal economic democracy

'It has been said that democracy is the worst form of government except for all those other forms that have been tried from time to time.'

(Sir Winston Churchill, House of Commons, 1947)

Sat on the moon with our all-seeing telescope we have already organised many of the things we see by reference to categories of thought – to theories and theory-related conceptions. We observe human action and, for convenience and simplicity, divide human action into systems – groups of individuals being 'social', 'political' 'religious', 'economic', 'ethical', etc. There is potentially an infinite number of ways in which human action might be categorised and analysed. Liberal economics and democracy is only one such way, but in the 100–200 years of its growth and maturity it has become by far the most influential and ubiquitous in the so-called developed world and, globally, in business, management, accounting and finance thought.[4]

At its simplest, the liberal economic democratic conception envisages a world of equal individuals, free to act (liberal)[5] and to express choice through actions in markets (economic) and actions in the political arena (democratic). The state (the government and its organs and institutional structures) is presumed to be small, to act to maintain freedom and, most importantly, to be *neutral* with respect to serving particular groups' interests.[6] The liberal economic democracy conception is both a **positive** conception (i.e. an attempted description of how the world *is* presumed to be), and a **normative** conception (i.e. a conception of how the world *should be*).[7] The essence is that the individual's freedom is paramount, that we all come to economic exchange equally able and free to express our personal economic choices. For those choices which cannot be expressed through economic exchange, we are presumed to be able to express them through either individual or group social action or through exercising equal power through the ballot box or other political action. As power, the ability to influence others, is assumed to be equally distributed, no one individual or group can systematically dominate, or impose their preference upon, any other. However, occasionally, it may be necessary for individuals to form, disband and reform, probably temporary, groups as coalitions to make a political point – for example,

---

[4]Democracy has a history of over 2500 years (see, for example, Held, 1987). Liberal economics and its relationship with democracy (hence 'liberal economic democracy') has a much more recent history and is usually dated from the work of Jeremy Bentham (1748–1832) and James Mill (1773–1836) which itself grew from the work of Hobbes, Locke and Rousseau and provided the foundation for the much richer work of John Stuart Mill (1806–1873). By way of contrast, this intellectual tradition also provided the basis for the very different interpretations of the world offered by Marx (1818–1883) and Engels (1820–1895). It is predominantly the modern, 19th and, particularly, 20th/21st century manifestations of liberal economic democracy that will be discussed (and painfully simplified) here.

[5]The word 'liberal' tends to cause problems for those not versed in political thought. It refers to the freedom of action of the agent and, in the modern context, the economic agent. It often bears little correspondence with modern manifestations of political parties with 'liberal' in their titles and more general notions of 'liberality' which tend to be associated with being tolerant towards others – allowing them their freedom as it were.

[6]In large part, this occurs 'naturally' in the conception because it is assumed that there are no systematic or systemic conflicts of interests between identifiable groups – itself because there are assumed to be no systematic or systemic groupings of 'classes'. That is, the model is 'atomistic' – a conception of a social world which consists entirely of individuals who may coalesce into groups (see below) but then fly apart again – constantly moving.

[7]See, for example, Hayek (1960, 1982); Friedman (1962); Nozick (1974).

political parties, pressure groups, representative groups, etc. Everyone is assumed to be free to join these groups, or to form alternative groups. This 'refinement' of liberal economic democracy to allow for 'groups' as opposed to just 'individuals' is, at its simplest, what is known as **pluralism**.[8]

The claimed analytical power of the liberal economic democratic conception develops when each agent is presumed to be acting in their own self-interest. The sum total of all these individual social, political and, especially, economic actions of self-interested agents does, it is claimed, produce maximum economic efficiency unfettered by social and political interference. The self-interested pursuit of economic efficiency seeks out the 'best' economic choices and ensures that finance, labour, know-how, physical capital and materials are put to the 'best' economic uses. As a result, it is claimed, this generates maximum profits and economic growth via maximum efficient output from scarce resources. Thus, it is concluded, an economy which is generating more financial wealth must also make the society better off and thus make everyone within that society better off. Minor inequalities arise either through choice (e.g. leisure versus work) or can be eroded through political action (e.g. pressure groups of the disadvantaged).

The individual is thus free to be rich, free to starve, free to be politically active or inactive. In this world, the institutional framework – and thus the law and government – represent the wishes of the actively choosing people (the demos). The institutional framework is brought into existence because the majority – acting freely, rationally and in their own self-interest – choose that institutional framework.

Finally, this pen-picture of liberal economic democracy has avoided any explicit reference to emotive things like ethics and morals. This is because embedded, implicitly, in the assumed workings of liberal economic democracy is a version of the ethic of **utilitarianism**. This ethic states that every action should be judged by the consequences of that action and, in particular, by reference to the consequences to the agent – the change in his/her utility. In the liberal economy of recent history, this utility is to be measured by cash flows, profit and GNP and thus, the consequences (and thus the 'rightness') of an action are captured in profit. A profitable action is thus a good action. This matter of *ethics* we will return to later.

This is a brief a statement of the 'pristine' liberal economic democratic (what we refer to as PLED) position associated with the 'New Right'.[9]

So *if* all agents were equal and *if* markets were information efficient and *if* this led to allocative efficiency and *if* this led, in turn, to economic growth and *if* this ensured maximum social welfare and *if* maximum social welfare is the aim of the society *then* conventional accounting, the pursuit of profit and the actions of self-interest in finance are morally, economically and socially justifiable and may lay claim to an intellectual framework.

---

[8]See, for example, Barach and Baratz (1962); Dahl (1970, 1972); Lukes (1974); Held (1987); and for a brief introduction see, for example, Speake (1979); Abercrombie *et al.* (1984); Robertson (1986).

[9]Experience suggests that students would normally have little appreciation of concepts like 'left' and 'right' wing in political terms. We have just outlined the right wing position – pristine liberal economic democracy (the extreme of the 'right' wing is assumed to be Fascist). Pristine liberal economic democracy does not exist (and probably cannot exist) anywhere in the world but is presumed to have come closest in the USA and possibly (but only possibly) in pre-Chinese Hong Kong and Singapore. The other end of the left/right political spectrum is usually attributed to a form of Marxism and, in particular, socialism. The pure form of socialist utopia similarly does not exist in the world (its extreme counterpart is usually assumed to be the State Communism of the ex-USSR, Cuba and pre-liberalisation China). In the socialist (left wing) model, the state plays a greater role in (putting it simply) protecting and supporting the less able and less privileged, but the 'cost' of this is the reduced freedom of economic agents. Most countries exist somewhere along this spectrum with a very noticeable move towards the 'right' during the 1980s. A third dimension to the spectrum has been introduced by the deep greens to which we will return shortly.

Of course this does not appear to be the case.

None of these 'ifs' can be shown unequivocally to hold and most of them can be shown *not* to hold. This leaves pristine liberal economic democracy, and business, finance and accounting which are implicitly justified by reference to it with no obvious moral, economic or logical theoretical foundation (Jacobsen, 1991). Similarly, it may even leave social accounting without a moral and theoretical foundation also. If the pristine liberal economic democracy conception was reliable *and* desirable then we could assume that the casualties of economic activity – species extinction, exploitation, pollution, poverty, community destruction – were voluntary choices – the eggs which had to be broken in the making of the developed country omelette. Such is the implicit assumption in most business, finance and accounting teaching, research and theory. It assumes that the business system is amoral and its responsibility is to develop and enhance the workings of the economic system for the benefit of the wealthy. In such a world, there is little need for such soft-hearted ideas like social accountability except perhaps to legitimise the system, to perhaps paper over or explain away occasional excesses or as an occasional tool for use by corporate management (Walton, 1983; Den Uyl, 1984; Mulligan, 1986). This is the position taken by Benston (see, for example, 1982a, 1982b, 1984) in his liberal economic analysis of social accounting. It is *also* the conclusion reached by Puxty (1986, 1991) from a critical theoretic, anti-liberal point of view (see also Tinker and Okcabol, 1991) and by writers such as Neu and Everett (Everett and Neu, 2000; Everett, 2004) from a more postmodern perspective. In a pristine liberal economic democracy world, social accounting is largely irrelevant at best and damaging at worst because it interferes with freedom and makes no contribution to liberalism. In a critical theoretic world, in a 'radical' conception,[10] social accounting is also largely irrelevant because it cannot achieve any change of substance, because it is essentially part of liberalism and controlled by corporations (capital). Both views have some substance, but we find neither view conclusively persuasive. We will return to this issue in some detail later in this chapter (see also Chapter 3).

In the meantime, whilst pristine liberal economic democracy may be the conceptual model that is widely assumed and adopted, the world in which we live cannot be shown to be consistent with that model. There are so many empirical failings that the pristine version of the model is no longer descriptive. What we may have, however, is a *version* – albeit a badly perverted and twisted version – of liberal economic democracy. (And do note that whilst we may be able to come to conclusions about aspects of the world in which we live, in which accounting operates and in which we wish to place social accounting, it is probably the case that the *actual* form of the power distribution, how economies and societies *actually* operate, etc., is virtually unknowable in any ultimate, 'factual', sense.) Our assumptions about societies are just that – assumptions, faiths, in fact. If we do, in fact, face some version of liberal economic democracy then we have to come to some conclusion on how 'bad' or 'good' this is. Will sticking-plaster solutions work? Do we need something more substantial in the way of encouraging social change? Or should we just attempt to get rid of the system and replace it with something hopefully better?[11] What role does, can and should social and environmental

---

[10]When we use the term 'radical' without any qualifying adjective, we will be referring to thought stimulated by Marx and post-Marxist critical theory and most usually associated with the radical left wing. Benston and his ilk are also radical – right wing radicals.

[11]The possibilities with which one might replace liberal economic democracy are infinite and limited only by imagination. They include variants of socialism, feudalism, anarchy, monarchy, dictatorship, theocracy, etc. Democracy may be the best of a bad bunch we have so far tried, but it may not be enough to undo the appalling social and environmental conditions that exist in the world today. If we do not allow democracy – or at least a version which empowers the choice of the people – and we have no wholly accepted form of authority (e.g. a deity) then whose vision of utopia and justice should be allowed to dominate? (see also, for example, Marcuse, 1964, 1986; Adorno and Horkheimer, 1972; Weston, 1986; Dobson, 1990; Sunstein, 1997; Burchell, 2002).

accounting and accountability play in this? This proves to be a major theme within discussions of social accounting.

2.5 The failings of liberal economic democracy

The PLED model can be severely criticised on a number of important issues.

In the first place, all individuals are clearly not equal economically, equal politically, or free to act in abstraction from their background, experience and the system in which they operate (Marcuse, 1964, 1986). The differences in economic and political power and freedom between even the average professional household and the homeless, the destitute, the mentally ill, the victimised and the abandoned, etc. in, for example, Australia, New Zealand, the USA, Canada, Germany and Holland are clearly enormous. Suggestions that an individual chooses to be born to a homeless, unemployed parent in a major city, with whatever 'social disadvantages' one may care to acknowledge, or else to a stable, wealthy and privileged middle-class household in 'pleasant' suburban surroundings is an essential tenet of liberal economic democracy, and clearly it is nonsense.[12] If that professional household in a developed country is then compared with an Ethiopian peasant family caught in a famine then some idea of the full extent of the inequities starts to become apparent.

Second, individuals cannot act independently of their framework and it is *not* generally individuals that exercise the real power but institutions – states, governments, corporations, etc. That is, there is 'power asymmetry' between the actors – Gray, Owen, Adams and General Motors are four actors in the liberal economic democratic world acting with equal power, wealth and freedom? Hardly! But far more reaching than this is *how* that power is used by corporations and by states. The Marxian critique, simply stated, suggests that:

- power is held by 'capital' and exercised on its behalf:
- conflict is the natural state in capitalism and, in particular, conflict between capital and labour;
- the state is 'captured' by capital and operates on its behalf to protect, reinforce and support capital's expansion and to maintain the suppression of labour;
- the emergence of a middle-class does not necessarily change anything in that it (and especially professionals such as accountants and managers) acts on behalf of capital, is privileged by it (as long as it serves capital) and is beholden to it.

One does not need to be a 'Marxist' or any other kind of 'ist' – nor does one have to accept the whole panoply of Marxism[13] to be persuaded that Karl Marx was, at a minimum, one of the brightest and wisest minds commenting on human conditions in recent centuries. One does not need to accept 'capital' and 'labour' as discrete and identifiable groups in a constant state of conflict to recognise that companies and their owners hold vast un-elected power

---

[12]Whilst it is clearly possible for some especially able, tenacious or lucky individuals to 'claw their way out' or substantially 'advance themselves' and it is probably the case that 'social mobility' has increased in certain parts of the world, this does alter the basic premise that privilege generally encourages privilege and deprivation tends to lead to further deprivation. To give a simple example to which most readers of this text can relate, the probability of someone with a 'professional' background being able to go to university and subsequently earn a 'professional' salary is very significantly higher than for someone from a 'manual' or 'unemployed' background. To believe that one attained university entrance simply because one was *intrinsically* more able or hard working – rather than because one had a background which encouraged opportunity – is the worst kind of arrogance. All peoples, even in the affluent West, are far from equal.

[13]'Marxism' comes in myriad different forms and so to talk of just 'Marxism' is misleading. It is probably more accurate to refer to 'Marxian' meaning forms of thought influenced by Karl Marx's insights and arguments.

throughout the globe and exercise that power in pursuit of organisational goals (however defined). One does not need to believe that the state works entirely on behalf of capital to be well aware of the many, many examples of governments acting – willingly and under pressure – to protect companies and their shareholders.[14] Furthermore, one does not need to be convinced of the ubiquity of the bourgeoisie to recognise that many professionals act to support and strengthen 'capital' – after all, what else is financial accounting? When was the last time any country's Companies Acts established the information rights of the poor, the starving, the unemployed, the homeless, environmentalists, employees, communities, etc. above the rights of the financial community?[15]

Third, the model of liberal economic democracy has many internal contradictions. Some of the most important are:[16]

● the links between individual self-interest ('greed' as it is manifested in conventional economic, finance and accounting literature) and social welfare cannot be demonstrated. It is vaguely possible that such a link exists – and, in very specific circumstances, might actually do so – but it cannot be shown to hold for everyday western economies;

● increases in income, whether measured by, for example, personal income, profit or GDP, whilst typically related to an increase in the consumption of certain things, do not measure quality of life, health, happiness or the 'consumption' of other things such as pleasurable activities, spiritual experience or the quality or quantity of the individual's relationship with community or nature;[17]

● the measure of society's wealth (gross domestic product – GDP) has many anomalies in it such that, for example, an increase in road accidents and environmental degradation count as *increases* in wealth rather than decreases (see, for example, Anderson, 1991; Ekins, 1992b);

● increases in financial wealth say nothing about the distribution of that wealth. Although total financial wealth as measured is generally increasing, the gap between rich and poor within and between nations is increasing in many places and the increased wealth in the 'developed' nations can normally be assumed to have been achieved at the expense of the 'lesser developed' nations (see, for example, Daly and Cobb, 1990; Ekins, 1992a,b; Korten, 1995; Eden, 1996; Bakan, 2004; and see also Collison *et al.*, 2007)

Fourth, the PLED model makes no allowance for environmental matters except in so far as they are represented in price. Nature is assumed to have no worth independent of its provision of economic facilities and environmental, ecological or nature-centred values can find

---

[14]There are far too many documented examples of this to make it really contentious. One of the most obvious would be the constant lobbying by European and North American companies to persuade governments to forbear from legislation. In the 1990s, such pressure was best illustrated in the attempts of companies to avoid environmental legislation. On the other side of the argument, the Companies Acts of most countries are amongst the most detailed examples of legislation and are unique in being focused entirely upon the companies' right to operate and providing the only real means of power and control in the hands of a select group – the shareholders. No other group in society has such an enabling piece of powerful legislation.

[15]There *are* examples of some moves in this direction but, whilst not trivial, they are not radical either and certainly do not challenge the apparently immutable rights of the shareholder. Note also that the standard counter argument is obviously that it is only through liberalism and economic growth that poverty, injustice and environmental degradation can be prevented. It all gets a bit circular doesn't it?

[16]For a more detailed analysis see, for example, Galbraith (1973, 1991); Hahn (1984a,b).

[17]This is particularly striking in measures such as the Index of Sustainable Welfare and the Genuine Progress Indicator (Donovan and Halpern, 2002; Meadows *et al.*, 2004; Layard, 2005) which claim to demonstrate that well-being is now travelling in the opposite direction to conventional measures of 'wealth' – at least for the more 'developed countries' (see also Jackson, 2009).

no space within the conception. Just as social desecration is possible – and even encouraged – within liberal economic democracy, so ecological desecration is an inevitable consequence of a model of the world based upon the liberal economic democratic view (Gray, 1990).[18]

Fifth, ethical problems are also crucial and significant. In the first place, 'more' is not necessarily 'better' (see, for example, Gorz, 1989; Power, 1992). Gorz's widely repeated statement is very apposite:

> Accountancy is familiar with the categories of 'more' and 'less' but doesn't know that of enough.
>
> (Gorz 1989: 112)

The pursuit and reward of 'more profit' is not amoral – it is a deliberate moral choice. Equally, the decision to judge the desirability of action by reference to cash flow and/or profit follows most of conventional economics where the desirability of an action can be judged by its (financial) consequences. This is a *moral* position based upon a particular interpretation of utilitarianism (J. S. Mill, 1962; Jacobsen, 1991). Choice always has some moral element – some suggestion that the action is better or worse, more or less desirable, is essentially more good or more bad. That conventional economics, accounting and much of business studies have attempted to strip the explicitly moral from decisions should not blind us to the fact that decisions are still moral choices. Even the attempt to make financial and economic decision-making non-moral is itself a moral choice (Cartwright, 1990; Malachowski, 1990).[19] It is a form of intellectual dishonesty to claim, as much of business, accounting, finance and economics does, that moral issues are 'nothing to do with me' (see, for example, McPhail, 1999).

Within, the PLED vision, self-interested utilitarianism *is* moral because its consequences – economic efficiency, growth, maximal social welfare – *are* assumed to be morally desirable. It is essential to know that the shareholder's hold–sell–buy decision or the determination of a discounted cash flow-based investment decision, for instance, are *moral* acts. They are implicitly justified in terms of the assumed economic (and therefore the assumed social welfare) consequences of the actions and the criteria (self-interested utility maximising – i.e. greed) on which they are based.

However 'consequentialism' (of which utilitarianism is one part) has a fatal flaw – not all consequences can be identified or known. Of course finance, accounting and economics avoid this by completely ignoring all non-financial (human, social, environmental) consequences and using statistical analysis to deal with the known elements of the *identified* consequences. That is, the unidentified consequences are ignored.[20] Furthermore, consequentialism is only

---

[18]It is worth noting, however, that Marxian analysis is also guilty of ignoring the environmental issues (Gray, 1992). Accounting (and social accounting if derived from it) which is based upon either the liberal economic democratic vision or the Marxian vision is almost certain to encourage and reward ecological desecration (Maunders and Burritt, 1991; Gray, 1992; Tinker and Gray, 2003).

[19]The attempt to strip the explicitly moral from economics was an essential element in leading to economics' undoubted analytical power. This is associated, primarily, with the work of Bentham and James Mill. It *does not*, however, make human action free of moral choice. The difficulties arising from this attempt are clear in John Stuart Mill's refinement of Bentham and James Mill's work. Furthermore, the 'invisible hand' which amorally achieves this mythical rise in social welfare derives from Adam Smith's work which was known to Bentham and Mill. However, Smith's 'self-interest' is wholly predicated upon the assumption of a moral and spiritual individual acting in a reasonable, moral, caring and thoughtful way. His understanding of 'self-interest' was much different from that of Bentham and wholly different from that of modern conventional economics (see for example, McKee, 1986; Raines and Jung, 1986; Coker, 1990; Reilly and Kyj, 1990; Jacobson, 1991).

[20]Even within a simplistic accounting world, this has problems as research work on the post-audit (or lack of it) of investment decisions adequately show (see, for example, Scapens and Sale, 1981; Pike, 1984; Neale, 1989).

Figure 2.1 Example of conflicts in ethical codes

The decision to invest in a new plant and where to site it is a situation potentially fraught with ethical conflict.

Which consequences should be taken into account and if so how? This may lead to the sort of ethical conflict *within* a criteria that can occur with cost-benefit analysis. That is, which costs (e.g. damage to life, communities, environment) and what benefits should be recognised and, if recognised how should they be introduced to the decision? Should they be 'valued' and if so whose valuation? Are the consequences to the corporation taking precedence over those to the local community? Why? Is it ethical?

The siting of a plant may involve bribery, for example. Is that ethical? In consequential terms it may well be so but for many people, bribery is intrinsically wrong.

Why is the company (and why are you) considering this investment? What are the motives? Is making profit an ethical motive?

Is the plant an intrinsically 'good' thing? By what criteria? Whilst it may be consequentially desirable it may not be deontologically ethical. Thus, given a particular frame of reference we might find an energy efficient plant of elegant and non-intrusive technology located in a sympathetic way an intrinsically good act. What is the plant to produce? Is producing deodorants or computer toys for rich western children an intrinsically good thing to be doing?

And so on. (For more detail, see, for example, Donaldson, 1988; Mintz, 1990.)

one way to assess the rightness or wrongness of actions. It is also possible to assess an action by reference to the motives of the actor (motivism) or by reference to the intrinsic rightness or wrongness of the act itself (deontological).

There can be no unequivocally correct way in which to judge an action and, one might suppose, a truly good action would satisfy all three criteria by reference to some 'higher' set of moral standards deduced from some conception of the actor's relationship with society, the natural environment and the metaphysical (e.g. a deity). More usually, however, the criteria will conflict. A common example of this is provided by questions about whether ends (consequences) justify means (motive and deontology). Under utilitarianism as practised by accounting and economics, the answer must be 'yes'. Under other frames of reference, for example the pacifist deep green position, they may not (see Figure 2.1).

No human action can be wholly amoral. The world of accounting and finance, within a world of conventional economics, based as it is, in essence, on liberal economic democracy has recourse to moral justification purely in terms of financial consequences – the rise in financial utility generated. This, in turn, rests upon the assumptions about the mechanisms within liberal economic democracy producing social welfare consequences. This link cannot be shown to exist.

So can organisations operating within liberal economic democracy deliver ethically justifiable actions and desirable outcomes? Of course it is possible that they can but, to get a little closer to understanding how accountability and social accounting might be important, we need to have a brief look at the broader system within which all this operates – the system of capitalism.

2.6 Capitalism and corporations

Capitalism is a broad term for the dominant way in which increasing parts of the globe organise their economic activity. Bannock *et al.* (1987), for example, define capitalism as '*[a] social and economic system in which individuals are free to own the means of production and maximize profits and in which resource allocation is determined by the price system*'. It is based very largely upon freedoms – the freedom to consume, to produce and to maximise profits (and hence is typically associated with liberalism). Its central tenet is that those with capital should be free to seek out returns on that capital. Capital is the central focus of the system, and profit – the return to capital – is the central measure of success. To all intents and purposes, when we talk of PLED we are assuming a system of capitalism, and Freidman, for example, would be seen as a champion of both: somebody who wishes all systems to be capitalistic and all capitalism to be PLED.

However, although the claim might prove to be a bit contentious, not all capitalism is the same. Consider, for the sake of illustration, three different arrangements. The first is a system of small, often one-person, businesses. Many high streets, towns and markets across the world are still organised like this. The important factors are that there are, in all probability, only one or two owners of each business; the owner(s) also works in the business, and no business is big enough, on its own, to dominate any market or the political and social processes of a locality. This might be called 'Adam Smith capitalism'. What makes this category of capitalism worth noting is that it is both potentially controllable (by customers, the community and/or legal and state bodies) and it can be stable (at least in the short to the medium term) as there is no inherent need for growth. The second category, for illustration, is the capitalism of the social democratic countries of Northern Europe, Scandinavia and (at the risk of stretching a point too far) Japan. Although now we are looking at a very wide range of organisations from the very smallest to some of the world's largest and, additionally, ownership may often be very diverse, the practice of social, cultural and government control is pre-eminent. That is, there exists an, at times, explicit recognition that society, environment and economy co-exist and the economy exists to serve its society. In such economies, there are relatively reasonably low disparities of wealth between the richest and the poorest. In such economies, explicit social goals, reasonably high taxation and a plethora of laws protecting the weak are considered normal and desirable (see, for example, Collison *et al.*, 2007).

It is, however, the third category of capitalism that, in view of its seemingly inexorable growth steadily squeezing out the first two categories, mostly exercises the world and will mostly occupy us here. This is the capitalism of international financial expansion, often thought of as Anglo-American capitalism, which is most typically associated with international financial markets, diverse and distant shareholding, voracious growth and dominance through size. It is typified by multi-national companies. These are the organisations which are frequently bigger than countries (Korten, 1995); whose leaders are paid increasingly alarmingly large amounts of money and whose power is akin to baronial princes of old; which can only be controlled by markets, which they often dominate, and shareholders, who are either too diverse, too distant, too uninformed and/or too well-rewarded to actively interfere, and whose influence on the global political economy (including the social and environmental state of the planet) is vastly in excess of any other body (see, for example, Bailey *et al.*, 1994, 2000; Eden, 1996; Hirst and Thompson, 1996; Bakan, 2004).

This modern international financial capitalism has been enormously successful as the engine of economic growth and globalisation but has also, in the deployment of, frequently inappropriate, argument in support of PLED, had the effect of dominating much debate throughout the developed and developing world about such matters as liberalisation, privatisation, lower taxes, removal of trade barriers and, consequently, on the whole of societies' structures and values. It is with these organisations that Joel Bakan's (2004) influential analysis of corporations as 'psychopaths' is concerned and where our concern for organisations which are bigger and more important than people really takes shape. These organisations are a manifestation of a capitalism which is – literally – potentially out of control. It is here that the need for social, environmental and sustainability accountability is at its most acute and offers the most potential.

2.7 Reformism or radical change?

This brings us to three of the central tensions that arise in any study of capitalism and accountability . . . and, like the study of any area of social science, they involve trying to clarify key assumptions and values underlying our arguments. We need to address these briefly before we can proceed.

First, there is the theoretical – but also importantly practical – argument over whether or not there are indeed three (or more) categories of capitalism or whether in fact all forms naturally gravitate towards the voracious international financial capitalism we now have. *Second*, there is a widespread range of views as to how the current version(s) of capitalism actually work(s) and, more particularly, whether we judge the system as we understand it to be a 'good' thing or not. *Finally*, if we decided that current forms of capitalism do not represent the best of all possible worlds, there is the hotly contested debate over whether the system can be reformed. If such reform is simply not possible and the consequences of our current means of organising are so profound, it may be that it is capitalism itself which must be destroyed before it destroys society and the environment.

We shall not dwell on the first of these to any great degree. There is substantial argument around whether all capitalism is ultimately voracious and/or successfully efficient and over whether different forms can exist and the more extreme elements of the system controlled for human benefit. To seriously discuss the 'categories of capitalism' would take us further into the heart of political theory than we have space for here. As with many arguments, it is probably the case that both 'sides' have some truth: capitalism has a built in engine of growth and appropriation *and* some societies control that nature more successfully than others. At a minimum, we should always be aware that it takes consistent moral courage by owners, governments and civil society to ensure that the 'bad' (the avaricious, growth-orientated and ruthless individuals, groups and organisations) does not drive out the 'good' (the smaller, community-based, values-based, innovative grass-roots and personal organisations). It is not obvious that such moral courage is in excess supply.

The other two tensions are more obviously central to any concerns about social accountability. That is, much of the optimism that underlies the view that social accounting has real potential and real value lies in a reformist assumption – namely that organisations and especially corporations can be redirected, re-designed and controlled . . . and that this is a 'good thing'.

These are matters explored in much more detail in **Chapter 3**. Suffice to say at this stage that there is a very wide diversity of possible views about the relationships between organisations, business, society and the environment and the extent to which gradual reform is both possible and desirable. Indeed, such views are based on nothing less than beliefs about what it is to be human and how humanity can and should organise itself. These often very complex views vary from pristine capitalism through socialism and feminism to postmodern and deep ecological worldviews. They represent a diversity of beliefs about how the world works and what we might need to do about it that reflects both understandings of the world as it is (the **positive**) and the ethical desiderata that we all pursue (the **normative**). It helps us place the PLED views in context and points to values and perceptions through which we might come to disagree with the PLED view. This range of worldviews will help explain some of the arguments and debates with which any examination of social accounting must be concerned. And, although we will continue with the idea that accountability is a reformist notion, we would like to suggest that, in all probability, it may transpire that it has enormous radical potential (Owen, 2008).

Worldviews, along with a rich understanding of social responsibility, require much deeper examination (see Buhr and Reiter, 2006; and **Chapter 3**) but, for the time being, we need to 'take a position' which doesn't exclude other, equally validly held, views. To do this, we will adopt something called a 'neo-pluralist' point of view.

2.8 A neo-pluralist vision of the world

In this section, we will attempt to explain the neo-pluralist conception of the world. The 'plural' in 'pluralism' refers to the idea that there are many sources of power and influence in a society. As we also saw above, neo-classical economics, and thus a great deal of business, accounting and finance, implicitly assumes that power is widely distributed between all individuals – that all individuals are equal. Whilst we might wish this to be the case *in principle*, it is clearly not so *in fact*. '**Neo-pluralism**' recognises this and assumes that, whilst power is not located in a single individual or group, for example the state, capital, a ruling elite, or whoever, neither is it evenly distributed (Shell, Oxfam and Reuters are all clearly more powerful than you are and therefore have more influence in political, economic and social matters than you do). Whilst we can develop this idea a lot further, for our present purposes, we can summarise the main elements in the visualisation presented in Figure 2.2 (for more detail, see Held, 1987).

Figure 2.2 A neo-pluralist visualisation of society

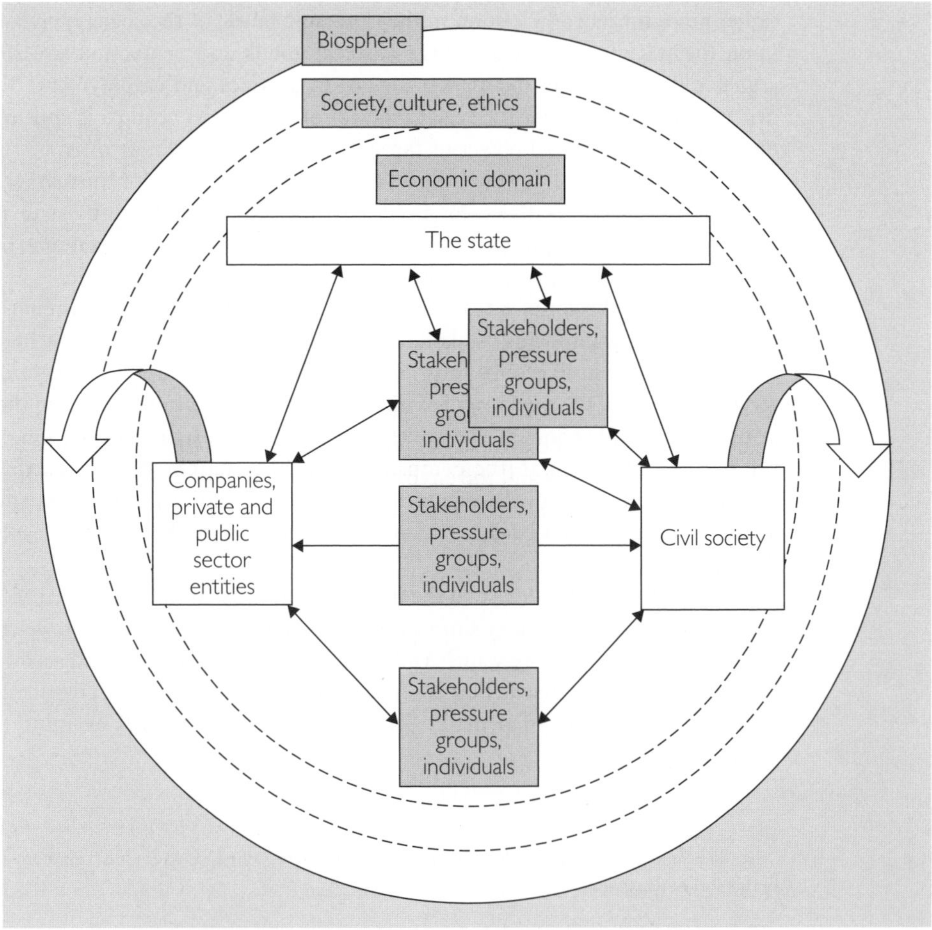

That is, again using the systems perspective which we introduced earlier, Figure 2.2 presents the 'economic domain' as being located within the 'societal, cultural and ethical domain'. The society, its culture and ethics determine, to a considerable degree, the structure and the acceptable modes of behaviour in the economic domain. The society is, itself, located within – and perhaps should be thought as indistinguishable from – the natural environment. It is a very modern, western conceit to think of society and economics as distinct from each other and both as distinct from the natural environment in which they are located. (The dotted lines in the figure are suggestive of a permeable relationship which we have not developed here.)

Lying within these systems, it is normal to think of society as having three principal components: the state (government and its organs), the market (dominated by business) and civil society (being individuals and the smaller, often socially – or community-based, organisations such as self-help groups).[21] Figure 2.2, also seeks to indicate that we need to recognise the interaction between these three systems – or subsystems if you prefer – and the social, cultural, ethical and environmental domains. The diagram is a serious simplification of a complex world.

The three principal systems interact in a myriad of ways: purchases, sales, advertising, laws, taxes, employment, education, etc., etc. None can act independently of the others. Putting it naively simply: business depends upon civil society for customers and employees and upon the state for laws, protection and legitimacy; the state depends upon civil society for its support and taxation and upon business for, to a degree, its taxes but probably rather more for its contribution to the economic fabric of the society; civil society depends upon the state for its social contract, social goods and protection whilst it depends, to arguable degrees, upon the market for goods, services and employment. The point of the neo-pluralist view is that the relative power of these three groups is shifting and contestable. Whilst the potential power of the state is recognised, it may often be that it cannot set laws, for example, without the support of the civil society and/or the market. The dominant power of the market is also recognised, but this can be tempered by state and civil society action. So, with a neo-pluralist view there is recognition, that, for example, both the pristine capitalist and the socialist have points worthy of attention and respect.

What makes the actual situation so much more dynamic is the existence of the plurality of individuals, groups and 'stakeholders' who act upon the dominant bodies in the economic domain. These groups form, disband and reform and act as pressures upon the different elements of the system and as mediations between different parts of the system. Thus, again being simple about it, the power that has slowly leached away from civil society and the state towards capital (the companies and their shareholders) is partially balanced by NGO and protest groups (Greenpeace, the anti-globalisation protests and so on) which attempt to bring power back to the society and to require the state to act other than simply for capital.

This finally brings us close to the nub of the matter. In essence, social accounting is concerned with a few apparently simple questions: What sort of society do we currently live in? What sort of society do we want? And what role does information and accountability play in the understanding, construction and maintenance of that society? To offer a way of making sense of all this, the final major section of this chapter will examine democracy and information.

---

[21]See, for example, Gray *et al.* (2006); Unerman and O'Dwyer (2006); O'Dwyer (2007) for more detail and references to this sort of discussion.

2.9 Democracy and information

A key aspect of PLED is that for freedom, choice and democracy to work, the elements that make up the system – ideally assumed to be the people or the 'demos' – must know about things which affect them and have the time and knowledge to take decisions and make choices. Information is, therefore, key to any examination of capitalism and/or PLED.

It is therefore potentially surprising to discover that traditionally capitalism has favoured the provision of information to only one major group – capital itself. As a result of law, those with the most power and influence (the shareholders) are given a unique service of information (the financial accounts or financial statements) which has been expensively and specifically prepared by a state monopoly of highly trained and regulated individuals (accountants) in order that these shareholders might make decisions about how they should invest to make yet more money. This is odd, because there is no other group in democratic society which has anything like such a privileged and legally required supply of information to support *their* decisions and preferences.

Furthermore, this information, this financial account, more or less ignores all social and environmental matters except in so far as aspects of society or the environment are reflected in costs and prices. As a consequence, those who use financial accounting information to make their decisions are highly likely to make a decision with little or no explicit concern for social or environmental effects. And yet, it is quite apparent that economic decisions have an enormous range of effects, not all of which are negative of course, on society and the natural environment.

Therefore, we can begin to see that the uneven distribution of information can, to a considerable degree, be taken as reflecting an uneven distribution of power. Financial accounts contribute to that, essentially un-democratic, situation. Similarly therefore, a change in information – including accounting and social accounting – can be taken as reflecting a change in society and can even perhaps be used either to reinforce or change the distribution of influence in a society. Thus if some information on polluting discharges to water (say) is provided to society by companies, there may have been a shift in the relative power of these two groups which brought this about *and* that switch in the levels of information asymmetry may reinforce that change. That is, the information is never neutral. The sort of information we have tells us what sort of democracy we have, if indeed we have one at all. A full democracy would have full accountability based on full information flows.

Democracy is a very broad term and, at its simplest,

> means little more than that, in some undefined sense, political power is ultimately in the hands of the whole adult population, and that no smaller group has the right to rule.
>
> (Robertson, 1986: 80)

There are, very crudely, three forms of democracy. **Representative democracy** is the sort of democracy which most of us in the (so-called) developed Western nations associate with the term. The 'will of the people' is operated through the election of representatives to speak and act on their behalf. It is usually operated through a party-political system which, whilst being politically expedient, vastly reduces the choices of the demos to the selection between two or three political parties every few years. Representative democracy is thus a basically fairly passive form of democracy and wide open to abuse. **State democracy** was conceived by Marx and developed by Lenin as an intermediate step towards a 'truer', socialist democracy. In this conception, the state first removed all of the inequalities in the nation which prevent the essential democratic ingredient of equality of power being exercised. This fairness is an essential element of socialism and, as we saw, one of the major flaws in liberal democratic economy. The state then acted on behalf of the people, rising above individual exercise of power, with the assumption, in Marx's conception anyway, that an equality of power would slowly be handed back to the people in a socialist utopia. Versions, albeit arguably somewhat perverted versions, of such a model have operated in the past, for example, in the USSR, Cuba and China, and in these nations the major disadvantages – as well as the far from trivial advantages – were so clearly observable. Finally, there is something called **participative (or participatory) democracy** which requires a much higher level of personal involvement of the demos in the political process through, normally, a major devolution of power to 'local level' politics. Such was the basis of the Athenian model and, very importantly, is the sort of democracy assumed in liberal democratic economics where 'active votes' are expressed through 'markets' by *informed* and active individuals. As we have seen, the modern economics version does not exist and Held (1987), for example, shows that the Athenian version was only a myth.[22] But its attractions are considerable and have led Macpherson (see, for example, 1973, 1977, 1985) – one of the leading modern political commentators – to see it as the last untried possibility for democratic organisation of society.[23]

Each model of democracy has its share of strengths and weaknesses. However, modern political thinking suggests that characteristics such as fairness and justice can be married with other desirable characteristics like freedom and opportunity through a re-democratisation of society. This can only be achieved through the return of power to the people. A necessary – but by no means sufficient – requirement for this is that information flows are themselves more 'democratic'. Such a flow would be thought to be necessary but not sufficient because, on the one hand, we have the considerable imbalance in current power distributions whilst, on the other hand, there must be scepticism as to whether simply informing people is sufficient to bring about change.

The importance of participatory information flows is emphasised by the considerable emphasis placed on them by environmentalists. Their argument is that the planetary ecology is so complex that no-one can possibly know how to solve it and no one individual or group can possibly have the right to take decisions which affect the planet and the life-expectancy of large swathes of the world's population. Only through empowering the individual to make informed and caring choices might there be any chance of ameliorating the environmental crisis – or at least permitting the species to make itself extinct in an informed way (Gray, 1992).

Such a view is, however, also highly contentious. In essence, this view, which we will continue to maintain for convenience if nothing else, has a rather naïve, wide-eyed optimism about it that may, in fact, be dangerous. It is by no means certain that an informed demos will still 'do the right thing'. Evidence is fairly clear that even when people know what is a 'good' action they still have difficulty overcoming habit, self-interest and inertia. Furthermore, given that the environmental crisis is a pressing need and that democracy tends to be a fairly slow system through which change can be brought about, even a well-informed and well-meaning demos may fail to change quickly enough. These are complex political matters that we have no space to develop further here (see, for example, Dobson, 1990; Zimmerman, 1994; Baxter, 1999; Burchell, 2002; Kovel, 2002), but it is essential to remain aware that how change might come about – and whether in fact accountability is a sufficient condition for change – are moot points.

---

[22]'Mythical' because the franchise (the right to vote) was only given to males of wealth and thus *disenfranchised* the very system which supported it – the women, slaves and freemen without wealth.

[23]See, for example, Held (1987) and Macpherson (1977). There is also a growing interest in something called 'communitarianism' which is developing a notion of a more active society which seems to us to bear a more than passing resemblance to aspects of the participatory democracy envisaged by Macpherson (see, especially Taylor, 2001). In the arena of social accounting, this position is developed well by Glen Lehman, (Lehman, 2006, 2007).

So, returning to our current attachment to the attractions of a participative democracy, it is essential that there must be flows of information in which those controlling the resources provide accounts to society of their use of those resources. This is **accountability**, the development of which we see to be the major potential for social accounting. That is a key component of our view of social accounting and is developed in much more detail in Gray *et al.* (1996) and in **Chapter 3**.

2.10 Summary and conclusions

This chapter has sought to develop a broad understanding of the context within which our concerns about justice, sustainability and social accountability might be considered. Systems thinking has given us a broad intellectual heuristic within which we can try to frame problems and the unpicking of pristine liberal democracy has provided some insights into how this world of ours is assumed to work. This understanding is necessary if we are to start to discuss how we might go about dealing with issues that concern us and, in particular, why social accountability might be desirable and the likely implications it may have.

To move us forward, we have introduced a neo-pluralist visualisation of society which conceives of power and influence as widely – but not equally – spread. Neither political power (the power of 'votes') nor economic power (the power to 'vote with dollars in the market place') is distributed with equality, justice or fairness. The distribution of and access to information in general and accounting in particular also follows this asymmetry. Information reflects, reinforces and/or helps create those inequalities. In a participatory democracy, those inequalities would be less pronounced and they certainly would be more visible: the demos would have rights to information and the actions taken on their behalf would be more transparent. **Accountability** would be more developed and more widely discharged (see Gray *et al.*, 1996; and **Chapter 3**).

References

Abercrombie, N., Hill, S. and Turner, B. S. (1984) *Dictionary of Sociology*. Harmondsworth: Penguin.

Ackoff, R. L. (1960) Systems, organisations and interdisciplinary research, *General Systems Theory Yearbook*, **5**: 1–8.

Adorno, T. W. and Horkheimer, M. (1972) *Dialectic of Enlightenment* (trans J. Cummings). New York: Herder and Herder [originally published 1947].

Anderson, V. (1991) *Alternative Economic Indicators*. London: Routledge.

Bailey, D., Harte, G. and Sugden, R. (1994) *Making Transnationals Accountable: A significant step for Britain*. London: Routledge.

Bailey, D., Harte, G. and Sugden, R. (2000) Corporate disclosure and the deregulation of international investment, *Accounting, Auditing and Accountability Journal*, **13**(2): 197–218.

Bakan, J. (2004) *The Corporation: the pathological pursuit of profit and power*. London: Constable and Robinson.

Bannock, G., Baxter, R. E. and David, E. (1987) *Dictionary of Economics*. London: Penguin.

Barach, P. and Baratz, M. S. (1962) Two faces of power, *The American Political Science Review*, **56**: 947–52.

Baxter, B. (1999) *Ecologism: An introduction*. Edinburgh: Edinburgh University Press.

Beishon, J. and Peters, G. (1972) *Systems Behaviour*. London: Open University/Harper & Row.

Benston, G. J. (1982a) An analysis of the role of accounting standards for enhancing corporate governance and social responsibility, *Journal of Accounting and Public Policy*, **1**(1): 5–18.

Benston, G. J. (1982b) Accounting & corporate accountability, *Accounting. Organizations and Society*, **7**(2): 87–105.

Benston, G. J. (1984) Rejoinder to 'Accounting and corporate accountability: An extended comment', *Accounting Organizations and Society,* **9**(3/4): 417–19.

Boulding, K. E. (1966) The economics of the coming Spaceship Earth, in Jarratt, H. (ed.), *Environmental Quality in a Growing Economy,* pp. 3–14. Baltimore, MD: John Hopkins Press.

Bryer, R. A. (1979) The status of the systems approach, *Omega,* **7**(3): 219–31.

Buhr, N. and Reiter, S. (2006) Ideology, The Environment and One World View: A discourse analysis of Noranda's environmental and sustainable development reports, in Freedman, M. and Jaggi, B. (eds), *Advances in Environmental Accounting and Management* lss. 3, pp. 1–48.

Burchell, J. (2002) *The Evolution of Green Politics.* London: Earthscan.

Carter, R., Martin, J. Mayblin, B. and Munday, M. (1984) *Systems, Management and Change: a graphic guide.* London: Harper Row.

Cartwright, D. (1990) What price ethics? *Managerial Auditing Journal,* **5**(2): 28–31.

Checkland, P. B. (1981) *Systems Thinking. Systems Practice.* Chichester: Wiley.

Coker, E. W. (1990) Adam Smith's concept of the social system, *Journal of Business Ethics,* **9**(2): 139–42.

Collison, D., Dey, C., Hannah, G. and Stevenson, L. (2007) Income inequality and child mortality in wealthy nation, *Journal of Public Health,* March, **29**(2): 114–17.

Dahl, R. A. (1970) *Modern Political Analysis.* Englewood Cliffs, NJ: Prentice Hall.

Dahl, R. A. (1972) A prelude to corporate reform, *Business and Society Review,* Spring: 17–23.

Daly, H. E. and Cobb, J. B. Jr. (1990) *For the Common Good: Redirecting the economy towards the community, the environment and a sustainable future.* London: Greenprint.

Den Uyl, D. J. (1984) The new crusaders: The corporate social responsibility debate, *Studies in Social Philosophy and Policy* No. 5 (Ohio: Social Philosophy and Policy Center).

Dobson, A. (1990) *Green Political Thought.* London: Unwin Hayman.

Donaldson J. (1988) *Key Issues in Business Ethics.* London: Academic Press.

Donovan, N. and Halpern, D. (2002) *Life Satisfaction: The state of knowledge and implications for government.* London: UK Government Strategy Unit, Cabinet Office.

Eden, S. (1996) *Environmental Issues and Business: Implications of a changing agenda.* Chichester: Wiley.

Ekins, P. (1992a) *A New World Order: Grassroots movements for global change.* London: Routledge.

Ekins, P. (1992b) *Wealth Beyond Measure: An atlas of new economics.* London: Gaia.

Emery, F. E. (ed.) (1969) *Systems Thinking.* Harmondsworth: Penguin.

Everett, J (2004) Exploring (false) dualisms for environmental accounting praxis, *Critical Perspectives on Accounting,* **15**(8): 1061–84.

Everett, J. and Neu, D. (2000) Ecological modernization and the limits of environmental accounting, *Accounting Forum,* **24**(1): 5–29.

Feyerabend, P. (2011) *The Tyranny of Science.* Cambridge: Polity.

Friedman, M. (1962) *Capitalism and Freedom.* Chicago, IL: University of Chicago.

Galbraith, J. K. (1973) *Economics and the Public Purpose.* Harmondsworth: Penguin.

Gorz, A. (1989) *Critique of Economic Reason* (trans Handyside, G. and Turner, C.). London: Verso.

Galbraith, J. K. (1991) Revolt in our time: the triumph of simplistic ideology, in Kaldor, M. (ed.), *Europe from below,* pp. 67–74. London: Verso.

Gray, R. H. (1990) *The Greening of Accountancy: The profession after Pearce.* London: ACCA.

Gray, R. H. (1992) Accounting and environmentalism: an exploration of the challenge of gently accounting for accountability, transparency and sustainability, *Accounting Organisations and Society,* **17**(5): 399–426.

Gray, R., Bebbington, J. and Collison, D. J. (2006) NGOs, Civil Society and Accountability: Making the people accountable to capital, *Accounting, Auditing and Accountability Journal,* **19**(3): 319–48.

Gray, R. H., Owen, D. L. and Adams, C. (2010) Some Theories for Social Accounting?: A review essay and tentative pedagogic categorisation of theorisations around social accounting, *Advances in Environmental Accounting and Management,* **4**: 1–54.

Hahn, F. (1984a) Reflections on the invisible hand, in Hahn, F. (ed.), *Equilibrium and Macroeconomics,* pp. 109–33. Oxford: Blackwell.

Hahn, F. (1984b) Economic theory and Keynes' Insights, *Empirica,* **11**(1): 7–22.

Hayek, F. A. (1960) *The Constitution of Liberty.* London: Routledge.

Hayek, F. A. (1982) *Law, Legislation and Liberty Vol. 3.* London: Routledge.

Held, D. (1987) *Models of Democracy*. Oxford: Polity.

Hirst, P. and Thompson, G. (1996) *Globalization in Question*. Cambridge: Polity.

Hopper, T. and Powell, A. (1985) Making sense of research into the organisational and social aspects of management accounting: A review of its underlying assumptions, *Journal of Management Studies*, **22**(5): 429–65.

Jackson, T. (2009) *Prosperity Without Growth? The transition to a sustainable economy*. London: Sustainable Development Commission.

Jacobsen, R. (1991) Economic efficiency and the quality of life, *Journal of Business Ethics*, **10**: 201–9.

Kast, F. E. and Rosenzweig, J. E. (1974) *Organisation and Management: A systems approach*. Tokyo: McGraw Hill Kogakusha.

Korten, D. C. (1995) *When Corporations Rule the World*. West Hatford/San Francisco, CA: Kumarian/Berrett-Koehler.

Kovel, J. (2002) *The Enemy of Nature: The end of capitalism or the end of the world?* London: Zed Books.

Laughlin, R. C. and Gray, R. H. (1988) *Financial Accounting: method and meaning*. London: Van Nostrand Reinhold.

Layard, R. (2005) *Happiness: lessons from a new science*. New York: Penguin.

Lehman, G. (2006) Perspectives on language, accountability and critical accounting: An interpretative perspective, *Critical Perspectives on Accounting*, **17**(6): 755–79.

Lehman, G. (2007) Ethics, communitarianism and social accounting, in Gray, R. H. and Guthrie, J. (eds), *Social Accounting, Mega Accounting and beyond: A Festschrift in Honour of M. R. Mathews*, pp. 35–42. St Andrews: CSEAR Publishing.

Lowe, A. E. (1972) The finance director's role in the formulation and implementation of strategy, *Journal of Business Finance*, **4**(4): 58–63.

Lowe, A. E. and McInnes, J. M. (1971) Control of socio-economic organisations, *Journal of Management Studies*, **8**(2): 213–27.

Lukes, S. (1974) *Power: A radical view*. London: Macmillan.

Macpherson, C. B. (1973) *Democratic Theory: Essays in retrieval*. Oxford: Oxford University Press.

Macpherson, C. B. (1977) *The Life and Times of Liberal Democracy*. Oxford: Oxford University Press.

Macpherson, C. B. (1985) *The Rise and Fall of Economic Justice and Other Papers*. Oxford: Oxford University Press.

Malachowski, A. (1990) Business ethics 1980–2000: an interim forecast, *Managerial Auditing Journal*, **5**(2): 22–7.

Marcuse, H. (1964) *One-Dimensional Man*. Boston, MA: Beacon.

Marcuse, H. (1986) *Reason and Revolution*. London: Routledge [originally published 1955].

Maunders, K. T. and Burritt, R. (1991) Accounting and ecological crisis, *Accounting, Auditing and Accountability Journal*, **4**(3): 9–26.

McKee, A. (1986) The passage from theology to economics, *International Journal of Social Economics*, **13**(3): 5–19.

McPhail, K. (1999) The threat of ethical accountants: An application of Foucault's concept of ethics to accounting education and some thoughts on ethically educating for the other, *Critical Perspectives on Accounting*, **10**(6): 833–66.

Meadows, D. (2009) *Thinking in Systems: A primer*. London: Earthscan.

Meadows, D. H., Randers, J. and Meadows, D. L. (2004) *The limits to growth: The 30-year Update*. London: Earthscan.

Mill, J. S. (1962) *Utilitarianism* (edited by M. Warnock). London: William Collins [first published 1863].

Mintz, S. M. (1990) *Cases in Accounting Ethics and Professionalism*. New York: McGraw Hill.

Mulligan, T. (1986) A critique of Milton Friedman's essay 'the social responsibility of business is to increase its profits', *Journal of Business Ethics*, **5**: 265–9.

Neale, C. W. (1989) Post-auditing practices by UK firms: Aims, benefits and shortcomings, *British Accounting Review*, **21**(4): 209–328.

Nozick, R. (1974) *Anarchy, State and Utopia*. Oxford: Blackwell.

O'Dwyer, B. (2007) The nature of NGO accountability: motives, mechanisms and practice, in Unerman, J., Bebbington, J. and O'Dwyer, B. (eds), *Sustainability Accounting and Accountability*, pp. 285–306. London: Routledge.

Owen, D. (2008) Chronicles of wasted time? A personal reflection on the current state of, and future prospects for, social and environmental accounting research, *Accounting. Auditing and Accountability Journal*, **21**(2): 240–67.

Pike, R. H. (1984) Sophisticated capital budgeting systems and their association with corporate performance, *Managerial and Decision Economics*, **5**(2): 91–7.

Power, M. (1992) After calculation? Reflections on critique of economic reason by Andre Gorz, *Accounting, Organizations and Society*, **17**(5): 477–500.

Puxty, A. G. (1986) Social accounting as immanent legitimation: A critique of a technist ideology, *Advances in Public Interest Accounting*, **1**: 95–112.

Puxty, A. G. (1991) Social accountability and universal pragmatics, *Advances in Public Interest Accounting*, **4**: 35–46.

Raines, J. P. and Jung, C. R. (1986) Knight on Religion and Ethics as agents of social change, *American Journal of Economics and Sociology*, **45**(4): 429–39.

Reilly, B. J. and Kyj, M. J. (1990) Economics and ethics, *Journal of Business Ethics*, **9**(9): 691–8.

Robertson, D. (1986) *Dictionary of Politics*. London: Penguin.

Scapens, R. W. and Sale, J. T. (1981) Performance measurement and formal capital expenditure controls in divisionalised companies, *Journal of Business Finance and Accounting*, **8**(3): 389–419.

Speake, J. (ed.) (1979) *A Dictionary of Philosophy*. London: Pan.

Sunstein, C. R. (1997) *Free Markets and Social Justice*. New York: Oxford University Press.

Taylor, C. (2001) A tension in modern democracy, in Botwinick, A. and Connolly, W. (eds), *Democracy and Vision*, pp. 79–99. Princeton, NJ: Princeton University Press.

Tinker, A. M. and Okcabol, F. (1991) Fatal attractions in the agency relationship, *British Accounting Review*, **23**(4): 329–54.

Tinker, T. and Gray, R. (2003) Beyond a critique of pure reason: from policy to politics to praxis in environmental and social research, *Accounting Auditing and Accountability Journal*, **16**(5): 727–61.

Unerman, J. and O'Dwyer, B. (2006) Theorising accountability for NGO advocacy, *Accounting, Auditing and Accountability Journal*, **19**(3): 349–76.

von Bertalanffy, L. (1956) General Systems Theory, *General Systems Yearbook*, **1**: 1–10.

von Bertalanffy, L. (1971) *General Systems Theory: Foundations, Development, Applications*. Harmondsworth: Penguin.

von Bertalanffy, L. (1972) General Systems Theory – a critical review, in Beishon, J. and Peters, G. (eds), *Systems Behaviour*. London: Open University/Harper & Row.

Walton, C. W. (1983) Corporate social responsibility: The debate revisited, *Journal of Economics and Business*, **3**(4): 173–87.

Weston, J. (ed.) (1986) *Red and Green: the new politics of the environment*. London: Pluto Press.

Zimmerman, M. E. (1994) *Contesting Earth's Future: Radical ecology and postmodernity*. London: University of California Press.

CHAPTER **3**

Corporate social responsibility and accountability

3.1 Introduction

Chapter 2 sought to lay out a broad theoretical backdrop against which we might set our study of social accounting. We offered a few, albeit brief, views of how the economic, business and organisational world might be thought to work. We introduced the notion of a neo-pluralist view of the world through which we might study organisational activity whilst remaining properly sensitive to the insights of (say) Marxian analysis and the claims of (say) liberalism. We also chose, perhaps a little reluctantly, to assume (for the time being at least) a more *reformist* rather than a more overtly *radical* approach to addressing social and environmental issues primarily through our understanding of the operations of financial and economic organisation.[1] Those social and environmental issues we set within the concerns for social justice, environmental stewardship and sustainability from Chapter 1 and conceived of the inter-connections between social desiderata, environmental exigencies and organisational behaviour via *systems thinking* through which inter-dependences and inter-actions might be more dynamically understood.

For those more immersed in the theories of social science, in the work of Marx, Foucault, Bourdieu, Giddens, Latour and others, this may seem a little superficial. However, for many this outline might have been one of their first introductions to the basic elements of social theory. (Chapter 4 provides a somewhat more detailed introduction to theory.) Either way, we hope that Chapter 2 has provided a sufficiently interesting and clear landscape to allow us to move closer to the organisational and corporate issues of social accounting that will occupy us for much of the rest of the book.

This mental framework – *meta-theory*[2] of a sort – prepares the way for us to start to focus on the key *meso*-theoretic ideas which will be more recognisable to us and which we will use as our framework for much of what follows. The two major themes which we explore in this chapter are those of corporate social responsibility (CSR) and of accountability.

---

[1]It is worth repeating the concern that the tension between the views of reform – the gradual and incremental change in society – versus radical change – the root and branch destruction and rebuilding of a society – is significant and should not be trivialised. Equally, reform via information and the very organs in which current power resides may well have the seeds of its own destruction built into it. Much of what we consider is based on the potential for goodwill and action from both civil society and the state. This may be naive.

[2]Theory is conveniently thought of as occurring at three levels. Meta-theory concerns 'grand' theory which tries to offer a broad explanation of the major sweeps of influence that structure and are structured by our societies, economies and cultures. Meso-theory (or, crudely, 'middle' theory) works at a higher level of resolution and deals with a more recognisable level of theory wherein we might talk about elements in society, organisations and groupings. Finally, there is 'micro-theory' which is focused and specific. Chapter 2 dealt with meta-theory, this chapter deals with meso-theory and Chapter 4 will introduce a range of meso/micro-theories relevant to our study. (The distinctions are not always precise!)

Figure 3.1 Top 10 benefits to (US) companies of becoming socially responsible

The Aspen Institute (2003) Survey of MBA students

- Enhanced public image/reputation (75%)
- Greater customer loyalty (51%)
- More satisfied and productive labour force (37%)
- Fewer regulatory or legal problems (37%)
- Long-term viability in marketplace (36%)
- A stronger and healthier community (34%)
- Increased revenue (6%)
- Lower cost of capital (2%)
- No benefit (2%)
- Easier access to foreign markets (2%)

Source: Adapted from Carroll and Buchholtz (2006).

Social responsibility has an intrinsic importance for any person who wants to know what it means to be a good person or to lead a good life. That is not, however, the only reason for emphasising it here. CSR has occupied an important place in the discussion of what it is to be a corporation for nearly 50 years. And whilst, in that time, very little firm conclusion has been reached, the discussion throws up a great many interesting and crucial insights into organisational life in the late 20th and early 21st centuries. But, more important to our present concerns, as we shall see later, accountability largely depends upon some idea of responsibility. So we will need to understand responsibility before we can talk sensibly about what forms of accountability are necessary and why.

The broad outline of the argument we will develop is this. Accountability derives from responsibility – to say anything about accountability we need to understand responsibility. Understanding what is meant by responsibility is key to so much of life – including organisational life. We will suggest that responsibility is a function of relationships between people, organisations, groups, societies and so on. To understand responsibility requires that we understand relationships, but here we hit a snag: there are a great many views (we will call them 'worldviews') about the nature of these relationships – and especially the relationships between organisations and society. The worldview one holds directly affects how one perceives these relationships. And the matter is made more complex because we are caught up in a world of un-sustainability which raises yet further questions about relationships – especially between societies, justice and the planet. It is into this complex melange that we will place accountability as a way of seeing through some of this confusion.

Consequently, this fairly long chapter is organised as follows. Section 3.2 introduces CSR and suggests why it is important. In Section 3.3 we consider definitions of CSR and then link those definitions with different worldviews in Section 3.4. Section 3.5 provides a brief initial examination of the link between sustainability and social responsibility before we try and offer, in Section 3.6, explanations as to why CSR is such a confused area. Section 3.7 introduces the notion of accountability and provides the basis for the model of accountability in Section 3.8. Sections 3.9 and 3.10, respectively, explore some of the practical components of accountability and then return to the matter of defining CSR. On this basis, we then return and undertake a more careful consideration of the relationship between social responsibility and sustainability in Section 3.11. Section 3.12 summarises the chapter and draws a few conclusions. There is a brief Appendix to the chapter in which the limitations of the accountability are explored.

3.2 Why is (corporate) (social) responsibility important?

Understanding personal responsibility is one of the ways in which we can try to comprehend what it is to be human and how we need to act to be good members of society. But trying to get to the bottom of what it is to be responsible is never simple.[3] It has probably ever been thus – and, because of the apparent inextricable link between power and responsibility, it seems to be that the more power an individual has, the more difficult it becomes for them to understand their responsibility in that the moral choices appear to be more complex. However, individuals do seek to articulate their responsibility for themselves: but who sorts it out for organisations?

For organisations, the nature of responsibility becomes yet more elusive. For *some* organisations this may not be an issue requiring especial attention. That is, for (very?) small organisations it may well be that we can often see a congruence between the personal values of those owning/running the organisation, what society wants from the organisation and the responsibility (whatever we mean by this) of the organisation itself. Equally, for non-profit organisations it may well be that again there is a congruence between the organisational objectives and the values held for it by the society whilst those who come to work for such organisations may do so because they share those perceived values.[4]

More especially, social responsibility might well not be a problem at all if we could accept (as the pristine capitalists claim – see below) that the only responsibility of a for-profit organisation is to make money – i.e. to make profit. The problems arise because many of the world's most influential and important organisations seek profit and growth such that their actions, their exercise of power and the way in which they behave increasingly influences the non-economic aspects of life (Thielemann, 2000). In addition, much of the behaviour of these large organisations seems to be increasingly thought of as 'not responsible' in the sense that most societies would understand it. This seems to be especially true for profit-seeking organisations and especially multi-national corporations (MNCs).[5] To what extent, then, can we consider organisations responsible for the impact of their actions and, relatedly, to what extent can a complex organisation have a simple – or even single – responsibility like maximising share price or profit?

Indeed, Bakan (2004) has argued that corporations need to be increasingly seen as *psychopathic*. Moreover, he argues, any suggestion that such organisations can exhibit human characteristics – like 'responsibility' – must be, at best, misleading. Any large corporation, he demonstrates, can only pursue its own self-interest and will be 'responsible' only to the

---

[3]The longevity of the debate around responsibility and such matters is illustrated nicely by the recognition that Aristotelian arguments are still a key part of the discussion (see, for example, Chryssides and Kaler, 1993). This is not the place to embark upon a detailed study of ethical theory but see, for example, Donaldson (1988); Matravers (2007); Shafer-Landau (2007).

[4]Of course, this is also an over-simplification and may even be naive. There is a complex world of understanding concerning the sense of self, individuals and organisations through the organisation studies literature (see, for example, Knights and Willmott, 2007). For introductions to the NGO issues see, for example, Edwards and Hulme (1995); Gray *et al.* (2006); Unerman and O'Dwyer (2006).

[5]Although it is worth emphasising that these concerns can also be equally true of monolithic and enormous State and other not-for-profit bodies whose cumbersome bureaucracies can swamp their ability to serve the purposes for which they were intended.

extent that such 'responsibility' is commensurate with its self-interest or, more danger-ously, when it is in the organisation's self-interest to appear to be thought of as responsi-ble. Given the extent to which responsibility is used to legitimate and justify organisations and the extent to which appeals to 'responsibility' are so often used to justify avoidance of strict regulation, such arguments have an importance and urgency that all of us need to understand. To do that, we need to try and get underneath the skin of social responsibility and CSR.

3.3 What is CSR – can it be defined?

> A corporate executive . . . has a direct responsibility . . . to conduct the business . . . to make as much money as possible while conforming to the basic rules of society . . .
>
> (Friedman, 1970)

> Philanthropy does little or nothing to help companies make profits, while all CSR activities are linked to improving a company's bottom line.
>
> (MHCi MONTHLY FEATURE (pdf/e-journal) April 2004: 2)

These quotations alone are sufficient to indicate that views on CSR vary considerably. Let us start simply. One of the simplest – and most popular – places to start is Archie Carroll's 'Pyramid of Social Responsibility' which appears here as Figure 3.2.

Figure 3.2 The pyramid of corporate social responsibility

Source: Adapted from Carroll A. B. (1991) The pyramid of corporate social responsibility: Towards the moral management of organizational stakeholders. *Business Horizons,* July-August: 42.

Figure 3.3 Top 10 reasons (US) companies are becoming more socially responsible

PWC (2002) Sustainability survey report

- Enhanced reputation (90%)
- Competitive advantage (75%)
- Cost savings (73%)
- Industry trends (62%)
- CEO/Board commitment (58%)
- Customer demand (57%)
- SRI demand (42%)
- Top-line growth (37%)
- Shareholder demand (20%)
- Access to capital (12%)

Source: Adapted from Carroll and Buchholtz (2006).

The simple suggestion in Carroll's pyramid of four domains of concern to organisations has long appealed – especially to those seeking to manage such organisations. But Carroll's pyramid fails to address such issues as: can one be permitted to act illegally and unethically if one needs to make a profit? why is being profitable the first concern of business? what is it to be ethical? and can contribution to quality of life be found other than through philanthropy? So Carroll's useful start only gets us so far.

A potentially more telling analysis is offered by Frederick's work (see, for example, Frederick, 1978, 1986, 1994) where he identified three principal eras in the development of CSR. He calls these CSR_1, CSR_2 and CSR_3. CSR_1 was concerned with **social responsibility** which referred to the encouragement to organisations to adopt key principles and seek to adopt the obligation to work for social betterment – beyond law, economics and shareholders. Businesses were encouraged to solve the problems that they created. This is too static, commentators argued, and made it difficult for a manager to manage. Managers needed something that engaged them and their organisations. Furthermore, on what moral foundation would these principles be based and justified? This vagueness prompted management scholars to seek out a more tractable notion which Frederick calls CSR_2 or **social responsiveness**. Social responsiveness was more concerned with the processes of management and, in particular, how organisations respond to social pressure. It involved careful understanding and consequent management of the organisation's stakeholders. The notion

Figure 3.4 7 reasons to abandon social responsibility

- Origins suspect, derived from economics, ignores culture, history, religion, etc.
- All accept the terms of the debate set by Friedman – the company is a profit maximizer
- Based on 'capitalism – love it or leave it'
- Conservative – takes status quo, fixes unintended consequences
- Takes managers out of their areas of competence
- Sees business and society as separable – linked by responsibilities
- Rights and responsibilities are limiting and irrelevant to practising manager

Source: Adapted from Freeman and Leidtka, (1991).

could be reactive, proactive or interactive and was much more attractive to organisations in that it placed companies' interests first. The notion failed, however, to consider how organisations exercised their impact on public policy and it still failed to articulate the moral basis of action. Frederick, in seeking out how to articulate the moral basis of organisational action, posited CSR_3 which addresses the nature of **social rectitude**. Social rectitude was concerned with the 'moral correctness of programmes and policies' and embedding the values underlying right actions and policies (rectitude) in the organisational culture. How this was to be done and how an organisation was to choose which values remained elusive.

And that is the key to much of the CSR literature – a long and rather circular attempt to nail down a concept and to do so in terms which are acceptable to an increasingly powerful and avaricious corporate sector. There seems to be no way in which to square this circle. Indeed, one review of the CSR literature (Mohan, 2003 as reported in de Bakker *et al.*, 2005: 288) identifies a 50 year struggle for meaning using terms from business ethics and philanthropy (in the 1950s) through 'stakeholder model', 'sustainable development', 'triple bottom line' and 'corporate citizenship' in addition to CSR_1, CSR_2 and CSR_3 – all of which appear to be inconclusive. And whilst, as we shall see, all of these terms plus the widely embraced notion of social performance all have currency – they all turn out to be empty at best and tautological at worst.[6]

But all is not lost. We will show that the reasons for this confusion are really quite simple and are solved by understanding that: (i) you cannot address the principal organs of capitalism without addressing capitalism itself and (ii) responsibility is a societal, not an organisational notion and the issue is solved through the application of a full and rigorous accountability. But more of this later.

To give this discussion a little more depth and focus we will find it useful to start to map differing views of CSR with political and other worldviews.

3.4 Views of the world and views of CSR

One's views about the nature and extent of organisational social responsibility derive from one's views about how you believe the world to be and how you would *like* the world to be. There are widely different views on this – as we saw in Chapter 2. For illustration, we will run through a few general ways in which different groups in society might envisage the organisation–society relationship and, thus, how they might see the nature of social responsibility.[7]

1 The '*pristine capitalists*' are those who see liberal economic democracy (see Chapter 2) as a good approximation of how the world works and also as the way in which the world *should* work. Consider the following quotations:

> In a free enterprise, private property system, a corporate executive is an employee of the owners of the business. He has a direct responsibility to his employers. That responsibility is to conduct the business in accordance with their desires, which

---

[6]One might try and speculate that there needs to be at least one further step in the evolution – perhaps we might call it CSR_4 – and dub it social reconciliation or reconstruction? This is to recognise that we are entering (perhaps have entered) a phase in which the notions of social responsibility are being *appropriated* by some elements of the business world. This phase appears to be characterised by the 'reasoning' that companies *must* deliver social goals (because no one else can) and therefore *will* deliver social goals. The world is safe in the hands of business?

[7]For more detail on these issues see, for example, Kempner *et al.* (1976); Tinker (1985); Lehman (1992); Chryssides and Kaler (1993); Mathews (1993); Perks (1993); Bailey *et al.* (1994a, b); Kovel (2002); Bakan (2004); Jones *et al.* (2005); Blowfield and Murray (2008).

generally will be to make as much money as possible while conforming to the basic rules of society, both those embodied in law, and those embodied in ethical custom.

(Friedman 1970)

In so far as 'socially responsible' businesses find that their new role is bringing with it higher costs and lower profits, they have a strong interest in having their unregenerate rivals compelled to follow suit, whether through public pressure or government regulation. The effect of such enforced uniformity is to limit competition and hence to worsen the performance of the economy as a whole. The system effects of CSR, as well as the enterprise effects, will tend to make people in general poorer . . . [T]he case for private business derives from its links with competition and economic freedom. [It is mistaken to] identify defence of the market economy with making businesses more popular and respected, through meeting 'society's expectations'. Whether all this is responsible conduct is open to doubt.

(Henderson 2001: 18)

If you subscribe to these views then any notion of social responsibility is dominated by the need to make money for shareholders, to grow, make profits and seek economic efficiency. It should be obvious from Chapters 1 and 2 that this view is widely held but often implicit.[8] Indeed, it is the very ubiquity (and implicitness) of the view that ensures that there are still good educational reasons for examining the extent to which the current social, economic and political systems do, in fact, work in this way.

2 Alternatively, you might subscribe to an '*expedient*' point of view. Such a view considers that long-term economic welfare and stability can only be achieved by the acceptance of certain (usually minimal) wider social responsibilities:

The important issues involve accommodation between different and often conflicting values. There are the values associated with the market economy – efficiency, freedom, innovation, decentralisation, incentive, individual achievement. And there are the values associated with political and social rights – equality of opportunity, the right of an individual to participate in important decisions affecting his or her life, the right to standards of health, education, personal privacy and personal dignity. What we are constantly faced with are the difficult choices and trade-offs needed to achieve balance among all these values.

(J. G. Clarke, Exxon Corporation, 1981)

Sustainable development has increasingly come to represent a new kind of world, where economic growth delivers a more just and inclusive society, at the same time as preserving the natural environment and the world's non-renewable resources for future generations.

(BT Social and Environmental Report: Summary and Highlights, 2005: 19)

Such views might be thought of as 'enlightened self-interest'. An expedient or enlightened self-interest view will usually recognise that current economic systems *do* generate unacceptable excesses and so some additional moral content must be added to the organisation–society relationship. This will typically manifest itself in a preference for, for example, some fairly limited social legislation plus a recognition of the place of 'business ethics' (see also Donaldson, 1988; Schmidheiny, 1992; O'Dwyer, 2003).

---

[8]For more detail see, for example, Benston, (1982a), b; Owen *et al*, (1992); Gray, (1994); Henderson, (2001).

3 The '*proponents of the social contract*', on the other hand, tend to consider that companies and other organisations exist at society's will and therefore are beholden (to some degree) to society's wishes:

> Any social institution – and business is no exception – operates in society via a social contract, expressed or implied, whereby its survival and growth are based on:
>
> 1) The delivery of some socially desirable ends to society in general and
>
> 2) The distribution of economic, social or political benefits to groups from which it derives its power.
>
> In a dynamic society, neither the sources of institutional power nor the need for its services are permanent. Therefore an institution must constantly meet the twin tests of legitimacy and relevance by demonstrating that society requires its services and that the groups benefitting from its rewards have society's approval.
>
> (Shocker and Sethi, 1973: 97)

> . . . every large corporation should be thought of as a social enterprise; that is as an entity whose existence and decisions can be justified in so far as they serve public or social purposes.
>
> (Dahl, 1972)

The social contract is a powerful image and, indeed, a powerful strand in social and democratic theory. The contract suggests the notion of mutual dependence and mutual responsibility and places the principle of accountability at its heart. Those who subscribe to this view still have to decide on how distant our current systems of economic and institutional organisation are from this state of explicit mutual dependence and, most especially, how unbalanced the current power distribution is. That is, are organisations dependent upon society or is society dependent upon the organisation? The decision on these factors determines the extent of the recognised responsibilities and, subsequently, the accountability that is implied by this contract (see, for example, Parkinson, 1993; Sunstein, 1997; Tozer and Hamilton, 2007).

4 We could use the term '*social ecologists*' to describe those who are concerned for the human environment (in the widest sense), who see serious problems developing if nothing is done about organisation–environment interactions soon, and who consider that large organisations (in particular) have been influential in creating the social and environmental problems and so could be equally influential in helping eradicate them:

> The principal defect of the industrial way of life with its ethos of expansion is that it is not sustainable. Its termination within the lifetime of someone born today is inevitable – unless it continues to be sustained for a while longer by an entrenched minority at the cost of imposing great suffering on the rest of mankind. We can be certain, however, that sooner or later it will end (only the precise time and circumstances are in doubt) and that it will do so in one of two ways: either against our will, in a succession of famines, epidemics, social crises and wars; or because we want it to – because we wish to create a society which will not impose hardship and cruelty upon our children – in a succession of thoughtful, humane and measured changes.
>
> (Goldsmith *et al.*, 1972: 15)

The rapid succession of crises which are currently engulfing the entire globe is the clearest indication that humanity is at a turning point in its historical evolution. The way to make doomsday prophecies self-fulfilling is to ignore the obvious signs of perils that lie ahead. Our scientifically conducted analysis of long term world

development based on all available data points out quite clearly that such a passive
course leads to disaster.

(Mesarovic and Pestel 1975, as quoted in Robertson, 1978: 22–3)

Human activity is putting such a strain on the natural functions of Earth that the ability of
the planet's ecosystems to sustain future generations can no longer be taken for granted.

(United Nations Millennium Ecosystem Assessment, 2005: 2)

Much of the early impetus for the original social responsibility and social accounting
debates of the 1970s as well as initial stimulus for the 'greening of business' and the environmental debates of the 1990s (in particular, see Chapter 7) would derive from these
beliefs. The implicit element here is that something has gone wrong with the economic
system and, at a minimum, the economic processes that lead to resource use, pollution,
waste creation and so on must be amended if the quality of human life is to be improved –
or held constant (see also, Owen, 1992; Gray and Bebbington, 2001).

5 The '*socialists*' – although this is a very general term covering a wide range of views – tend
to believe that the present domination of social, economic and political life by capital is
inimical. There, thus, needs to be a significant re-adjustment in the ownership and structure of society:

We shrink back from the truth if we believe that the destructive forces of the modern
world can be 'brought under control' simply by mobilising more resources – of
wealth, education, and research – to fight pollution, to preserve wildlife, to discover
new sources of energy, and to arrive at more effective agreements on peaceful
co-existence. Needless to say, wealth, education research, and many other things are
needed for any civilisation, but what is most needed today is a revision of the ends
which these means are meant to serve. And this implies, above all else, the
development of a life-style which accords to material things their proper, legitimate
place, which is secondary and not primary.

(Schumacher, 1973: 290)

From the political point of view it is important to emphasize that the problems
associated with advanced technology cannot be framed merely in terms of economic
categories, concerning solely the ownership and control of the means of production,
but challenge the political nature of our social and cultural institutions, that of the
concept of the nature of man to which they have given rise, and the technological
practices which have been based on them. Institutions that promote social hierarchies
must be confronted with demands for the recognition of the equality and shared
collective experience of all men. Not only must the division of society into oppressors
and oppressed be broken down, but so too must the barriers that separate mental
activity from manual labour, and abstract theory from concrete practice. Only
through such changes can we create a situation that will enable us to reintegrate all
aspects of social life and experience and to establish a situation in which man can be
liberated to fulfil his full potential as a sensitive, creative and social being.

(Dickson, 1974: 203)

Such views have a strong echo in the views expressed by critical management and
accounting scholars (see, for example, Tinker, 1985; Puxty, 1986, 1991; Kovel, 2002;
Grey and Willmott, 2005). In general terms, views such as these are likely to be associated
with a deep intellectual suspicion of accounting, business management and capitalism
and, as we shall see, much of social and environmental accounting.

6 There is also an increasingly influential voice from feminism and, in particular, the *'radical feminists'*. The essence of the view expressed here is that our economic, social, political and business systems – and thus the language of business and accounting – are essentially 'masculine' constructs which emphasise, for example, aggression, traditional success, achievement, conflict, individualism, competition and so on. Our world thus denies a proper voice to, for example, compassion, love, reflection, cooperation and other 'feminine' values:

> . . . most people and societies in the present planetary order are dominated by the un-balanced Masculine or Yang worldview. In order for a balancing, an emergence of the Universal Feminine or Yin is necessary, so that a genuine union or integration may occur between the two: an integration of thinking and reasoning with intuition and feeling; a balancing of active and productive doing with stillness and contemplation; a preparedness to receptively wait as well as aggressively confront; a blending of material concerns with spiritual realization; a dilution of the respect for analysis, discourse and argument with a love of silence; a contemplating of dualistic thought with intuitive holistic seeing; a softening of the attachment to logic with a receptivitly to imagination and dreams; a turning of the coin, to see that the other side . . .
>
> (Hines, 1992: 337)

> Mas(k)culine gestures based on the notion of unified, self-present subjects of rationality are made here and there in accounting. But there is no recognition of the masculine fear of thinking the unthinkable, giving up power and control and struggling for difference. Calls are made for corporate social reporting, 'participative budgeting', accountability (in terms of masculine power or agency relationships). They are all in the end doomed to the masculine proliferation of the self-same (what is like me and hence is good/safe) – a feminine affective economy of gift, affirmation and love would be more disruptive. . . . At present, environmental accounting too, being founded on phallogocentric understandings, will be unable to bring about revolutionary change despite the high hopes and desires of some of its proponents.
>
> (Cooper, 1992: 28)

Such a view then suggests that there is an essential 'sickness' in much of what we take for granted. Even the way in which we think about apparently benign concepts like 'social responsibility' (for example) completely misses the point that this is no way for wise and compassionate human beings to organise their world.

7 The *'deep ecologists'*, in many regards, hold views closer to the *'radical feminists'* than their more obvious bed-fellows, the *'social ecologists'*. This is because the deep ecologists hold that human beings do not have any greater rights to existence than any other form of life (and indeed, may have actually forfeited their rights as a consequence of the human race's mistreatment of the rest of life). Such views inform animal rights activists and organisations such as Earth First!:

> . . . the world is an intrinsically dynamic, interconnected web of relations in which there are no absolutely discrete entities and no absolute dividing lines between the living and the nonliving, the animate and the inanimate, or the human and the non-human. This model of reality undermines anthropocentrism . . . which may be seen as a kind of ecological myopia or unenlightened self-interest that is blind to the ecological circularities between the self and the external world . . .
>
> (Eckersley, 1992: 49, 52)

This is not to claim that accounting information plays one of the primary roles in either creating or inhibiting the solution of ecological problems. Such primacy can, arguably, be attributed to cultural factors – nationalism/parochialism, anthropocentricism, selfishness and ideologically induced attitudes towards the desirability of economic growth, efficiency and property rights. . . . [S]ome of the attributes of conventional accounting information and its uses act to exacerbate or reinforce the effects of these primary factors.

(Maunders and Burritt, 1991: 10)

The essence of these views is that the very foundation – even the existence of – our economic (and social) system is an anathema. Put at its simplest, our economic system can – and does – contemplate trade-offs between, for example, the habitat of threatened species and economic imperatives. To a deep ecologist it is inconceivable that a trade-off could have any form of moral justification. Such a view therefore challenges virtually every aspect of taken-for-granted ways of human existence – especially in the Western developed nations.

8 Finally, we might hold a view more consonant with *postmodernism*. Not a simple – or indeed a single – perspective, postmodernism would typically reject both the 'grand narrative' offered by (especially) Marxian views and express itself in term of the failures of modernity – many of whose characteristics exercise the radical feminists and the deep ecologists. A postmodernist view would also, typically, reject much of the structure and paraphernalia of modern (mostly) western life. Typically associated by scholars in the business, management and accounting literature with theorists such as Derrida, Lyotard, Baudrillard, Rorty and, especially, Foucault, the diversity represented here ranges from the profoundly radical to the potentially very conservative (Zimmerman, 1994). In its diversity of view it does not, in our view, help us develop social accounting at this stage of the proceedings. But postmodern thought offers a fundamental critique of modernity and, in doing so, it significantly adds to the challenges that social accounting must address (see, for example, Everett, 2004; Gray, 2010).

3.5 Clarifying responsibility in the interests of sustainability?

Whilst we are contemplating this range of views, it is an appropriate point at which to revisit the concept of sustainability which we introduced briefly in Chapter 1. We will develop this further still in Section 3.11, but we need to remind ourselves again of the 'elephant in the bedroom' at this stage. In large part this is because – to put it simply – it is difficult to consider a position of 'responsibility' which did not acknowledge the exigencies of 'sustainability'. And yet, although it may appear so in much business writing, the two terms are not interchangeable (Gray and Milne, 2004; Milne *et al.*, 2007; 2009).

Sustainability (or, as it is often called, *sustainable development*) is most commonly defined as a system of development which:

meets the needs of the present without compromising the ability of future generations to meet their own needs.

(The Brundtland Report, United Nations, 1987)

The essential idea behind sustainability is that current modes of behaviour – especially in the developed world – are *un-sustainable* and therefore threaten current and future ways of life. Sustainability is important for at least two major reasons: *first,* it brings firmly onto the political and business agendas of the world that the present ways of doing things do, indeed, have 'externalities'; and *second,* it is an almost universally accepted principle (Bebbington and Thomson, 1996; Dresner, 2002).

Sustainability, however, is something of a wolf in sheep's clothing. First, although it is generally used to refer only to planetary environmental issues, it is also an essentially social concept as well and concerned as much with wealth distribution (**intra-generational equity**) as well as the well-being of future generations (**inter-generational equity**) (see, for example, Gladwin, 1993; Milne, 2007; Unerman *et al.*, 2007). This raises major questions about such matters as current distributions of wealth, current levels of consumption in the West and population growth. Unsurprisingly, there are widely differing attitudes to these things in much the same way as there are on matters of social responsibility. Second, there is widespread dispute about how big is the current gap between our current patterns of consumption and production on the one hand and a sustainable way of life on the other. *This disagreement naturally leads to very different views on how substantial a change is necessary to bring us round to sustainability.*

So, although sustainability has provided a new, demanding and potential holistic basis on which to discuss such matters as organisational responsibility and society, it has also brought out into the open very many differences of belief which will remain sources of continuing conflict. *Indeed, each of the views expressed above would offer a different interpretation of what comprises sustainability and how it could be reached.* What seems to be almost universally accepted, however, is that western ways of economic and business organisation are certainly not sustainable (and that applies only to the West: how much more true it must be when the global population aspires to western levels of production and consumption). Any change intended to make developed countries' organisations more sustainable will have a major impact on organisational life, and the way in which the changes come about will significantly depend on which of the above groups' opinions finally prevail (for more detail see, for example, Jacobs, 1991; Bebbington and Thomson, 1996; Gray and Bebbington, 2001; Atkinson *et al.*, 2007; Unerman *et al.*, 2007).

As a final point at this stage on sustainability, it is worth highlighting that many of the '*deep ecologists*' have very serious reservations about sustainability. This is because it is concerned about the continued viability of human species – not about all species. At one extreme, a very plausible argument can be made that humanity has had its chance. It is now the chance of other species and life-forms to escape from the domination of people and if this involves the extinction of humanity, this is a reasonable and natural consequence. Such views are increasingly widely held (Maunders and Burritt, 1991; Gray, 1992).

There are, of course, yet other points of view (we have not, for example, considered religious perspectives here). The range we have identified here is probably sufficient for our purposes as it highlights that the organisation–society relationship is far more complex, and is continually becoming more complex, than we are inclined to think of it. This complexity means that the roles we may envisage for social and environmental accounting and accountability are far from simple.

3.6 Why does so much confusion remain on CSR?

There seems every likelihood that confusion over CSR is likely to continue in the business and the management literature. We can see a number of key reasons for this continuing confusion and, in identifying them, we believe that we can perhaps offer a way out.

Attempts have been made to chart the confusion. For example, De Bakker *et al.* (2005) offer three reasons for a lack of closure over CSR: a *progressive view* which sees the early vagueness over concepts giving way to an optimistic and increasingly sophisticated research approach; a *variegational view* in which the continual introduction of new terms, ideas and concepts (as we saw itemised above by Mohan, 2003, for example) constantly muddies the water; and, finally, what they call a *normatist view* which would believe that getting bogged down in normative, values-based, issues simply confuses those involved in management practice and research. Whilst, in a particularly careful analysis, the authors find support for the first two reasons, their paper illustrates the points we wish to make here.

In the first place, lively interest and academic coherence around a subject do not necessarily mean that we are any closer to a grounded and firm understanding of the issue. Whilst more and more organisations and scholars apparently embrace, debate and investigate CSR the incidence of organisational irresponsibility continues to grow (see, for example, Vogel, 2005). That is, the apparent increase in the acceptance of sustainability and its importance appears to be accompanied by an increase in un-sustainability!

Second, and we see this more clearly when the debate turns to embrace social *performance* (for more detail also see Chapter 8 where we look at the financial performance and social performance debate), the lack of a firm and independent description of what CSR actually means can so easily result in a key tautology. That is, if you wanted to show, for example, that the more responsible organisations are also the more successful ones, you might find yourself selecting only those social responsibilities which are win–win in nature – i.e. those which accord with good business practice. That is, for example, it is clearly responsible to not kill or injure your employees, to minimise wastes and not to annoy either consumer or environmental groups. It is also clearly good business practice to do these things. That the two sets of things might be correlated tells us almost nothing about responsibility; it simply tells us that only stupid companies waste money/resources and annoy their powerful stakeholders. A truly responsible company on the other hand might, however: ensure that it never had an accident; refuse to employ any threatened environmental resource; adopt United Nations standards for all its employees; ensure that any activity it undertook was approved – *independently* – by the whole community affected; refuse to ever lie or persuade people to buy things that they did not need; and so on, and so on. Such a company could actually exist, we guess, but it could not exist in international financial markets and it would almost certainly not be a multi-national corporation. Put more simply, a profit-seeking organisation which behaved to the standards that we would admire in a great human being would almost certainly not be a profitable organisation for long at all (for more detail, see, for example, Estes (1976); Schwartz and Gibb (1999); Hertz (2001); Klein (2001); Bakan (2004)).

And this brings us to the third point about CSR. Corporations – and especially large organisations whose shares are quoted on international stock markets – are creatures of capitalism and entirely governed by its strictures. Capitalism is *not* a 'responsible' system in and of itself. If you wanted the poor fed, the homeless housed and the environment protected you would be quite mad to wait for unfettered international financial capitalism to deliver it. It won't. There is no money in it. Capitalism requires that organisations only do something, on the whole, if it makes a profit. It is a strange notion of 'responsibility' that says a responsible action is one that you are paid for. Corporations will normally only do things if they are paid to do it. Under capitalism, it is almost certain that on the whole corporations *cannot* be responsible. The only acts of 'responsibility' that financial capitalistic organisations can undertake are those that are profitable (Bakan, 2004).

We conclude that to expect corporations to be 'responsible' in some absolute sense is both ridiculous and improper. *But,* by the same token, we need our organisations *not* to be irresponsible and, indeed, we need to know which *responsibilities* they are meeting. And here is the key to unpicking CSR. As Walters stated it in 1977:

> [S]ocial responsibility is not telling society what is good for society but responding to what society tells the firm the society wants and expects from it.
>
> (Walters, 1977: 44, as quoted in Mintzberg, 1983: 13)

There may well be practical difficulties with such a view but, as a starting point, such an analysis seems entirely plausible: society decides what social responsibility is. The individual, group or organisation who is to be held responsible does not make that decision themselves – or if they do so (and many of us may morally decide on our own responsibility) they must face the consequences of not meeting the social requirements of the society.[9] So we can establish that the specification of responsibilities is properly the domain of society and that the responsibility will be required of organisations by society. The key, then, is to determine:

- which responsibilities; and
- the extent to which these required responsibilities are being met.

This is the task of a full and enforced accountability.

3.7 Accountability

Accountability is a widespread, even ubiquitous, phenomenon that arises, in some form or other, in nearly all relationships. It can be simply defined as:

> The duty to provide an account or reckoning of those actions for which one is held responsible

Accountability has two crucial components: it arises as a result of a relationship between two or more parties (be they individuals, loose associations or organisations) and its nature is determined by the social and moral context in which the relationship is manifest. Each individual will have a myriad of relationships – with, for example, friends, family, workmates, colleagues, merchants, local community, pen-pals, casual acquaintances as well with the state, companies, NGOs and other organisations. Each of these relationships will be different in character – some emotional, some economic, some casual, some intense and so on. What is considered reasonable behaviour within the relationship and thereby governs behaviour of the parties is the context in which it takes place. Different cultures, different generations, have different acceptable standards as to how (say) a child treats a parent, a man treats a woman, a student treats a teacher, a friend treats a friend, a company treats an employee and a merchant treats a customer. This combination of the nature of the relationship and the context and 'rules' governing that relationship leads to what Dillard calls an 'ethic of accountability' (Dillard, 2007).

That is, each relationship has a moral dimension determined by the nature of the relationship, the actions expected and/or required in the relationship and the community context. One aspect of this moral dimension is the requirement to offer 'accounts' – to explain oneself, to articulate one's intentions and aspirations, to offer detailed explanations of one's actions. This is accountability – and we undertake it all the time with varying degrees of success, formality and transparency.

These accounts will, on a personal basis, often be very informal accounts – casual comments, brief stories about my day at work, my evening out, my weekend, my efforts for the group, my borrowing of your resources, etc. Such informal accounts arise due to what Rawls (1972) calls 'closeness'. There is an intimacy, a physical, values and/or moral proximity between the parties such that formal accounts are an anathema, an insult even ('pay me back

---

[9]The decision to be (say) a conscientious objector in a time of war is clearly a moral decision and clearly goes against the social decision about responsibility. In making such a moral decision, one explicitly accepts the consequences of society's judgement.

when you can'). The greater the closeness, the less the need for formality. By contrast, the less the closeness the more the need for formality – and this is where what we normally think of as formal accountability really starts to matter. Non-close relationships can generally be assumed to require some sort of formal 'accounts' which, in turn, would typically require some formality over the rules governing, for example, public statements and, hence, a more formal accountability (see, for example, Gray *et al.*, 2006). *Crucially,* the very essence of financial and economic relationships is that they do not encourage – actually they discourage – closeness. Thus the more a relationship is dominated by the economic, the less closeness there is likely to be and the more the relationship will require formality in the accountability (see also Thielemann, 2000).

3.8 A model of accountability

So if we begin to formalise our understanding of accountability, we may start by identifying that it involves two broad responsibilities or duties: the responsibility to undertake certain actions, or forbear from taking actions, and the responsibility to provide an account of those actions. The simplest, most familiar, *formal* case of accountability is that of the shareholders and a company. The directors of a company have a responsibility to manage the resources, both financial and non-financial, entrusted to them by the shareholders *and* a responsibility to provide an account of this management. We can therefore see the annual report and the financial statements as this account – as the mechanism for discharging this accountability. How well the statements succeed in discharging the accountability is, of course, another matter. The essential elements of this process arise from a *relationship* between the directors and the shareholders; the relationship is almost *exclusively economic **and** it lacks closeness*; it is a relationship which is *defined by society,* in this case through, *inter alia,* the Companies Acts, and which provides the shareholders with a **right to information**.

This specific case of the accountability model can be generalised to apply to all relation-ships and rights to information in Figure 3.5.[10]

Figure 3.5 is an extremely simplified model but can be used to explore many complex situations. This basic version of the model hypothesises a simple two-way relationship between an accountee (who we might call the 'principal' and in conventional financial accounting would be the shareholder) and accountor (who we might call the agent and in conventional financial accounting would be the director of a company). The terms of the flows between the parties and the actions and accountability required will be a function of the relationship, which might be thought of as a 'contract', between the parties. This, in turn, will reflect the social context of that relationship (e.g. the importance that society places on the flow of capital to the company in return for the privileges of limited liability and rights to information).

The model is a great deal more flexible than it might look. For example, the accountee and accountor may be individuals, organisations or groups. A particular pair of accountor

---

[10]Please note that this model bears only the slightest resemblance to the Principal–Agent or Agency Model so popular in accounting and finance research in the 1980s and 1990s (see, for example, Jensen and Meckling, 1976). The Agency Model is a two-person model based upon Coase's Theorem and makes a whole raft of assumptions about the motivations and characters of the principal and agent – most notably that they are so-called 'rational' economic actors who are wholly selfish and wholly greedy. These are the sort of people you hope never to meet and certainly wouldn't want to invite to your house for tea (see, for example, Gambling, 1978; Tinker and Puxty, 1995). The present model makes none of these assumptions, predates 'Agency Theory' by several centuries and is subject to none of the fatal criticisms that make Agency Theory so intellectually unappealing (see, for example, Christenson, 1983; Noreen, 1988; Arrington and Francis, 1989; Arrington, 1990; Armstrong, 1991; Tinker and Okcabol, 1991).

Figure 3.5 A simple model of accountability

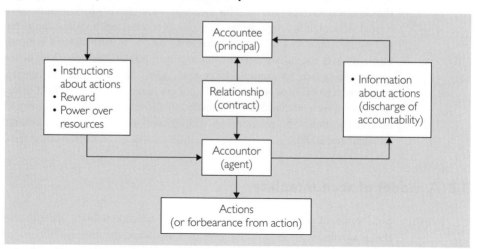

and accountee might have several different relationships and thus be accountor in one and accountee in another – thus an employee may be accountable to the management of an organisation for his/her work performance whilst the management may well be accountable to the employee for the extent to which the company is complying with health and safety at work legislation. The employee may also be a shareholder and thus able to hold the management to account whilst both employee and management, as members of the community, may wish to hold the company to account for its pollution record.

The essence of the model is the relationship between the parties and the role that society – or elements of society – ascribes to that relationship. It is this relationship through which notions of responsibility are established and through which associated rights to information (accountability) are determined. So the crucial issue, once we move into more formal, less close relationships, is how we should establish when a relationship – a 'contract' – **exists** and whether we need to recognise different types of contracts.

What we are envisaging here is that a society may be thought of as a series of sets of relationships – for example, between individuals, between organisations, between the state and the individual or organisation and between individuals and the rest of the natural environment and so on. In essence, a society can then be thought of as a series of individual relationships or 'social contracts' between the members of the society and the society itself.[11] If we formalise these into contracts – rather than the more general notion of relationships – we need to examine how the contracts are formed. Generally, the formation can be thought of as both legal and non-legal with the latter arising from the general context of society and normal social intercourse plus the more compelling moral or natural contracts. That is, some relationships and parts of some relationships are governed by law whereas other relationships – and some parts of all relationships – are governed by the ruling ethics, values and principles of the society (Dillard, 2007). These 'contracts' provide the basis for the rights of the parties in that relationship – including rights and responsibilities relating to information flows – i.e. accountability.

---

[11]This 'social contractarian' tradition is well established in political philosophy and whilst it is, like most models, an over-simplification, it captures a richer set of actual and potential social relationships than are present in, say, the atomistic view of pristine liberal economic democracy. We should stress that we acknowledge a greater intellectual debt in this model to Rawls than we do to Coase. A richer and more explicitly theorised approach to these issues is present in Lehman (1999, 2001, 2002).

3.9 Some practical components of accountability

To see what this means for the more formal examination of accountability, we therefore need to distinguish between *legal* and non-legal, or *moral* or *natural*, rights and responsibilities (see, for example, Likierman and Creasey, 1985; Likierman, 1986).

The most obvious rights and responsibilities are those established within law. Whatever one feels about the justice of laws of a country or about the processes that generate the laws,[12] they are certainly the 'rules of the game' by which each of us – including organisations – is supposed to play.[13] The law lays down the *minimum* level of responsibilities and rights and thus the *minimum* level of legal accountability at any given time in any given country (Tinker *et al.*, 1991).

Whilst law frequently identifies responsibility for actions, it rarely specifies the responsibility to account for those actions – the accountability. So, for example, in the EU a company is legally responsible (*inter alia*) to protect the health and safety of its workforce, to give equal opportunities to all members of the workforce and to give special attention to the employment of the disabled. In the UK, only in the case of employment of the disabled is that legal responsibility accompanied by a legal responsibility to disclose. Even then, the information the company is required to disclose – the discharged accountability – is trivial and certainly does not provide information to permit any assessment of whether the responsibility has been met. Therefore, we can note that the legal responsibility for action and the legal responsibility for accountability are not equal – the *legal* responsibility for action brings a *moral* responsibility to account which is only partially discharged by the *legal* responsibility to account. If we were content to leave accountability to only legal forces and voluntary initiatives, the demands of accountability would rarely be satisfied. This represents one of the major reasons why social accounting, if it is to be a meaningful activity, *must* be mandatory. Evidence is quite clear that the encouragement of voluntary social and environmental accounting (SEA) has little lasting or substantive influence on reporting practices (see Adams *et al.*, 1995; Gray *et al.*, 1995; for summaries of this evidence).

Indeed, the requirement to report to shareholders (financial accounting) is one of the very few instances of explicit accountability being established within the law itself[14] – and thus, one of the very few examples of where there is *any* sort of congruence between an organisation's defined responsibility and its discharged accountability. So the first role for social accounting therefore, is to fill this gap – to develop means for the moral responsibility for accountability required in law to be satisfied.[15]

Now, it is probably true that we could relatively simply establish organisational responsibility as enshrined in law. If we were then able to enforce a full discharge of the accountability associated with those legal responsibilities, a considerable improvement in transparency, accountability and democracy would probably result. But that is not the whole story by any means. Not only does the law not represent a full and complete manifestation of society's

---

[12]See, for example, Dowling and Pfeffer (1975); Stone (1975); Lindblom (1993).

[13]Nothing we say here should be taken to imply that we believe there is a moral imperative to obey the law. Indeed, civil disobedience and direct action (e.g. demonstrations and the actions of Greenpeace and Earth First!) may be essential democratic mechanisms for the maintenance of accountability. This is a complex matter, but space prevents us dwelling upon it here. It is re-examined to a degree when we look at the external social audits in Chapter 10.

[14]Although there is the potential for this to change with the development of Freedom of Information Acts in a number of countries.

[15]In addition, accountability requires the establishment of apposite *channels* of accountability – the means through which accountability information might flow from accountor to accountee. The existence of a real channel of accountability is rather more important than what the channel is actually used for.

values (mainly due to power asymmetry and delays in the legal process, Stone, 1975), but also there are a wide range of additional responsibilities which are placed upon the organisation. What these might be, and the extent to which an organisation should or could respond to them, is a matter subsumed within *corporate social responsibility* – which is why it is such an important (if difficult) area.[16] (More detail on the model and its limitations are provided in the Appendix to the chapter).

3.10 Defining corporate social responsibilities

Accepting that we are inevitably going to simplify things somewhat (for more detail see, for example, Parkinson, 1993; Matravers, 2007), we can start to put some flesh on the bones of the responsibilities that a full accountability would seek to discharge. The key is the conclusion from Walters (1977): the activities for which an organisation must be held accountable are determined by society – not exclusively by the organisation itself. For convenience, you will recall from the preceding two chapters, in addition to the notion of a society itself, we tend to think of society as comprising a range of stakeholders, and it is these stakeholders to whom we look to determine accountability. So what responsibilities will a stakeholder impose upon an organisation – how will we know and which of these will be legitimate?

As Lehman (1999, 2001), Owen and Swift (2001), Adams (2004) and Thomson and Bebbington (2005) have all argued in different ways, what one is seeking here, ideally, is a much greater *closeness* between stakeholder and organisation coupled with a mechanism (or mechanisms) that will ensure that the obvious power asymmetry between large organisations and their stakeholders is overcome to some degree or other. A greater intimacy and dialogue and understanding between the parties might then bring us back to some notion of the more personal and informal accountabilities we discussed earlier. This is the potential of social accounting. It is crucial to keep in mind that social accounting has this potential to move us towards a world in which such vastly changed accountability and information flows could, in turn, potentially bring about really substantial change in the organisational and economic structure of 21st century financial capitalism. In the meantime, however, in a world of distance and power differential, what sorts of *formal* accounts should we aspire to?

As we will reconsider later in the book, in order to ensure that all responsibility is formally considered and any attendant accountability is potentially complete and meaningful, we need to consider *each* of the relationships that exists between the organisation and *each* of its stakeholders (see Gray *et al.*, 1997). For each relationship, full accountability then needs:

---

[16]**A Compliance-with-Standard Report** (Gray *et al.*, 1986, 1987; and see later in the book) would provide much of the information necessary to assess the extent to which organisations had met their basic responsibilities – that is, it would discharge the organisation's accountability for legal responsibility. If legal rights have the advantage of being relatively easy to identify (if not to enforce), non-legal or moral and natural rights and responsibilities are far more difficult to establish in any unique way. It is therefore useful to split these non-legal rights and responsibilities into *quasi-legal* and other, *philosophical* rights and responsibilities. The quasi-legal rights and responsibilities are those enshrined in codes of conduct (e.g. the United Nations' Global Compact or the various protocols on air pollution and greenhouse gases), statements from authoritative bodies to whom the organisations subscribe (Industry Associations or The International Chamber of Commerce, for example), plus other 'semi-binding agreements' – possibly from the organisations themselves – such as mission statements, published social and environmental policy statements, statements in speeches from chief executives or statements of objectives. So, in effect, a 'contract' is established by an authoritative body, by an organisation to which the 'accountable' organisation subscribes or by the 'accountable' organisation itself. The 'philosophical' rights and responsibilities are the most tricky but probably the most important.

- *descriptive* information which expresses the parameters of the relationship and allows a reader to understand it (e.g. numbers of employees, wage rates, diversity, etc., location, turnover);

- *accountability* information which expresses the information required by context and contract. This is, basically, information expressed in both legal and *quasi-legal* aspects of the relationship – where *quasi-legal* refers to anything from voluntary codes of practice, organisational policies and industry statements of good practice to such widely accepted codes as SA8000 on use of child labour and the United Nations Millennium Development Goals;

- *stakeholder voice* information is an expression of those activities which the stakeholder would wish to hold an organisation accountable for – whether or not the organisation chooses to account for it or not;

- the *organisational voice* must also be heard – and, whilst this is not a problem (organisations have any number of channels through which they can and do speak about their activities), to ignore such a voice would undermine any claim to completeness.

This then permits responsibility to be articulated by each element of society, the stakeholders and the organisation itself and, consequently, a complete accountability would be an account of the performance by the organisation against each aspect of the relationship.

The key point that then flows from this is that a full social accountability would *not* specify what it was that an organisation should do but rather require information about the extent to which the organisation did – or more likely did not – meet the expectations of its stakeholders and society. *Full accountability may be more about what an organisation cannot do than what it has done.*

Inevitably, there is much more than could be said here,[17] but we will restrict ourselves to some of the key issues. Perhaps the most important of these at this stage is the notion of the 'responsibility' of the stakeholders themselves and the level of self-awareness that they may have. That is, do stakeholders always know what it is they need? Do they have the information, knowledge, experience and attention span to focus onto the most important issues? To what extent do stakeholders really only know about what companies tell them and, from the organisational point of view, it is clearly impossible for any organisation to respond entirely to each and every stakeholder demand. Such conflict can be well illustrated when we consider the 'responsibility' to be 'sustainable'.

3.11 Social responsibility and sustainability

Sustainability, as we have already seen, is quite possibly the most important issue any of us can consider: it is the 'elephant in the bedroom' in that it is so big and dominant that we must always address it first and any other issue pales beside it. Sustainability, as we shall see, both clarifies – and confuses – some of the social responsibility issues.

In the first place, there is the question of whether or not organisations in general and companies in particular can and should be held responsible for sustainability. We shall see shortly that the answer to this turns out to be 'yes' but *a priori* would we think this reasonable? Well, there is a fairly strong *a priori* case that whilst the un-sustainability of the planet is clearly a function of population, technology and consumption (as we saw in Chapter 1, Section 1.4), it is also fairly clear that the enormous advances in material well-being,

---

[17]The additional references given throughout the text develop many of these points more fully.

technology and consumption are, in large part, due to the growth of global capitalism. The principal engines of our economic and business organisation are MNCs and they, in particular, must be implicated in some substantial ways with the consequent level of social justice and material and environmental consumption. (Such a case is developed in, for example, Beder, 1997; Kovel, 2002; Meadows *et al.*, 2004; Porritt, 2005; Gray, 2006a,b; Unerman *et al.*, 2007.) So the idea of there being a responsibility linking organisations and sustainability is plausible.

There is not, however, any law (that we know of) requiring organisations to act sustainably[18] – despite the range of law governing health and safety, consumer protection and various aspects of environmental protection these do not sum to a total of sustainability. So how would responsibility for sustainability manifest itself at the organisational level?

The obvious thought is that any responsibility would be expressed through the voices of stakeholders. We then realise three important things which are more clearly seen in the case of sustainability. First, for much of the life of 'sustainability' as a dominant concept, relatively few stakeholders would have the impetus, knowledge and understanding to raise it as an issue. Consequently, the question is raised: how many stakeholders does it take to make a responsibility(!). Second, most stakeholders especially at the turn of the century would not have had much understanding of the elements of sustainability and, in particular, would know little about how a particular organisation's activities influenced and were influenced by sustainability. Indeed, for many stakeholders such knowledge would almost certainly have been created by the information and publicity produced by the organisation itself![19] And third, the obvious important stakeholders in the case of sustainability are the disenfranchised and future generations: it is far from clear how such voices can be heard and understood under normal circumstances.

It transpires that a major key to determining responsibility with regard to sustainability and organisations lies in the realms of the quasi-legal and the organisations' own voices. Since sustainability rose to prominence on the global agenda, organisations have been closely involved. NGOs have been active since the beginning (Beder, 1997; Welford, 1997; Bendell, 2000); local authorities, for example, have been closely involved through such initiatives as Local Agenda 21 (see, for example, Ball, 2002); and MNCs in particular have been active in promoting, managing and suppressing the sustainability agenda (which depends on your interpretations). Business (in so far as such a collective noun makes sense), through such organs as the World Business Council for Sustainable Development (WBCD) and the International Chamber of Commerce (ICC), has greatly influenced the pronouncements of the United Nations and others on such matters as the definition of sustainable development, the role of business in its delivery, climate change, ozone depletion and such like (see, for example, Bruno, 1992; Eden, 1996; Beder, 1997; Mayhew, 1997; Hertz, 2001).[20] Such involvement by business and business organisations suggests, at the very least, the implication of large business organisations in the achievement of (or deviation from)

---

[18]Although the original Earth Summit conference in Rio de Janeiro in 1992 issued guidance to this effect – most famously captured in Agenda 21 (see, for example, Welford, 1997).

[19]There is an anecdote – which may actually be apocryphal as the evidence is elusive – that a company in the early 1990s (let us call it Company A) sought the views of its customers on the extent to which Company A used animal testing on its products. The point is that the customers could not know this (although Company A could) but could only know the extent to which they perceived animal testing to be used – and this, in turn, was likely to be a direct function of Company A's own publicity. Asking customers about animal testing was actually a way of testing the quality of Company A's own information diffusion.

[20]Business' influence on the Earth Summit (Mayhew, 1997) and the implicit contract which derives from both Rio and, more recently, both the Millennium Development Goals and the Global Compact speak eloquently of a presumption of MNC involvement in the paths to and away from sustainability.

sustainability. But, more tellingly, many individual businesses and business organisations have spoken directly of their involvement in sustainability – many going as far as to produce 'Sustainability Reports'.[21]

On this basis, it seems that much of the international business community could be thought to be accepting the responsibility to seek out sustainability and, as a consequence, the extent to which these corporations are genuinely seeking out sustainability has become a genuine and legitimate concern for accountability. As Milne *et al.* (2003, 2006) have shown, not only are MNCs not sustainable, they probably cannot be so and, most especially, their claims in this direction are not matched by their accountability. (We explore this in Chapter 9. See also, Gray, 2006b.)

It is at this point that one realises that accountability (and, of course, responsibility) is a moving target, constantly changing and, very importantly, constantly negotiated with (typically) those with the power trying to persuade the demos that either (a) they don't need to be accountable on this issue ('trust us') or (b) they are already fully accountable (Tinker *et al.*, 1991). This would typically be a task for the state but there is much evidence that the state too frequently takes the side of capital in matters of responsibility and accountability. In such cases, it falls to *civil society* through grass-roots movements and, as we shall see in Chapter 10, the *external social audits* to constantly challenge the accepted notions of responsibility and, consequently, to push and challenge the current levels of accountability.

3.12 Conclusions and implications for accountability and responsibility

This chapter has sought to develop a broad argument about how accountability derives from responsibility. Any understanding of accountability needs to be grounded in an understanding of responsibility. Responsibility is an elusive but central notion in personal as well as organisational life. We argued that responsibility is set within and determined by the relationships that obtain (and which could obtain) between people, organisations, groups, societies and so on. Understanding those relationships – and consequently understanding responsibility – is dependent on the worldview we hold. Each worldview sees the nature of these relationships – and especially the relationships between organisations and society – rather differently. In any circumstances, such worldviews matter – as do the relationships between organisations and society – but in a world of un-sustainability these concerns become really very critical indeed as we have to reconsider and renegotiate not just our accountability relationships but our most basic relationships between societies, justice and the planet. It was into this complex set of ideas that we placed accountability as a way of seeing through some of this confusion.

Parkinson (1993) has famously argued that a civilised society either controls actions (responsibility) or controls the information relating to those actions (accountability). Within a 'close', active democracy, much of this is achieved by personal and societal interaction itself (Lehman, 1995, 1999, 2001; Dillard, 2007). It must be this vision that we strive towards but, in a complex, changing and imbalanced world in which the opportunities for oppression and manipulation appear to be so gleefully accorded to power, having active mechanisms to define and articulate responsibility and to pursue the attendant accountability is probably essential. Only in this way – through mandated and regulated accountability – can the demos begin to learn a more balanced picture about itself and about what the organs of societies wishes – mainly its organisations – actually do and do not do; can and cannot achieve.

---

[21]For more detail see, for example, Kolk (1999, 2003); Gray (2006b); Milne (2007).

Responsibility is constantly changing, is articulated through a complex array of means. Accountability is the information related to that stated responsibility and is as much about what has *not* been done (e.g. failure to be sustainable) as it is about what *has* been done. So, although we have sought to present accountability as reasonable, middle ground between the more extreme views, and accountability has indeed been accused of conservatism as a consequence (Tinker *et al.*, 1991), this is to misunderstand the potential nature of accountability. As Owen (2008) argues, accountability is a profoundly radical notion because it requires the powerful (large corporations and the state for example) to be accountable to those with rights but (individually at least) little power – civil society. It is this profoundly democratic notion and the extent to which it is opposed that provides us with some illustration of how important it might actually be.

Appendix

Some limitations and extensions of accountability

The principle of accountability holds out the possibility for the development of information flows – 'accounts' and accounting – in such a way that it both contributes to – and reflects – the sort of democratic society in which individuals are better informed and more empowered, in which the inequalities of wealth are potentially exposed and the inequalities of power are somewhat reduced.

Accountability is based upon an ideal and an abstraction – that of participative democracy. Its assumptions can be, however, made transparent, and the attractiveness or otherwise of accountability does not depend upon any particular assumptions about the ways in which society is currently organised. So, although the neo-pluralist conception of society is usefully illustrative, it is not a pre-requisite for the development of the accountability model. The accountability model can be used to analyse current practice under whatever assumptions one chooses to make about the organisation of society. In the pro-active rather than analytical sense, accountability is essentially a mechanism – the development of which contributes to the normative position of a more justly organised and better informed democracy.

Establishing how information in general and social and environmental accounting and accountability in particular will actually contribute to the desired social change must, inevitably, be a little speculative. However, two things are apparent. *First,* information is a prerequisite of an active (participative) democracy and, virtually by definition, social and environmental accounting and accountability are necessary – but not sufficient – conditions for greater democracy. *Second,* any increase in organisational transparency through more formal social and environmental accounting and accountability will have a number of effects:

- the increased and *different* information will help socially (re)construct the organisation (see, for example, Hines, 1988). More aspects of organisational life will be made visible[22] and the consequences of organisational activity and the actions of society with respect to the organisation will become more transparent. We will think of our organisations and the organisation of economic life differently;

- the increased and *different* information will tend to encourage *information inductance* (Prakash and Rappaport, 1977) whereby the type of information one is required to report

---

[22]And whilst total visibility will probably increase, each increase in visibility, in focusing attention on some matters, decreases attention on others and, thus, inevitably makes other things 'invisible'. See, for example, Hopwood (1990); Broadbent *et al.* (1991); Hines (1992).

tends to influence the behaviour of not just the recipient of the information (e.g. society) but the creator and transmitter of the information (e.g. company management). Nobody likes reporting data on something about which either one is ashamed or which one believes one will attract criticism. Generally, therefore, organisations can be expected to act differently (to a degree at least) in order to reduce the necessity to report such negative information;

- the transparency engendered by the accountability can have the effect of bringing the organisation and the results of the actions of the organisation into closer conjunction.[23] There is some evidence to suggest that 'closeness' between actors themselves – as well as between actors and the results of their actions – decreases conflict of interests and increases the exercise of responsibility. That is, as we discussed above, accountability is a result of responsibility and, in turn, increases responsibility;[24]

- and, finally, if information exposes myths about organisational life (e.g. that all organisations are just and sustainable), that will change the political economy of the society encouraging – one might hope – increased activism in civil society and increased regulation and control by the state. Thus if the organisation does not (probably cannot) change then the 'rules of the game' will have to be changed.[25]

Such influences on the business, society and environment relationships and consequent development of more democratic relationships are an essentially emancipatory (Gallhofer and Haslam, 1997a, b) and evolutionary process. That is, the process is driven by the rebalancing of power relationships through a changing of control over, and access to, information. This will produce change which, whilst starting from and grounded upon present practice, will be a constant and developing process. More information usually leads to different actions and demands for even more, rather than less, information. The demos becomes increasingly empowered to demand more democratic relationships.

However, for some commentators, accountability still has major problems. These are principally to do with 'power'. For example, we have already seen that a case might be made for a particular set of rights and responsibilities – let us say, for illustration, the preservation of whales. Those making the case believe an accountability is due to them. Those from whom the accountability is owed do not. Who is right? Tricker (1983) and Stewart (1984) argue that unless the principal can enforce the accountability, then no accountability is due. This might be called 'positive accountability' – what *should be* is considered to be identical to *what is*. Tricker goes on to argue that if the agent voluntarily chooses to disclose some or all of the information that (an assumed) accountability might demand and the principal is unable to enforce that disclosure then we should think of this as '*ex gratia* disclosure'. That is, the agent, for reasons of his/her own (see Chapter 4) in placing information in the public domain should not be thought of as acknowledging and discharging a (non-existent) accountability. This theme can be further developed. Many of the types of information with which SEA is concerned can be, indeed often are, made available to the state. That is, the state can

---

[23]This relates to a reduction in what is known in Marxian analysis as 'alienation' and to an increase in what is called (as we saw above) in the work of John Rawls 'closeness' (see, for example, Rawls, 1972). For more detail in an accounting context see, for example, Lehman (1999, 2001, 2002).

[24]Some evidence to this effect is quoted in Gray (1992). Indeed, a major element in 'deeper green' thinking is the recognition that being accountable, the giving of an account, is often a morally sound and spiritually uplifting thing to do. There are frequently times in normal human relationships when an 'agent' wants, even demands, to give an account of themselves. It is only habit, social convention and the conditioning effects of large organisations that make the idea of freely giving an account so bizarre.

[25]Of course there is considerable evidence that organisations, and especially large corporations and their representative bodies, will actively seek to prevent such changing of the rules of the game (see, for example, Parkinson, 1993).

(and occasionally does) enforce the accountability, but the demos is unable to enforce the next step in the accountability chain onto either the agent or the state. In this, we certainly learn something about the nature of the state – not least that it is not accountable to the demos and, perhaps more disturbing, it is perhaps not a democratic institution itself (see, for example, Tinker 1984a, b).

Be that as it may, accountability can still be due, even when it cannot be enforced. A moral or natural right to information may flow from an established (legal or non-legal) responsibility. The fact that this accountability is not discharged at least reaffirms our (reluctant) arguments about the lack of democracy in modern western industrialised society. Roberts and Scapens (1985) take this further. They argue, as we have done above, that accountability is essentially a reflection of a social process. However, they go further and, following Giddens (1976), deduce that this suggests that power rests with the principal who is able to impose his/her social values onto the agent. The accountability relationship is thus potentially exploitive.[26] The irony, in the present context, is that in CSR we are generally concerned with principals (e.g. society, employees, etc.) unable to enforce their accountability onto the agents (e.g. the state or companies). There is, therefore, no chance in these situations to exercise exploitation even if they wished to. Furthermore, the desire for accountability is driven by a desire to actually *reduce* the levels of (assumed) exploitation of the powerful over the weak. So whilst accountability may well reflect power asymmetry – full accountability is designed to seek something closer to power equality.

This is where the flexibility of the accountability model comes into play. Remember that we said that the accountor and accountee (agent and principal) can change and swap places. Take the society–state–company relationship. The accountability relationships are far from simple. We might think that the society elects the state which controls and enfranchises the company. Certainly, one set of relationships runs that way. But the company can control the state (the simplest illustration is by threatening to move overseas and create unemployment and a reduction in tax revenue, etc.) and the state can control society to a fair degree as well. The relationship can thus run the other way. So, each party can be both an agent and a principal to the other. When the explicit, actual and perhaps economic power outstrips the moral or natural rights (as seems to be increasingly the case in a less and less moral world), then only those relationships in which the principal can enforce the accountability are positively observable. And 'yes', such relationships will impose one group's view on another – whether that is better or worse, good or bad will depend upon your worldview and moral position.[27]

Similarly, there may well be situations in which the principal is able to enforce the accountability but chooses not to – through trust, ignorance, lack of concern, laziness, stupidity or whatever.[28] Does the accountability also fail here? Stewart (1984) argues that it does, but we are unconvinced. Certainly, principals who fail to exercise their rights are also failing to exercise a *duty* of control (an important argument in the shareholder–director relationship) but it seems that the *principle* of accountability still obtains. This is the argument of Hedlund and Hamm (1978) and Greer *et al.* (1978) in which a natural or moral right does not wither through lack of use. It may well encourage a self-interested agent *not* to discharge accountability, but it need not suggest that the agent *should* not discharge it. Indeed, Gray (1978) has argued that accountability can be discharged by the *existence* of a channel of

---

[26]There is no *a priori* reason that we can see why this need be the case.

[27]For example, do you believe it is necessarily a good thing that companies can impose their views on society but not the other way round?

[28]Indeed, it is frequently argued that only through constant struggle and vigilance can democracy ever be maintained – it is never static (see, for example, Walker 2002).

accountability – a means whereby the information can be obtained – and that this is far more significant than the existence of the account itself.

So, although charges can be brought against the accountability model (and these are only some of them), it is nevertheless a useful means for analysing information in a society which claims to be a democracy. The model allows analysis. It has some positive (descriptive) power (as with, for example, the shareholder–director relationship), but it is predominantly a *normative model* (a model of the world as it should be from a particular point of view). Such normative models are essential, not just for planning one's route (where do we want to get to?), but also for evaluating the steps along the way (how good is our present situation? do we wish to be here? how well are we progressing in our development towards our goal – whatever that is?).

References

Adams, C. A. (2004) The ethical, social and environmental reporting-performance portrayal gap, *Accounting, Auditing and Accountability Journal*, **17**(5): 731–57.

Adams, C. A., Coutts, A. and Harte, G. (1995) Corporate equal opportunities (non)disclosure, *British Accounting Review*, **27**(2): 87–108.

Armstrong, P. (1991) Contradiction and social dynamics in the capitalist agency relationship, *Accounting, Organizations and Society*, **16**(1): 1–26.

Arrington, E. (1990) Intellectual tyranny and the public interest: the quest for the holy grail and the quality of life, *Advances in Public Interest Accounting*, **3**: 1–16.

Arrington, E. and Francis, J. (1989) Letting the chat out of the bag: deconstruction, privilege and accounting research, *Accounting. Organizations & Society*, **14**(1/2): 1–28.

Atkinson, G., Dietz, S. and Neumayer, E. (2007) *A Handbook of Sustainable Development*. Cheltenham: Edward Elgar.

Bailey, D., Harte, G. and Sugden, S. (1994a) *Making Transnationals Accountable: A significant step for Britain*. London: Routledge.

Bailey, D., Harte, G. and Sugden, R. (1994b) *Transnationals and Governments: Recent policies in Japan, France, Germany, the United States and Britain*. London: Routledge.

Bakan, J. (2004) *The Corporation: the pathological pursuit of profit and power*. London: Constable and Robinson.

Ball, A. (2002) *Sustainability Accounting in UK Local Government: An agenda for research*. London: ACCA.

Bebbington, K. J. and Thomson, I. (1996) *Business Conceptions of Sustainability and the Implications for Accountancy*. London: ACCA.

Beder, S. (1997) *Global Spin: The corporate assault on environmentalism*. London: Green Books.

Bendell, J. (ed.) (2000) *Terms for Endearment: Business, NGOs and Sustainable Development*. Sheffield: Greenleaf Publishing / New Academy of Business.

Benston, G. J. (1982a) An analysis of the role of accounting standards for enhancing corporate governance and social responsibility, *Journal of Accounting and Public Policy*, **1**(1): 5–18.

Benston, G. J. (1982b) Accounting & corporate accountability, *Accounting, Organizations and Society*, **7**(2): 87–105.

Blowfield, M. and Murray, A. (2008) *Corporate Responsibility: A critical introduction*. Oxford: Oxford University Press.

Broadbent, J., Laughlin, R. and Read, S. (1991) Recent financial and administrative changes in the NHS: A Critical Theory analysis, *Critical Perspectives on Accounting*, **2**(1): 1–30.

Bruno, K. (1992) *The Greenpeace Book of Greenwash*. Washington, DC: Greenpeace.

Carroll, A. B. (1991) The pyramid of corporate social responsibility: Towards the moral management of organizational stakeholders, *Business Horizons*, July/August: 42.

Carroll, A. B. and Buchholtz, A. K. (2006) *Business and Society: Ethics and Stakeholder Management*. Thomson: London.

Christenson, C. (1983) The methodology of positive accounting, *The Accounting Review*, January: 1–22.

Chryssides, G. D. and Kaler, J. H. (1993) *An Introduction to Business Ethics*. London: Thomson.

Cooper, C., Dunn, J. and Puxty, A. (1992) *Anxious Murderers? – Death Drives in Accounting*, paper presented at British Accounting Association Scottish Area Group Conferences, Dundee, 10 September.

Dahl, R. A. (1972) A prelude to corporate reform, *Business and Society Review*, Spring: 17–23.

De Bakker, F. G. A., Groenewegen, P. and Den Hood, F. (2005) A bibliometric analysis of 30 years of research and theory on corporate social responsibility and corporate social performance, *Business and Society*, **44**(3): 283–317.

Dickson, D. (1974) *Alternative Technology and the Politics of Technical Change*. Glasgow: Fontana.

Dillard, J. (2007) Legitimating the social accounting project: an ethic of accountability, in Unerman, J., Bebbington, J. and O'Dwyer, B. (2007) (eds), *Sustainability Accounting and Accountability*, pp. 37–54. London: Routledge.

Donaldson, J. (1988) *Key Issues in Business Ethics*. London: Academic Press.

Dowling, J. and Pfeffer, J. (1975) Organisational legitimacy: social values and organisational behaviour, *Pacific Sociological Review*, January: 122–36.

Dresner, S. (2002) *The Principles of Sustainability*. London: Earthscan.

Eckersley, R. (1992) *Environmentalism and Political Theory: Towards an ecocentric approach*. London: UCL Press.

Eden, S. (1996) *Environmental Issues and Business: Implications of a changing agenda*. Chichester: Wiley.

Edwards, M. and Hulme, D. (eds) (1995) *Non-Governmental Organisations – Performance and Accountability: Beyond the magic bullet*. London: Earthscan.

Estes, R. W. (1976) *Corporate Social Accounting*. New York: Wiley.

Everett, J. (2004) Exploring (false) dualisms for environmental accounting praxis, *Critical Perspectives on Accounting*, **15**(8): 1061–84.

Frederick, W. C. (1986) Towards CSR3: Why ethical analysis is indispensable and unavoidable in corporate affairs, *California Management Review*, **28**(2): 126–41.

Frederick, W. C. (1994) From CSR1 to CSR2: the maturing of business-and-society thought, *Business and Society*, **33**(2): 150–67 (originally 1978 Working Paper).

Freeman, E. R. and Liedtka, J. (1991) Corporate social responsibility: a critical approach, *Business Horizons* (July–August): 92–8.

Friedman, M. (1970) The social responsibility of business is to increase its profits, *The New York Times Magazine*, September 13: 122–6.

Gallhofer, S. and Haslam, J. (1997a) The direction of green accounting policy: critical reflections, *Accounting, Auditing and Accountability Journal*, **10**(2): 148–96.

Gallhofer, S. and Haslam, J. (1997b) Beyond accounting: The possibilities of accounting and critical accounting research, *Critical Perspectives on Accounting*, **8**(1/2): 71–96.

Gambling, T. (1978) The evolution of accounting man, *Accountants' Weekly*, 10 November: 30–31.

Giddens, A. (1976) *New Rules of Sociological Method*. London: Hutchinson.

Gladwin Thomas, N. (1993) Envisioning the sustainable corporation, in Smith, E. T. (ed.), *Managing for Environmental Excellence*. Washington, DC: Island Press.

Goldsmith, E. *et al.* (1972) *Blueprint for Survival*. Harmondsworth: Penguin.

Gray, V. (1978) Accountability in policy process: An alternative perspective, in Greer, S., Hedlund, R. D. and Gibson, J. L. (eds), *Accountability in Urban Society*. London: Sage.

Gray, R. H. (1992) Accounting and environmentalism: an exploration of the challenge of gently accounting for accountability, transparency and sustainability, *Accounting Organisations and Society*, **17**(5): 399–426.

Gray, R. H. (1994) Corporate reporting for sustainable development: accounting for sustainability in 2000AD, *Environmental Values*, **3**(1): 17–45.

Gray, R. (2006a) Social, Environmental, and Sustainability Reporting and Organisational Value Creation? Whose Value? Whose Creation?, *Accounting, Auditing and Accountability Journal*, **19**(3): 319–48.

Gray, R. H. (2006b) Does sustainability reporting improve corporate behaviour? Wrong question? Right time?, *Accounting and Business Research* (International Policy Forum), **36**: 65–88.

Gray, R. (2010) Is accounting for sustainability actually accounting for sustainability . . . and how would we know? An exploration of narratives of organisations and the planet, *Accounting, Organizations and Society*, **35**(1): 47–62.

Gray, R. H. and Bebbington, K. J. (2001) *Accounting for the Environment*, 2nd Edition. London: Sage.

Gray, R. H. and Milne, M. (2004) Towards reporting on the triple bottom line: mirages, methods and myths, in Henriques, A. and Richardson, J. (eds), *The Triple Bottom Line: does it all add up?*, pp. 70–80. London: Earthscan.

Gray, R. H., Owen, D. L. and Maunders, K. T. (1986) Corporate social reporting: the way forward?, *Accountancy*, December: 6–8.

Gray, R. H., Owen, D. L. and Maunders, K. T. (1987) *Corporate Social Reporting: Accounting and accountability*. Hemel Hempstead: Prentice Hall.

Gray, R. H., Kouhy, R. and Lavers, S. (1995) Corporate social and environmental reporting: A review of the literature and a longitudinal study of UK disclosure, *Accounting, Auditing and Accountability Journal*, **8**(2): 47–77.

Gray, R. H., Collison, D. J. and Bebbington, K. J. (1997) Environmental and social accounting and reporting, in Fisher, L. (ed.), *Financial Reporting Today: Current Trends and Emerging Issues*, 1998, pp. 179–214. London: Accountancy Books.

Gray, R., Bebbington, J. and Collison, D. J. (2006) NGOs, civil society and accountability: making the people accountable to capital, *Accounting, Auditing and Accountability Journal*, **19**(3): 319–48.

Greer, S., Hedlund, R. D. and Gibson, J. L. (eds) (1978) *Accountability in Urban Society*. London: Sage.

Grey, C. and Willmott, H. C. (2005) *Critical Management Studies: A Reader*. Oxford: Oxford University Press.

Hedlund, R. D. and Hamm, K. E. (1978) Accountability and political institutions, in Greer, S., Hedlund, R. D. and Gibson, J. L. (eds), *Accountability in Urban Society*, pp. 34–47. London: Sage.

Henderson, D. (2001) *Misguided Virtue: False Notions of Corporate Social Responsibility*. London: The Institute of Economic Affairs.

Hertz, N. (2001) *The Silent Takeover: Global capitalism and the death of democracy*. London: Heinemann.

Hines, R. D. (1988) Financial accounting: In communicating reality, we construct reality, *Accounting, Organizations and Society*, **13**(3): 251–61.

Hines, R. D. (1992) Accounting: filling the negative space, *Accounting, Organizations and Society*, **17**(3/4): 313–42.

Hopwood, A. G. (1990) Ambiguity, knowledge and territorial claims: some observations on the doctrine of substance over form: a review essay, *British Accounting Review*, **22**(1): 79–88.

Jacobs, M. (1991) *The Green Economy: Environment, sustainable development and the politics of the future*. London: Pluto Press.

Jensen, M. and Meckling, W. (1976) Theory of the firm: managerial behaviour, agency costs and ownership structure, *Journal of Financial Economics*, October: 305–60.

Jones, C., Parker, M. and Ten Bos, B. (2005) *For Business Ethics*. London: Routledge.

Kempner, T. K., MacMillan, K. and Hawkins, K. H. (1976) *Business and society*. Harmondsworth: Pelican.

Klein, N. (2001) *No Logo: No Space, No Choice, No Jobs*. London: Flamingo/Harper Collins.

Knights, D. and Willmott, H. (2007) *Introducing Organizational Behaviour and Management*. London: Thomson Learning.

Kolk, A. (1999) Evaluating corporate environmental reporting, *Business Strategy and the Environment*, **8**(4): 225–37.

Kolk, A. (2003) Trends in sustainability reporting by the Fortune Global 250, *Business Strategy and the Environment*, **12**(5): 279–91.

Kovel, J. (2002) *The Enemy Of Nature: The end of capitalism or the end of the world?*, London: Zed Books.

Lehman, C. (1992) *Accounting's Changing Roles in Social Conflict*. London: Paul Chapman.

Lehman, G. (1995) A legitimate concern for environmental accounting, *Critical Perspectives on Accounting*, **6**(6): 393–412.

Lehman, G. (1999) Disclosing new worlds: a role for social and environmental accounting and auditing, *Accounting Organizations and Society*, **24**(3): 217–42.

Lehman, G. (2001) Reclaiming the public sphere: problems and prospects for corporate social and environmental accounting, *Critical Perspectives on Accounting*, **12**(6): 713–33.

Likierman, A. (1986) *Rights and Obligations in Public Information*. Cardiff: University College Cardiff Press.

Likierman, A. and Creasey, P. (1985) Objectives and entitlements to rights in government financial information, *Financial Accountability and Management*, **1**(1): 33–50.

Lindblom, C. K. (1993) *The Implications of Organizational Legitimacy for Corporate Social Performance and Disclosure*, Paper presented at the Critical Perspectives on Accounting Conference, New York.

Mathews, M. R. (1993) *Socially Responsible Accounting*. London: Chapman Hall.

Matravers, M. (2007) *Responsibility and Justice*. Cambridge: Polity.

Maunders, K. T. and Burritt, R. (1991) Accounting and ecological crisis, *Accounting, Auditing and Accountability Journal*, **4**(3): 9–26.

Mayhew, N. (1997) Fading to grey: the use and abuse of corporate executives' 'representational power', in Welford, R. (ed.), *Hijacking Environmentalism: Corporate response to sustainable development*, pp. 63–95. London: Earthscan.

Meadows, D. H., Randers, J. and Meadows, D. L. (2004) *The Limits to Growth: The 30-year Update*. London: Earthscan.

Milne, M. (2007) Downsizing Reg (Me and You)!: Addressing the 'real' sustainability agenda at work and home, in Gray, R. H. and Guthrie, J. (eds), *Social Accounting, Mega Accounting and Beyond: A Festschrift in Honour of M. R. Mathews*, pp. 49–66. St Andrews: CSEAR Publishing.

Milne, M. J., Kearins, K. N. and Walton, S. (2006) Creating adventures in wonderland? The journey metaphor and environmental sustainability, *Organization*, **13**(6): 801–39.

Milne, M. J., Tregigda, H. M. and Walton, S. (2009) Words not actions! The ideological role of sustainable development reporting, *Accounting Auditing and Accountability Journal*, **22**(8): 1211–57.

Mintzberg, H. (1983) The case for corporate social responsibility, *The Journal of Business Strategy*, **4**(2): 3–15.

Mohan, A. (2003) *Strategies for the management of complex practices in complex organisations: A study of the transnational management of corporate responsibility*. Unpublished doctoral dissertation, University of Warwick.

Noreen, E. (1988) The economics of ethics: A new perspective on agency theory, *Accounting Organizations and Society*, **13**(4): 359–70.

O'Dwyer, B. (2003) Conceptions of corporate social responsibility: the nature of managerial capture, *Accounting Auditing and Accountability Journal*, **16**(4): 523–57.

Owen, D. (2008) Chronicles of wasted time? A personal reflection on the current state of, and future prospects for, social and environmental accounting research, *Accounting, Auditing and Accountability Journal*, **21**(2): 240–67.

Owen, D. L., Gray, R. H. and Adams, R. (1992) A green and fair view, *Certified Accountant*, April: 12–15.

Owen, D. L. and Swift, T. (2001) Social accounting, reporting and auditing: beyond the rhetoric, *Business Ethics: A European Review*, **10**(1): 4–8.

Parkinson, J. E. (1993) *Corporate Power and Responsibility*. Oxford: Oxford University Press.

Perks, R. W. (1993). *Accounting and Society*. London: Chapman & Hall.

Porritt, J. (2005) *Capitalism: as if the World Matters*. London: Earthscan.

Prakash, P. and Rappaport, A. (1977) Information inductance and its significance for accounting, *Accounting, Organizations and Society*, **2**(1): 29–38.

Puxty, A. G. (1986) Social accounting as immanent legitimation: A critique of a technist ideology, *Advances in Public Interest Accounting*, **1**: 95–112.

Puxty, A. G. (1991) Social accountability and universal pragmatics, *Advances in Public Interest Accounting*, **4**: 35–46.

Rawls, J. (1972) *A Theory of Justice*. Oxford: Oxford University Press.

Roberts, J. and Scapens, R. (1985) Accounting systems and systems of accountability, *Accounting, Organizations and Society*, **10**(4): 443–5.

Robertson, J. (1978) *The Sane Alternative*. London: James Robertson.

Schmidheiny, S. (1992) *Changing Course*. New York: MIT Press.

Schumacher, E. F. (1973) *Small is Beautiful*. London: Abacus.

Schwartz, P. and Gibb, B. (1999) *When Good Companies Do Bad Things: Responsibility and Risk in an Age Of Globalization*. New York: Wiley.

Shafer-Landau, R. (2007) *Ethical Theory: An anthology*. Oxford: Blackwell.

Shocker, A. D. and Sethi, S. P. (1973) An approach to incorporating societal preferences in developing corporate action strategies, *California Management Review*, Summer: 97–105.

Stewart, J. D. (1984) The role of information in public accountability, in Hopwood, A. and Tomkins, C. (eds), *Issues in Public Sector Accounting*, pp. 13–34. Oxford: Philip Allen.

Stone, C. D. (1975) *Where the law ends*. New York: Harper & Row.

Sunstein, C. R. (1997) *Free Markets and Social Justice*. New York: Oxford University Press.

Thielemann, U. (2000) A brief theory of the market – ethically focused, *International Journal of Social Economics*, **27**(1): 6–31.

Thomson, I. and Bebbington, J. (2005) Social and environmental reporting in the UK: a pedagogic evaluation, *Critical Perspectives on Accounting*, **16**(5): 507–33.

Tinker, A. M. (ed.) (1984a) *Social Accounting for Corporations*. Manchester: Manchester University Press.

Tinker, A. M. (1984b) Theories of the state and the state of accounting: economic reductionism and political voluntarism in accounting regulation theory, *Journal of Accounting and Public Policy*, **3**: 55–74.

Tinker, A. M. (1985) *Paper Prophets: a social critique of accounting*. Eastbourne: Holt Saunders.

Tinker, T., Lehman, C. and Neimark, M. (1991) Corporate social reporting: Falling down the hole in the middle of the road, *Accounting, Auditing and Accountability Journal*, **4**(2): 28–54.

Tinker, A. M. and Okcabol, F. (1991) Fatal attractions in the agency relationship, *British Accounting Review*, **23**(4): 329–54.

Tinker, T. and Puxty, T. (1995) *Policing Accounting Knowledge*. London: Paul Chapman.

Tozer, L. and Hamilton, F. (2007) Re-energising the social contract in accounting: The case of James Hardie Industries, in Gray, R. H. and Guthrie, J. (eds), *Social Accounting, Mega-accounting and Beyond: A Festschrift in Honour of M. R. Mathews*, pp. 107–26. St Andrews: CSEAR Publishing.

Tricker, R. I. (1983) Corporate responsibility, institutional governance and the roles of accounting standards, in Bromwich, M. and Hopwood, A. G. (eds), *Accounting Standards Setting – An International Perspective*, pp. 27–41. London: Pitman.

Unerman, J., Bebbington, J. and O'Dwyer, B. (eds) (2007) *Sustainability Accounting and Accountability*. London: Routledge.

Unerman, J. and O'Dwyer, B. (2006) Theorising accountability for NGO advocacy, *Accounting, Auditing and Accountability Journal*, **19**(3): 349–76.

United Nations Millennium Ecosystem Assessment, (2005) *Living Beyond Our Means: Natural Assets and Human Well-Being: Statement from the board* (http://www.millenniumassessment.org/en/Products.BoardStatement).

United Nations World Commission on Environment and Development (1987) *Our Common Future (The Brundtland Report)*. Oxford: Oxford University Press.

Vogel, D. (2005) *The Market for Virtue: The potential and limits of corporate social responsibility*. Washington, DC: Brookings Institute.

Walker, P. (2002) *We the People: Developing a new democracy*. London: New Economics Foundation.

Walters, K. D. (1977) Corporate social responsibility and political ideology, *California Management Review*, **19**(3): 40–51.

Welford, R. (ed.) (1997) *Hijacking Environmentalism: Corporate response to sustainable development*. London: Earthscan.

Zimmerman, M. E. (1994) *Contesting Earth's Future: Radical ecology and postmodernity*. London: University of California Press.

CHAPTER 4

Description, development and explanation of social, environmental and sustainability accounting and reporting[1]

4.1 Introduction

Any initial introduction to social accounting will provide some of the parameters and definitions of the topic. Our approach has been to then try to provide a context within which we might see the role, the potential, even the necessity of social accounting as a major mechanism in promising the species any kind of civilised future in the face of rapacious (albeit massively successful) development.[2] We are now in a position to flesh out the bones of that initial introduction to social accounting. This is what we attempt in this chapter. To do this, we are going to consider three principal themes in this chapter: descriptions of what social accounting is (and can be); brief histories of the development in social accounting; and, finally, an overview of some of the theoretical interpretations and explanations that are used to make sense of the activity.[3]

It is possible to view 'social accounting' as the *universe of all possible accountings* – that is, that social accounting can embrace any possible way in which we can imagine that individuals and/or groups/organisations might choose to request, give and receive accounts from one another. However, as we remarked there, it is not obvious that such an all-embracing definition actually helps us much (see, for example, Owen, 2008). Even if we draw our boundaries a little tighter, we will still find that social accounting is a complex, diverse, amorphous and constantly changing craft (see, for example, Unerman *et al.*, 2007). We will do what we can to limit the frustration that must inevitably flow from this diversity but, as we work through this chapter, it should become apparent why social accounting is not precise or definable and how it came to be so. Indeed, this imprecision is very likely to remain the case until such time as social accounting is subject to demanding international

---

[1]We very gratefully acknowledge comments from Jan Bebbington and David Collison on earlier versions of this chapter. A substantial part of this chapter – the sections on theory – has previously been published in a similar form as Gray, Owen and Adams (2010) 'Some theories for social accounting?: A review essay and tentative pedagogic categorisation of theorisations around social accounting', *Advances in Environmental Accounting and Management* 4:1–54.

[2]See Gray *et al.*, (1996) for one illustration of this. More support is provided in Chapters 1–3.

[3]We strongly recommend that you make yourself familiar with examples of social accounting and, if you have not already done so, that you now start obtaining and consulting such accounts. The Appendix to this chapter gives guidance on how to go about this.

regulation – and even then evolution may still produce the new and the unexpected.[4] To a considerable extent, it is the voluntary – one might even say wilful – nature of much of social accounting practice that produces the diversity and the lack of any systematic or organised development. And this is something that becomes both more acute and more bothersome when we look at the explanations (the theories) offered for social accounting practice.

The chapter is organised as follows. The next section considers the diversity of social accounting. Section 4.3 offers a brief history of corporate social responsibility (CSR) and is followed in Section 4.4 by a brief history of some of the milestones in the development of social accounting. Section 4.5 starts us on our journey into theory and outlines a justification for our need to consider theory in social accounting. Section 4.6, examines systems-level theories and then offers a tentative schema which we can use to (cautiously) organise our theorising about social accounting. Sections 4.7 and 4.8 cover, respectively, (what we have called) sub-systems-level theories and organisational-level theories of social account-ing. These sections comprise most of the most popular (and influential) theories of social accounting but, as we shall see, these theoretical perspectives do not provide as much insight as we need. So in Section 4.9 we look inside the organisation and examine the perspectives employed and offered by the growing number of field-based studies of social accounting practice. We try and pull this all together in the conclusions in Section 4.11. The chapter concludes with an Appendix in which we provide a 'study guide' to obtaining and using examples of organisational disclosure to illustrate and challenge the (potentially) vague and general comments that a chapter on theory inevitably produces.

4.2 The diversity of social accounting

Social accounting is only partially regulated and such regulations as exist vary considerably from country to country (see, for example, Guthrie and Parker, 1990; Hibbitt and Collison, 2004). SEA is, as a result, a virtually limitless area of potential activity. This diversity can be illustrated by considering just a few of the characteristics that any account might possess.

The subject matter of the account: Social accounting tends to focus on areas such as employees, community, consumers and the natural environment. But even this is too tidy: responsibility and sustainability (or sustainable development) are crucial aspects of social accounting and they do not always fall neatly into these categories. Further still, any review of social accounting prac-tice is likely to unearth examples of accounting and reporting on/for other matters such as: ethics; religion; taxation; standards; political involvement; lobbying; human rights; international relations; and the characteristics of organisational investment. In essence, we can expect to find within social accounting virtually *any* element of *any* relationship between the organisation and its stakeholders.

Who, then, are the stakeholders?: Stakeholders are any person or group who can influence and/or is influenced by the organisation under consideration (see, especially, Friedman and Miles, 2006, for more detail as well a range of analyses of this statement). This suggests that everybody in some way or other is a stakeholder in almost any major organisation. We won't worry about the practicalities of this at this stage. However, we must keep a constant

---

[4]There have been attempts to address standardisation in social accounting both academically and professionally. For an introduction see, for example, Parker (1986) and Gray *et al.* (1997).

vigilance to recognise that those whom the organisation may *wish* to recognise as the key stakeholder groups and those whom society may require to be recognised as key stakeholders are frequently not the same thing at all. With that said, in broad terms it is common to consider the principal stakeholder groups as including: management of the organisation; shareholders; other providers of finance; government; employees; local communities; customers; suppliers; the environment itself; and future generations. But this is *not* – nor is it ever likely to be – comprehensive.

The audience for any account: The stakeholders and the audience need not be the same thing. Whilst some examples of social accounting are closely defined for a specific target audience (for example, employee reports, information for collective bargaining, local environmental information, information for use in schools or internal documents for management use), it may often be difficult to establish for whom a report is principally intended. Certainly, as much of the disclosure which *is* regulated appears in Annual Reports it must, therefore, be assumed to be intended primarily for shareholders. This suggests that, although the *general* or social accountability of companies may still be relatively underdeveloped, social disclosure to shareholders *has* shown a steady increase over time. Later in the chapter, this issue of intended audience will re-emerge as important when we consider the range of theoretical explanations of social accounting.

What content will comprise the account?: Whilst we may now know what to expect from a social account in terms of its subject matter, what detail, form, medium or completeness might we expect? This is where the real diversity of social accounting reveals itself. A 'social account' might comprise such things as: a brief assertive statement from the Chairperson or Directors; beautifully crafted reports on organisational involvement in the local community; intensely focused tables of data and targets; or the far rarer attempts at full (perhaps even financial) social or environmental accounts. A few minutes spent with reports and/or websites of organisations (as we suggested above) will provide you with examples from the full spectrum (see also Adams and Frost, 2004).

What is the organisation's motivation for producing a report?: As we will see later in this chapter, it would be both interesting and useful to know why an organisation chooses to produce a social/environmental account and report. But motives are notoriously tricky to infer with any fairness or accuracy.[5] Equally, it is unlikely that an individual, let alone a complex organisation, operates with a single motive. As we will see, research has speculated, inferred and shown that motives for social and environmental disclosure may be as diverse as: individual commitment; idealism; competitive advantage; manipulation of public perception; forestalling of legislation; keeping up with competitors; inducing change; public image; pressure from ethical investors; communication of risk management; legitimation and so on. Rarely, however, does the motive appear to be the *discharge of full accountability*.

How reliable is the account?: With any information, it is necessary to know to what extent one may trust it to present (say) a complete, fair, balanced or reliable picture of the issues and/or organisation in question – insofar as this is possible at all. Of course, a report prepared by an organisation and un-audited does not mean it is a pack of lies anymore than a report prepared by an external party can be assumed to be a full and balanced picture. However, the source and reliability of the preparer of the information should be taken into account when reading the social and environmental reports. As a general guide, does the

---

[5]This comment is something of a personal statement in that to simply assume a motive – especially a motive of simple 'self-interest' – is both trite and, potentially, deeply offensive to the individuals concerned. Inevitably, however, as we consider the theoretical explanations we will need to try and infer motive later.

report cover the relationship with all its stakeholders? Is it actually a *complete* account? And, again as we shall see later, just because the data appears to be attested to, audited or assured by an apparently independent third party may not mean much either (see, for example, Ball *et al.*, 2000; O'Dwyer and Owen, 2005; Owen, 2007).

To what extent is the account governed by law, codes or guidelines? It would be natural to expect more from an account which was governed by extensive law and supported by a well-trained profession like accounting.[6] Whether this would be entirely rational is another question, but we simply want to emphasise that whilst many jurisdictions do require disclosure of social/environmental data (most typically data about employment numbers and conditions and maybe other items such as environmental contingencies and liabilities), the bulk of the data that does and should comprise an external social account will be either voluntarily disclosed or not disclosed at all.

Finally, in this section, it is necessary to dwell briefly on the distinction between 'internal' and 'external' accounting and reporting. Indeed, we should note distinctions between internal and external *preparation* and internal and external *consumption* of the accounts and the reports. These sources and destinations tend to result in different forms of social accounts. Figure 4.1 seeks to illustrate this.

An organisation's first concern might normally be expected to be its day-to-day management plus strategy and plans. So organisations will typically prepare information for their own internal use – through such things as management accounting, management information systems, reports, research and the like. This internal information will, in all likelihood, be augmented by information which is prepared for the use of the regulatory authorities[7] and information provided under contract by external consultants.[8] Only rarely does such information reach the public domain and so relatively little tends to be known about it outside the organisation itself. The principal purpose of this information will be to help guide management decisions within the organisation (for more detail see Gray and Bebbington, 2001; Bebbington, 2007; Chapters 5, 6, and 7 of this text).

The difficulty that the general public (and researcher) have in obtaining information about what is really happening inside the organisation is just one of the reasons why the emphasis in social accounting tends to be on published material for external consumption – although this may be only the tip of a social and environmental accounting iceberg. (We will seek to both justify this emphasis as well as – perhaps contradictorily – redress the balance as far as possible as the book progresses.)

Organisations have the full range of possible approaches to social accounting from which to choose if they decide to proceed with a published social and environmental account. But,

---

[6]As you are aware, the information disclosure by an organisation tends to be dominated by the, principally regulated, information provided to the owners and providers of funds in an organisation. Whilst it is possible to trace the early motivations for this disclosure to what might be thought of as 'social' and well as economic reasons, most current discussion of conventional financial accounting tends to ignore the role and purpose of Companies Acts in favour of concentration on 'explaining' behaviour in terms of short-term, self-interested gamblers in stock markets. This is one of the major reasons that social accounting tends to place less emphasis on providers of financial funds (for a discussion see, for example, Owen *et al.*, 1987; Cooper, 1988; Gray, 2006).

[7]Including 'Quangos' – quasi-autonomous non-governmental bodies such as a country's Health and Safety Inspectorate or Environmental Protection Agency.

[8]As was the case with environmental consultants reports in the early stages of environmental debate of the late 1980s and early 1990s. In these cases, environmental consultants might help design waste-minimisation programmes, give guidance to the organisation on areas of toxic waste or contaminated land or, more proactively, help the company towards the establishment of environmental quality management or environmental audit and management systems such as ISO 14000 (see, for example, Gray *et al.*, 2001a; Gray and Bebbington, 2001 for more detail).

Figure 4.1 Different elements of social accounting

Report for use/consumption of:	Report compiled by	
	Internal Organisational Participants	**External Organisational Participants**
Internal Organisational Participants	• Health + Safety Data • Environmental Management Systems Reports • Environmental Accounting • Strategic Environmental Assessment • Stakeholder Assessment • Compliance Audit • SWOT Analysis	• Government Body Reports (e.g. environmental protection agency or other regulators reports); • Consultants for (e.g.) waste; energy; environmental management; risk assessment • Due Diligence Assessment
External Organisational Participants	• Social, Environmental, Responsibility or Sustainability Stand Alone Reports • Annual Report Disclosures • Corporate Website • Employee Reports • Mission Statements • Press Releases • Analysts' Briefing Documents	• Government Body Reports (e.g. advertising standards reports) • The External Social Audits • Civil Society Reports (e.g. NGOs; consumer protection bodies etc.) • Investigative Journalism • Ethical Investment Reports • Academic Reports (e.g. Reporting Portrayal Gaps; Silent Accounts)

increasingly, various independent bodies have developed the ability to produce and publish information about other organisations. So a multi-national company (say) which is reluctant to produce data on (for example) its environmental performance may find that the public, its employees, shareholders and others are increasingly well-informed about that company's performance as a result of the ferreting activities of bodies such as the Council on Economic Priorities, Friends of the Earth, Greenpeace, CorporateWatch or AdBusters (for more detail see Chapter 10). This, provides yet another pressure upon organisations to undertake the disclosure of their social and environmental performance themselves before 'less sympathetic' organisations do it for them![9]

The foregoing, then, provides some idea of the range of possibilities that can fall within the ambit of social accounting. And this is just the *formal* accountings – this is an indication of the range of social accounts when we keep to the language of reporting, disclosure and accountants. As we said earlier, social accounting might be more effective and attractive if it were less formal, perhaps involving deep discussion and exchange of ideas (something which is sometimes called *dialogics* – see, for example, Thomson and Bebbington, 2005) or something which involved a much deeper and closer sense of community and connectivity (as Lehman, 1995, 1999, 2001, 2007 has developed in his writing). These (and other) possibilities will be matters for later study.

We now turn in the next section to provide a brief historical context before we go on to examine this thing we know as social accounting.

---

[9]The emergence of Freedom of Information Acts – initially in the USA and now elsewhere – has certainly exercised organisations which now know that outside bodies have access to data that the organisation might prefer was kept from the public domain.

4.3 A brief history of social responsibility

> Those who cannot remember the past are condemned to repeat it.
>
> (George Santayana, *Life of Reason* 1905)[10]

We need to recognise that social responsibility, social justice and environmental sustainability are not uniquely 21st century concerns. Indeed, anxiety about responsibility and accountability are probably as old as humanity; concerns over greed, wealth, property and trade run through most religions; and environmental concerns appear variously in ancient Babylon, the mystery of Easter Island and species extinction by early Maori settlement in New Zealand (see, for example, Ponting, 1991). Equally, deep concern about mercantilism and capitalism were the backbone of the 18th and 19th century writings of, respectively, Adam Smith and Karl Marx. Conflict over the treatment and exploitation of people – and especially issues like slavery – have occupied mankind throughout history. So the issues are not new. What *is* new is the scale of the issues and the inexorable intertwining of the issues in the growth of business organisations. From about the middle of the 20th century when the opportunities offered during WWII provided the means for the first modern MNCs to emerge (see, for example, Korten, 1995), the relationship between societies (in the widest sense) and business organisation has become a major critical concern. A rapid tour through some of the milestones in each of the last few decades will provide a useful – if seriously basic – background to what follows.

Modern thinking about business and its (actual or potential) social responsibility is normally dated from Bowen (1953) who was one of the first business commentators[11] to identify a separate – and largely new and previously unrecognised – potential for responsibility for western business and its managers.[12] Post-WWII scarcity and then prosperity coupled with both greater internationalisation and significantly larger corporations provided a basis upon which the modern debates were built.

By the 1960s, concerns over social responsibility were very much live issues – especially in the USA – and were widely debated in business, government and business school circles. By the mid-1960s, Peter Drucker, a well-respected establishment figure in the field of management, was able to report that young, white, middle-America was increasingly disappointed with the level of social responsibility exercised by the American corporation (Drucker, 1965). Although these concerns were far from radical, they represented the beginnings of a major shift in thinking where it mattered – in the business and government.[13] No longer could western middle-class society and business assume that the corporations of

---

[10]*Life of Reason, Reason in Common Sense*, Santayana, G., Seribner's, 1905, p. 284. Critical Edition, MTT Press, 2011, p. 172. Reproduced with Permission.

[11]Radical critique of business has a much longer history – at least as far back as social reformers like the Webbs and the seminal work of Marx and Engels in the mid-19th century. However, the 'social responsibility of business' debate which Bowen is credited with originating was much less concerned with the structural inequities in society and more with a recognition of the new power and ubiquity of corporations and the implications that this had for the way in which managers conducted the affairs of the organisation. For more penetrating analyses of these issues see, for example, Dickson (1974); Kapp (1978); Tinker (1984a,b, 1985); Held (1987); Neimark (1992); Parkinson (1993); Kovel (2002); Bakan (2004); Buhr (2007).

[12]Do note that both paternalism and undertaking business with a strong moral/religious emphasis have a long history as illustrated by, for example, the Quakers and influential reformers such as Robert Owen, the Lever brothers and Titus Salt.

[13]It was some years before the middle classes of the 'developed' nations began to wake up to the structural inequities, environmental degradation, exploitation of labour and irresponsibility of dealings with 'developing nations' and repressive regimes arising from 'normal' business practice.

market capitalism were unequivocally benign institutions. Other key developments of the 1960s included a growing anxiety about the financial accountability of increasingly large businesses and, of crucial importance to what we now know as the environmental movement, the publication of Rachel Carson's seminal book, *Silent Spring,* in 1962.[14] *Silent Spring* was a shocking book which chronicled, probably for the first time, the systematic and systemic impact on natural ecosystems of the modern chemical industry.

The 1970s[15] was a particularly important decade for CSR – even if it ultimately represented a series of lost opportunities. Here were laid down the foundations of the modern environmental movement through, for example, the United Nations Stockholm Conference in 1972 (United Nations Environment Programme, 1972), the seminal writings of Schumacher (1973, 1980) and the publication of both *Blueprint for Survival* (Goldsmith *et al.,* 1972) and the first *Limits to Growth* report (Meadows *et al.,* 1972). General systems theory (GST) was developed during this decade (see, for example, von Bertalanffy, 1971) and directly influenced the environmental movement in turn (Boulding, 1966; Ward and Dubos, 1972). The energy crisis of the mid to late 1970s focused a lot of minds at the time. The contest over social responsibility really gained momentum through Friedman's seminal contributions (Friedman, 1970) and business-led explorations of the concept (Confederation of British Industry, 1971; Hargreaves and Dauman, 1975; Kempner *et al.,* 1976). The status of, employees and employment gained (what is probably) an all-time high in the West, and law governing employees, health and safety, unions and such matters advanced rapidly throughout Europe, North America and Australasia (see, for example, Foley and Maunders, 1977). Interest by the professional accounting bodies in the UK, USA and Europe reached an all-time high (Accounting Standards Steering Committee, 1975; American Institute of Certified Public Accountants, 1977) and what is probably the first textbook on social accounting was published (Estes, 1976).

The 1980s is a decade usually typified as one in which the western right-wing (liberal) values came to dominate the world and the backbones of globalisation, liberalisation and privatisation were put in place. This triumph of market mentality certainly devastated the social justice advances made during the 1970s, but to see the decade as so simply bleak would probably be too trite. The decade saw 'social responsibility' give way to 'business ethics', employee rights give way to consultation and environmental liabilities enter law. The United Nations continued to try to control multi-national corporations – with relatively little success it has to be said (see, for example, Rahman, 1998). But the United Nations had much more success with the utterly pivotal Brundtland Commission (United Nations World Commission on Environment and Development, 1987) which still stands as the cornerstone of most modern thinking on sustainability. Such changes were set against a backdrop of disasters which were both more closely associated with industry and more widely reported than ever before: the famine in Ethiopia, the devastating chemical explosion in the Union Carbide plant in Bhopal, India and the catastrophe of the oil spills from the ship *Exxon Valdez* in Prince William Sound in 1989 all provided a backdrop to this unqualified triumph of capitalism. A triumph (incorrectly) claimed as both the cause of and as a manifestation of the fall of the Berlin Wall in 1989.

---

[14]Although Carson (1962) remains the most frequently cited initiation of the environmental movement, there are other preferred candidates. At least as influential is Aldo Leopold's *A Sand County Almanac* first published in 1949 (Leopold, 1989).

[15]Selecting only a few events and writers from a whole decade is obviously a dangerous thing to do – especially a decade that produced writings from as diverse a range of authors as Bourdieu, Kuhn and Rawls – but we intend to give only a flavour (and a very specific and personal history) here. We hope nobody is too offended by the oversimplification.

By contrast, the 1990s, although still witnessing disaster, catastrophe and greater acts of systematic alienation in the so-called developed nations, was more difficult to give a simple character to. The 1990s saw the official end of the Cold War and the end of apartheid in South Africa. Significantly, it was the period when environmental issues in particular entered the mainstream. It was the period when business, at least superficially, noisily embraced both social responsibility and environmental management. The environmental movement advanced through the United Nations Summit at Johannesburg in 1992 whilst the Treaty of the European Union in 1992 created a 'single market' to rival that of North America. Under the surface of the apparently calm and increasingly successful global economy in which India and China were beginning to emerge as crucial elements, the battle between the environmental and social justice movements from the 1970s and the liberal free-market movements of the 1980s was becoming, despite superficial appearances to the contrary, much, much more serious.

In the early 21st century, the bleakness of the global picture is now in the starkest contrast to the triumphalism of much of big business. Never have social responsibility, social justice and sustainability been so widely debated and applauded; never have the conditions of the planet and the oppression of its peoples been so stark. It is in this context that social accounting becomes a development of such crucial potential.

4.4 A brief history of social accounting

There are many relatively easily accessible 'histories' of social accounting[16] and so we can be correspondingly brief here. Our aim here is simply to note that social accounting (and its analogues) is not an entirely recent phenomena and that there is a wealth of experience from which to draw – one does not need to re-invent wheels. Examples of social accounting and details on historic examples can be found elsewhere (see, for example, Estes, 1976; Johnson, 1979; Gray *et al.*, 1987, 1996).

Although Guthrie and Parker (1989), for example, track social disclosure back to 1885 in BHP (a major Australian company) and Adams and Harte (1998), for example, track social disclosure concerning women back to 1935, we can, again, trace the genesis of modern social accounting from the 1960s. The pioneering work of such organisations as the Council on Economic Priorities and Ralph Nader (in the USA) and, a little later, The Consumers Association, Social Audit Ltd and Counter Information Services (in the UK) laid a challenge to corporate behaviour that was to be instrumental in prompting the experimentation and self-disclosure that laid the foundations of social accounting.[17]

The range and diversity of experimentation in the 1970s was inspiring. Clark C. Abt and Associates (Abt and Associates, 1972 *et seq*) reported a series of attempts at a fully financially quantified statement of social and environmental impacts. (This experiment was, incidentally, replicated by the Cement Corporation of India in 1981 and also tried by a Dutch company BSO Origin in the 1990s. Sadly, the latter exhibited no apparent awareness of the earlier Abt experiments.) Eastern Gas and Fuel Associates (1972) and The Phillips Screw Company (1973) reported on their performance against standards. Atlantic Richfield (1974–1977) and 1st National Bank of Minneapolis (1974) provided embryonic attempts at what we would now call stakeholder dialogue. Reports to employees ('Employee Reports')

---

[16]Such histories can be found in Bloom and Heymann (1986); Lehman (1992); Neimark (1992); Gray *et al.* (1996, Chapter 4); Mathews (1977); Gray (2002); Parker (2005); Deegan (2007); Buhr (2007); Owen (2008).

[17]A much more intensive examination of a number of these initiatives is presented in Chapter 10.

were widely produced and many large companies produced reports on their employees ('Employment Reports'). All this was accompanied by the growth in the Value Added Statement (Burchell *et al.*, 1985) and experimentation in both 'energy accounting' (see, for example, Dick-Larkham and Stonestreet, 1977; Hewgill, 1977, 1979; CIMA, 1982; Odum, 1996) and 'Human Resource Accounting' (see, for example, Flamholtz, 1974; Preston, 1981; Roslender and Dyson, 1992).[18]

The decade that followed seemed much less innovative by comparison – not least because the global political climate also seemed less accommodating. The 1980s saw a considerable increase in the range and type of 'External Social Audits' (Gray *et al.*, 1996) whilst the extent of both regulated and voluntary corporate disclosure in the Annual Report rose steadily. In the USA, such disclosure was dominated by environmental liability data whilst, elsewhere, it was data about employees that tended to be the more widespread.

It was the arrival of the 'Stand Alone Report' in the 1990s that seemingly transformed both the debate and the practice of social disclosure. Led by the innovative steps of a selection of companies like Noranda (Canada), Norsk Hydro (Norway) and British Airways (UK), the prospect of substantial *environmental* reporting seemed again realistic. These pioneer reports, along with experiments from the Danish Steel Works (based on something called the eco-balance) and BSO-Origin (based on the Abt experiments from the 1970s), set a standard that has rarely been met since (see, for example, Gray and Bebbington 2001). Environmental Reporting became a major part of large corporate activity throughout the 1990s, but it was not until the mid 1990s that social issues were permitted to re-enter the disclosure debate.[19] This in turn led to the rise of the Triple Bottom Line (Elkington, 1997; Henriques and Richardson, 2004) and, eventually, the production of Responsibility Reports and Sustainability Reports on a global scale (see, for example, KPMG, 1999, 2002, 2005, 2008, 2010).

If you have supplemented this brief trip through recent history with visits to the web and the websites of both key reporters and those who are monitoring the disclosure as we recommended at the start of this chapter (see the Appendix to this chapter), you will be more or less up to date with reporting practice – at least at a very general level. So now, if that is broadly what social accounting practice has looked like over the last 50 years or so, we can turn to the altogether more difficult question of how we might explain, interpret and evaluate such practice. That is the job of theory.

4.5 Some theory for social accounting

Theory is a tricky thing. Theory is something we use all the time – but typically implicitly. Cooking a meal, going to the shops, commenting on the football or discussing the behaviour of friends, colleagues or family members all involve theory – although we mostly would not bother to explicitly identify and consider that theory. Theory is, at its simplest, *a conception of the relationship between things*. It refers to a mental state or framework and, as a

---

[18]Human resource accounting experienced something of a revival in the early 21st century in areas such as Intellectual Capital Accounting and related matters (see, for example, Mouritsen, 2006; Roslender *et al.*, 2006).

[19]It is probably apposite to recall that social issues were being deliberately excluded from the agenda of the early 1990s by corporations as well as by accountants. Despite this, innovations by companies such as Traidcraft (see, for example, Dey *et al.*, 1995; Gray *et al.*, 1997; Dey, 2007) and by SbN Bank had begun to develop novel and penetrating approaches to reporting on social issues. Much like the early pioneers in environmental reporting, these early examples of substantial social disclosure have rarely been surpassed.

result, determines, *inter alia*, how we look at things, how we perceive things,[20] what things we see as being joined to other things and what we see as 'good' and what we see as 'bad'. If we are going to try and explain social accounting practice, make sense of its potential and its impacts (interpret it) and evaluate its effectiveness, we are going to need some theory.[21]

You will, of course, already have met some theory. Within social accounting, we have used *systems theory* as a way of seeing the world, and we employ the *theory of accountability* as an articulation of why social accounting was potentially important and as an indication of what it actually could achieve (see **Chapter 3**). Equally, we introduced in Chapter 3 notions of *meta-theory* and *meso-theory* in order to try to understand why different views of organisation, economic activity and social accounting might arise from different (political, social, moral, cultural or religious) views of the world.

There is a whole body of work related to the philosophy of science on the formation of theory, what makes a good theory and so on (see, for example, Burrell and Morgan, 1979; Morgan and Smircich, 1980; Borg and Gall, 1989; Ghauri *et al.*, 1995; Laughlin, 1995),[22] but we can be a little more relaxed about such matters here. For social accounting we need theory to help us observe, organise and explain a range of things. These 'things' might include (but not be restricted to): What is (and what is not) social accounting? Why do organisations undertake (or not) social accounting? Why do we see the social accounting practice that we do? Why does that change over time? What effects does social accounting have? What effects could (and should) social accounting have? What makes a good or a bad social accounting practice?

The preferred theoretical framework we have adopted in this book has elements which are each *positive, normative* and *pragmatic*. The **positive** (i.e. descriptive)[23] elements are two-fold. First, we take as given that the current ways of human organisation are not sustainable and that planetary balance, social justice (however defined) and eco-system stability are under serious threat. This statement arises from a reading of the data and is a *positive* (i.e. descriptive) conception. Second, we observe from the research that organisations in general – and large corporations in particular – are not accountable for most of their activity. The **normative** (i.e. explicitly value-laden)[24] elements are that: first, we believe that organisations *should* be accountable and that accountability is a *good thing*; secondly, we believe that democracy is a *good thing* (and, positively, that accountability is an essential

---

[20]There is a widely held view that all perception is theory laden. This means that what we see, recognise and/or respond to is a function of our predispositions, beliefs and understandings – our theories in fact. See, for example, Wilber (2000) for a stimulating if eccentric approach to these issues.

[21]Theory is the term we will most generally hear, but 'framework', 'hypothesis' and 'model' are also terms which *can* mean similar things. We will not be worrying much about the difference in these terms here.

[22]In fact, almost any good textbook on research methods will help in understanding theory and the role that it plays. However, the range of references given here will help in beginning to understand the range of conflict there is over theory. For illustration, there are a range of social science researchers (usually called positivists) to whom theory is only valuable if it is 'scientific' (whatever that means) and permits prediction. If you were not a 'positivist', you would not necessarily believe this to be either true or necessary.

[23]'Positive' in this context means descriptive: a statement of 'what is' which is not sullied by concerns of 'what should be' (which is 'normative') and, for many scholars, is a matter of observation and facts. It does *not* mean 'good' in the sense of 'feeling positive'. The word then lends itself to 'positivism' which is a belief set that the only good research is that which relies entirely upon observation free from values – at its simplest, this is something we know better as 'the scientific method'. You can (and indeed do) use positive statement and method without being a positivist. (It all gets quite confusing really.)

[24]'Normative' means a statement of values: a statement of 'what should be' not 'what is'. Any normative statement derives from your values, ethics, morals, principles. There is a widely held view that, although it is essential to distinguish the normative from the positive in argument, no positive statement can ever be entirely free of the normative. That is, our ways of seeing the world reflect our theories and these reflect out values. (We said it got quite confusing.)

component of democracy); and, finally, we believe strongly that sustainability in the Brundtland sense (United Nations World Commission on Environment and Development, 1987) is also a *good thing*. These are contestable statements of values and, therefore, normative. Finally there are **pragmatic** elements to the framework of this book. The adoption of a *neo-pluralistic* framework is entirely strategic – it is a place where most political theories can meet and dispute; it is not *totalising* in that it does not close down any voices as far as we can see. Equally, there are pragmatic choices in that (for example) the notion that an accountable democracy is capable of delivering sustainability is entirely un-examined (and may be actually quite wrong). This is pragmatic in that one value (that of the self-evident desirability of accountability) is assumed to be complementary with the self-evident desirability of sustainability. It may well not be.[25]

These elements combine to provide the framework within which this book sits. The 'theory' such as it is though, is both *underspecified* (it does not deal with all and every eventuality nor does it deal with each and every element in the human experience of social accounting and the planet) and *loosely coupled* (in that the linkages between the elements are not all explicated fully). This is a fairly common occurrence with theoretical frames, in that few people (if any) fully understand everything. Theories, typically, as a result are either (i) wide-ranging and consequently underspecified and loosely coupled (like ours); or (ii) are very narrowly focused and, in overcoming our limitations, exclude a considerable range of potentially important elements from the theoretical framework or model. (As we shall see shortly, these latter qualities are both the strength and the weakness of traditional economic models.) In essence, we suggest that you assume that theory is always incomplete in the social sciences.[26] More especially, each of the theoretical lenses we are going to now introduce will have insights and understandings to offer, but they each and every one of them will fail to fully explain the phenomena of social accounting that interests us.[27]

With this *caveat* in mind, we will now review some of the theoretical frames that have been used in and around social accounting. Our review is inevitably selective and partial. Its purpose is *not* to be comprehensive but to be illustrative of the range of approaches, perspectives and views we might bring to bear upon the matters that concern social and environmental accounting.

One of the major things we learn from **general systems theory (GST)** is the importance of *level of resolution*. That is, problems, issues and solutions vary depending upon how widely or narrowly we spread our perception and the range and level of 'things' we allow (or invite) into our perception and, therefore, into our theoretical framing. We will therefore attempt to offer an (albeit crude) organisation of theories by their level of resolution. We do this, not to close down or simplify debate – quite the opposite – but rather to provide an instrumental pedagogic framework with which to begin this discussion without getting entirely bogged down in theories and theories of theories, etc. Equally, we are going to

---

[25]Indeed, one would need to be very careful indeed if you wanted to suggest that a full and free-flowing democracy would deliver sustainability as there is not (and probably cannot be) any direct evidence on this – not least because current democracies are mostly influenced by a media which itself is supportive of – and a creation of – the international financial capitalism that may well be the source of much of the problem.

[26]Indeed, it may actually be that the essential nature of social accounting – given its emergent and political nature – must always remain a problem which is not amenable to simple framing and which must remain unstructured in order to retain its radical edge. Furthermore, if this is the case (and it is certainly true for some methodological positions) then it may be that we should stop trying to test theories (other than in a positivistic frame) and get used to the idea of employing a range of underspecified theory to help as guides, lenses and aids to explanation.

[27]For more detail on the role of theory, see Laughlin (1995, 2004) and for a discussion of the failings of theory in social accounting see Adams (2002).

crudely allocate theories to three *dominant metaphors*. We again do this entirely for convenience – if it works for you that is really nice and if it doesn't please ignore it. These three metaphors are: the *biological*; the *political/sociological*; and the *economic/rationalist*.[28] (There is a fourth metaphor – that of the other or 'the *Other*' – which we also find useful and which is developed in Gray *et al.* (2010). This metaphor captures those theories which our categories miss and includes those theories more obviously associated with postmodernity, radical feminism and the recognition of difference. This metaphor is touched upon in what follows but it is not developed here.)

Our basic mental framework – *with considerable simplification* – therefore looks like Figure 4.2. In its avoidance of many of the principal theories in social science for the last century or so, the figure might even be thought trivial. However, it allows us to place some of the theoretical lenses more commonly employed in social accounting into a context; to illustrate where other theoretical insights may be sought; and permits our conversations and thoughts about theory to be explicitly conscious about some of their limitations. Using Figure 4.2, we turn to look at meta-level theories.

4.6 Social accounting and system-level/meta-theories

On the (albeit contestable) basis that we maintain/seek a coherence in our theorisation, our approach to selecting theory will be pre-determined (to some degree at least) by our views about how we believe the world to be and how we would *like* the world to be (see also, O'Dwyer, 2003). There are widely different views on this and deeply held, pre-empirical, notions like religious and spiritual values; views on the nature of mankind and ecology; and beliefs about the purpose of existence; which will all directly and indirectly impinge upon our worldviews.[29] These worldviews will, consequently, have profound influence on how we view, for example, organisation-society relationships and, consequently, the functions (potential and actual) of social accounting.[30] At a minimum, such views are likely to determine with which theorisations we feel intellectually, spiritually and emotionally comfortable. In this section, we begin by briefly reviewing some of the key systems-level/meta-theories, categorised around our metaphors.[31]

Biological

The dominant biological metaphor we have employed so far is that of systems theory. This meta-theory has little in the way of the political or the economic automatically embedded in it. It does, however, allow us to bring into our conception such systems as we decide are important – hence it provides part of our pragmatic frame for the book. So, systems theory directs our attention to, for example, the notion that societies, organisations, economics, accounting, ecology are all systems and they interact and affect, and are affected by, each other. Simply assuming that an activity is unrelated to societal or environmental desecration does not make it so.[32]

---

[28]These three metaphors are derived principally from casual empiricism but are a categorisation that we have found notably persuasive. Much more detail is give in the article based on this section – Gray *et al.* (2010).

[29]A religious pacifist is unlikely to see the world in the same way as an aggressive solipsist.

[30]For more detail on these issues see, for example, Kempner *et al.* (1976); Kovel (2002); Tinker (1985); Lehman (1992); Chryssides and Kaler (1993); Mathews (1993); Perks (1993); Bailey *et al.* (1994a,b); Bakan (2004).

[31]We reiterate that any adoption of any metaphor should invoke our caution at the meta-theoretical level.

[32]See, for example, Thielemann (2000), for an interesting development of this approach.

Figure 4.2 A tentative and highly speculative, *non-discrete*, categorisation of a selection of theorisations around social accounting

THEORY LEVEL (of resolution)	METAPHOR			
	Biological	Political/ Sociological	Economic/ Rationalist	Other
Meta-Theory (System Level)	• General Systems Theory • Deep Ecology	• Marxian Political Economy • Communitarianism • Discourse • Habermas	• Friedman's Liberal Economics	• Post-Modernity
Meso/ Sub-Systems Level	• Autopoiesis • (Neo) Institutional	• Bourgeois Political Economy • Social Contract • Accountability • Media Agenda Setting • Cultural Conceptions	• Efficient Capital Markets Hypothesis	• Foucault • Radical Feminism • Actor-Network
Micro I/ Organisational	• Stakeholder • (Neo) Institutional • Resource Dependence • Contingency	• Legitimacy • Stakeholder	• Decision Usefulness • Signalling • Principal-Agent • Transaction Costs	• Emerging New Conceptions of Enterprise
Micro II/Internal to Organisation	• Autopoiesis • Organisational Change (multiple) • Boundary Management	• Structuration • Discourse • Group/Identity	• Positive Accounting • The Business Case	
Micro III/ Individual		• Values driven • Motivation	• Principal-Agent	• Feminism • Identity

Source: Adapted from Gray *et al.,* 2010: 12.

One of the principal contributions of *general systems theory* (GST) is a physical conception of planetary and ecological systems (for much more detail, see Meadows, 2009). It sees integration and self-regulating systems, myriad species and eco-systems interacting with each other. It is the basis of virtually all deep green and radical ecological perceptions. As such, we might view current human interactions as malignant and view modernity itself as a profound failure to live within the principles of ecology and nature. The problems of conflict between ecology and humanity are deep and are almost certainly not solved by the application of more pseudo-modern curatives like social accounting (see, for example, Lamberton, 1998; Andrew, 2000). It is this set of concerns which lead us into **deep ecology** and the profound political and sociological implications that this holds for mankind (see, for example, Goldblatt, 1996; for an introduction).

Political/sociological

By far the most influential of the political/sociological theorisations is the **classical political** economy of Marx. *Political economy* is a useful phrase that considers the way in which power and economic organisation work in a society and the influences that they have.[33] Marx directed our attention to the big picture (the lower level of resolution) to examine the role of the state, the role of capital (investors, management, companies and their supporting structures and institutions) and the role of labour and the bourgeoisie (what we normally think of as the middle classes).[34] In essence, capital held the power, the state was 'captured' by capital and could be expected to do its bidding – aided and abetted by the bourgeoisie.

Labour (pretty much the rest of society) was conceptualised as oppressed and its wealth (the value that labour created through its efforts) was appropriated (stolen, really) by capital. From this perception, injustice, structural conflict and power are essential to any understanding of how society works. Injustice can only be remedied by the removal of power from capital – something which cannot be expected of the state (as it is controlled by capital). It is therefore assumed that such structural change must come, if at all, from labour movements.[35] The ethical foundation of this position – socialism in essence – is that justice is more important than freedom (the liberal economic perception is the opposite of this) and, consequently, nobody in a civilised society should have the 'freedom' to be without basic amenities. The corollary of this is that nobody should have the freedom to be ridiculously wealthy and/or to control the basic elements through which societies provide for themselves (the 'means of production' in Marx).

It is crucial to note, as a consequence, that one does not need to be a 'Marxist' to be stimulated by the insights offered by Marx. Equally importantly, it is essential to note that the concerns that exercised Marx are not ones which can be solved through marginal adjustment of our present world order – nothing less than complete structural change (of capitalism in his case) can possibly begin to address these issues.

Whilst classical political economy might be the most substantive meta-theory, other (possible) meta-theory level conceptions such as **communitarianism** and **discourse theory** are increasingly influential. Communitarianism is an explicitly normative conception of political organisation with an explicit preference for fairness and locally determined need rather than the more typical socialist preference for equality and egalitarianism (see, for example, Gray 1996). This has been taken up by Lehman (1999, 2001, 2007) who has articulated a social accounting which might emerge with an emphasis on dialogue and local democracy. Here, we have the beginnings of something like a meta-theory in which social accounting has an explicit place.[36] Discourse theory assumes that '*it is language, signs, images, codes and signifying systems that organise the psyche, society and everyday life. Meaning is socially constructed. . . .*' (Friedman and Miles, 2006: 69). If, then, language is all, and pre-existing

---

[33]Political economy was the term which would previously have referred to what we now know as economics – although modern economics has largely abandoned the political and sociological in its analysis.

[34]The most obvious source of further reading is Marx's work itself, but more accessible help can be gained from, for example, Held (1980); Tinker (1984); Kovel (2002).

[35]Post Marxist writing widened the sources of structural change (Marcuse, 1964), and other theorists, such as Gramsci, led to the formation of the Frankfurt School from which Critical Theory emerged. Theorists as diverse as Foucault and Habermas are direct descendants of Marxian thinking, and their influence in accounting, business and social accounting research is considerable (see, for example, Held, 1980).

[36]In this, the communitarianism vision has a strong affinity with much work in the 'third sector' and social enterprise movements where social accounting is developing with relatively little direct engagement with – or by – the dominant academic literature. See, for example, Pearce (1996); Doane (2000); Ball and Seal (2005); Gray *et al.*, (2011).

structure elusive or even illusory, then our view of what a society is, what is desirable and what is not, is both constructed by language and entirely understood through it. This can make for an interesting range of questions about whether (for example) our earlier claims of planetary crisis or injustice have any content (see, for example, Zimmerman, 1994; Gray, 2010). Social accounting is itself a manifestation of language (in that accounts are language) and might then be seen as both manifestation of *and* a construction of the society itself.[37]

This only scratches the surface of course, and there is an array of important theorists who have yet to be fully integrated into social accounting (although see, for example, Shenkin and Coulson, 2007; Spence, 2009). But, broadly, areas such as critical theory and developments like critical discourse analysis offer considerable potential within social accounting.[38]

Economic/rationalist

The economic/rationalist view might be best typified by **neo-liberal economics** and its most feted exponent, Milton Friedman (see, for example, Jacobsen, 1991). It is worth emphasising that the liberal economic position, much like the Marxist position, has little to say about social accounting except that it is either an impediment to profit making or is likely to be manipulated by corporations and capital to foster a climate in which liberalism can flourish. Such observations are worth bearing in mind as they resonate strongly with a lot of the empirical work we will meet.

Other

Finally, we stressed that we would not be dwelling here on 'the Other', but if one is to have any kind or wide understanding of theorising then one must be aware that **postmodernity** is a powerful lens which would typically reject both the 'grand narrative' offered by (especially) Marxian views and, rather, would express itself in terms of the failures of modernity. A postmodern view would typically reject much of the structure and paraphernalia of modern (mostly) western life. Postmodernism (drawing from scholars such as Derrida, Lyotard, Baudrillard, Rorty and, especially, Foucault) offers a fundamental critique of modernity and, in doing so, it significantly adds to the challenges that social accounting must address (see, for example, Everett, 2004; Gray, 2010). The **radical feminist** world view suggests that our economic, social, political and business systems – and thus the language of business and accounting – are essentially 'masculine' constructs which emphasise, for example, aggression, traditional success, achievement, conflict, competition and so on. Our world thus denies a proper voice to, for example, compassion, love, reflection, cooperation and other 'feminine' values. A radical feminist would almost certainly challenge this 'masculine' attempt to organise and categorise (see Hines, 1992; Shearer, 2002).

In brief, then, our choice of meta-theory can be assumed to reflect our worldview. (Whether or not the worldview can be assumed to be chosen in a disinterested and informed way is quite another matter.) Whilst that worldview is unlikely to have anything directly to say about social accounting, it is highly likely to have implications for some or all of: responsibility, information, communication, justice, organisations, power, systems, accountability and so on. In so doing, the worldview provides a frame within which our *meso-theories* gain credence

---

[37]For a critique see Everett (2004); Spence (2009). For an application of discourse theory in social accounting, see Milne *et al.* (2006); Tregidga and Milne (2006).

[38]There is a myriad of omissions from the chapter; from the detail of critical theory via the work of Habermas (for example, Thielemann, 2000; Unerman and Bennet, 2004) to theorising more normally associated with postmodernism (such as in the writings of Foucauld, Baudrillard, Bourdieu and Laclau) or new emerging themes (such as the use of Dean's notions of 'governmentality' to consider information, disclosure and reporting as mechanisms within a 'mentality' of governance (Dean, 1999; Gouldson and Bebbington, 2007). That said, we have not reviewed theories of theology, psychology or anthropology either.

and coherence. Each of our foci in research thinking and policy-making might be thought to have a reflexive relationship with the theories 'above it' (the meta-theories) and those 'below it' (the micro-theories). Any choice of theory about aspects of social accounting might, therefore, be assumed to ultimately reflect the world view of the person concerned.

4.7 Increasing resolution – sub-system level/meso-theories

A major problem that besets much discussion in management, business and accounting is that the meta-theory level is typically excluded from the discussion (see, especially Chwastiak, 1996, for a stimulating example of this phenomenon). We find time and time again that, for example, the sustainability of corporations is debated in the absence of any discussion of the sustainability of the planet; the responsibility of organisations is debated in the absence of any discussion of the responsibility of capitalism; claims to serve the public interest do not ground their claims in any notion of society or justice . . . and so on. As a consequence, most theorising about corporations – and, it then follows, about social accounting – too often only starts at (what we have called in Figure 4.2) the 'sub-systems level'.[39]

Now the most interesting thing about 'sub-system level' theories is whether or not their proponents think of them as embracing the system level or whether they recognise that they are considering only a subset of the system. This is classically illustrated in the central and ancient political notion of the social contract and bourgeois political economy (which we consider below) where, for example, the distribution of power is examined without considering how that distribution came about in the first place and is now maintained – i.e. without considering the meta-theoretical level.

Biological

As many theories can be employed at different levels of resolution, our locating them at any point in Figure 4.2 may well be a little arbitrary. Arbitrary categorisation is illustrated well by the theory of **autopoiesis** which is another biological metaphor that can be used at individual, organisational or systems levels. Autopoiesis as it is applied to social science is most usually associated with Luhmann (1989) and was introduced to the accounting literature by Power (1994). At the risk of over-simplification, autopoiesis can be thought of as a property of systems whereby they only permit into their architecture those elements that 'code' with the system itself: intrusions (or threats) which are not recognised by the system – which do not 'code' to the system – will be rejected. (The parallel with cell biology is fairly obvious as a cell's immune system learns to reject alien bodies but accepts what it recognises as benign elements.) Autopoiesis is an elegant metaphor for the way in which systems (of, for example, information flow, disclosure, corporate behaviour, financial markets or whatever) will 'reject' any invasion or other development which does not accord with the design archetype of the system itself. Consequently, we can hypothesise that any social accounting which might be seen as a threat to (say) capitalism will be rejected by it and only social accounting which 'codes' to the system will be accepted by the system.[40] Such a conception offers us a useful

---

[39]It is worth recording again that the more focused the theory (the higher the level of resolution), the more likely it is to be able to address and, indeed, say something specific about social accounting and social accounting practice. This trade-off between scope and specificity seems to inevitably bedevil all theoretical speculation.

[40]This is what Kirman (1999) found when seeking explanations for why substantive social accounting regulation which had been effectively promised in New Zealand in the 1990s was eventually rejected. Laughlin's (1991) model of change (see below) also bears a notable resemblance to this way of thinking.

explanation of the way in which (for example) serious accountability and/or sustainability reporting is rejected by western economies.

Political/sociological

The most common theorisations about corporations in general and social accounting in particular adopt (often unwittingly) a 'bourgeois' political economy.[41] Whilst, as we saw, classical political economy places structural conflict, inequality and the role of the state at the heart of the analysis, **bourgeois political economy** tends to take the '*status quo*' as given and thus excludes them from the analysis. As a result, the bourgeois political economists tend to be concerned with interactions between groups in an essentially pluralistic world (for example, the negotiation between a company and an environmental pressure group, or between a local authority and the state). Whilst this produces useful analysis, it does, according to the classical political economists, entirely miss the more important point of how those relative differences in power, wealth, etc. were generated and maintained by the system in the first place. In essence, what happens with a bourgeois political economic viewpoint is that we examine social accounting when it is (say) legitimating specific elements of the system, of a company, of an industry or of a practice (say), and thereby fail to see that the issue being legitimated is actually systemic: the issue under consideration is a direct consequence of the system within which it arises. This would mean that when studies throw up 'irresponsible' behaviour by a corporation such as (for example) Union Carbide, Exxon, BP, Premier Oil, Nike, Nestle or whoever, the only thing of real interest to a classical political economist is not that such behaviour took place but simply that they got caught. The behaviour itself is expected from a system (like capitalism) under which irresponsibility is, it can be argued, encouraged.

Bourgeois political economy provides us with a subsystem-level context within which most of the theories we are going to briefly review are typically located. Its relatively restricted perspective allow us to focus on theories which tell us more about – or at least give us more direct insights into – social accounting.

The **social contract** is most usually associated with 17th and 18th century writers such as Hobbes and Rousseau. It considers that, in essence, each individual undertakes to contract with society for the benefit they derive from being part of that society – defence, laws, mutual support, etc. More formally (as Tozer and Hamilton, 2007: 108, put it) the contract is derived between those who are empowered (typically the government) and those who grant that power (by election, abstention or submission). From this then arises the frequently stated position that an organisation exists at the will of a society to the extent that it continues to provide society with benefits. This in turn brings us to an analysis of rights and responsibilities (see, for example, Donaldson, 1988) which, in turn, leads into the conception of **accountability** which, as we have already seen, is widely employed in social accounting (see, for example, Gray *et al.*, 2006). Much of the use of accountability in social accounting echoes the social contract although accountability allows us to ask how the rights, responsibilities and accountabilities are established and maintained (see, for example, Mathews, 1993; Dillard *et al.*, 2005; Tozer and Hamilton, 2007).

---

[41]Bourgeois political economy is most usually associated with Adam Smith and John Stuart Mill and subsequent economists. It is explained in a little more detail in Gray *et al.* (1996: 48 *et seq.*). In fact, it predates the separation of 'politics' and 'economics' that we currently take for granted in our schools and universities. Until relatively modern times – the late 19th and increasingly through the 20th century – one would have studied political economy on the understanding that society, politics and economics were inseparable. This makes our classification of metaphors the more obviously artificial.

Media agenda setting theory was introduced to the literature by Craig Deegan (see, for example, Brown and Deegan, 1998) and it focuses our attention on the way in which issues are constructed through popular culture in general and the media in particular. Put simply, there is always pollution, but pollution is only recognised as an issue worthy of attention (and corporations only then respond) when the (largely corporate-owned) media finds out and is able and willing to make a fuss about it. (In a sense that echoes discourse theory, 'pollution' doesn't socially exist until it is defined, managed and communicated by the media.) One of the ways in which a company, a series of companies and/or an industry might respond (and by which we might, therefore, learn about what affects organisations) is through their voluntary disclosure. The systemic production of issues (such as pollution, injustice and so on) plus the way in which a society comes to rely upon a media to inform it about substantive matters is something on which the meta-theory would have something to say. A classical political economy would direct us to how those issues are created and manipulated whilst media agenda setting theory would look at how the issue was then re-constructed, managed and manipulated through media and organisational interaction.

A related concern for context is offered by, for example, Mathews and Perera (1991), Adams (2002), Haniffa and Cooke (2005), and Perera (2007) who direct our attention to culture as a key systemic variable that will influence social accounting practice. We already know that things like size of the organisation and the profile of the industry in which it operates have a major impact on an organisation's predisposition to formally disclose (see, for example, Murray et al., 2006). What culture captures is a range of important aspects such as: attitudes to disclosure; attitudes to the accountability of organisations; the expectations and reactions of civil society; and the likely response by the organs of the state. Our theories of why social accounting does (or does not) take place, the form it takes and its regulation are, therefore, going to be clearly culturally dependent and, in the same way that an understanding of culture has helped understand organisations and indeed accounting practice (see, for example, Hofstede, 1984; McSweeney, 2002), it will also add to our understanding of social accounting (see also Adams et al., 1995). Indeed, Islam, for example, has frequently been cited as a major influence on disclosure regimes in a number of countries[42] (see, for example, Hanafi and Gray, 2005; Kamla et al., 2006; Belal and Owen, 2007).

Economic/rationalist

There are many subsystems level theories within neo-classical economics. Few of these are *explicitly* employed in social accounting but one exception is the **efficient capital market hypothesis** (ECMH). The ECMH (and the variants that surround it, see, for example, Hines, 1984; Belkaoui, 1997) operates at the level of financial markets[43] and suggests that (typically) the prices of shares in stock markets respond rapidly and unbiasedly to new information. The constraints of the theory are clear: it looks at economic actors in a specific (but exceptionally) powerful subsystem of the society. The theory operates around the fascinating tautology that information is that which affects share prices; that which does not affect share prices is not 'information' and any reaction to 'non-information' is itself irrational (see, for example, Hines, 1984). Therefore we can study social accounting and discover whether or not it has 'information content' to actors in the stock market and, from

---

[42]Reasons as to why this might be vary but amongst the suggestions are that Islam encourages a modesty and a disinclination to speak of one's virtues and successes.

[43]The place (sometimes a physical space, more usually an electronic place) where financial 'products' and, most especially, the shares of companies are traded.

there, infer whether investors do or do not interest themselves in social accounting data.[44] Generally speaking, we find that investors who are pursuing their own wealth and self-interest are generally uninterested in social accounting unless it relates to risk and/or future earnings.[45]

Other

To round off this section of the chapter we should briefly recognise some of the notable omissions – a number of which might fall loosely under our heading of 'the *Other*'. We do not revisit radical feminism here except to note that Shearer (2002) neatly articulates an accountability deeply embedded in feminist perspectives of intersubjective relationships and, in so doing, not only warns us of the dangers and limitations of economic theory but offers us a more context-sensitive and ethically explicit approach to the interpretation of giving and receiving accounts. (See also Dillard, 2007, for an important development of this idea.) No review could be complete without an acknowledgement – however brief – of Foucault. He is one of the 20th century's most influential thinkers and his work is foundational throughout management academe but, as yet, remains relatively under-used in social accounting (although see Everett and Neu, 2000; Lehman 2006). The essence of Foucault's work relates to forms of knowledge, discipline and power, and (what he calls) the practices of the self. Broadly speaking, whether we are concerned to understand the resistance that social accounting can offer, the difficulties we face engaging with modernity and/or the postmodern critique of 'conventional' social accounting approaches, there is much yet to be drawn from Foucault's work.

And finally we should briefly mention **actor network theory** (ANT) which has at its heart a simple idea: that our conception of issues, problems, sites of research enquiry should be based around the notion of dynamic and interacting networks. At this level, it looks a lot like a child of GST, but the key to ANT is its claim for the heterogeneous nature of networks which contain many dissimilar elements. With its attachment to ethnomethodology, ANT then distinguishes itself from other theories employing networks in that an actor-network contains people, objects, ideas and organisations: collectively known as actors or actants (Knights and Willmott, 2007: 428). Networks are transient and maintained through constant performance of the relationship between the actants and hold out the potential to make visible the infrastructure within which events and actions take place (Callon and Law, 1997; Lukka, 2004; Callon, 2009). Placing social accounting – whether as a technology or a semiotic category – within ANT could suggest ways of seeing social accounting and disclosure in novel and more dynamic guises.

At a very general level of analysis, we can perhaps see that these subsystems theories relate to social accounting in terms of the emergent properties of social accounting as well as the functions that social accounting might be expected to serve and (to the extent that the system is purposive) how the system might use social accounting for its own ends. However, by far the greatest volume of research and theorising around social accounting occurs not at the level of the system or even at the subsystem but at the level of the organisation itself and, generally, seeks to answer questions about why organisations do (or do not) produce social, environmental and sustainability disclosures.

---

[44]There is considerable work done in this field and the implications of these issues roll over into 'ethical investment' and socially responsible investment (SRI) (see, for example, Kreander, 2001).

[45]The situation is, of course, not quite this simple. Although we are taught that all investors are selfish and greedy, for many institutional and 'ethical' investors this is simply too trite a view. For a broad introduction to the issues see, for example, Owen (1990); Margolis and Walsh (2003); Murray *et al.* (2006).

4.8 Micro-level/theories of social accounting and organisations

For some time now, the theories most widely employed in the social accounting literature have involved constrained conceptions of the 'organisation' and its interactions with a (usually partially defined) substantive environment. It will come as no surprise to find that the theories at this level also are not neatly accommodated by our four metaphors (as might be suggested by Figure 4.2) and that they, in particular, will often combine elements of the biological, the social/political and the economic.

Biological

Stakeholder theory[46] (along with legitimacy theory – see the next section) has been one of the most widely-employed theories in the social accounting literature at this level of resolution. Stakeholder theory could be located quite easily under either the political metaphor (mainly as a result of its link with the social contract) or the rationalist metaphor (as a result of its rational management link – see below), but we find it most valuable as a more organic – and hence biological – metaphor. A 'stakeholder' of an organisation is any (human) agency that can be influenced by, or can itself influence, the activities of the organisation in question (see, in particular, Freeman, 1984, 1994). An organisation therefore has very many stakeholders including as diverse a range as employees, management, communities, society, the state, future generations and non-human life.[47] It is, thus, an explicitly systems-based view of the organisation and its environment. And it is a view which recognises the dynamic and complex nature of the interplay between the organisation and its environment. There are two major variants of stakeholder theory and this general perception applies to both.

The first variant of stakeholder theory relates directly to the accountability model that we have used elsewhere (Gray *et al.*, 1996; **Chapter 3**) and perceives the organisation-stakeholder interplay as a series of socially-grounded relationships which involve responsibility and accountability. Thus, the organisation owes an accountability to all its stakeholders. The nature of that accountability is determined by the relationship(s) of that stakeholder with the organisation. Thus, to all intents and purposes, this is the *normative* approach to accountability. It has little descriptive or explanatory power in a social accounting context (Gray *et al.*, 1997).

The second variant of stakeholder theory relates more closely to Tricker's (1983) concern over *empirical accountability*. That is, stakeholder theory may be employed in a strictly organisation-centred way. Here, the stakeholders are identified *by the organisation of concern* (and not by society as they would be in the accountability framework), by reference to the extent to which the organisation believes the group needs to be managed in order to further the interests of the organisation (what Mitchell *et al.*, 1997, call 'salience'). The more important (salient) the stakeholder to the organisation, the more effort that will be exerted in managing that relationship. Information – including financial accounting and social accounting – is a major element that can be deployed by the organisation to manage (or manipulate) the stakeholder in order to gain their support and approval (or to distract their opposition and disapproval). It is quite possible to interpret a proportion of social accounting and disclosure as commensurate with an organisation operating in accordance with stakeholder theory. Furthermore, stakeholder theory encourages us to interpret examples of

---

[46]For an introduction to stakeholder theory see, for example, Ullmann (1985) and Roberts (1992) and, for more detail, Friedman and Miles (2002, 2006) and Donaldson and Preston (1995).

[47]Although this would raise some issues about the 'human agency' requirement – an important and difficult debate in its own right that (for example) the deep ecologists have major issues over.

voluntarily disclosed social accounting as indicative of which stakeholders matter most to an organisation and, thus, those which the organisation may be seeking to influence (Roberts, 1992; Mitchell *et al.*, 1997).

Social accounting in the 21st century has seen a growing interest in institutional (or more properly **neo-** or **new institutional**) **theory** as a promising alternative theoretical frame.[48] It concerns itself with organisations and *organisational fields*. Organisational fields comprise '*both cultural and network systems* [which give] *rise to a socially constructed arena within which diverse, interdependent organizations carry out specialized functions. It is within such fields that institutional forces have their strongest effects and, hence, are most readily examined*' (Scott, 2004: 7). Fields are thus socially constructed space arising from interactions, shared interests, common concerns, joint activities and so on. Larrinaga-González (2007) identifies a number of such spaces in the area of social accounting including the Global Reporting Initiative and the Environmental Audit and Management Scheme. The process of *institutionalisation* is primarily a process of homogenisation – or *isomorphism* – in which organisations converge in their behaviours to give a field stability and (eventually) inertia. Broadly speaking, this process of institutionalisation is presumed to occur through a combination of *coercion* (e.g. regulations, laws or major market changes), *normative* mechanisms (shared and converging values through, for example, education or professionalisation) and *mimetic* mechanisms (typically imitation of behaviours that appear to be successful). From such perspective, one will be able to explain part of social accounting behaviour through a combination of (say) increasingly shared values (about, for example, the capacity and responsibility of organisations) and a mimetic tendency to imitate others in the field (Bansal and Roth, 2000; Larrinaga-González, 2007).

Institutional theory has a close relationship with both stakeholder theory (where the web of stakeholders and their interactions and relative strengths might be thought of as fields) and legitimacy theory – which we consider below. (Larrinaga-González, 2007, argues that legitimacy theory is a special case of institutional theory.) Institutional theory also has direct relationships with theories such as resource dependency theory (RDT).[49] RDT is a derivation of systems theory and a close relation of contingency theory (see below) and maintains a dynamic relationship between an organisation and its dependency on (and hence vulnerability to) unpredictable resource supplies. Uncertainty and hostility are key components of the organisation's environment. Consequently, the demands placed upon it by agents who control the key supplies, are a major explanation of organisational choice and action. The resources upon which an organisation is dependent need not be only finance, labour, supplies, markets, etc., but may well also include legitimacy, reputation and so on (Deegan, 2002). The potential for a disclosure regime (and thus the use of social accounting) to operate in such a climate is obvious.

Contingency theory posits that any organisation, to function well, will adopt the structures, postures, missions, activities and suchlike that best fit its environment and circumstances.[50] Thus there is no single, ideal type of organisation and organisation structure. Neither will there be (say) any single ideal position on social responsibility or any single ideal system

---

[48]For an introduction see Larrinaga-González (2007). To offer institutional theory as an organisational-level theory and as a biological metaphor may be thought misleading by some. Institutional theory is most typically associated with DiMaggio and Powell (see, for example, 1983) and Scott (see, for example, 2004).

[49]See Pfeffer and Salancik (1978) and, for a brief introduction, see Knights and Willmott (2007: 215).

[50]The biological connection lies in the way in which there is assumed to be some ideal form(s) of an entity given the environment in which it operates and so, to the extent that environmental conditions can be generalised, it can be assumed that the more successful the organisation (organism) the more closely it will approximate this ideal. For more detail see, for example, Knights and Willmott (2007: 208–9).

of information flows and disclosure regimes. The best for the organisation will depend upon its circumstances (Otley, 1980; Thomas, 1986). It would be possible to suggest that social responsibility and social accounting may be *contingent variables* – i.e. variables which are dependent upon key environmental and organisational factors (see, for example, Adams, 2002). Indeed, there is a whole body of literature which explores the association between organisational factors and such matters as political exposure, industry affiliation, company size and so on, and this literature can be thought of as having a link to contingency theory (see, for example, Husted, 2000; and see also Gray *et al.*, 2001b).

Political/sociological

Legitimacy theory basically takes the second variant of stakeholder theory above and adds conflict and dissension to the picture (Guthrie and Parker, 1989; Patten, 1992; Lindblom, 1993; Deegan, 2002). At its simplest, the theory argues that organisations can only continue to exist if the society in which they are based perceives the organisation to be operating to a value system which is commensurate with the society's own value system (i.e. if they are perceived as legitimate by the 'relevant publics'). Organisations can face many threats to their legitimacy (e.g. a serious accident, a major pollution leak or a financial scandal) and, in consequence, may employ broad *legitimation strategies* to counter that threat. Lindblom (1993) identifies four such strategies: 'educate' its stakeholders; change the stakeholders' perceptions of the issue; distract (i.e. manipulate) attention away from the issue of concern; or seek to change external expectations about its performance.

Legitimacy theory, in this general form, offers important insights into social accounting practice. Many major social accounting initiatives can be traced back to one or more of Lindblom's suggested legitimation strategies. For example, the general tendency for social and/or environmental disclosure to emphasise the positive points of organisational behaviour, rather than the negative elements, may be explained as commensurate with a legitimation action on the part of the organisation (see Deegan, 2002, 2007; O'Donovan, 2002).

But legitimacy theory also has two principal variants. The first tends to be concerned with the legitimacy of individual organisations – for example, a company which is involved in a major oil spill or a charity caught up in a financial scandal may find its legitimacy threatened. The literature offers examples of where social accounting has been used to try and close 'legitimacy gaps' (O'Donovan, 2002). The second variant of legitimacy theory, however, takes a wider perspective (a lower level of resolution) and, principally informed by Marxian thinking, raises questions about the legitimacy of the *system* (e.g. capitalism) as a whole. Such a perspective might lead one to ask, for example, why shareholders have the dominant role in external information provision, or why companies are permitted to act in ways that most individuals would find unacceptable in their private lives.[51] Under this perspective, social accounting is more subtly employed. It might be used by a range of organisations[52] to (say) either 'explain' about changing organisation–employee relationships which may look, on the surface, like an attempt to educate stakeholders, but which might be interpreted as an attempt to cover moves towards the emasculation of trade unions. Similarly, we can see trends in (especially) sustainability disclosure which can be interpreted as attempts

---

[51]One illustration of this in the UK arises from the Church of England which, as a Christian Church, is committed to the principle of 'thou shalt not kill' and yet did, for many years, have a substantial number of financial investments with weapons manufacturers. The matter is clearly a complex one if weapons manufacturing is seen as a legitimate form of business to people who are sworn to uphold the sanctity of life.

[52]Or, more likely, supported and encouraged by and through an industry representative body, a pseudopolitical body or a 'front' organisation – sometimes referred to as Astroturf organisations.

to maintain public perception of the importance of a company, an industry and a system in the 'creation' of 'wealth' and 'jobs'. Such uses of social accounting can be interpreted as attempts to continue the legitimacy of the system rather than of individual organisations.[53] None of this, however, really tells us very much about why organisations might choose *not* to disclose at all or necessarily tell us why disclosure might be so selective.

Economic/rationalist

That stalwart of accounting theorisation – **decision-usefulness** – has also been used to help explain social accounting. This theory simply suggests that information (like social accounting) will be produced if appropriate decision-makers find it useful in their decisions. The theory is a useful heuristic, but it fails to expose which decision-makers concern us and why – and, consequently, the theory concerns itself with the powerful decision-makers like management and investors and thereby implicitly ignores most other decision-makers. However, the theory is also confused over the *normative* and the *positive*. As a descriptive theory, it does not help a great deal in the sense that almost anything can be useful. (A teaspoon is useful in digging a hole if that is all you have.) On the other hand, it does not tell us who *should* receive information (investors are assumed but not justified in the theory) and so it ducks the normative question (which is why accountability works so well in this vein). So we can study social accounting and discover that investors and financial participants in companies find social information 'quite useful' information (see, for example, Firth, 1978; Epstein and Freedman, 1994), but such information tends only to be central to a minority of 'ethical investors'. How social information would influence all the decisions of all corporate stakeholders if it were complete, direct and fairly stark is entirely another matter and remains largely untested (see, for example, Guthrie and Parker; 1989; Chan and Milne, 1999; Milne and Chan, 1999).

One interesting variant on decision usefulness is caught by the notion of signalling which suggests that management might produce social accounting as a signal to (primarily) their financial stakeholders that they are keeping an eye on (for example) social and environmental risks. Consequently, the investors can be persuaded that the organisation is both well-run and relatively free from unexpected social (de-legitimating) shocks. This would certainly go someway towards explaining why so many organisations would produce largely vacuous stand alone reports – they are not directed at informed members of civil society but are intended for management, investors and the media as a signal of the organisation's competence (Neu *et al.*, 1998).

Agency (or principal-agent) theory is both an exceptionally closely focused theory and an exceptionally popular one. It conceives of the world as comprising pairs of individuals – a principal and an agent – who contract together under assumptions of short-termism, utter selfishness and utility maximisation.[54] The principal (e.g. a manager or a shareholder) seeks to induce the agent (e.g. an employee or a director of the shareholder's firm) to do things that are in the best interest of the principal and thereby overcome the agent's own preferences (known as 'moral hazard') and any likelihood of the agent making the wrong choice ('adverse

---

[53]Such a view is commensurate with the work of, for example, Tregidga and Milne (2006) which examines how the language around sustainability is taken by corporations and their representative bodies and stripped of meanings (like zero or negative growth, equity and so on) that might be seen to challenge current business hegemony. That these authors use discourse theory as the key to unlocking this issue demonstrates the fluidity of theory and its categorisation.

[54]The basic language and structure of agency theory sound a lot like the theory of accountability: contract, principal and agent. This is as far as the similarity goes. In essence, if one stripped accountability theory of all its humanity, context and relationships and assumed narrow selfish motives then one might have ended up with agency theory.

selection'). The principal achieves this through monitoring the agent (typically via information) and offering financial inducements for correct behaviour. The theory can be most comfortably employed at the personal level but, following Jensen and Meckling's (1976) argument that the firm is no more than a 'nexus of contracts', it is now widely applied at the organisational level. The theory is used to model manager–employee behaviour and company management–stockholder/market relationships and used, for example, to explain incentives and control.[55] The direct use of agency theory in social accounting is relatively scarce (although see Ness and Mirza, 1991, for one exception) although its underlying assumptions and reasoning are widely used in the statistical analyses of social accounting disclosure which, typically, might be concerned with isolating and understanding investor-relevant financial effects. Broadly, investors seem relatively uninterested in social accounting information (see, for example, Chan and Milne, 1999). Agency theory is relatively unpopular in mainstream social accounting largely because something as individualistic and self-serving as agency theory sits uncomfortably with the more expansive, liberationist and even emancipatory ethical basis that most bring to social accounting.

Equally, social accounting has not yet fully embraced the potential of institutional economics, markets and hierarchies and, particularly, **transaction cost theory** (see, for example Williamson, 1979). This branch of theory begins from an explanation as to why organisations exist: mainly because it costs too much to transact each and every action in the market place and so these actions are more easily and efficiently undertaken *within* an organisational setting. In doing so, organisations (it is argued) are able to overcome problems of transaction terms between agents who must all have imperfect information. This, in turn, leads organisations to be able to more easily overcome difficulties in maintaining reliability and quality in goods and services. (The trend towards out-sourcing is a reverse of this process.) MNCs can then be seen as massive mechanisms for minimising transactions costs worldwide (see also, Korten, 1995; Agmon, 2003). The role that social accounting might play in such a conception is not immediately obvious, but we might see social accounting used internally in the organisation to maintain culture and ease internal transaction costs whilst the larger organisations can employ disclosure to influence their negotiations (and therefore their transactions) over cost, regulations and market advantages.

For completeness, we should note that although the social accounting literature has been slow to embrace 'the *Other*' at an organisational level, there has been an enthusiastic engagement with *new conceptions of organisational life* and an associated imagining of social accounting. Faced with the stresses of social and environmental un-sustainability we need to (re-) imagine what an organisation which embraced nature and/or sustainability might look like. Experiments with social enterprises, fair trade and 'values-based' corporations and cooperatives are the tip of this iceberg (Gladwin *et al.*, 1995b; Dauncey, 1996; Johnson and Bröms, 2000; Young and Tilley, 2006; Barter and Bebbington, 2010) and this, in turn, is stimulating new approaches to social accounting (see, for example, Gray *et al.*, 1997; Evans, 2000; Cooper *et al.*, 2005; Dey, 2007; and see **Chapters 12 and 13**).

One thing that unites most (if not all) of the theories considered in this section is that they are underspecified, they really do not explain 'why' organisations do what they do regarding CSR and social accounting in any consistently thorough or convincing way. They are, in fact, outside-looking-in theories – theories which observe organisations from the outside and speculate on what is happening. Much more penetration on the detail of what organisations are doing is acquired through the inside-looking-in theories – or theories which derive from field work research conducted inside the organisation itself.

---

[55]It is also a theory which attracts considerable criticism. See, for example, Christenson (1983); Arrington (1990); Armstrong (1991); Tinker and Okcabol (1991).

4.9 Social accounting inside the organisation (micro-theory II)

Although there has always been fieldwork-based enquiry[56] in social accounting, by the turn of the century it had not dominated research to the same extent as had the more arms-length enquiries such as the study of organisation's disclosure or even the use of postal questionnaires. Consequently, theorisation about social accounting within the organisation – how it comes about; why it happens; why it doesn't happen; why it takes the form that it does; and so on – had been relatively less well-developed. Given the considerable volume of management and management accounting theorisation based, primarily, upon fieldwork this is actually a bit of a surprise (see, for example, Emmanuel and Otley, 1985; Puxty, 1986; Knights and Willmott, 2007).

There is currently no dominant organisational theory of (or for) social accounting inside the organisation. There are, however, a number of themes that seem to stand out – regardless of the theory employed.[57] For example, research continues to show the diversity and complexity of both individual and organisational motivation for social accounting. Whilst there may well be times when social accounting might be undertaken for a simple, singular direct reason, it would be contestable to assume this was always the case.[58] Additionally, studies increasingly identify the importance of the role of key individuals in the developing of social accounting as well as the problems that an individual might experience in the conflict between personal and organisational values regarding social (non) disclosure (Antal, 1985; Jones, 1986; Gray *et al.*, 1997, 1998; Buhr, 1998, 2007; de Villiers, 1999; Gray and Bebbington, 2000; Adams, 2002; Miles *et al.*, 2002; Norris and O'Dwyer, 2004; Rahaman *et al.*, 2004; Dey, 2007; Spence and Gray, 2008).

Biological

One major area of theorising over social accounting within the organisation relates to models of *organisational* change. For example Gray *et al.* (1995) employed an adapted form of Laughlin's (1991) model of organisational change to provide a framework within which to study the emergence of social accounting in a number of institutions (see also Larrinaga-González *et al.*, 2001; Larrinaga-González and Bebbington, 2001). In a manner similar to the discussion of autopoiesis (which we briefly revisit below), the study found that whilst environmental accounting (in that case) was both a result of external pressures and a potential source of change itself, the range of influences that the organisation 'recognised' and responded to was limited to those that accorded with the design archetype of the organisation. The model was further extended to embrace Llewellyn's (1994) approach to organisational **'boundary management'**. That model suggests that issues (such as the natural environment, climate change or social responsibility) will, to a degree at least, be absorbed

---

[56]Fieldwork is a general term that refers to research in which the researcher leaves the office/university and studies the phenomena of interest in the context in which it arises – they 'go into the field'. This contrasts with other research methods which might involve study of data sets (e.g. share prices) or analysing documents (for example, annual reports or CSR reports) away from the setting where the data or the documents were created.

[57]There is a parallel here with a number of dominant observations in the more 'positivistic', arms-length research. For example, regardless of theoretical position, social disclosure is more likely to happen in bigger organisations in certain countries and in certain industries.

[58]The range of potential influences on the disclosure decision in organisations can verge on the bewildering. For example, unpublished PhD theses involving fieldwork from countries such as Bangladesh and Egypt have identified culture, the role of civil society, Islam, relationships with communities, the importance of overseas investors, the international financial community and the influence and attitude of other western companies all as active issues in the disclosure decision.

by the organisation and/or that the organisation will extend its boundaries to embrace 'outside' issues. That is, organisations can be said to 'manage' the boundaries of their entity and to determine what is, and what is not absorbed or recognised by it – in effect, what is or is not 'part' of the organisation and consequently part of the business of the organisation. Social and environmental issues and the management of and accounting for them is just such an issue and will be embraced, absorbed or rejected to the extent that it seems to be in accord with the organisation and its sense of itself.

Political/sociological

An influential illustration of how to approach the use of theory is offered by the intensive case study of Buhr (1998). Buhr employs a range of theoretical lenses through which to explore how issues (like pollution or social accounting) actually emerge as issues within organisations. The paper concludes that the dominance of engineers in the company leads to a predominantly technological approach to both solutions and explanations of the issues (in this case emissions) and their solution. The paper then concludes that, in this case at least, a social constructionist/legitimacy theory perspective offers the more powerful explanation of events.

By contrast, Buhr (2002) formally theorises her examination of two different organisations and *their* very different reactions to and involvement with environmental reporting through Giddens' **structuration theory**. At its simplest, structuration theory is an articulation of the way in which individuals influence and are in turn influenced by the structures around them. That is, the relationship between individuals (agency) and structure is *reflexive*. The theory argues that, on the one hand, what we know as social life cannot be understood as a simple sum of all individual/micro-level activity, but neither can all social activity be completely explained from a structural/macro perspective. The middle way between the extremes sees agents' repetition of acts both producing and re-producing structure – but, importantly, all social structures are understood to be neither inviolable nor permanent. Buhr uses this framework to study how pressures, issues and concerns were perceived, interpreted and then responded to by two separate organisations. The paper contains recognition of the roles played by key agents, the possibilities offered and the restrictions placed by structure and culture, the influence (or not) of stakeholders as a function of either agency and/or structure and the long-term process through which reporting practices change or revert to type.

The organisation theory literature (as we have already seen) is rich in theoretical perspectives, and many of these have the potential for greater insights into social accounting. One further illustration might suffice for now – that of discourse theory. Discourse theory (which you will recall we also met under meta-theories above) is concerned with how the ebb and flow of communication both reflects and creates both meaning and reality. The way we describe something reflects how we think of it and, depending upon our individuality and relative power, may have a major influence on how we and others come to think of the issue in question. For example, Livesey and Kearins (2002) and Tregidga and Milne (2006) explored how language is used in disclosure to 'construct' the sense of the relationship between 'sustainability', the organisation and its traditional pursuits (see also Buhr and Reiter, 2006).

Finally, we should just flag up the role of groups (and teams) and identity. The psychological and business management literatures are replete with explorations of the role that groups play in organisation, in the completion of tasks, the contribution of effort but also on the formation of an individual's identity and on acceptable norms of behaviour, thought and language. Consequently, shared beliefs (such as, for example, 'we are an ethical company') become reified and cannot be examined. An individual who might want to challenge such views is likely to find it very difficult indeed and so here (as with culture and a whole host of other factors – see below) may be another way in which (non) disclosure and (non) accountability decisions around social accounting can be explored (see, for example, Knights and Willmott, 2007, for an introduction).

Economic/rationalist

It seems to be relatively unusual for micro-studies, typically based upon fieldwork, to seek out and/or apply economic explanations for social accounting practice (but see Miles *et al.*, 2002). One exception is Spence and Gray (2008) which examines the language used when officers of organisations explain their organisation's (non) engagement with both CSR and social and environmental reporting.[59] The monograph infers that there appears to be a prevailing necessity for organisational participants to articulate most things through a version of *the business case* – there is little space for something which is not a business case and anything that is to be adopted within the organisation must be expressed as part of a business case (regardless of any economic or other 'reality'). Thus issues like sustainability and CSR, which are increasingly being pressed upon organisations, must be (and can only be) re-articulated into terms commensurate with a business case, otherwise, they cannot 'code' to the organisation. So CSR and sustainability end up trivialised (Shamir, 2004). Such reasoning resonates with autopoiesis as we saw it earlier – only at a higher level of resolution. It seems to be the case that any initiative concerning social, environmental or sustainability accounting and/or reporting must be stated in tune with (must 'code' to in the autopoietic sense) the overall (economic) mission of the organisation itself.

As should now be apparent, explanations of organisational behaviour (especially regarding social accounting) need to be complex if they are to catch the range of issues at work. One illustration of how theory construction might be taken forward is offered by Adams (2002). That paper (augmented by interviews conducted in the UK and Germany) drew from the extant literature a potentially bewildering array of factors that had a potential impact on the form and content of the social accounting. Adams categorises these influences as: characteristics of the corporation; issues internal to the organisation; and general contextual issues. The resultant model is shown in Figure 4.3.

4.10 Individual-level theories (micro-level III)

The individual-level (sometimes referred to as 'agency' as opposed to 'structure') theories that are relevant to our discussion here are primarily explanations of why individuals do things and, in particular, why they might initiate or resist the development of social accounting. There are, as you might expect, a myriad of theories about human motivation, agency and resistance, etc. (up to and including agency theory which we met above). Research has identified an enormous array of personal motivations and concerns behind agency in social accounting. Perhaps the most interesting thing to emerge from this has been the twin influence of the role of key individuals (sometimes called 'champions') and the way in which the social and environmental agendas have enabled such champions to merge their personal and their organisational values – something which is considered relatively rare in most profit-centred organisations (see, for example, Gray *et al.*, 1995, 1998; Adams, 2002; Buhr, 2002; Spence and Gray 2008). However, it remains largely unclear why individuals do (and do not) support and develop social accounting (and accountability); how salient issues are selected

---

[59]The two ideas were frequently seen as identical/interchangeable by respondents.

Figure 4.3 Diagrammatic portrayal of the influences on corporate 'social' reporting

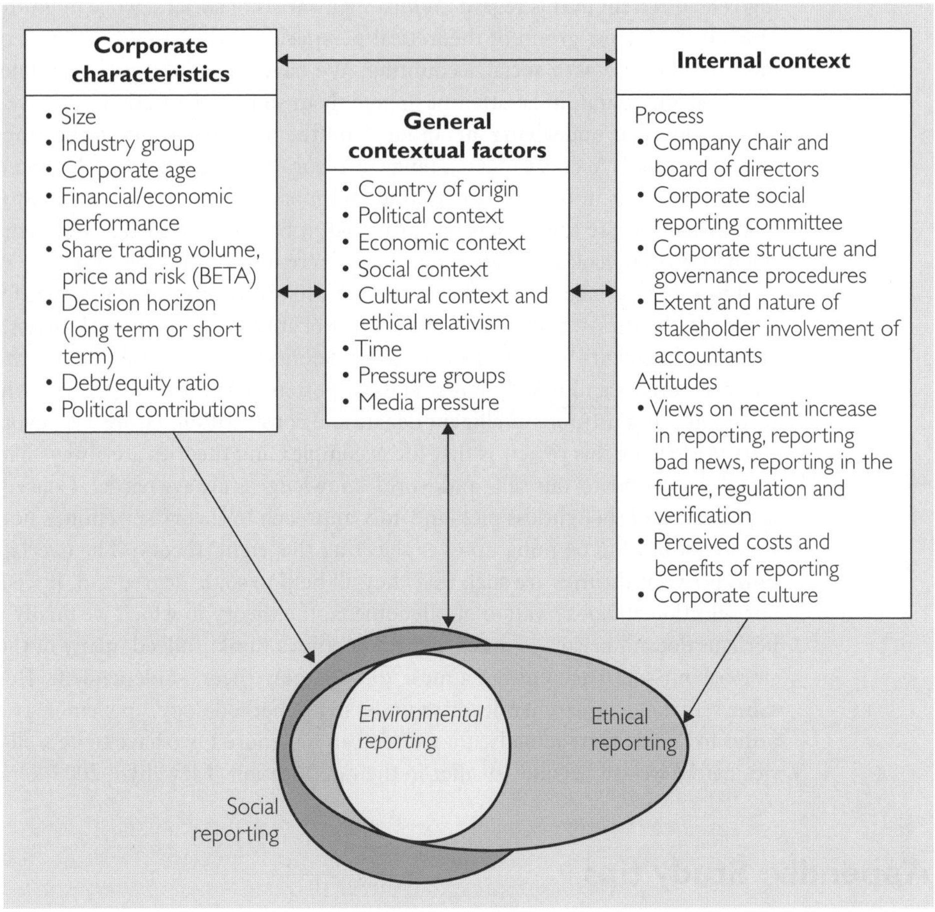

Source: Adams, (2002).

Note: Arrows show direction of influence.

and managed; why some things cannot be discussed and how initiatives are developed or opposed. Organisational theory, psychology and anthropology have much to show us here.

4.11 Summary and conclusions

In this (unfortunately rather long) chapter we have sought to achieve two very basic things.

First, we have tried to provide some overall sense of what social accounting is, can be and might be. The diversity of the area and the largely *ad hoc* manner in which social accounting has developed means that we have a potentially enormous array of potentially incoherent practices to examine in our study of social accounting. One of the major themes that emerge from this overview is that the *importance* of social accounting is not necessarily determined by either the often weak, even trivial, practice or by its frequency.

Second, we have tried to provide a brief tasting of some of the range of theoretical lenses that might be brought to bear in our attempts to make some sense of social account-ing. Theory is essential for any act of organising, analysing, understanding, evaluating, making proposals about or even trying to predict the future of any practice. Social accounting is no different in this regard. Whilst many areas of social science might be dominated by a relatively narrow group of theoretical perspectives, this need not – and probably should not – be the case with social accounting. We have seen that there is an enormous array of potential lenses and, in an attempt to give them some organisation, we have categorised the theories by their underlying metaphor[60] and the level of resolution they most conveniently appear to adopt. Access to such a range of theories offers, we hope, the potential for a wider range of insights into social accounting and more imaginative research into its (non) practice. Whatever else this review has provided, it has *not* provided a complete specification of the best ways to look at social accounting. There may be no such thing.

The things to take away from this review are probably quite liberating. First, identifying our meta-theories – our worldviews in all probability – helps unpick biases, influences and even disagreements because it seems to be definitionally the case that our theories influence our perceptions and our disagreements are often at the level of theory. More clarity over theory may lead to more intelligent debate and conversation. However, beyond this, what we start to see from this review is that life is complex and theories are always likely to be imperfect and incomplete: our self- and world-knowledge is always partial. Consequently, whilst a haphazard and thoughtless pick-and-mix approach to theory selection is not going to work, we do not need to be hung up over selecting the 'right' theory. The overlaps and intersections between theories are such that they all help to some degree and, if we are careful with our selection and observation of phenomena, the theory to which we finally nail our colours perhaps doesn't matter as much as we sometimes think. Indeed, many of the major conclusions about social accounting appear to be largely theory-independent. But in exactly the same way that a researcher needs to understand methodology, epistemology and ontology in order to decide that, actually, they often can be ignored, we have to be well-versed in theories before we can become cavalier in their application (Llewellyn, 2003; Gray *et al.*, 2010).

Appendix: Study tips

Consulting organisational reporting

As with any form of study that entails engagement with a variety of real-world phenomena (and very little does not), the educational experience can be enhanced by regularly consulting the phenomena under consideration. Throughout your study, we know that you will gain enormous benefit from consulting various manifestations of social, environmental and sustainability accounting and reporting.

By this stage in the text you should be getting into the habit of regularly consulting organisations' reporting practices. (Later on we will direct you to, for example, externally prepared documents and output from the social investment community.)

Who?

Our principal focus is large companies and so it is to large company reporting that we would first direct you. Choose a company that interests you – you buy their products, you have worked for them, you are secretary of a campaign group that is trying to get them to change their practices, you would like a job with them later, etc. Similarly, you may have a stronger

---

[60]See the excellent Smircich (1983) which offers a range of metaphors of organisation and Morgan (1986) for a more detailed and in-depth approach to this activity.

interest in public or third sector organisations – then seek out reporting by your local authority or the local health organisation for example. (Guidance on which large companies produce stand-alone reports can be obtained from websites such as that of **Global Reporting Initiative (GRI)** or **Corporate Register** or **The Association of Chartered Certified Accountants (ACCA) Reporting Awards Scheme**.

Where?

Generally speaking, you will make more sense of documents that talk about issues that you understand and therefore choosing an organisation from the country in which you are studying makes the most sense and may well be the easiest. (As you have already seen, context is all.)

What?

By now you will be aware that self-reporting primarily takes place in (a) the organisation's Annual Report and (b) in stand-alone reports (which are often called things like *Responsibility Reports, Sustainability Reports* or *Reports on the Environment* and so on. Equally, an increasing number of organisations report through either (a) hard copy documents (actual physical reports) or (b) the website (electronic reports). Some do both.

How?

Having chosen your organisation, you can obtain most information by visiting their websites. There you will find any electronic reporting (noting how easy or difficult it was to locate the material you wanted) plus, in all likelihood, information on whether hard copy is produced and how to obtain such documents if you want them.

Why?

Having obtained the report(s) and/or regular access to them, you should get into the habit of consulting them as you consider issues. You will start to relate what we talk about to the reports. So, for example, which of the range of crisis issues are touched upon in the report? To what extent are the views of the world and stakeholders recognisable from the report? Which social responsibilities, if any, are reflected in the report you are looking at. Given the range of issues that social accounting might cover: to what extent does your report do this?

Further guidance?

If you need further guidance on this or wish to extend your research, then we would direct you to **The Centre for Social and Environmental Accounting Research (CSEAR) website** where the article by Adams and Laing (2000), 'How to Research a Company' will give you considerable help. (It is located in the 'Introductory Materials' section of the website.) The website also has an 'Approaches to Practice' section that should be helpful.

References

Abt, C. C. and Associates (1972) *Annual Report and Social Audit 1972.*
Accounting Standards (formerly Steering) Committee (1975) *The Corporate Report*. London: ICAEW.
Adams, C. A. (2002) Internal organisational factors influencing corporate social and ethical reporting, *Accounting, Auditing and Accountability Journal*, **15**(2): 223–50.

Adams, C. A. and Frost, G. (2004) *The Development of Corporate Web-Sites and Implications for Ethical, Social and Environmental Reporting Through These Media.* Edinburgh: ICAS.

Adams, C. A. and Harte, G. (1998) The changing portrayal of the employment of women in British banks' and retail companies' corporate annual reports, *Accounting, Organizations and Society,* **23**(8): 781–812.

Adams, C. A., Hill, W. Y. and Roberts, C. B. (1995) *Environmental, Employee and Ethical Reporting in Europe.* London: ACCA.

Adams, C. A. and Laing, Y. (2000) "How to research a company" *Social and Environmental Accounting Journal* **20**(2) (pp. 6–11).

Agmon, T. (2003) Who gets what: The MNE, the national state and the distributional effects of globalization, *Journal of Industrial Business Studies,* **34**(5): 416–27.

American Institute of Certified Public Accountants (1977) *The Measurement of Corporate Social Performance.* New York: AICPA.

Andrew, J. (2000) The accounting craft and the environmental crisis: Reconsidering environmental ethics, *Accounting Forum,* **24**(2): 197–222.

Antal, A. B. (1985) Institutionalizing corporate social responsiveness: Lessons learned from the Migros experience, *Research in Corporate Social Performance and Policy,* 7: 229–49.

Armstrong, P. (1991) Contradiction and social dynamics in the capitalist agency relationship, *Accounting, Organizations and Society,* **16**(1): 1–26.

Arrington, E. (1990) Intellectual tyranny and the public interest: The quest for the holy grail and the quality of life, *Advances in Public Interest Accounting,* **3**: 1–16.

Bailey, D., Harte, G. and Sugden, S. (1994a) *Making Transnationals Accountable: A significant step for Britain.* London: Routledge.

Bailey, D., Harte, G. and Sugden, R. (1994b) *Transnationals and Governments: Recent policies in Japan, France, Germany, the United States and Britain.* London: Routledge.

Bakan, J. (2004) *The Corporation: the pathological pursuit of profit and power.* London: Constable and Robinson.

Ball, A. and Seal, W. (2005) Social justice in a cold climate: could social accounting make a difference?, *Accounting Forum,* **29**(4): 455–73.

Ball, A., Owen, D. L. and Gray, R. H. (2000) External transparency or internal capture? The role of third party statements in adding value to corporate environmental reports, *Business Strategy and the Environment,* **9**(1): 1–23.

Bansal, P. and Roth, K. (2000) Why companies go green: A model of ecological responsiveness, *Academy of Management Journal,* **43**(4): 717–36.

Barter, N. and Bebbington, J. (2010) *Pursuing Environmental Sustainability Research Report 116.* London: ACCA.

Bebbington, J. (2007) *Accounting for Sustainable Development Performance.* London: CIMA.

Belal, A. and Owen, D. (2007) The view of corporate managers on the current state of, and future prospects for social reporting in Bangladesh, *Accounting, Auditing & Accountability Journal,* **20**(3): 472–94.

Belkaoui, A. (1997) *Accounting Theory,* 3rd Edition. London: Dryden Press.

Bloom, R. and Heymann, H. (1986) The concept of social accountability in accounting literature, *Journal of Accounting Literature,* **5**: 167–82.

Borg, W. R. and Gall, M. D. (1989) *Educational Research: An Introduction,* 5th Edition. London: Longman.

Boulding, K. E. (1966) The economics of the coming Spaceship Earth, in Jarratt, H. (ed.), *Environmental Quality in a Growing Economy,* pp. 3–14. Baltimore, MD: John Hopkins Press.

Bowen, H. R. (1953) *Social Responsibilities of the Businessman.* New York: Harper & Row.

Brown, N. and Deegan, C. (1998) The public disclosure of environmental performance information – A dual test of media agenda setting theory and legitimacy theory, *Accounting and Business Research,* **29**(1): 21–42.

Buhr, N. (1998) Environmental performance, legislation and annual report disclosure: the case of acid rain and Falconbridge, *Accounting, Auditing and Accountability Journal,* **11**(2): 163–90.

Buhr, N. (2002) A structuration view on the initiation of environmental reports, *Critical Perspectives on Accounting,* **13**(1): 17–38.

Buhr, N. (2007) Histories of and rationales for sustainability reporting, in Unerman, J., Bebbington, J. and O'Dwyer, B. (2007) (eds), *Sustainability Accounting and Accountability*, pp. 57–69. London: Routledge.

Buhr, N. and Reiter, S. (2006) Ideology, the environment and one world view: a discourse analysis of Noranda's environmental and sustainable development reports, in Freedman, M. and Jaggi, B. (eds), *Advances in Environmental Accounting and Management ss. 3*, pp. 1–48.

Burchell, S., Clubb, C. and Hopwood, A. (1985) Accounting in its social context: Towards a history of value added in the United Kingdom, *Accounting, Organizations and Society*, 10(4): 381–413.

Burrell, G. and Morgan, G. (1979) *Sociological Paradigms and Organizational Analysis*. London: Heinemann.

Callon, M. (2009) Civilizing markets: carbon trading between in vitro and in vivo experiments, *Accounting, Organizations and Society*, 34(3/4): 535–48.

Callon, M. and Law, J. (1997) After the individual in society: lessons on collectivity from science, technology and society, *Canadian Journal of Sociology*, 22(2): 165–82.

Carson, R. (1962) *Silent Spring*. Boston, MA: Houghton Mifflin.

Chan, C. C. C. and Milne, M. J. (1999) Investor reactions to corporate environmental saints and sinners: An experimental analysis, *Accounting and Business Research*, 29(4): 265–79.

Chartered Institute of Management Accountants (1982) *The Evaluation of Energy Use: Readings*. London: CIMA.

Christenson, C. (1983) The methodology of positive accounting, *The Accounting Review*, January: 1–22.

Chryssides, G. D. and Kaler, J. H. (1993) *An Introduction to Business Ethics*. London: Thomson.

Chwastiak, M. (1996) The wrong stuff: the accounting review's forum for defense clients, *Critical Perspectives on Accounting*, 7(4): 365–81.

Confederation of British Industry (1971) *The Responsibilities of the British Public Company*. London: CBI.

Cooper, D. J. (1988) A social analysis of corporate pollution disclosures: a comment, *Advances in Public Interest Accounting*, 2: 179–86.

Cooper, C., Taylor, P., Smith, N. and Catchpowle, L. (2005) A discussion of the political potential of social accounting, *Critical Perspectives on Accounting*, 16(7): 951–74.

Dauncey, G. (1996) *After the crash: The emergence of the Rainbow economy*, 2nd Edition. Rendlesham: Greenprint.

De Villiers, C. J. (1999) The decision by management to disclose environmental information: a research note based on interviews, *Meditari Accountancy Research*, 7: 33–48.

Dean, M. (1999) *Governmentality: Power and Rule in Modern Society*. London: Sage.

Deegan, C. (2002) The legitimising effect of social and environmental disclosures: a theoretical foundation, *Accounting Auditing and Accountability Journal*, 15(3): 282–311.

Deegan, C. (2007) Organizational legitimacy as a motive for sustainability reporting, in Unerman, J., Bebbington, J. and O'Dwyer, B. (eds), *Sustainability Accounting and Accountability*, pp. 127–49. London: Routledge.

Dey, C. (2007) Developing silent and shadow accounts, in Unerman, J., Bebbington, J. and O'Dwyer, B. (eds), *Sustainability Accounting and Accountability*, pp. 307–26. London: Routledge.

Dey, C., Evans, R. and Gray, R. H. (1995) Towards social information systems and bookkeeping: A note on developing the mechanisms for social accounting and audit, *Journal of Applied Accounting Research*, 2(3) December.

Dick-Larkham, R. and Stonestreet, D. (1977) Save it! The accountants' vital role, *Accountants' Weekly*, 15 April: 22–3.

Dickson, D. (1974) *Alternative Technology and the Politics of Technical Change*. Glasgow: Fontana.

Dillard, J. (2007) Legitimating the social accounting project: an ethic of accountability, in Unerman, J., Bebbington, J. and O'Dwyer, B. (eds), *Sustainability Accounting and Accountability*, pp. 37–54. London: Routledge.

Dillard, J., Brown, D. and Marshall, S. (2005) An environmentally enlightened accounting, *Accounting Forum*, 29: 77–101.

DiMaggio, P. D. and Powell, W. W. (1983) The iron cage revisited: institutional isomorphism and collective rationality in organizational fields, *American Sociological Review*, 48(2): 147–60.

Doane, D. (2000) *Corporate Spin: The troubled teenage years of social reporting.* London: NEF.

Donaldson, J. (1988) *Key Issues in Business Ethics.* London: Academic Press.

Donaldson, T. and Preston, L. (1995) The stakeholder theory of the corporation: Concepts, evidence and implications, *Academy of Management Review,* **20**(1): 65–91.

Drucker, P. (1965) Is business letting young people down?, *Harvard Business Review,* Nov/Dec: 54.

Elkington, J. (1997) *Cannibals with Forks: the triple bottom line of 21st century business.* Oxford: Capstone Publishing.

Emmanuel, C. R. and Otley, D. T. (1985) *Accounting for Management Control.* London: Van Nostrand Reinhold.

Epstein, M. J. and Freedman, M. (1994) Social disclosure and the individual investor, *Accounting, Auditing and Accountability Journal,* **7**(4): 94–109.

Estes, R. W. (1976) *Corporate Social Accounting.* New York: Wiley.

Evans, R. (2000) *Corporate Ethical Accounting: (How) Can Companies Tell the Truth?* Cambridge: Grove Books.

Everett, J. (2004) Exploring (false) dualisms for environmental accounting praxis, *Critical Perspectives on Accounting,* **15**(8): 1061–84.

Everett, J. and Neu, D. (2000) Ecological modernization and the limits of environmental accounting, *Accounting Forum,* **24**(1): 5–29.

Firth, M. (1978) A study of the consensus of the perceived importance of disclosure of individual items in corporate annual reports, *The International Journal of Accounting Education and Research,* **14**(1): 57–70.

Flamholtz, E. G. (1974) *Human Resource Accounting.* California: Dickenson.

Foley, B. J. and Maunders, K. T. (1977) *Accounting Information Disclosure and Collective Bargaining.* London: Macmillan.

Freeman, R. E. (1984) *Strategic Management: A stakeholder approach.* Boston, MA: Pitman.

Freeman, R. E. (1994) The politics of stakeholder theory: Some future directions, *Business ethics Quarterly,* **4**(4): 409–21.

Friedman, A. and Miles, S. (2002) Developing stakeholder theory, *Journal of Management Studies,* **39**(1): 1–21.

Friedman, A. L. and Miles, S. (2006) *Stakeholder: Theory and practice.* Oxford: Oxford University Press.

Friedman, M. (1970) The social responsibility of business is to increase its profits, *The New York Times Magazine,* September 13: 122–6.

Ghauri, P., Grônhaug, K. and Kristianslund, I. (1995) *Research Methods in Business Studies: A practical guide.* London: Prentice Hall.

Gladwin, T. N., Kennelly, J. J. and Krause, T.-S. (1995a) Shifting paradigms for sustainable development: Implications for management theory and research, *Academy of Management Review,* **20**(4): 874–907.

Gladwin, T. N., Krause, T.-S. and Kennelly, J. J. (1995b) Beyond eco-efficiency: Towards socially sustainable business, *Sustainable Development,* **3**: 35–43.

Goldblatt, D. (1996) *Social Theory and the Environment.* Oxford: Blackwell.

Goldsmith, E. *et al.* (1972) *Blueprint for Survival.* Harmondsworth: Penguin.

Gouldson, A. and Bebbington, J. (2007) Corporations and the governance of environmental risk, *Environment and Planning C: Government and Policy,* **25**(1): 4–20.

Gray, J. (1996) *After Social Democracy.* London: Demos.

Gray, R. H. (2002) The social accounting project and accounting organizations and society: privileging engagement, imaginings, new accountings and pragmatism over critique, *Accounting Organizations and Society,* **27**(7): 687–708.

Gray, R. (2006) Social, environmental, and sustainability reporting and organisational value creation? Whose value? Whose creation?, *Accounting, Auditing and Accountability Journal,* **19**(3): 319–48.

Gray, R. (2010) Is accounting for sustainability actually accounting for sustainability . . . and how would we know? An exploration of narratives of organisations and the planet, *Accounting, Organizations and Society,* **35**(1): 47–62.

Gray, R. H. and Bebbington, K. J. (2000) Environmental accounting, managerialism and sustainability: Is the planet safe in the hands of business and accounting?, *Advances in Environmental Accounting and Management,* **1**: 1–44.

Gray, R. H. and Bebbington, K. J. (2001) *Accounting for the Environment*, 2nd Edition. London: Sage.

Gray, R. H., Owen, D. L. and Maunders, K. T. (1987) *Corporate Social Reporting: Accounting and accountability*. Hemel Hempstead: Prentice Hall.

Gray, R. H., Kouhy, R. and Lavers, S. (1995) Corporate social and environmental reporting: A review of the literature and a longitudinal study of UK disclosure, *Accounting, Auditing and Accountability Journal*, 8(2): 47–77.

Gray, R. H., Owen, D. L. and Adams, C. (1996) *Accounting and Accountability: Changes and challenges in corporate social and environmental reporting*. London: Prentice Hall.

Gray, R. H., Dey, C., Owen, D., Evans, R. and Zadek, S. (1997) Struggling with the praxis of social accounting: stakeholders, accountability, audits and procedures, *Accounting, Auditing and Accountability Journal*, 10(3): 325–64.

Gray, R. H., Bebbington, K. J., Collison, D. J., Kouhy, R., Lyon, B., Reid, C., Russell, A. and Stevenson, L. (1998) *The Valuation of Assets and Liabilities: Environmental law and the impact of the environmental agenda for business*. Edinburgh: ICAS.

Gray, R. H. and Collison, D. J. with J. French, K. McPhail and L. Stevenson (2001a) *The professional accountancy bodies and the provision of education and training in relation to environmental issues*. Edinburgh: ICAS.

Gray, R. H., Javad, M., Power, D. M. and Sinclair, C. D. (2001b) Social and environmental disclosure and corporate characteristics: A research note and extension, *Journal of Business Finance and Accounting*, 28(3/4), 327–56.

Gray, R., Bebbington, J. and Collison, D. J. (2006) NGOs, civil society and accountability: making the people accountable to capital, *Accounting, Auditing and Accountability Journal*, 19(3): 319–48.

Gray, R. H., Owen, D. L. and Adams, C. (2010) Some theories for social accounting?: A review essay and tentative pedagogic categorisation of theorisations around social accounting, *Advances in Environmental Accounting and Management*, 4: 1–54.

Gray, R., Dillard, J. and Spence, C. (2011) A brief re-evaluation of 'the social accounting project': Social accounting as if the world matters, in Osborne, S. and Ball, A. (eds), *Social Accounting and Public Management: Accountability for the Public Good*, pp. 11–22. London: Routledge.

Guthrie, J. and Parker, L. D. (1989) Corporate social reporting: a rebuttal of legitimacy theory, *Accounting and Business Research*, 9(76): 343–52.

Guthrie, J. and Parker, L. D. (1990) Corporate social disclosure practice: a comparative international analysis, *Advances in Public Interest Accounting*, 3: 159–76.

Hanafi, R. A. and Gray, R. H. (2005) Collecting social accounting data in developing countries: A cautionary tale from Egypt, *Social and Environmental Accounting Journal*, 25(1): 15–20.

Haniffa, R. M. and Cooke, T. E. (2005) The impact of culture and governance on corporate social reporting, *Journal of Accounting and Public Policy*, 24: 391–430.

Hargreaves, B. J. A. and Dauman, J. (1975) *Business Survival and Social Change: A practical guide to responsibility and partnership*. London: Associated Business Programmes.

Held, D. (1980) *An Introduction to Critical Theory: Horkheimer to Habermas*. London: Hutchinson.

Held, D. (1987) *Models of Democracy*. Oxford: Polity.

Henriques, A. and Richardson, J. (2004) *The Triple Bottom Line: does it add up?* London: Earthscan.

Hewgill, J. (1977) Can energy become the new currency?, *Accountants' Weekly*, 9 September: 13.

Hewgill, J. (1979) New frontiers: Into the unknown, *Accountants' Weekly*, 7 February: 11.

Hibbitt, C. and Collison, D. J. (2004) Corporate environmental disclosure and reporting developments in Europe, *Social and Environmental Accounting Journal*, 24(1): 1–11.

Hines, R. (1984) The implications of stock market reaction (non-reaction) for financial accounting standard setting, *Accounting and Business Research*, 15(57): 3–14.

Hines, R. D. (1992) Accounting: filling the negative space, *Accounting, Organizations and Society*, 17(3/4): 313–42.

Hofstede, G. (1984) *Culture's Consequences: International differences in work-related values*. Beverly Hills, CA: Sage.

Husted, B. W. (2000) A contingency theory of corporate social performance, *Business and Society*, 39: 24–38.

Jensen, M. and Meckling, W. (1976) Theory of the firm: managerial behaviour, agency costs and ownership structure, *Journal of Financial Economics*, October: 305–60.

Johnson, H. L. (1979) *Disclosure of Corporate Social Performance: survey, evaluation and prospects*. New York: Praegar.

Johnson, H. T. and Bröms, A. (2000) *Profit Beyond Measure: Extraordinary Results Through Attention to Work and People*. New York: The Free Press, Simon Schuster.

Jones, C. (1986) Corporate social accounting, *Occasional Papers in Sociology* No. 3, Bristol Polytechnic, April.

Kamla, R., Gallhofer, S. and Haslam, J. (2006) Islam, nature and accounting: Islamic principles and the notion of accounting for the environment, *Accounting Forum*, **30**(3): 245–65.

Kapp, K. W. (1978) *The Social Costs of Business Enterprise*. Nottingham: Spokesman [first published 1950].

Kempner, T. K., MacMillan, K. and Hawkins, K. H. (1976) *Business and Society*. Harmondsworth: Pelican.

Kirman, C. E. (1999) *A Luhmann-informed account of the attempt to introduce environmental reporting in New Zealand*. PhD Thesis: University of Dundee.

Knights, D. and Willmott, H. (2007) *Introducing Organizational Behaviour and Management*. London: Thomson Learning.

Korten, D. C. (1995) *When Corporations Rule the World*. West Hatford/San Francisco, CA: Kumarian/Berrett-Koehler.

Kovel, J. (2002) *The Enemy of Nature: The end of capitalism or the end of the world?* London: Zed Books.

KPMG (1999) *KPMG International Survey of Environmental Reporting 1999*. Amsterdam: KPMG/WIMM.

KPMG (2002) *KPMG 4th International Survey of Corporate Sustainability Reporting*. Amsterdam: KPMG/WIMM.

KPMG (2005) *KPMG International Survey of Corporate Responsibility 2005*. Amsterdam: KPMG International.

KPMG (2008) *KPMG International Survey of Corporate Responsibility Reporting 2008*. Amsterdam: KPMG International.

KPMG/UNEP/GRI/UCGA (2010) *Carrots and Sticks-Promoting Transparency and Sustainability*. Amsterdam: KPMG.

Kreander, N. (2001) An Analysis of European Ethical Funds, *Research Paper* Number 33. London: ACCA.

Lamberton, G. (1998) Exploring the accounting needs of an ecologically sustainable organisation, *Accounting Forum*, **22**(2): 186–209.

Larrinaga-González, C. (2007) Sustainability reporting: insights from neoinstitutional theory, in Unerman, J., Bebbington, J. and O'Dwyer, B. (eds), *Sustainability Accounting and Accountability*, pp. 150–67. London: Routledge.

Larrinaga-González, C. and Bebbington, J. (2001) Accounting change or institutional appropriation? A case study of the implementation of environmental accounting, *Critical Perspectives on Accounting*, **12**(3): 269–92.

Larrinaga-González, C., Carrasco-Fenech, F., Javier Caro-González, F., Correa-Ruiz, C. and Maria Páez-Sandubete, J. (2001) The role of environmental accounting in organisational change: an exploration of Spanish companies, *Accounting, Auditing and Accountability Journal*, **14**(2): 213–39.

Laughlin, R. C. (1991) Environmental disturbances and organisational transitions and transformations: some alternative models, *Organisation Studies*, **12**(2): 209–32.

Laughlin, R. C. (1995) Methodological themes: empirical research in accounting: alternative approaches and a case for 'middle-range' thinking, *Accounting, Auditing and Accountability Journal*, **8**(1): 63–87.

Laughlin, R. C. (2004) Putting the record straight: A commentary on 'Methodology choices and the construction of facts: Some implications from the sociology of knowledge', *Critical Perspectives on Accounting*, **15**(2): 261–77.

Lehman, C. (1992) *Accounting's Changing Roles in Social Conflict*. London: Paul Chapman.

Lehman, G. (1995) A legitimate concern for environmental accounting, *Critical Perspectives on Accounting*, **6**(6): 393–412.

Lehman, G. (1999) Disclosing new worlds: a role for social and environmental accounting and auditing, *Accounting, Organizations and Society*, **24**(3): 217–42.

Lehman, G. (2001) Reclaiming the public sphere: problems and prospects for corporate social and environmental accounting, *Critical Perspectives on Accounting*, **12**(6): 713–33.

Lehman, G. (2006) Perspectives on language, accountability and critical accounting: An interpretative perspective, *Critical Perspectives on Accounting*, **17**(6): 755–79.

Lehman, G. (2007) Ethics, communitarianism and social accounting, in Gray, R. H. and Guthrie, J. (eds), *Social Accounting, Mega Accounting and Beyond: A Festschrift in Honour of M. R. Mathews*, pp. 35–42. St Andrews: CSEAR Publishing.

Leopold, A. (1989) *A Sand County Almanac: And sketches here and there*. Oxford: Oxford University Press [first published 1949].

Lindblom, C. K. (1993) *The Implications of Organizational Legitimacy for Corporate Social Performance and Disclosure*, Paper presented at the Critical Perspectives on Accounting Conference, New York.

Livesey, S. M. and Kearins, K. (2002) Transparent and caring corporations? A study of sustainability reports by The Body Shop and Royal Dutch/Shell, *Organization and Environment*, **15**(3): 233–58.

Llewellyn, S. (1994) Managing the boundary: How accounting is implicated in maintaining the organisation, *Accounting, Auditing and Accountability Journal*, **7**(4): 4–23.

Llewellyn, S. (2003) What counts as 'theory' in qualitative management and accounting research? Introducing five levels of theorising, *Accounting, Auditing and Accountability Journal*, **16**(4): 662–708.

Luhmann, N. (1989) *Ecological Communication* (trans J. Bednarz). Cambridge: Polity.

Lukka, K. (2004) How do accounting research journals function? Reflections from the inside, in Humphrey, C. and Lee, B. (eds), *The Real Life Guide to Accounting Research*, pp. 433–48. London: CIMA/Elsevier.

Marcuse, H. (1964) *One-Dimensional Man*. Boston, MA: Beacon.

Margolis, J. D. and Walsh, J. P. (2003) Misery loves companies: rethinking social initiatives by business, *Administrative Science Quarterly*, **48**: 268–305.

Mathews, M. R. (1993) *Socially Responsible Accounting*. London: Chapman Hall.

Mathews, M. R. (1997) Twenty-five years of social and environmental accounting research: Is there a silver jubillee to celebrate?, *Accounting, Auditing and Accountability Journal*, **10**(4): 481–531.

Mathews, M. R. and Perera, M. H. B. (1991) *Accounting Theory and Development*. London: Chapman Hall.

McSweeney, B. (2002) Hofstede's model of national cultural differences and their consequences: A triumph of faith – a failure of analysis, *Human Relations*, **55**(1): 89–118.

Meadows, D. (2009) *Thinking in Systems: A primer*. London: Earthscan.

Meadows, D. H., Meadows, D. L., Randers, J. and Behrens, W. H. (1972) *The Limits to Growth*. London: Pan.

Miles, S., Hammond, K. and Friedman, A. L. (2002) Social and environmental reporting and ethical investment, *ACCA Research Report No. 77*. London: ACCA.

Milne, M. J. and Chan, C. C. (1999) Narrative corporate social disclosures: How much difference do they make to investment decision-making?, *British Accounting Review*, **31**(4): 439–57.

Milne, M. J., Kearins, K. N. and Walton, S. (2006) Creating adventures in wonderland? The journey metaphor and environmental sustainability, *Organization*, **13**(6): 801–39.

Mitchell, R., Agle, B. R. and Wood, D. J. (1997) Toward a theory of stakeholder identification and salience: defing the principle of who and what really counts, *Academy of Management Review*, **22**(4): 853–86.

Morgan, G. (1986) *Images of Organization*. London: Sage.

Morgan, G. and Smircich, L. (1980) The case for qualitative research, *Academy of Management Review*, 8(4): 491–500.

Mouritsen, J. (2006) Problematising intellectual capital research: ostensive versus performative IC, *Accounting, Auditing and Accountability Journal*, **19**(6): 820–41.

Murray, A., Sinclair, D., Power, D. and Gray, R. (2006) Do financial markets care about social and environmental disclosure? Further evidence and exploration from the UK, *Accounting, Auditing and Accountability Journal*, **19**(2): 228–55.

Neimark, M. K. (1992) *The Hidden Dimensions of Annual Reports*. New York: Markus Wiener.

Ness, K. E. and Mirza, A. M. (1991) Corporate social disclosure: A note on a test of agency theory, *British Accounting Review*, 23(3): 211–18.

Neu, D., Warsam, H. and Pedwell, K. (1998) Managing public impressions: Environmental disclosures in annual reports, *Accounting Organizations and Society*, 23(3): 265–82.

Norris, G. and O'Dwyer, B. (2004) Motivating socially responsive decision making: the operation of management controls in a socially responsive organisation, *The British Accounting Review*, 36(2): 173–96.

O'Donovan, G. (2002) Environmental disclosure in the annual report: Extending the applicability and predictive power of legitimacy theory, *Accounting, Auditing and Accountability Journal*, 15(3): 344–71.

O'Dwyer, B. (2003) Conceptions of corporate social responsibility: the nature of managerial capture, *Accounting, Auditing and Accountability Journal*, 16(4): 523–57.

O'Dwyer, B. and Owen, D. L. (2005) Assurance statement practice in environmental, social and sustainability reporting: a critical evaluation, *British Accounting Review*, 37(2): 205–30.

Odum, H. T. (1996) *Environmental Accounting: Energy and environmental decision making*. New York: Wiley.

Otley, D. T. (1980) The contingency theory of management accounting: achievement and prognosis, *Accounting, Organizations and Society*, 5(4): 413–28.

Owen, D. L. (1990) Towards a theory of social investment: a review essay, *Accounting, Organizations and Society*, 15(3): 249–66.

Owen, D. L. (2007) Assurance practice in sustainability reporting, in Unerman, J., Bebbington, J. and O'Dwyer, B. (eds), *Sustainability Accounting and Accountability*, pp. 168–83. London: Routledge.

Owen, D. (2008) Chronicles of wasted time? A personal reflection on the current state of, and future prospects for, social and environmental accounting research, *Accounting, Auditing and Accountability Journal*, 21(2): 240–67.

Owen, D., Gray, R. and Maunders, K. (1987) Researching the information content of social responsibility disclosure: A comment, *British Accounting Review*, 19(2): 169–76.

Parker, L. D. (1986) Polemical themes in social accounting: A scenario for standard setting, *Advances in Public Interest Accounting*, 1: 67–93.

Parker, L. D. (2005) Social and environmental accountability research: A view from the commentary box, *Accounting, Auditing and Accountability Journal*, 18(6): 842–60.

Parkinson, J. E. (1993) *Corporate Power and Responsibility*. Oxford: Oxford University Press.

Patten, D. M. (1992) Intra-industry environmental disclosures in response to the Alaskan oil spill: A note on legitimacy theory, *Accounting, Organizations and Society*, 17(5): 471–5.

Pearce, J. (1996) *Measuring Social Wealth: A study of social audit practice for community and cooperative enterprises*. London: New Economics Foundation.

Perera, H. (2007) The international and cultural aspects of social accounting, in Gray, R. H. and Guthrie, J. (eds), *Social Accounting, Mega-accounting and Beyond: A Festschrift in Honour of M. R. Mathews*, pp. 91–100. St Andrews: CSEAR Publishing.

Perks, R. W. (1993) *Accounting and Society*. London: Chapman & Hall.

Pfeffer, J. and Salancik, G. (1978) *The External Control of Organizations: A resource dependence perspective*. New York: Harper & Row.

Ponting, C. (1991) *A Green History of the World: The Environment and the Collapse of Great Civilizations*. London: Sinclair-Stevenson.

Power, M. (1994) Constructing the responsible organisation: Accounting and environmental representation, in Teubner, G., Farmer, L. and Murphy, D. (eds), *Environmental Law and Ecological Responsibility: The concept and practice of ecological self-organisation*, pp. 370–92. London: Wiley.

Preston, L. E. (1981) Research on corporate social reporting: Directions for development, *Accounting, Organizations and Society*, 6(3): 255–62.

Puxty, A. G. (1986) Social accounting as immanent legitimation: A critique of a technist ideology, *Advances in Public Interest Accounting*, 1: 95–112.

Rahaman, A. B., Lawrence, S. and Roper, J. (2004) Social and environmental reporting at the VRA: institutionalised legitimacy or legitimation crisis?, *Critical Perspectives on Accounting*, 15(1): 35–56.

Rahman, S. F. (1998) International accounting regulation by the United Nations: a power perspective, *Accounting, Auditing and Accountability Journal*, 11(5): 593–623.

Roberts, R. W. (1992) Determinants of corporate social responsibility disclosure, *Accounting, Organizations and Society*, 17(6): 595–612.

Roslender, R. and Dyson, J. R. (1992) Accounting for the worth of employees: a new look at an old problem, *British Accounting Review*, 24(4): 311–29.

Roslender, R., Stevenson, J. and Kahn, H. (2006) Employee wellness as intellectual capital: an accounting perspective, *Journal of Human Resource Costing & Accounting*, 10(1): 48–64.

Schumacher, E. F. (1973) *Small is Beautiful*. London: Abacus.

Schumacher, E. F. (1980) Buddhist economics, *Resurgence*, 1(11) Jan–Feb 1968, reprinted in Daly, H. E. (ed.), *Economy, Ecology, Ethics: Essays toward a steady state economy*, pp. 138–45. San Francisco, CA: W. H. Freeman.

Scott, W. R. (2004) Institutional theory: contributing to a theoretical research paradigm, in Smith, K. G. and Hitt, M. A. (eds), *Great Minds in Management: The Process of Theory Development*. Oxford: Oxford University Press (http://www.si.umich.edu/ICOS/Institutional%20Theory%20Oxford04.pdf).

Shamir, R. (2004) The de-radicalization of corporate social responsibility, *Critical Sociology*, 24: 669–90.

Shearer, T. (2002) Ethics and accountability: from the for-itself to the for-the-other, *Accounting, Organizations and Society*, 27(6): 541–74.

Shenkin, M. and Coulson, A. B. (2007) Accountability through activism: learning from Bourdieu, *Accounting, Auditing and Accountability Journal*, 20(2): 297–317.

Smircich, L. (1983) Concepts of culture and organisational analysis, *Administrative Science Quarterly*, 28: 339–58.

Spence, C. (2009) Social accounting's emancipatory potential: Gramscian critique, *Critical Perspectives on Accounting*, 20(2): 205–27.

Spence, C. and Gray, R. (2008) *Social and Environmental Reporting and the Business Case*. London: ACCA.

Thielemann, U. (2000) A brief theory of the market – ethically focused, *International Journal of Social Economics*, 27(1): 6–31.

Thomas, A. P. (1986) The contingency theory of corporate reporting: some empirical evidence, *Accounting, Organizations and Society*, 11(3): 253–70.

Thomson, I. and Bebbington, J. (2005) Social and environmental reporting in the UK: a pedagogic evaluation, *Critical Perspectives on Accounting*, 16(5): 507–33.

Tinker, A. M. (ed.) (1984a) *Social Accounting for Corporations*. Manchester: Manchester University Press.

Tinker, A. M. (1984b) Theories of the state and the state of accounting: economic reductionism and political voluntarism in accounting regulation theory, *Journal of Accounting and Public Policy*, 3: 55–74.

Tinker, A. M. (1985) *Paper Prophets: a social critique of accounting*. Eastbourne: Holt Saunders.

Tinker, A. M. and Okcabol, F. (1991) Fatal attractions in the agency relationship, *British Accounting Review*, 23(4): 329–54.

Tozer, L. and Hamilton, F. (2007) Re-energising the social contract in accounting: The case of James Hardie Industries, in Gray, R. H. and Guthrie, J. (eds), *Social Accounting, Mega-accounting and Beyond: A Festschrift in Honour of M. R. Mathews*, pp. 107–26. St Andrews: CSEAR Publishing.

Tregidga, H. M. and Milne, M. J. (2006) From sustainable management to sustainable development: A longitudinal analysis of a leading New Zealand Environmental reporter, *Business Strategy and the Environment*, 15(4): 219–41.

Tricker, R. I. (1983) Corporate responsibility, institutional governance and the roles of accounting standards, in Bromwich, M. and Hopwood, A. G. (eds), *Accounting Standards Setting – An International Perspective*, pp. 27–41. London: Pitman.

Ullmann, A. E. (1985) Data in search of a theory: a critical examination of the relationships among social performance, social disclosure and economic performance of US firms, *Academy of Management Review*, 10(3): 540–57.

Unerman, J. and Bennet, M. (2004) Increased stakeholder dialogue and the internet: towards greater corporate accountability or reinforcing capitalist hegemony?, *Accounting, Organizations and Society*, **29**(7): 685–707.

Unerman, J., Bebbington, J. and O'Dwyer, B. (eds) (2007) *Sustainability Accounting and Accountability*. London: Routledge.

United Nations Environment Programme (1972) *Declaration of the United Nations Conference on the Human Environment*. Stockholm: United Nations.

United Nations World Commission on Environment and Development (1987) *Our Common Future (The Brundtland Report)*. Oxford: Oxford University Press.

von Bertalanffy, L. (1971) *General Systems Theory: Foundations, Development, Applications*. Harmondsworth: Penguin.

Ward, B. and Dubos, R. (1972) *Only one Earth: The care and maintenance of a small planet*. Harmondsworth: Penguin.

Wilber, K. (2000) *A Theory of Everything: An integral vision for business, politics, science and spirituality*. Dublin: Gateway.

Williamson, O. E. (1979) Transaction-cost economics: the governance of contractual relations, *Journal of Law and Economics*, **22**(2): 233–61.

Young, W. and Tilley, F. (2006) Can business move beyond efficiency? The shift toward effectiveness and equity in the corporate sustainability debate, *Business Strategy and the Environment*, **15**: 402–15.

Zimmerman, M. E. (1994) *Contesting Earth's Future: Radical ecology and postmodernity*. London: University of California Press.

CHAPTER 5

Social and community issues

5.1 Introduction

The interactions between an organisation and society are many and complex and we take many of them for granted. This is one of the many reasons why accounting (which guides the economic elements of many organisational decisions) and social accounting procedures (which might seek to introduce non-economic issues into decision-making and try to hold organisations to account) are so important.

A systems view (see Chapter 2) allows us a way of thinking about – and perhaps then organising our thoughts around – the social and community issues of organisational life and the associated accounting and accountability. We introduced some of the basic ideas involved here in Chapter 2. Section 5.2 introduces the ideas of the stakeholders. This may be a simplifying notion, but it is a useful mechanism to help us organise our thoughts.

Thinking of a small organisation – say, a bicycle repair shop or a community hall for example – one can immediately imagine the sorts of stakeholders that such an organisation will have. It will have financial stakeholders – perhaps shareholders or trustees or partners or the funding agency or other ownership structure and possibly the bank (see Chapter 8). It will have an implicit stakeholder in the form of the environment which it will use, belong to and have impact upon (see Chapter 7). It may very well also have employees (see Chapter 6).

In addition, the organisation will have a wide range of other stakeholders, including those who use its services, and perhaps we would think of these as customers. Through the use of the products and services, lives are changed and possibilities enhanced or reduced. Then there are those who supply and support it with goods and services – the suppliers and the supply chain itself. Where the organisation chooses to obtain its inputs and the speed with which it pays its bills, for example, all have small but important impacts elsewhere. (Think of movements like Fairtrade and Foodmiles and you start to get the picture.) Then, of course, there are the far more important (although much more elusive) impacts that the organisation will have on the immediate community around it and – in all probability – on a wider community through its interaction with (e.g. local) politics, clubs, associations, advertising and so on. How might the organisation account for all of this – and for what, if anything, might we consider the organisation accountable? These are not easy questions – indeed, it is probably as good a time as any to recognise that it is probably impossible to produce a wholly complete account equally relevant to all sections of society. We can, as we shall see, however, get close.

Size is crucial when considering organisational stakeholders, and we will return to these issues once we have expanded this simple image a little. Imagine how complex our brief analysis will become when we start to consider medium-sized companies operating in a number of different locales; or a local government and its impacts; or a large non-governmental organisation (NGO) whose members and campaigns are spread all over the world. We would need to begin to recognise a range of regional issues, including such things as local employment concerns, political sensitivities in the area and issues to do with the relative well-being (or otherwise) of those living in the region.

And, of course, it becomes more complex still when we look at the multi-national corporations (MNCs). They are of a size that can swamp the countries in which they operate. More especially, MNCs which are based in (say) the Netherlands or Japan can be operating in countries which have little direct involvement with that company and its products, services and operations. Equally, as is increasingly apparent, organisations – and especially large organisations – have an immense influence on government policy in both host and home countries and, consequently, can be amongst the most influential entities in an area, a region or a nation – regardless of the product or service with which they are associated (Bailey *et al.*, 1994; Korten, 1995; Rahman, 1998; Banerjee, 2007).

In this chapter, we begin to try and address some of this complexity. The chapter is organised as follows. The next section looks at how we might think about what is meant by 'society' before we move on to look, in Section 5.3, at some of the developments and trends in social accounting and reporting. Section 5.4 briefly explores the organisation's own point of view in these matters and we contrast that, in Section 5.5, with the stakeholder view and raise the challenging issues of stakeholder engagement. Section 5.6 undertakes an examination of community, philanthropy and the role of NGOs with an especial focus on the concept of corporate community investment. In Sections 5.7 and 5.8 we touch upon a number of the enormous but very challenging issues that raise their heads when we consider the wider context of (especially) MNC operations. These later sections concentrate on, respectively, lesser developed countries and human rights. This brings us to a concluding section where we re-iterate the complexities of trying to understand the business–society relationship and introduce some of the limitations to our discussion. Indeed, it is crucial to note from the outset that there are as many ways of approaching the notion of society as there are worldviews. For instance, this chapter could have had a much stronger emphasis on such central matters as conflict (as the political science literature might do) or ethics (as the business ethics and management literature might emphasise) and/or the role and emancipation of women (as many literatures outside accounting might emphasise). All of these appear throughout this chapter (and, indeed, in other chapters) and are key themes rather than central *motifs*. It is for the reader to select what is, for them, the essential matter(s) through which an understanding of society should be framed.

5.2 Society? Social issues? Stakeholders?

If *social* accounting is about anything, it is probably about society and social issues. Many of the early approaches to social accounting took a relatively simple view of what was meant by society and social issues and, to a degree at least, the tendency to equate these terms with corporate social responsibility (CSR) and stakeholders continues that simplification. Does this matter? Well, to a degree it does. The holistic view we try to initiate in **Chapter 2** gives us a sense of the importance of trying to maintain the 'big picture' whilst simultaneously dealing with details. A review of worldviews would show that there are then different ways of understanding the notions of society and social issues (see Chapter 3). A review of theories (see Chapter 4) would identify, in part at least, how different perspectives embrace, emphasise or exclude different elements of the world with which we are concerned.[1]

---

[1]This is a good point to emphasise that different academic disciplines approach and frame these issues often very differently. Our intention is to try to synthesise the literature from both the accounting + finance and the business + management (including business ethics) literatures. In broad terms, whilst both have critical branches in their literatures, the latter has a much deeper and longer (if managerial) engagement with social responsibility whilst accounting would, to the extent that it cares at all, rather emphasise reporting issues. There is a real danger that the literatures might sometimes fail to talk to each other and learn from each others' experience. This would be genuinely unfortunate.

Inevitably then, most discussions and approaches are likely to be partial, but one way to guard against this is to adopt a concept of 'society' that minimises that risk. One such concept is that derived from Gramsci which articulates society as comprising: the **state**; the **market**; and **civil society** (Abercrombie *et al.*, 1984; Bendell, 2000a; Moon, 2002; Vogel, 2005). The state comprises the government, government bodies and other central and local government functions: the public sector if you will. The market comprises private commercial activities and therefore is dominated by companies, businesses, banks and the whole panoply of finance. Civil society is the rest – people as individuals, families, community groups, charities and other NGOs and, under most specifications, religious bodies. Our focus, as in social accounting generally, as we have said, tends to be on organisations and the organisations we mainly focus on are in the 'market' – companies and other businesses. The market focus also takes in suppliers and customers: either as other organisations or as individuals from civil society who enter the market through economic relationships with the organisation of concern. Other organisations are of interest in both the state sectors (e.g. local government – see, for example, Ball and Osborne, 2011) and in civil society (notably NGOs – see, for example, Unerman and O'Dwyer, 2006) although less attention is given to them in social accounting (but see Chapter 12). From a holistic and neo-pluralist perspective we can see organisations as interacting with other elements of the market (economic interactions with social consequences), the state (social, political and economic interactions) and civil society (mostly social interactions). It is these interactions which give us our social issues.

But two things require immediate notice.

- *First*, this conception makes no especial mention of the 'environment' (see Chapter 7). Equally, although we might suggest that employees and the world of 'work' might be thought of as interaction between civil society and the organisation, this is only one such conception and it is traditional to look at employees, trades unions and employment as a separate (although clearly related) issue (see Chapter 6).

- *Second*, the notion of 'social issues' is potentially infinite. The range of issues is a function of how we articulate them (Henriques, 2010). We might want to include: wealth, oppression, children, education, poverty, housing, consumption, well-being, happiness, . . . the list goes on and on (see Figures 5.6 and 5.9 below). Consequently, thinking of social issues in terms of CSR and stakeholders becomes that much more attractive: and while these are not perfectly substitutable notions, they will broadly serve our purpose here. So any reasonably large organisation will have a complex range of social stakeholders, including finance providers, employees, customers, suppliers, the state, communities, other governments, the media, civil society as a whole, . . . and so on. We deal with these to varying degrees in this chapter (and throughout the text).

5.3 Developments and trends in social reporting and disclosure

As we have already suggested, the reporting of non-financial social information by organisations has a long history: both Guthrie and Parker (1989) and Maltby (2005) find illustrations in the late 19th and early 20th centuries. However, the emergence of **social reporting** and **social disclosure** as recognised phenomena is typically dated from the 1960s as a direct outgrowth of the increasing concern over CSR that emerged (especially in the USA) after the Second World War. Whilst employee/employment and the natural environment were key elements in this disclosure, the predominantly social concerns in these early years tended to be around community involvement, consumers and some broad (but ill-defined) notion of total engagement with society.

Buhr (2007) actually identifies the 1970s as the decade of *social* reporting, and there is little question that interest in the field died in the 1980s and was subsequently swamped by environmental, triple bottom line (TBL) and sustainability (sic) reporting (see Chapter 9). As Buhr (2007) goes on to note, the international accounting firm, Ernst and Ernst (as it then was), produced a regular survey of USA corporate reporting from 1971 to 1978. This plotted the developments in the (largely voluntary) reporting by large companies in their annual reports of such matters as fair business practices, community involvement and products. Over this period, the levels of disclosure grew so that by the late 1970s about 90% of the Fortune 500 had some social disclosure in their reports. However, generally speaking, this disclosure would typically only relate to a single item and cover less than half a page of the report (Ernst & Ernst, 1971 *et seq*; Gray *et al.*, 1987). Broadly similar patterns were reported across Europe (Lessem, 1977; Preston, 1978; Brockhoff, 1979; Schreuder, 1979), Australia (Trotman, 1979), New Zealand (Robertson, 1978), India (Singh and Ahuja, 1983) and Malaysia (Teoh and Thong, 1984), for example, although Estes (1976) was not alone in condemning much of this reporting as being 'incomplete, defensive[,] . . . sprinkled with propaganda . . . and blatantly self-serving' (p. 55). It is difficult to argue that much has changed really.

That aside, in addition to the 1970s being the period in which social reporting began to enter the mainstream, the decade was of especial interest for the experimental social reports that they engendered. These experimental reports embraced a range of non-financial approaches to reporting (see Figure 5.1) as well as a range of early attempts to integrate financial information with social data (see Figure 5.2).[2]

Instances of innovative, experimental, social (and other) accounts are not confined to the early years of social accounting and they continue to the present. So, for example, the 1990s also saw an impressive range of innovative experiments including such gems as the UK's Traidcraft Exchange 1996 accounts (which are also on the CSEAR website), the North American ice cream gurus, Ben and Jerry's *Social Assessment* in the 1990s, the Danish SbN Bank's *Ethical Accounting Statement* and the Canadian VanCity Credit Union social audits.

In the main, social reporting has continued to develop – initially through the annual report and then, increasingly, through the organisation's websites and the production of (what have become known as) standalone reports. As with most voluntary reporting, the topics covered vary over time as issues like (for example) apartheid, community investment, human rights, dealing with repressive regimes, ethical supply chains, internal governance arrangements and engagement with stakeholders have been perceived as more or less important (Gray *et al.*, 1995; SustainAbility 1996 *et seq*; Palenburg *et al.*, 2006; Pricewaterhouse, 2010). But by far the biggest trends in reporting over the years have been (a) the rise and fall of employee and employment reporting (Roberts, 1990; Adams *et al.*, 1995; Adams and Harte, 1998; Adams and McPhail, 2004) during the 1980s; (b) the emergence and dominance of environmental reporting in the 1990s; and then (c) the development of the (so-called) TBL and sustainability reporting this century. Inevitably, any demarcation lines between (say) social, environmental, sustainability and employee reporting have blurred considerably and organisations may label their non-financial reporting under headings as diverse as sustainable development, citizenship, social responsibility, report to society, and so on (KPMG, 2002, 2005, 2008). Social reporting – as distinct from environmental and sustainability reporting – is not that common and, somewhat strikingly, Kolk (2003) reported that in her study of 250 of the world's largest companies only 33 actually reported on social issues.

---

[2]There was a plethora of such reports – especially in the USA. Consult Estes (1976), Johnson (1979), Gray *et al.* (1987, 1996) and the CSEAR website 'Approaches to Practice' for further examples and illustrations. Linowes and other financial approaches are reconsidered in Chapter 9.

Figure 5.1 Summary of the Atlantic Richfield Company's social report, 1977

ASSETS	LIABILITIES
Minority Affairs	Most minorities and women who work for Atlantic Richfield hold low-level jobs. There is not a single black or female officer.
Atlantic Richfield has worked hard to provide job opportunities for minorities. Minority group members account for 13% of the total work force, a ratio that ranks Atlantic Richfield at the top of the petroleum industry.	
	More than 70 major U.S. companies have elected blacks to their boards of directors. Many have also named women directors. The petroleum industry has resisted this trend – and so has Atlantic Richfield. Its board is all-white, all-male, all-Christian.
Jobs formerly restricted to men – such as refinery work – have been opened up to women.	
The number of minorities and women in professional, managerial and sales positions has nearly doubled since 1970.	The Company has not been aggressive or innovative in its support of minority enterprise. Standard Oil of Indiana, for example, requires its purchasing agents to set goals and goes out of its way to help fledgling companies. Result: Indiana Standard spends four or five times what Atlantic Richfield spends in purchases from minority suppliers.
To aid minority economic development, Atlantic Richfield maintains deposits of over $1 million in minority-owned financial institutions across the country.	
Atlantic Richfield reported purchases of $3.2 million from minority suppliers in 1974. This was double its 1973 purchases.	
Contributions	
Its $5.5 million of charitable contributions in 1974 supported a large number of educational, health and cultural organisations in the United States.	To encourage charitable contributions, the Internal Revenue Service allows corporations a deduction of up to 5% on pretax profits. At least two companies – Dayton Hudson and Cummins Engine – takes this full deduction. Other companies – Aetna Life & Casualty, for example – have sharply increased their giving. Atlantic Richfield gives away 1.3% of pretax profits.
Atlantic Richfield matches, dollar for dollar, employee contributions to educational institutions.	
One unusual grant in 1974 was $10,000 to the Council on Economic Priorities, an organisation that monitors corporate social responsibility.	The pattern of Atlantic Richfield's giving is in the traditional mould, with most money going to old-line, established institutions. Of the $850,000 committed to education in 1973, for example, more than a quarter went to one school, the Massachusetts Institute of Technology.
Community organisations backed by Atlantic Richfield Foundation grants include the Boy Scouts, YMCA, Junior Achievement, Urban Coalition, American Red Cross, Salvation Army and Urban League.	Black colleges receive only minimal support.
Shareholder Information	
The firm's Form 10K financial report, which contains more detailed information than the annual report and which all corporations must file with the Securities and Exchange Commission, was offered free of charge to all shareholders in 1972 and 1973.	The Company's annual report has been niggardly in providing meaningful details of pollution control programmes or specific information about social responsibility activities. The tendency has been to substitute rhetoric for hard data. Shell Oil Company has consistently released far more information.

Figure 5.1 (*continued*)

ASSETS	LIABILITIES
Environment and Conservation	
Atlantic Richfield was the first company in the petroleum industry to announce that it would make a lead-free gasoline.	Atlantic Richfield was slow to comprehend the environmental problems connected with the Alaskan pipeline and for too long resisted protection measures later incorporated into the project.
In the interests of what it called 'America's natural beauty', the Company in 1972 cancelled its entire out-door advertising – 1,000 billboards in 36 states.	The Company, while paying its respects to the conservation ethic in solving our energy problems, persists in the view that more development and more growth can solve our energy problems.
Its Cherry Point refinery in the state of Washington has been recognised as a model nonpolluter.	
It has emphasised energy conservation in its own operations.	
Consumerism	
It was one of the first companies in the petroleum industry to post the octane levels of its gasolines at the pump.	At many US companies the concept of social responsibility has been institutionalised at least to the extent that new positions and/or committees have been created, some of them with high standing in the table of organisation. Atlantic Richfield has floundered through a series of organisational re-shuffles, with the social responsibility functions still scattered, relegated to lower levels of the Company and concerned largely with peripheral areas outside the mainstream activities.
Social Management	
The Company's public affairs programme in Alaska is outstanding, far surpassing any comparable effort by Atlantic Richfield in the lower 48 states both in the range and depth of activities. The Company has made its presence felt in Alaska as a concerned corporate citizen.	

CONCLUSION

As the youngest of the petroleum giants, the Company carries less baggage from the past. As a company still in transition, it is more conscious that its future lies ahead. And that is perhaps what is most hopeful; it is a company not yet fully formed. When oil from Alaska begins to flow and Atlantic Richfield becomes even bigger than it is today, it will have a splendid opportunity to demonstrate that social concerns can be built into the day-to-day operations of a petroleum company. More than most giant companies, it has its future in its hands. It need not relive or repeat the mistakes of the past.

Source: Participation II, Atlantic Richfield Company, (1977). Used with permission.

Attempts to make organisational social reporting more systematic have had varying degrees of commitment behind them and have achieved varying degrees of success over the years and across countries. Although research has identified a range of complex and diverse processes that encourage organisations to embrace forms of social responsibility, there is no very clear picture over which actually dominate and/or manifest themselves in a commitment to reporting (Hess *et al.*, 2002; Moon, 2002; den Hond and de Bakker, 2007; Matten and Moon, 2008).[3] What is absolutely clear is that the regulation and codification of social

---

[3] Again, here we might recognise the differences between the focus of the accounting and finance literature on the practice of reporting and the greater concern of the business and management literature with internal process and pressures.

Figure 5.2 Socio-economic operating statement

X Corporation

Socio-economic operating statement for the year ending December 31 19X1

I Relations with people

A Improvements

	1 Training programme for handicapped workers	$10,000
	2 Contribution to educational institution	4,000
	3 Extra turnover costs because of minority hiring programme	5,000
	4 Cost of nursery school for children of employees voluntarily set up	11,000
	Total improvements	$30,000

B *Less* Detriments

	1 Postponed installing new safety devices on cutting machines (cost of the devices)	$14,000

C Net improvements in people actions for the year $16,000

II Relations with environment

A Improvements

	1 Cost of reclaiming and landscaping old dump on company property	$70,000
	2 Cost of installing pollution control devices on Plant A smokestacks	4,000
	3 Cost of detoxifying waste from finishing process this year	9,000

A Total improvements $83,000

B *Less* Detriments

	1 Cost that would have been incurred to relandscape strip-mining site used this year	$80,000
	2 Estimated costs to have installed purification process to neutralise poisonous liquid being dumped into stream	100,000
		$180,000

C Net deficit in environment actions for the year $97,000

III Relations with product

A Improvements

	1 Salary of vice-president while serving on government Product Safety Commission	$25,000
	2 Cost of substituting leadfree paint for previously used poisonous lead paint	9,000
	Total improvements	$34,000

B *Less* Detriments

	1 Safety device recommended by Safety Council but not added to product	$22,000

C Net improvements in product actions for the year $12,000

Total socio-economic deficit for the year	$69,000
Add Net cumulative socio-economic improvements as of January 1, 19X1	$249,000
Grand total net socio-economic actions to December 31, 19X1	$180,000

Source: Linowes, D.F., An approach to socio-economic accounting, *Conference Board Record*, November 1972, p. 60. Reproduced by permission.

reporting is in no sense comparable with that exerted over financial reporting – which probably tells us something important. Beyond the general observation that many countries require their large companies to disclose information about employees, there is relatively little systematic regulation of social information disclosure. Whilst, for example, India and the UK[4] have very general requirements that corporations discuss their social (and environmental) performance and large Canadian banks are required to explain their contribution to society, the more detailed disclosure requirements laid down on French companies have generally been the exception (KPMG, 2008; Palenburg *et al.*, 2006 – although see KPMG *et al.*, 2010). Unfortunately, Kuasirikun and Sherer (2004) are not alone in finding that companies typically fail to comply with what little regulated disclosure there is.

By comparison with the seemingly lukewarm interest by governments in regulating social information, non-governmental bodies have been much more enthusiastic – although whether any more successful remains a moot point. Without question, the most consistent effort to bring system and regulation to the corporate world has been from the United Nation (UN). We will return to the UN below when we look at the role of MNCs in the lesser developed countries and the emergence of the UN Global Compact (UNGC). (We also meet the UN in **Chapters 7, 9 and 12**, when we touch upon its attempts to regulate environmental reporting and sustainability.) There is much that could be said about the serious and professional attempts by UN staff and delegates to develop proper accountability for social issues but, as Rahman (1998) so clearly shows, these attempts consistently failed in the face of opposition from (mainly) G7 countries and large MNCs. In this connection, Kamp-Roelands' (2009) review of her committee's largely derailed attempts to agree on and then apply a set of CSR indicators makes for poignant reading.

More successful, at least on the surface, has been the Global Reporting Initiative (GRI). This multi-stakeholder approach to establishing a framework of generally accepted reporting principles for environmental, social and sustainability reporting (similar to generally accepted accounting principles for financial reports) had been adopted by well over 1000 organisations worldwide by 2010[5] (Leipziger, 2010; Gray and Herremans, 2012). That this was a very small proportion of all the world's organisations – even of the world's largest organisations – illustrated the challenges of getting companies to adopt even a well-considered framework (Milne *et al.*, 2006, 2009; Milne and Gray, 2007).

Research into patterns of social reporting (as indeed, into patterns of environmental and sustainability reporting as we shall see later) has confirmed time and time again that voluntary reporting by companies is consistently related to the organisation's size, the industry in which it functions and the country of the company. For a variety of reasons, the larger companies are more likely to voluntarily produce more social information. Similarly, industry affiliation influences the likelihood and extent of reporting with, in general, those which are closer to the final customer tending to adopt disclosure[6] (Hackston and Milne, 1996; Al-Najjar, 2000; Cormier and Gordon, 2001; Brammer and Pavelin 2006). Research into the variation in disclosure between countries is not quite as well developed (Guthrie and Parker, 1990; Adams *et al.*, 1998; KPMG, 2008), although the countries that seem to lead in disclosure are not always the most obvious ones,[7] but culture, practice and attitudes of the state

---

[4]The UK also has requirements that companies disclose their political and charitable donations. This is a unique requirement as far as we are aware.

[5]The Guidelines are styled *Sustainability Reporting Guidelines*. This matter is considered in Chapter 9.

[6]The relationships with environmental disclosure are more related to perceived impact on the environment.

[7]So, for example, Japan was amongst the leading standalone reporters for many years and for a long time Spain had the largest number of reporters under GRI whilst China's reporting practice is altogether more difficult to interpret (Du and Gray, 2013).

and civil society all seem to play a part (Williams and Ho Wern Pei, 1999; Adams, 2002; Kolk, 2003, 2008). But it is Adams (2002) who particularly suggests that this research, which tends towards something of a black box approach, fails to capture the subtlety of the disclosure decisions by companies and that researchers need to spend more time inside the organisations to understand their motivations for social accounting and reporting. This is what we look at in the next section.

5.4 From the organisation's point of view

At the heart of the debate over what is meant by CSR is the essential conundrum of whether or not it is any business of business? Whilst the acts of social responsibility and/or philanthropy and/or the decision to undertake voluntary social reporting can be acts of genuine citizenship undertaken by an organisation or even the personal initiative of key individuals within the organisation, it is very unclear whether large, especially quoted, organisations have the moral or economic freedom to behave in this way unless such acts are clearly in the business interests of the organisation itself (Lantos, 2001; Margolis and Walsh, 2003; Blowfield and Murray, 2008). Thus, no matter what accountability civil society might want or might be owed, the firm itself has to march to a more pragmatic drum. Adams (2008) and Adams and Whelan (2009) see the adoption of both CSR and social reporting as strategic choices which are themselves part of the organisational strategy reflecting social and environmental objectives, targets and outcomes; risk management; stakeholder engagement; governance; and so on. Manifested in such matters as the employment of women, the employment of racial minorities, engagement with communities, the firm's posture on ethical matters and so on, the research literature continues to identify and explore those complex drivers of organisational reaction (Moon, 2002; Vogel, 2005).

It follows therefore that CSR and social reporting have to fit the logic of the organisation and sit comfortably in one form or other of a 'business case' (Spence and Gray, 2008). It is a truism therefore to say that such reporting must be (on the whole) managerialist, marginalist, predominantly in the interests of the organisation and understood differently by different organisations and by different industries (Wood, 1991; Herremans et al., 2008). How organisations understand their CSR and how they understand their stakeholders and the firm's effects on them will, in turn, determine how they report (Adams, 2008).

But *how* the firm understands its CSR and/or the purpose and function of its social reporting is, itself, a complex matter. Not only do the issues that civil society might hold as relevant to the organisation have to be translated into institutional logics, but they are then further mediated through national and international business bodies like the International Chamber of Commerce or the World Business Council for Sustainable Development and even the GRI. The message from these bodies is then mediated again into the sense-making and business case for the organisation itself (see, for example, Gray and Bebbington, 2000; Angus-Leppan et al., 2009). Add into the mix the need to sift through (what must appear to be) an almost infinite array of data on potential issues and the need to provide a reliable basis for the firm's own reporting and it is little wonder that understandings of social reporting and CSR vary so much and that such a gap exists between the desires of civil society and the actions of the corporations (Gray and Herremans, 2012).

Mediation by international bodies should not be underestimated. Not only did the GRI have a major influence on determining (misguidedly in our view) how organisations made sense of social, environmental and sustainability issues, but CSR itself is subject to international interpretation. For example, the European Union published a Green Paper on CSR in 2001 (European Commission, 2001) and defined CSR broadly sufficiently to engage all interested parties. Responses to the Green Paper provided a stark illustration of conflicting stakeholder views as, for instance, the Confederation of British Industry (CBI) was critical of its focus on engagement with employees and unions on matters such as work–life balance, equal opportunities and life-long learning. Trade unions and NGOs were, not surprisingly, strongly supportive (see Burchell and Cook, 2006). At least as influential was the issuance of the International Standards Organisation (ISO) guidance document 26000 concerning organisations and their adoption/management of social responsibility (ISO, 2010).[8] ISO 26000 'aims to be a first step in helping all types of organization in both the public and private sectors to . . . achieve the benefits of operating in a socially responsible manner'. And the adoption of CSR 'can influence, among other things: Competitive advantage [and] Reputation'.[9] The standard offers what O'Riordan and Fairbrass (2008) call a 'practical framework for CSR executives who face the challenge of responding in an effective manner to stakeholders' (p. 745). That organisations need guidance is not in question but, as we shall see below, it is not obvious that stakeholders are at all convinced that the two key principles of responsibility and accountability espoused by the standard have been much advanced in practice (Moratis and Cochius, 2011).

We will touch upon other attempts to help organisations adjust to CSR – such as AccountAbility's AA1000 standards and the UNGC – later. However, little of this guidance seems to recognise – as Adams and McNicholas (2007) and Adams and Whelan (2009) demonstrate – that an organisation addressing social reporting is undergoing what can be a very substantial degree of change. It is only through an understanding of the change process – or, more usually, the impediments to change – that organisations and those who work with them can really address the challenges of CSR and social reporting.

What it often all comes down to is 'why would an organisation change in order to adopt a practice whose benefits might appear elusory?' The key point from the perspective of most organisations is that their task is to manage the expectations and perceptions of the stakeholders in such a way as to ensure the successful continuation of the organisation itself (Polonsky and Jevons, 2006; den Hond and de Bakker, 2007). (This is quite the opposite of what we saw as the *normative* approach to stakeholder theory in which the preferences of civil society prevail.) Under these circumstances, the key point is for the organisation to only consider responsibility more seriously as it applies to the *salient* stakeholders – those which can most significantly influence the business (Mitchell *et al.*, 1997). Such a view must potentially move the organisation away from areas of philanthropy and/or activities which have no discernible benefit to the organisation – i.e. those which do not easily fit the business case. How the 'business case' is actually conceptualised is a whole other question (Spence and Gray, 2008), but it can certainly, under the right circumstances, transcend simple calculations of accounting profit or return. In fact, how organisations begin the process of developing their traditional control systems (including of course their accounting systems) so that more enlightened views towards CSR and social responsibility can prevail is an important – if resistant – issue. As far as we can tell, it seems to depend upon the culture of the organisation and its top management and the amount of freedom from financial markets it can maintain (Norris and O'Dwyer, 2004; Dey, 2007; Durden, 2008). Indeed organisations undertake social reporting for very many and complex reasons, but there is still much to be done to ensure that organisations are taking this as far as they can and that research is as supportive as it can be – and needs to be (Adams, 2002).

---

[8] The ISO have also been very influential on the matter of environmental management systems (see Chapter 7).
[9] http://www.iso.org/iso/iso_catalogue/management_and_leadership_standards/social_responsibility/sr_discovering_iso26000.htm (sampled 24 August 2011).

Indeed, Adams (2002) argues that we need to get beyond the current inadequate attempts to explain the reasons behind corporate voluntary disclosure and start to recognise more nuanced understandings that start to reflect the complexities that research in the field has revealed. Her suggestions are further developed in Figure 5.3. Figure 5.3 is derived from a wide review of both the accounting and the business/management literatures. The sheer diversity of potential influences on both the adoption and posture of social responsibility and the likely approach to reporting serve to illustrate how relatively underdeveloped is our understanding of organisational motivations in this field.[10]

5.5 Stakeholders' views

Perhaps one of the most striking developments in social reporting around the turn of the 21st century was the increasing attention demanded for – and given to – the views of the stakeholders. Although not everybody was convinced by this development (Owen *et al.*, 2000, 2001), any organisation which wished to understand CSR and which intended to take its social responsibility and social reporting seriously was advised to 'engage' its stakeholders in dialogue. Key to this development was the work of the independent organisation, AccountAbility,[11] whose AA1000 series of standards set the benchmark for stakeholder engagement and dialogue. In essence, it was argued that any organisation would come to better understand and respond to its stakeholders through systematic discussion and the regular seeking of views. This makes obvious sense but, even more than this, it was suggested that stakeholders and organisations would come to better understand each other through this process and there would follow – in an ideal world – a convergence of interests and needs. AccountAbility (1999) provided guidance for both organisations and societal groups about what would constitute robust dialogue (emphasising the importance of listening, understanding and responsiveness). The *AA1000 Stakeholder Engagement Standard* (AccountAbility, 2008) further developed the requirements for quality stakeholder engagement. It proposed that stakeholder engagement should be bound by three principles: **materiality** (knowing the stakeholders' and the organisation's material concerns), **completeness** (understanding stakeholder concerns, that is, views, needs and performance expectations and perceptions associated with their material issues) and **responsiveness** (coherently responding to stakeholders' and the organisation's material concerns).

We mentioned earlier the degree of change that embracing CSR and social reporting might require in an organisation. Adams and Whelan (2009) and Adams and McNicholas (2007) go as far as to suggest that there is the potential for stakeholder dialogue and stakeholder lobbying to act as the stimulus and catalyst for actually initiating that process of change. Although consulting stakeholders will (inevitably really) expose clashes between stakeholder views and needs, there is some evidence that it can be a successful vehicle for changing company practices (Burchell and Cook, 2006). The question of the extent of such change and whether such change goes deeply enough into the organisation core is another question entirely (Laughlin, 1991).

There is a long history of sensible organisations consulting with their stakeholders (see, for example, the First National Bank of Minneapolis social audits; Epstein *et al.*, 1977). Equally, the research literature has a long history of providing evidence that organisations

---

[10]One key theme here is that the explanations offered in the accounting literature could probably learn a great deal from a wider recognition of the research carried out in wider social science literatures.

[11]More (including the actual text of the standards) can be found on the organisation's website at http://www. accountability.org/.

Figure 5.3 Diagrammatic portrayal of the influences on corporate social responsibility and sustainability reporting

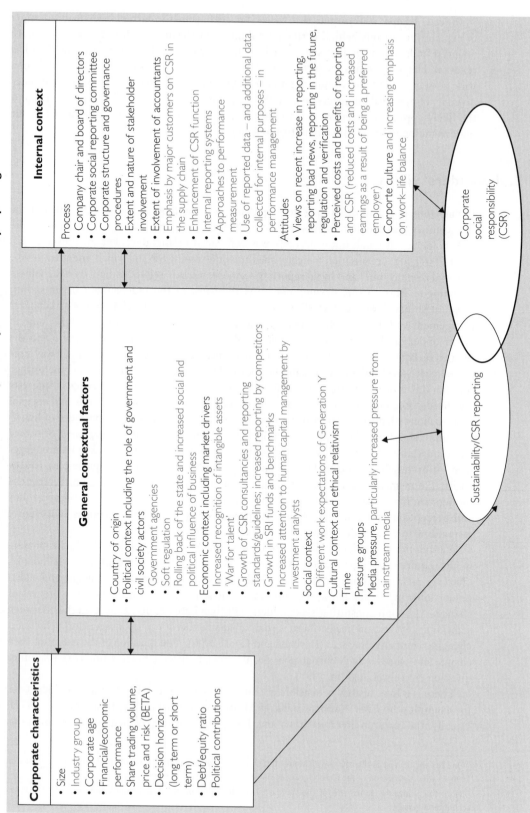

Corporate characteristics

- Size
- Industry group
- Corporate age
- Financial/economic performance
- Share trading volume, price and risk (BETA)
- Decision horizon (long term or short term)
- Debt/equity ratio
- Political contributions

General contextual factors

- Country of origin
- Political context including the role of government and civil society actors
  - Government agencies
  - Soft regulation
  - Rolling back of the state and increased social and political influence of business
- Economic context including market drivers
  - Increased recognition of intangible assets
  - 'War for talent'
  - Growth of CSR consultancies and reporting standards/guidelines; increased reporting by competitors
  - Growth in SRI funds and benchmarks
  - Increased attention to human capital management by investment analysts
- Social context
  - Different work expectations of Generation Y
  - Cultural context and ethical relativism
  - Time
  - Pressure groups
  - Media pressure, particularly increased pressure from mainstream media

Internal context

Process
- Company chair and board of directors
- Corporate social reporting committee
- Corporate structure and governance procedures
- Extent and nature of stakeholder involvement
- Extent of involvement of accountants
- Emphasis by major customers on CSR in the supply chain
- Enhancement of CSR function
- Internal reporting systems
- Approaches to performance measurement
- Use of reported data – and additional data collected for internal purposes – in performance management

Attitudes
- Views on recent increase in reporting, reporting bad news, reporting in the future, regulation and verification
- Perceived costs and benefits of reporting and CSR (reduced costs and increased earnings as a result of being a preferred employer)
- Corporate culture and increasing emphasis on work–life balance

Corporate social responsibility (CSR)

Sustainability/CSR reporting

Source: Adapted from Adams, (2002).

Notes: Arrows show direction of influence. Light grey text shows links between drivers for corporate social responsibility and corporate social disclosure in some cases using diversity issues as a specific example.

fail to supply the information needs of external and internal participants (see, for example, Benjamin and Stanga, 1977; Ingram and Frazier, 1980; Dierkes and Antal, 1985; Epstein, 1991, 1992; Tilt, 1994; Deegan and Rankin, 1999). As stakeholder consultation became more visible, further attempts to establish stakeholder information needs were undertaken and came to much the same view (see, for example, Adams and Kuasirikun, 2000; Adams, 2004; Thomson and Georgakopoulos, 2008; Benn *et al.*, 2009). The highly publicised SustainAbility/UNEP surveys *Engaging Stakeholders* series tried to be more upbeat about the situation but ended up telling much the same story (SustainAbility/UNEP, 1996, 1997, 1998, 1999). Similarly, an increasing attention to the information needs of NGOs reveals yet again that their information needs are not generally satisfied by organisational reporting (O'Dwyer *et al.*, 2005). So there are examples of what stakeholders would like – one illustration of which is provided by Pleon (2005). This survey was unusual in its international coverage with nearly 500 respondents in five different languages (although the English and German language responses dominated). The survey finds that 'CSR Reports are mainly aimed at shareholders and investors. But the financial community doesn't consider them useful' (p. 6). This is a fairly arresting (but not surprising) result. The survey goes on to identify that, apart from environmental issues, the top social concerns expressed by the stakeholders were human rights, corporate governance and standards in developing countries, followed by bribery and corruption and social issues in the supply chain (see also, Chenall and Juchau, 1977; Brooks, 1986; ICCR, 2011).

It is this continuing mis-match between what stakeholders want and what organisations voluntarily prefer to disclose that, in part, leads to the substantive critiques of the stakeholder consultation process (Henriques, 2007). Owen and Swift (2001) and Owen *et al.* (2000, 2001), for example, question whether the language and appearance of stakeholder dialogue have any real substance: do they amount to little more than corporate spin? Equally, the major differences in power between the organisation and its stakeholders are wilfully ignored and yet they must seriously distort any real attempt at communication, so much so that after 10 years of research, Cooper and Owen (2007) are still moved to note (p. 657): 'Clearly corporate governance mechanisms have not evolved in such a way that stakeholder accountability, as opposed to (enlightened?) stakeholder management, may be established.'

In summary, the idea of stakeholder consultation is attractive and may even be an essential component of a proper discharge of accountability (Gray *et al.*, 1997). Indeed, it is quite apparent that stakeholders have a need for accountability and, despite the language of initiatives such as ISO 26000, stakeholder engagement processes continue to lack robustness.

5.6 Community involvement and philanthropy

If there is one stakeholder that dominates discussions of the organisation's relationship with civil society it is the community. Indeed, in addition to the market stakeholders such as consumers and suppliers, it is commonplace to see an organisation's non-financial stakeholders identified as the environment, employees and the community. Now the community is an enormous concept covering not just the peoples who live near the organisation's sites – at home and overseas – but oftentimes also the elements of society from which the organisation draws its employees and customers as well as offering a hint of the wider civil society within which the organisation operates. Concern about how to understand organisational interaction with community is a serious matter and it is addressed, to a degree, in the Global Reporting Initiative (GRI) Guidelines. Disappointed with the response from reporting

Figure 5.4 Percentage of community topics reported in sample

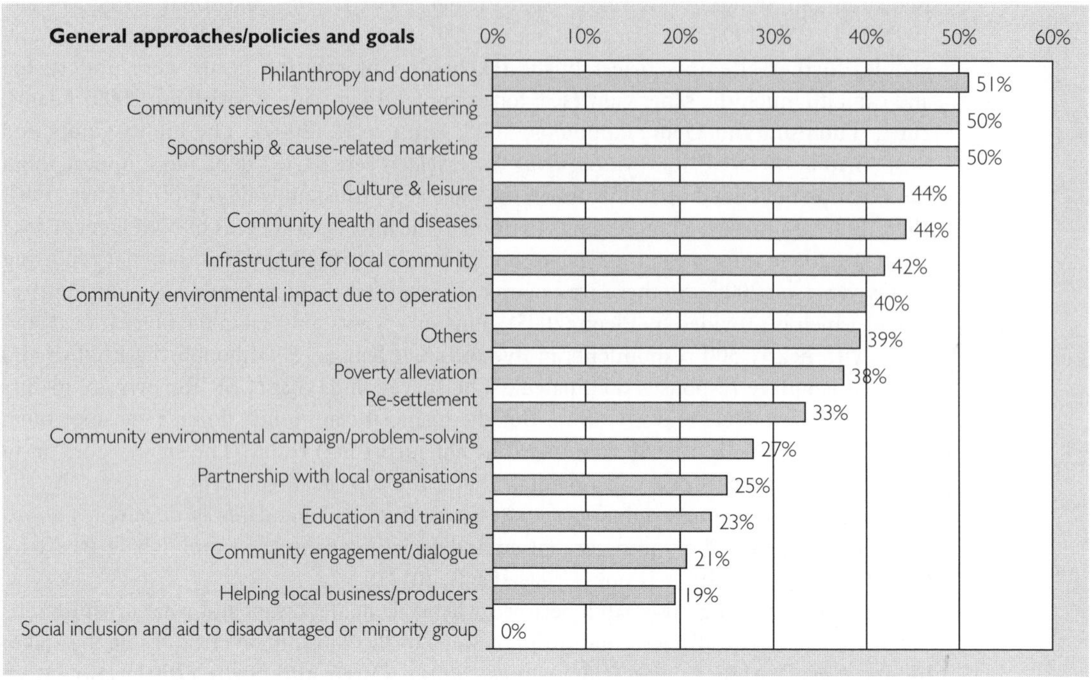

Source: Taken from GRI, (2008b): 20.

organisations, GRI undertook a study of 72 sustainability (sic)[12] reports and identified the following community-related topics therein (GRI, 2008b) (see Figure 5.4).

The range of issues covered even in such a relatively limited survey is daunting but they can, perhaps, be understood as comprising three main themes: philanthropy and corporate giving; community involvement and investment; and engagement with NGOs (and civil society organisations, CSOs). This section is structured around these three elements before we widen our scope in subsequent sections.

Philanthropy and corporate giving

Perhaps the simplest and most obvious interaction between an organisation and its community(ies) is in the form of corporate giving – typically through donations and sponsorship (Cowton, 1987). Corporate giving can vary from a simple response by the organisation to requests from the community through to strategic and carefully placed investments and the use of financial resources as part of reputation management. Simple corporate philanthropy is rarely simple. Carroll (1991) saw philanthropy as the peak of his CSR pyramid whilst Freidman saw it as an illegal and immoral use of shareholders' funds. For such a seemingly simple act, views are significantly diverse and, in the UK at least, corporations are required to disclose their charitable (and political) donations in their annual report.

Although many recipients of such donations may consider the amounts significant in their activities, and there are exceptional examples of corporations making major sponsorship deals,

---

[12]As explored more fully in Chapter 9, few if any so-called sustainability reports address the planetary issues of sustainability (see Milne *et al.*, 2009).

Figure 5.5 Charitable giving in the USA 2010 by source of contributions ($ 290.89 billion)

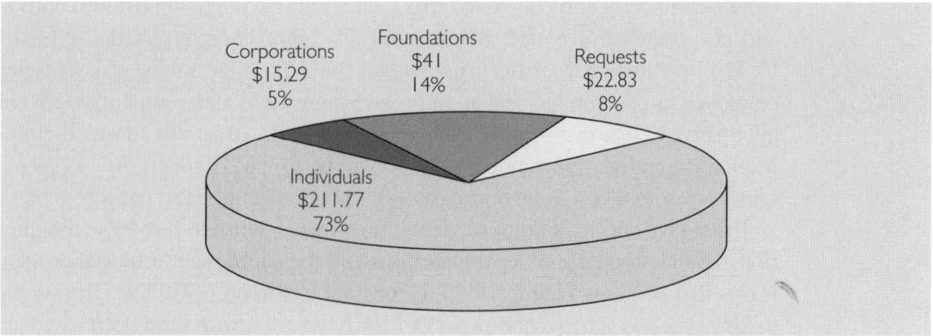

Notes: $ in billions. All figures rounded.

the amounts involved from the corporate perspective are relatively small. Overall, the amounts of philanthropic giving by the UK's largest companies, for example, tend to average out at about 0.5% of profit and, whilst US corporations tend to be a little more generous, corporate giving there is in decline (Campbell *et al.*, 2002). Indeed, in the USA, corporate giving is actually a fairly small percentage of philanthropy in total (see Figure 5.5).[13] Despite the relatively small amounts involved and in part prompted by the concerns that this is an improper use of stockholders' money (Bartkus *et al.*, 2002), more attention is being given to how the organisation might use the process of community philanthropy to advance their economic agendas. Porter and Kramer (2002) see such donations as strategic and as a source of competitive advantage to the firm, and a report by Deloitte (2011)[14] revealed that, indeed, more companies were thinking strategically about their community interactions. This is still not especially widespread though, and McKinsey (2011)[15] found that a fifth of respondents were now using corporate giving strategically with an emphasis on the perceptions and attitudes of consumers. At this point, philanthropy – the selfless provision of resources to those less fortunate – has given way to what is increasingly referred to as corporate community investment.

Community involvement and investment

An organisation's engagement with its communities – especially amongst the larger companies – is increasingly approached as one part of business decision-making. It is considered, in fact, as corporate community investment (CCI). CCI explicitly features in ISO 26000's understanding of CSR (Moratis and Cochius, 2011) and is highly implicit in the GRI's concerns about community. From the corporate point of view, this is, to all intents and purposes, an example of where the 'business case' meets stakeholder management. If one has a purely economic and amoral view of the organisation, it is quite unremarkable that a profit-seeking organisation will look to maximise the positive corporate impact from each dollar spent. Whether this might then continue to be thought of as CSR is quite another matter (Brammer *et al.*, 2009). Nevertheless, pragmatic charities and advisory groups increasingly seek to address potential

---

[13]This is taken from *Giving USA*, a report compiled annually by the American Association of Fundraising Counsel for 2010 as reported at http://www.nps.gov/partnerships/fundraising_individuals_statistics.htm.

[14]http://www.philanthropyuk.org/news/2011-07-15/corporate-philanthropy-more-strategic-deloitte-report-reveals.

[15]*The state of corporate philanthropy: A McKinsey Global Survey* (2011) at http://www.mckinseyquarterly.com/The_state_of_corporate_philanthropy_A_McKinsey_Global_Survey_2106 (sampled February 2012).

donor organisations through the language of a business case and investment rather than appealing to their 'better nature'.[16] ICCSR (2007) reported on a series of interviews with a range of British companies and concluded that CCI was increasingly linked into both business strategy and corporate governance. More especially, the report stressed the business benefits of CCI. These included greater trust by stakeholders, more robust risk management, increased employee motivation, as well as improved innovation and competitive advantage. Corporate giving may well be just another element of organisational activity to which capital budgeting and investment appraisal techniques can and will be applied. Philanthropy as such in these circumstances may well be a thing of the past – and that may not be an entirely good thing.

From civil society's point of view it now makes sense to recognise the uneven power relationship in such stakeholder transactions and the subsequent importance of accountability. It is this that prompts Hamil (1999) to consider examples of CCI activity (donations of cash or in kind, such as staff secondments) which have come with strings attached and/or which are otherwise linked to specific corporate benefits. It is important to recognise that corporate giving, despite its charitable appearance, is about what the organisation wishes to give to, not what the community needs. Hamil suggests that a requirement for disclosure by companies of their motivation for CCI involvement could 'make companies more accountable while at the same time retaining for the community the many benefits the activity delivers' (p. 23).

The GRI (2008b) report shows that globally, amongst leading edge reporters anyway, there is evidence of some movement in this direction. Figure 5.6 shows the topics on which the GRI sample reported – and whilst it would be stretching to suggest that all of these are recognised explicitly as 'investment', it is clear to see how much of the disclosure relates to what could so easily be covered by an investment business case.

Engagement with NGOs and CSOs

Defining NGOs is not as simple as it looks. NGOs are variously described as autonomous, non-profit-making, self-governing and campaigning organisations with a focus on the well-being of others (Gray et al., 2006). They have been characterised as organisations 'whose stated purpose is the promotion of environmental and/or social goals rather than the achievement or protection of economic power in the market place or political power through the electoral process' (Bendell, 2000a: 16; see also Edwards, 2000; Teegen et al., 2004). They represent one, major, element in CSOs, whose own growth appears to be a function of the increasing size, power and orientation of both the state and the market economy. This is ironic in that the state is supposed to represent civil society but seems to increasingly alienate it whilst the market economy has grown so virtual, large and hyper-real that it actively alienates the society from which it has sprung. Because of these complexities, it is increasingly commonplace (if incorrect) to equate the interests of civil society with those of NGOs (Bendell, 2000b; Chandhoke, 2002).

NGOs and CSOs matter because they can mobilise resources, undertake research, raise and develop campaigns and provide a focal point that communities, in the broadest sense, often cannot (Deegan and Blomquist, 2006; MacLeod, 2007). As a result, these organisations become an important, potentially salient, stakeholder from the organisation's point of view and a focal point for research into community needs and information demands (Unerman and Bennet, 2004). But, as we have seen, taken in the round, communities and civil society organisations do *not* get the accountability they need and inadequate forms of stakeholder dialogue are unlikely to change that. Consequently, whilst attempts are being

---

[16]The UK's Charities Aid Foundation is just one such example.

Figure 5.6 Percentage of reports in sample reporting on . . .

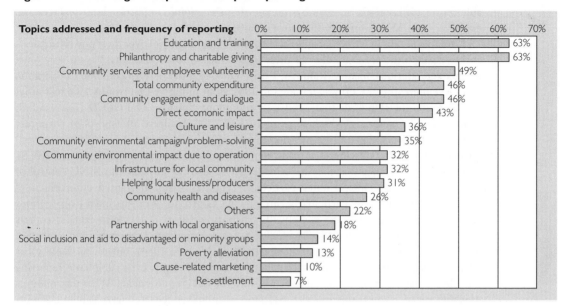

Source: Taken from GRI, (2008b):12.

made through CCI and such like to integrate the interests of organisations and communities, NGOs often find it necessary to adopt an increasingly adversarial position with companies. This, in turn, has contributed to calls for an increased accountability of NGOs – not just the accounting to NGOs (Gray *et al.*, 2006; Unerman and O'Dwyer, 2006).

The issues which can be at stake and the major imbalances in power and resources between organisations and communities are important enough when we are exploring relationships in developed countries. In the lesser developed or newly developed countries, the stakes are so very much higher with the conflicts going to the very heart of indigenous culture and the ability of peoples to continue living as they have done.

5.7 Accountability, MNCs and LDCs

The need for CSR strategies and programmes which are sensitive to the context of less developed countries (LDCs) and newly industrialised countries (NICs) is particularly acute given the large numbers of developed country multi-nationals operating in these countries. Further, foreign-owned multi-nationals must be accountable for their social and environmental impacts if host governments are to have any hope at all of exercising some degree of control over them (Gray and Kouhy, 1993). Briston (1984) directly addressed the issue of the control of MNCs by host countries and identified a range of information that would be an essential part of any social responsibility and control. This information included data on: the purchase of inputs locally; profit and capital repatriation; the extent of planned and actual local equity participation; the extent of local participation in top management; the level of employment provided; the obligation to train local personnel; environmental protection; and the construction of necessary infrastructure such as roads and housing. Some of these items would not, of themselves, in a western domestic context be considered as part of a CSR programme but, considered in light of the increasing north–south divide, most certainly are (Belal and Owen, 2007). Indeed, the questions of CSR and LDCs with respect to *tve impact*

MNCs are greatly complicated not just by the differences in social values and the consider-able power differentials but also by the enormous economic impacts – both positive and negative – that MNCs bring to their host countries.

Attempts to rein in MNCs, to give host countries some control over their powerful guests and to develop the accountability of these often enormous corporations, have a very long, but not very glorious, history (Rahman, 1998). The Organisation for Economic Cooperation and Development (OECD) was founded in 1961 as an initiative that would, amongst other things, guide the standards that MNCs should adopt globally. The *OECD Guidelines for Multinational Enterprises,* which were revised in 2000, 'address all aspects of corporate behaviour, from taxation and competition to consumer interests and . . . enhance the contribution to sustainable development made by multi-national enterprises' (Leipziger, 2010: 5). There is little evidence that they have made any substantial inroads into MNC control or accountability, and Leipziger notes, particularly, the lack of any substantive enforcement of the guidelines. The UN, particularly through the Centre for Transnational Corporations (UNCTC), has an even longer and equally frustrating history of trying to exercise control over MNCs (Rahman, 1998; MacLeod, 2007). For example, the UN established a range of minimum disclosure requirements in the 1980s (UNCTC, 1984), but global society has advanced little in moves towards such reporting and, strikingly, even such basic reporting guidelines have so far failed to find their way into the GRI requirements. What this amounts to is that very few governments have paid more than lip service to concerns about account-ability of MNCs for social and environmental impacts and all, but for a very limited amount of, disclosure remains voluntary (Aaronson, 2005).

Studies of actual MNC disclosure in this context do not offer a much more optimistic picture. Generally speaking, companies make more substantial disclosure in their home (western) country than they do in the host countries (UN, 1991), and Belal and Owen (2007) go further and highlight that standards developed for the benefit of stakeholders in western developed nations may have a negative impact on both social justice and the economic north–south divide as it relates to LDCs. Interestingly though, Chapple and Moon (2005), in a study of website reporting across seven Asian countries, found that, whilst MNCs are more likely to adopt CSR than specifically local companies, the profile of their CSR tended to reflect the country of operation rather than the country of origin. Similarly, Jamali (2008) found that the subsidiaries of MNCs in Lebanon were more likely than local companies to adopt CSR practices and engage with a broad range of stakeholders. That is, MNCs may be laggardly in exporting their social disclosure from home to host country but, generally speaking, their standards are still better than local companies.[17]

Optimism that voluntary initiatives will both improve corporate behaviour and increase serious accountability shows no signs of abating. The most striking initiative that follows in the wake of the OECD, UN and others is the *Global Compact.* Launched by the then UN Secretary-General, Kofi Annan, in 2000 as part of the Millennium Development Goals initiatives, the UNGC comprises 10 principles which companies are requested to publicly adopt, sign up to and then report upon annually as an indication of their progress in meeting these principles. The principles are themselves derived from prior codes and are shown in Figure 5.7. KPMG (2008) reports that 40% of their Global 250 sample claim to be reporting in line with the UNGC, although KPMG goes on to report that monitoring by the UNGC is a serious business and that 1000 companies had been de-listed from the signatories for failure to report on progress. Time will tell if this is the voluntary code that finally produces

---

[17]This is another example of a point made in the business and management literature but which is not obviously exploited in the accounting literature.

Figure 5.7 The UN Global Compact principles

Human Rights

Principle 1: Businesses should support and respect the protection of internationally proclaimed human rights; and

Principle 2: make sure that they are not complicit in human rights abuses.

Labour

Principle 3: Businesses should uphold the freedom of association and the effective recognition of the right to collective bargaining;

Principle 4: the elimination of all forms of forced and compulsory labour;

Principle 5: the effective abolition of child labour; and

Principle 6: the elimination of discrimination in respect of employment and occupation.

Environment

Principle 7: Businesses should support a precautionary approach to environmental challenges;

Principle 8: undertake initiatives to promote greater environmental responsibility; and

Principle 9: encourage the development and diffusion of environmentally friendly technologies.

Anti-Corruption

Principle 10: Businesses should work against corruption in all its forms, including extortion and bribery.

accountability to host countries. The UNGC principles in Figure 5.7 include both environmental issues (see Chapter 7) and labour issues (see Chapter 6). The following section touches upon a number of issues we have not yet addressed including that of human rights.

5.8 Indigenous people, repressive regimes, child labour and human rights

Underlying our discussions in this chapter so far are a range of implicit questions – questions we explored in earlier chapters. They include such fundamental concerns as what is it to be human and how, if at all, should we manage the relationship between the market and civil society? The heading of this section gives a brief glimpse of how those sorts of questions manifest themselves (a glimpse we will extend further in the final section of the chapter). To a considerable degree, the issues arise because of the essential difference in fundamental goals between organisation and communities (Brown, 2009). If we follow a liberal point of view, we see the goals as ultimately converging in that it is assumed that companies create wealth and wealth is spread around society. But, even if one subscribes to such a view, there are inevitably clashes between (say) the western democratic principles of civil society and the organisation's desire to do business. In doing business, organisations will work with regimes whose values we might consider offensive, they may need to mine land that is the ancestral home of an indigenous people, and so on. These clashes seem inevitable. They beg the question, what is a company doing in that country in the first place? If the company is already

Figure 5.8 Universal Declaration of Human Rights

Article 1: All human beings are born free and equal in dignity and rights. They are endowed with reason and conscience and should act towards one another in a spirit of brotherhood.

Article 2: Everyone is entitled to all the rights and freedoms set forth in this Declaration, without distinction of any kind, such as race, colour, sex, language, religion, political or other opinion, national or social origin, property, birth or other status.

Furthermore, no distinction shall be made on the basis of the political, jurisdictional or international status of the country or territory to which a person belongs, whether it be independent, trust, non-self-governing or under any other limitation of sovereignty.

Article 3: Everyone has the right to life, liberty and security of person.

there, and unlikely to withdraw, then the issue may be to negotiate their engagement through the contentious areas, in the least destructive way possible. Such matters considerably exercise the international and development organisations and the associated research literature (Bailey *et al.*, 1994, 2000; Korten, 1995; Munck and O'Hearn, 1999; Ebrahim and Weisband, 2007). For our purposes, we can illustrate the complexities by brief reference to just two issues – that of child labour and that of human rights. (Racial and gender equality, along with aspects of these two issues, will emerge again in Chapter 6.)

Human rights is a highly contested area that emerges from the United Nations Universal Declaration of Human Rights (UDHR) (UN, 1948). In brief, the central contested areas are the extent to which the notion of human rights is indeed universal (as many cultures are uncomfortable with it) plus the increasing role of corporations in an area which is traditionally and more sensibly a matter for states (Adams and Harte, 1999; Gray and Gray, 2011). Figure 5.8 illustrates the first three of the 30 articles which comprise the UDHR.

Concern for human rights occupies a central place in many of the initiatives we have already seen: the UNGC and the OECD Guidelines are two obvious examples of this. In addition, the International Finance Corporation (IFC) (www.ifc.org), a member of the World Bank Group, and the GRI (www.globalreporting.org), have established a joint global project to promote gender reporting whilst the GRI, UNGC and Realizing Rights are working together to improve human rights reporting.[18] The appointment of John Ruggie as the special representative of the UN Secretary-General for Business and Human Rights in 2005 brought the issue to central stage. States vary in their support of the declaration but, for example, in 2008 the Australian Human Rights Commission embarked on a project looking at the role of Australian companies in protecting and promoting human rights, a stated aim being to 'illustrate the relevance of human rights to all Australian companies'.[19] Whilst Amnesty International, along with a number of other NGOs and civil society groups, supports (for example) the *UN Norms on the Responsibilities of Transnational Corporations and other Business Enterprises with regard to Human Rights*, some states, businesses and business organisations actively oppose developments in this field (Warhurst *et al.*, 2004; Gray and Gray, 2011). Moreover, the matter is not one restricted to only for-profit organisations. Aaronson (2005) calls for governments to play a role by examining how government purchasing power might be used to improve human rights.

---

[18]Voluntary corporate reporting of aspects of human rights – typically related to employment – is extensive and particularly illustrated in mining companies like Rio Tinto and Xstrata. More detail is provided in Chapter 6.

[19]http://www.hreoc.gov.au/human_rights/corporate_social_responsibility/index.html (accessed 21/11/08).

Amongst the many key issues covered by human rights, that of the rights and roles of women has been of particular importance. Although much of this relates to the employment of women and has a number of interesting spill-overs into economic development and the role of micro-finance, the treatment, portrayal and status of women keenly reflects central *motifs* of a society (Tinker and Neimark, 1987; Cooper and Puxty,1996; Adams and Harte, 1999). Indeed, the unsettling question of the extent to which the very attempt to see these essential motifs of a culture through the lens of human rights also remains unresolved. As we have seen, there is an abiding question of whether western notions of human rights should properly be applied in some cultures as well as the extent to which such application simply reflects further cultural relativism (Pegg, 2003). But more pertinently, although it is quite clear that business is now seriously intertwined with human rights, it remains far from clear whether they are usurping – or being required to adopt – the duties of the state in an area of human experience where companies (labour issues aside) are relatively ill-equipped.

If human rights is a contested area, that of child labour is even more difficult still. Whilst the use of children as labour in oppressive ways is (almost) universally eschewed by the western democracies, there are cultures and circumstances under which there can seem to be little or no alternative. To pronounce absolutely on such a matter risks embarking upon the worst kind of imperialism (Mellahi *et al.*, 2010). Of course it might be thought that child labour would more properly sit in an analysis of employment, but a major concern that the issue brings to the fore is that combination of commercial pressures that globalisation forces upon countries and cultures coupled with the importance of organisations carefully analys-ing and scrutinising their supply chain – where the child labour is more likely to occur. It is here that the International Labour Organisation is especially active and where one of the earliest CSR standards, SA8000, comes into play. *Social Accountability 8000* covers more than child labour, but its clarity and focus have made it a popular standard with the leading companies seeking to improve their monitoring and performance in this area (Leipziger, 2010). The standard and compliance with it is also monitored and assessed and this gives the standard a credibility other guidelines may not have.

The reporting of these issues in organisations' disclosure appears to be relatively thin. In fact, although reporting on gender and race issues has a relatively long history (Adams and Harte, 1998; Adams and McPhail, 2004), the issues of both human rights and child labour have, it would seem, been slow to appear on the reporting radar. An initial review of surveys of reporting practices such as Trucost (2004), Palenberg *et al.* (2006), SPADA (2008), Martin and Hadley (2008) found either no reference or only very sparse reference to either human rights or child labour. ACCA/Corporate Register (2004) finds very little evidence of practice. However, human rights, but not child labour, features relatively prominently in KPMG (2008) which finds that 21% of the Fortune 250 sample explicitly make reference to UDHR and 40% of the sample explicitly reference the UNGC (which gives a high profile to human rights). By contrast, GRI (2008a) finds that only 7% of those companies reporting using the G3 Guidelines were actually complying with the Guidelines on the subject of human rights. One can expect the level of apparent interest to slowly develop, not least as these are issues about which elements of the financial community are increasingly exercised (Sullivan and Mackenzie, 2006). Increasingly, it seems, the growing institutionalisation of socially responsible investment (see Chapter 8), plus the growing awareness amongst many mainstream pension funds, has ensured that human rights is one of the major issues around which investors would seek to avoid ethical and financial risk (Coles, 2003). Human rights has a significant (if implied) position within the UNGC and the United Nations Principles of Responsible Investment (UNPRI) as well as in The Equator Principles and the (so-called) ethical indices like the Dow Jones Sustainability Index and the UK's FTSE4Good (see, for example, Collison *et al.*, 2009; and especially Oulton, 2006).

So, although the discharge of organisational accountability around social, community and human rights issues is a very long way from satisfactory, there is continuing evidence of a steady growth in awareness of the importance of these issues in both organisation logics and strategies as well as in organisational public disclosure. You do not have to look far to find MNCs producing relatively promising levels of disclosure (see, for example, companies like BASF, Rio Tinto and Xstrata), although here, as elsewhere, it seems unlikely that substantive accountability will ever be achieved in the absence of a well-policed regime of mandatory reporting.

5.9 Extensions, community and the social

It is quite apparent that organisational activity has a significant impact on a whole range of social and community issues. When those organisations are located in so-called developed countries, the issues seem to be more about equality and political economy. When those organisations are based in developing countries, the issues seem more acute, embracing, as they do, traditional ways of life, clashes of culture and levels of exploitation and oppression that the (so-called) western democracies find unacceptable. Two closely related things become apparent from even this brief review of social issues and organisational reporting: the range of issues that are relevant under the heading of 'social' is immense, perhaps even infinite; and the quality of reporting is relatively thin and patchy.

The range of issues is daunting – whether you are looking at the matter from the organisation's point of view or from that of civil society. Organisational accountability is always going to be complex (see Figure 5.9). It is important to realise just how sketchy our coverage of social and community issues has been in this chapter (and, in all probability, would always have to be).

Figure 5.9 A selection of social and community issues to be addressed by social reporting

● Advertising	● Human trafficking
● Anti-corruption	● Human rights
● Apartheid	● Inequality
● Bribery	● Infrastructure
● Child labour	● Lobbying
● Community investment	● Marketing
● Culture	● Media
● Diversity	● Philanthropy
● Eco-justice	● Poverty
● Education	● Propaganda
● Ethical trading	● Race
● Extortion	● Repressive regimes
● Fair trade	● Sexual exploitation
● Gender	● Slavery
● Happiness	● Stakeholder engagement
● Health	● Supply chain

And these are just the visible issues. Within such a list of issues sit a whole lot of what we should probably call 'political economy' issues. These are the systemic factors that provide the complex framing within which the relationships between the state, the market and civil society are manifest (Brown, 2009). Businesses clearly spend a lot of effort and resources in lobbying (SustainAbility, 2005). Reporting on such an activity is unlikely to capture the levels of influence that organisations exert over politicians, the law, the terms of the contract within society and even, especially, education (Mayhew, 1997; Beder, 2006). How we understand social justice, what we think of as the place of business, and so on is, to a not-insignificant degree, the result of influence from business itself. To really develop social reporting and to see its potential and its failures, it is crucial to imagine what the world might look like with a different information regime and a different set of power relationships.

So this only scratches the surface of the issues with which social reporting is concerned. Employees, the natural environment, the financial world and sustainability[20] are just another set of the components that make up the complex world we inhabit and which social accountability seeks to navigate.

References

Aaronson, S. A. (2005) 'Minding our business': what the United States Government has done and can do to ensure that U.S. multinationals act responsibly in foreign markets, *Journal of Business Ethics*, **59**: 175–98.

Abercrombie N., Hill, S. and Turner, B. S. (1984) *Dictionary of Sociology*. Harmondsworth: Penguin.

ACCA/Corporate Register.com (2004) *Towards Transparency: Progress on Global Sustainability Reporting*. London: Association of Chartered Certified Accountants.

AccountAbility (1999) *Accountability 1000: The foundation standard – an overview*. London: AccountAbility.

AccountAbility (2008) *Accountability 1000: AccountAbility Principles Standard*. London: AccountAbility.

Adams, C. A. (2002) Internal organisational factors influencing corporate social and ethical reporting, *Accounting, Auditing and Accountability Journal*, **15**(2): 223–50.

Adams, C. A. (2004) The ethical, social and environmental reporting-performance portrayal gap, *Accounting, Auditing and Accountability Journal*, **17**(5): 731–57.

Adams, C. A. (2008) A Commentary on: corporate social responsibility reporting and reputation risk management, *Accounting, Auditing and Accountability Journal*, **20**(3): 365–70.

Adams, C. A. and Harte, G. (1998) The changing portrayal of the employment of women in British banks' and retail companies' corporate annual reports, *Accounting, Organizations and Society*, **23**(8): 781–812.

Adams, C. A. and Harte, G. (1999) *Towards Corporate Accountability for Equal Opportunities Performance*, Occasional Paper 26. London: ACCA.

Adams C. A. and Kuasirikun, N. (2000) A comparative analysis of corporate reporting on ethical issues by UK and German chemical and pharmaceutical companies, *European Accounting Review*, **9**(1): 53–80.

Adams, C. A. and McNicholas, P. (2007) Making a difference: sustainability reporting, accountability and organisational change, *Accounting, Auditing and Accountability Journal*, **20**(3): 382–402.

Adams, C. A. and McPhail, K. (2004) Reporting and the politics of difference: (non)disclosure on ethnic minorities, *Abacus*, **40**(3): 405–35.

Adams, C. A. and Whelan, G. (2009) Conceptualising future changes in corporate sustainability reporting, *Accounting, Auditing and Accountability Journal*, **22**(1): 118–43.

Adams, C. A., Hill, W. Y. and Roberts, C. B. (1995) *Environmental, Employee and Ethical Reporting in Europe*. London: ACCA.

---

[20]These are dealt with in Chapters 6, 7, 8 and 9, respectively.

Adams, C. A., Hill, W. Y. and Roberts, C. B. (1998) Corporate social reporting practices in Western Europe: Legitimating corporate behaviour?, *British Accounting Review*, **30**(1): 1–21.

Al-Najjar, F. L. (2000) Determinants of social responsibility disclosure of US Fortune 500 firms: An application of content analysis, *Advances in Environmental Accounting and Management*, 1: 163–200.

Angus-Leppan, T., Metcalf, L. and Benn, S. (2009) Leadership styles and CSR practice: An examination of sense making, institutional drivers and CSR leadership, *Journal of Business Ethics*, **93**: 189–213.

Bailey, D., Harte, G. and Sugden, R. (1994) *Making Transnationals Accountable: A significant step for Britain*. London: Routledge.

Bailey, D., Harte, G. and Sugden, R. (2000) Corporate disclosure and the deregulation of international investment, *Accounting, Auditing and Accountability Journal*, **13**(2): 197–218.

Ball, A. and Osborne, S. P. (2011) *Social Accounting and Public Management: Accountability and the common good*. Abingdon: Routledge.

Banerjee, S. B. (2007) *Corporate Social Responsibility: The good, the bad and the ugly*. Cheltenham: Edward Elgar.

Bartkus, B. R., Morris, S. A. and Seifert, B. (2002) Governance and corporate philanthropy: Restraining Robin Hood?, *Business Society*, **41**(3): 319–44.

Beder, S. (2006) *Suiting Themselves: How corporations drive the global agenda*. London: Earthscan.

Belal, A. and Owen, D. (2007) The view of corporate managers on the current state of, and future prospects for social reporting in Bangladesh, *Accounting, Auditing & Accountability Journal*, **20**(3): 472–94.

Bendell, J. (ed.) (2000a) *Terms for Endearment: Business, NGOs and Sustainable Development*. Sheffield: Greenleaf/New Academy of Business.

Bendell, J. (2000b) Introduction: working with stakeholder pressure for sustainable development, in Bendell, J. (ed.), *Terms for Endearment: Business, NGOs and Sustainable Development*, pp. 14–30. Sheffield: Greenleaf Publishing/New Academy of Business.

Benjamin, J. J. and Stanga, K. G. (1977) Difference in disclosure needs of major users of financial statements, *Accounting and Business Research*, **27**: 187–92.

Benn, S., Dunphy, D and Martin, A. (2009) Governance of environmental risk: new approaches to managing stakeholder involvement, *Journal of Environmental Management*, **90**(4): 1567–75.

Blowfield, M. and Murray, A. (2008) *Corporate Responsibility: A critical introduction*. Oxford: Oxford University Press.

Brammer, S. and Pavelin, S. (2006) Voluntary environmental disclosure by large UK companies, *Journal of Business Finance and Accounting*, **33**(7/8): 1168–88.

Brammer, S., Pavelin, S. and Porter, L. A. (2009) Corporate charitable giving: multinational companies and countries of concern, *Journal of Management Studies*, **46**(4): 575–96.

Briston, R. J. (1984) Accounting standards and host country control of multinationals, *British Accounting Review*, **16**(1): 12–26.

Brockhoff, K. (1979) A note on external social reporting by German companies: a survey of 1973 company reports, *Accounting, Organizations and Society*, **4**(1/2): 77–85.

Brooks, L. J. (Jr) (1986) *Canadian Corporate Social Performance*. Hamilton: The Society of Management Accountants of Canada.

Brown, A. (2009) The milieu of reporting of Nacamaki and Nabuna villages of Koro Island, *Pacific Accounting Review*, **21**(3): 202–27.

Buhr, N. (2007) Histories of and rationales for sustainability reporting, in Unerman, J., Bebbington, J. and O'Dwyer, B. (eds), *Sustainability Accounting and Accountability*, pp. 57–69. London: Routledge.

Burchell, J. and Cook, J. (2006) Confronting the 'corporate citizen': shaping the discourse of corporate social responsibility, *International Journal of Sociology and Social Policy*, **26**(3-4), 121–37.

Campbell, D., Moore, G. and Metzger, M. (2002) Corporate philanthropy in the UK 1985–2000: some empirical findings, *Journal of Business Ethics*, **39**(1-2): 29–41.

Carroll, A. B. (1991) The pyramid of corporate social responsibility: towards the moral management of organizational stakeholders, *Business Horizons*, July/August: 42.

Chandhoke, N. (2002) The limits of global civil society, in Glasius, M., Kaldor, M. and Anheier, H. (eds), *Global Civil Society 2002*, pp. 35–53. Oxford: Oxford University Press.

Chapple, W. and Moon, J. (2005) Corporate social responsibility (CSR) in Asia: a seven-country study of CSR web site reporting, *Business and Society*, **44**(4): 415–41.

Chenall, R. H. and Juchau, R. (1977) Investor information needs: an Australian study, *Accounting and Business Research*, **26**: 111–19.

Coles, D. (2003) What is the attitude of investment markets to corporate performance on human rights?, in Sullivan, R. (ed.), *Business and Human Rights: Dilemmas and solutions*, pp. 92–101. Sheffield: Greenleaf.

Collison, D., Cobb, G., Power, D. and Stevenson, L. (2009) FTSE4Good: exploring its implications for corporate conduct, *Accounting, Auditing and Accountability Journal*, **22**(1): 35–58.

Cooper, C. and Puxty, A. (1996) On the proliferation of accounting (his)tories, *Critical Perspectives on Accounting*, **7**: 285–313.

Cooper, S. M. and Owen, D. L. (2007) Corporate social reporting and stakeholder accountability: the missing link, *Accounting, Organizations and Society*, **32**: 649–67.

Cormier, D. and Gordon, I. M. (2001) An examination of social and environmental reporting strategies, *Accounting, Auditing and Accountability Journal*, **14**(5): 587–616.

Cowton, C. J. (1987) Corporate philanthropy in the United Kingdom, *Journal of Business Ethics*, **6**(7): 553–8.

Deegan, C. and Blomquist, C. (2006) Stakeholder influence on corporate reporting: an exploration of the interaction between WWF-Australia and the Australian minerals industry, *Accounting, Organizations and Society*, **31**(4/5): 343–72.

Deegan, C. and Rankin, M. (1999) The environmental reporting expectations gap: Australian evidence, *British Accounting Review*, **31**(3): 313–46.

Deloitte (2011) *More Than Just Giving: Analysis of corporate social responsibility across UK firms*. London: Deloitte (http://www.philanthropyuk.org/news/2011-07-15/corporate-philanthropy-more-strategic-deloitte-report-reveals).

den Hond, F. and de Bakker, F. G. A. (2007) Ideologically motivated activism: how activist groups influence corporate social change activities, *Academy of Management Review*, **32**(3): 902–24.

Dey, C. (2007) Developing silent and shadow accounts, in Unerman, J., Bebbington, J. and O'Dwyer, B. (eds), *Sustainability Accounting and Accountability*, pp. 307–26. London: Routledge.

Dierkes, M. and Antal, A. B. (1985) The usefulness and use of social reporting information, *Accounting, Organizations and Society*, **10**(1): 29–34.

Du, Y. and Gray, R. (2013) The emergence of stand alone social and environmental reporting in Mainland China: an exploratory research note, *Social and Environmental Accountability Journal*.

Durden, C. (2008) Towards a socially responsible management control system, *Accounting, Auditing and Accountability Journal*, **21**(5): 671–94.

Ebrahim, A. and Weisband, E. (eds) (2007) *Global Accountabilities: Participation, pluralism and public ethics*. Cambridge: Cambridge University Press.

Edwards, M. (2000) *NGO Rights and Responsibilities: A new deal for global governance*. London: The Foreign Policy Centre/NCVO.

Epstein, M. J. (1991) What shareholders really want, *The New York Times*, April 28.

Epstein, M. J. (1992) The annual report report card, *Business and Society Review*, Spring: 81–3.

Epstein, M. J., Epstein, J. B. and Weiss, E. J. (1977) *Introduction to Social Accounting*. New York: National Association of Accountants.

Ernst & Ernst (1971 *et seq.*) *Social Responsibility Disclosure*. Cleveland, OH: Ernst & Ernst.

Estes, R. W. (1976) *Corporate Social Accounting*. New York: Wiley.

European Commission (2001) *Green Paper promoting a European framework for corporate social responsibility*, DG Employment and Social Affairs. Brussels: European Commission.

Gray, R. H. and Bebbington, K. J. (2000) Environmental accounting, managerialism and sustainability: is the planet safe in the hands of business and accounting?, *Advances in Environmental Accounting and Management*, **1**: 1–44.

Gray, R. and Gray, S. (2011) Accountability and human rights: a tentative exploration and a commentary, *Critical Perspectives on Accounting*, **22**(8): 781–9.

Gray, R. and Herremans, I. (2012) Sustainability and social responsibility reporting and the emergence of the external social audits: the struggle for accountability?, In Bansal, T. and Hoffman, A. (eds), *Oxford Handbook of Business and the Environment*, pp. 405–24. Oxford: Oxford University Press.

Gray, R. H. and Kouhy, R. (1993) Accounting for the environment and sustainability in lesser developed countries: An exploratory note, *Research in Third World Accounting*, **2**: 387–99.

Gray, R. H., Owen, D. L. and Maunders, K. T. (1987) *Corporate Social Reporting: Accounting and accountability*. Hemel Hempstead: Prentice Hall.

Gray, R. H., Kouhy, R. and Lavers, S. (1995) Corporate social and environmental reporting: a review of the literature and a longitudinal study of UK disclosure, *Accounting, Auditing and Accountability Journal*, **8**(2): 47–77.

Gray, R. H., Owen, D. L. and Adams, C. (1996) *Accounting and Accountability: Changes and challenges in corporate social and environmental reporting*. London: Prentice Hall.

Gray, R. H., Dey, C., Owen, D., Evans, R. and Zadek, S. (1997) Struggling with the praxis of social accounting: stakeholders, accountability, audits and procedures, *Accounting, Auditing and Accountability Journal*, **10**(3): 325–64.

Gray, R., Bebbington, J. and Collison, D. J. (2006) NGOs, civil society and accountability: making the people accountable to capital, *Accounting, Auditing and Accountability Journal*, **19**(3): 319–48.

GRI (with Roberts Environmental Center) (2008a) *Reporting on Human Rights*. Amsterdam: Global Reporting Initiative.

GRI (with University of Hong Kong and CSR Asia) (2008b) *Reporting on Community Impacts*. Amsterdam: Global Reporting Initiative.

Guthrie, J. and Parker, L. D. (1989) Corporate social reporting: a rebuttal of legitimacy theory, *Accounting and Business Research*, **9**(76): 343–52.

Guthrie J. and Parker, L. D. (1990) Corporate social disclosure practice: a comparative international analysis, *Advances in Public Interest Accounting*, **3**: 159–76.

Hackston, D. and Milne, M. (1996) Some determinants of social and environmental disclosures in New Zealand, *Accounting, Auditing and Accountability Journal*, **9**(1): 77–108.

Hamil, S. (1999) Corporate community involvement: a case for regulatory reform, *Business Ethics: a European review*, **8**(1): 14–25.

Henriques, A. (2007) *Corporate Truth: the limits to transparency*. London: Earthscan.

Henriques, A. (2010) *Corporate Impact: measuring and managing your social footprint*. London: Earthscan.

Herremans, I. M., Herschovis, M. S. and Bertels, S. (2008) Leaders and laggards: the influence of competing logics on corporate environmental action, *Journal of Business Ethics*, **89**: 449–72.

Hess, D., Rogovsky N. and Dunfee, T. W. (2002) The next wave of corporate community involvement: corporate social initiatives, *California Management Review*, **44**(2): 110–25.

ICCR (2011) *Building sustainable communities through multi-party collaboration*. New York: Interfaith Center on Corporate Responsibility.

ICCSR (2007) *The Role of Stakeholder Engagement in Corporate Community Investment*. Nottingham: Charities Aid Foundation/International Centre for Corporate Social Responsibility.

Ingram, R. W. and Frazier, K. B. (1980) Environmental performance and corporate disclosure, *Journal of Accounting Research*, **18**(2): 614–22.

ISO (2010) *Guidance on Social Responsibility (ISO 26000)*. Geneva: International Standards Organisation.

Jamali, D. (2008) A stakeholder approach to corporate social responsibility: a fresh perspective into theory and practice, *Journal of Business Ethics*, **82**: 213–31.

Johnson, H. L. (1979) *Disclosure of Corporate Social Performance: Survey, evaluation and prospects*. New York: Praeger.

Kamp-Roelands, N. (2009) Corporate responsibility reporting, in United Nations Conference on Trade and Development, *Promoting Transparency in Corporate Reporting: A Quarter Century of ISAR*, pp. 99–112. Geneva: United Nations.

Kolk, A. (2003) Trends in sustainability reporting by the Fortune Global 250, *Business Strategy and the Environment*, **12**(5): 279–91.

Kolk, A. (2008) Sustainability, accountability and corporate governance: exploring multinationals' reporting practices, *Business Strategy and the Environment*, 17(1): 1–15.

Korten, D. C. (1995) *When Corporations Rule the World*. West Hartford, CT/San Francisco, CA: Kumarian/Berrett-Koehler.

KPMG (2002) *KPMG 4th International Survey of Corporate Sustainability Reporting*. Amsterdam: KPMG/WIMM.

KPMG (2005) *KPMG International Survey of Corporate Responsibility 2005*. Amsterdam: KPMG International.

KPMG (2008) *KPMG International Survey of Corporate Responsibility Reporting 2008*. Amsterdam: KPMG International.

KPMG/UNEP/GRI/UCGA (2010) *Carrots and Sticks – Promoting Transparency and Sustainability*. Amsterdam: KPMG.

Kuasirikun, N. and Sherer, M. (2004) Corporate social accounting disclosure in Thailand, *Accounting, Auditing and Accountability Journal*, 17(4): 629–60.

Lantos, G. R. (2001) The boundaries of strategic corporate social responsibility, *Journal of Consumer Marketing*, 18(7): 595–630.

Laughlin, R. C. (1991) Environmental disturbances and organisational transitions and transformations: some alternative models, *Organisation Studies,* 12(2): 209–32.

Leipziger, D. (2010) *The Corporate Responsibility Code Book*, Revised 2nd Edition. Sheffield: Greenleaf.

Lessem, R. (1977) Corporate social reporting in action: an evaluation of British, European and American practice, *Accounting, Organizations and Society*, 2(4): 279–94.

MacLeod, M. R. (2007) Financial actors and instruments in the construction of global corporate responsibility, in Ebrahim, A., and Weisband E., (eds), *Global Accountabilities: Participation, pluralism and public ethics*, pp. 227–51. Cambridge: Cambridge University Press.

Maltby, J. (2005) Showing a strong front: corporate social reporting and the 'business case' in Britain, 1914–1919, *Accounting Historians Journal*, 32: 145–67.

Margolis, J. D. and Walsh, J. P. (2003) Misery loves companies: Rethinking social initiatives by business, *Administrative Science Quarterly*, 48: 268–305.

Martin, A. D. and Hadley, D. J. (2008) Corporate environmental non-reporting: a UK FTSE 350 perspective, *Business Strategy and the Environment*, 17: 245–59.

Matten, D. and Moon, J. (2008) 'Implicit' and 'explicit' CSR: a conceptual framework for a comparative understanding of corporate social responsibility, *The Academy of Management Review*, 33(2): 404–24.

Mayhew, N. (1997) Fading to grey: the use and abuse of corporate executives' 'representational power', in Welford, R. (ed.), *Hijacking Environmentalism: Corporate response to sustainable development*, pp. 63–95. London: Earthscan.

McKinsey (2011) *The State of Corporate Philanthropy: A McKinsey Global Survey* (http://www.mckinseyquarterly.com/The_state_of_corporate_philanthropy_A_McKinsey_Global_Survey_2106).

Mellahai, K., Morrell, K. and Wood, G. (2010) *The Ethical Business: Challenges and controversies*. London: Palgrave Macmillan.

Milne, M. and Gray, R. H. (2007) Future prospects for corporate sustainability reporting, in Unerman, J., Bebbington, J. and O'Dwyer, B. (eds), *Sustainability Accounting and Accountability*, pp. 184–208. London: Routledge.

Milne, M. J., Kearins, K. N. and Walton, S. (2006) Creating adventures in wonderland? The journey metaphor and environmental sustainability, *Organization*, 13(6): 801–39.

Milne, M. J., Tregigda, H. M. and Walton, S. (2009) Words not actions! The ideological role of sustainable development reporting, *Accounting, Auditing and Accountability Journal*, 22(8): 1211–57.

Mitchell, R., Agle, B. R. and Wood, D. J. (1997) Toward a theory of stakeholder identification and salience: defining the principle of who and what really counts, *Academy of Management Review*, 22(4): 853–86.

Moon, J. (2002) The social responsibility of business and new governance, *Government and Opposition*, 37(3): 385–408.

Moratis, L. and Cochius, T. (2011) *ISO 26000: The business guide to the new standard on social responsibility*. Sheffield: Greenleaf.

Munck, R. and O'Hearn, D. (eds) (1999) *Critical Development Theory*. London: Zed Books.

Norris, G. and O'Dwyer, B. (2004) Motivating socially responsive decision making: the operation of management controls in a socially responsive organisation, *The British Accounting Review*, **36**(2): 173–96.

O'Dwyer, B., Unerman, J. and Hession, E. (2005) User needs in sustainability reporting: perspectives of stakeholders in Ireland, *European Accounting Review*, **14**(4): 759–87.

O'Riordan, L. and Fairbrass, J. (2008) Corporate social responsibility (CSR) models and theories in stakeholder dialogue, *Journal of Business Ethics*, **83**(4): 745–58.

Oulton, W. (2006) The role of activism in responsible investment: the FTSE4Good indices, in Sullivan, R, and Mackenzie, C. (eds), *Responsible Investment*, pp. 196–205. Sheffield: Greenleaf.

Owen, D. L. and Swift, T. (2001) Social accounting, reporting and auditing: beyond the rhetoric, *Business Ethics: A European Review*, **10**(1): 4–8.

Owen, D. L., Swift, T., Bowerman, M. and Humphreys, C. (2000) The new social audits: accountability, managerial capture or the agenda of social champions?, *European Accounting Review*, **9**(1): 81–98.

Owen, D. L., Swift, T. and Hunt, K. (2001) Questioning the role of stakeholder engagement in social and ethical accounting, auditing and reporting, *Accounting Forum*, **25**(3): 264–82.

Palenberg, M., Reinicke, W. and Witte, J. M. (2006) *Trends in Non-financial Reporting*. Berlin: Global Public Policy Institute.

Pegg, S. (2003) 'An emerging market for the new millennium; Transnational corporations and human rights' in Frynas, J. G. and Pegg, S. (2003) (eds) *Transnational corporations and human rights* (Basingstoke: Palgrave MacMillan) pp. 1–32.

Pleon (2005) *Accounting for Good: The Global Stakeholder Report 2005*. Amsterdam: Pleon.

Polonsky, M. J. and Jevons, C. (2006) Understanding issue complexity when building a socially responsible brand, *European Business Review*, **18**(5): 340–49.

Porter, M. and Kramer, M. (2002) The competitive advantage of corporate philanthropy, *Harvard Business Review*, **80**(12): 5–16.

Preston, L. E. (1978) Analyzing corporate social performance: methods and results, *Journal of Contemporary Business*, **7**: 135–50.

Pricewaterhouse Coopers (2010) *CSR Trends 2010: Stacking up the results*. Toronto: Craib/PWC (www.pwc.com/ca/sustainability).

Rahman, S. F. (1998) International accounting regulation by the United Nations: a power perspective, *Accounting, Auditing and Accountability Journal*, **11**(5): 593–623.

Roberts, C. B. (1990) *International Trends in Social and Employee Reporting*, Occasional Research Paper 6. London: ACCA.

Robertson, J. (1978) *The Sane Alternative*. London: James Robertson.

Schreuder, H. (1979) Corporate social reporting in the Federal Republic of Germany: an overview, *Accounting, Organizations and Society*, **4**(1–2): 109–22.

Singh, D. R. and Ahuja, J. M. (1983) Corporate social reporting in India, *International Journal of Accounting*, **18**(2): 151–70.

SPADA (2008) *Environmental Reporting: Trends in FTSE 100 Companies* (http://www.spada.co.uk/wp-content/uploads/2008/11/environmental-reporting-spada-white-paper.pdf).

Spence, C. and Gray, R. (2008) *Social and Environmental Reporting and the Business Case*. London: ACCA.

Sullivan, R. and Mackenzie, C. (eds) (2006) *Responsible Investment*. Sheffield: Greenleaf.

SustainAbility (2005) *Influencing Power: Reviewing the conduct and content of corporate lobbying*. London: SustainAbility/WWF.

SustainAbility/UNEP (1996) *Engaging Stakeholders: The benchmark survey*. London/Paris: Sustainability/UNEP.

SustainAbility/UNEP (1997) *Engaging Stakeholders: The 1997 Benchmark survey*. London/Paris: SustainAbility/UNEP.

SustainAbility/UNEP (1998) *Engaging Stakeholders: The non-reporting report*. London/Paris: SustainAbility/UNEP.

SustainAbility/UNEP (1999) *Engaging Stakeholders: The internet reporting report*. London/Paris: SustainAbility/UNEP.

Teegen H., Doh, J. P. and Vachani, S. (2004) The importance of nongovernmental organizations (NGOs) in global governance and value creation: an international business research agenda, *Journal of International Business Studies*, **35**: 463–83.

Teoh, H. Y. and Thong, G. (1984) Another look at corporate social responsibility and reporting: an empirical study in a developing country, *Accounting, Organizations and Society*, **9**(2): 189–206.

Thomson, I. and Georgakopoulos, G. (2008) Social reporting, engagements, conflicts and controversies in an arena context, *Accounting, Auditing and Accountability Journal*, **21**(8): 1116–43.

Tilt, C. A. (1994) The influence of external pressure groups on corporate social disclosure - some empirical evidence, *Accounting, Auditing and Accountability Journal*, **7**(4): 47–72.

Tinker, T. and Neimark, M. (1987) The role of annual reports in gender and class contradictions at General Motors: 1917–1976, *Accounting, Organizations and Society*, **12**: 71–88.

Trotman, K. T. (1979) Corporate responsibility disclosure by Australian companies, *The Chartered Accountant in Australia*, **49**(8): 24–8.

Trucost (2004) *Environmental Disclosure in the Annual Report and Accounts of Companies in FTSE All-share*. Bristol: Environment Agency.

Unerman, J. and Bennet, M. (2004) Increased stakeholder dialogue and the internet: towards greater corporate accountability or reinforcing capitalist hegemony?, *Accounting, Organizations and Society*, **29**(7): 685–707.

Unerman, J. and O'Dwyer, B. (2006) Theorising accountability for NGO advocacy, *Accounting, Auditing and Accountability Journal*, **19**(3): 349–76.

UN (1948) *The Universal Declaration of Human Rights*. Geneva: United Nations.

UN (1991) *Accounting for Environmental Protection Measures*, Report of the UNCTAD Secretariat, January (E/C.10/AC.3/1991/5). New York: United Nations.

UNCTC (1984) *International Standards of Accounting and Reporting*. New York: United Nations.

Vogel, D. (2005) *The Market for Virtue: The potential and limits of corporate social responsibility*. Washington, DC: Brookings Institute.

Warhurst, A., Cooper, K., in association with Amnesty International (2004) *The 'UN Human Rights Norms for Business'*. Bradford on Avon: Maplecroft.

Williams, S. M. and Ho Wern Pei, C. (1999) Corporate social disclosures by listed companies on their web sites: an international comparison, *The International Journal of Accounting*, **34**(3): 389–419.

Wood, D. (1991) Corporate social performance revisited, *Academy of Management Review*, **16**(4): 691–718.

CHAPTER 6

Employees and unions

6.1 Introduction

Amongst the most complex – and, in some senses, the least understood – of the relationships that humankind develops are those which we hold with the world of **work**. Richard Donkin goes so far as to say: '[t]oday we seem to take the need to work for granted. Some would argue that it is a psychological necessity. . . . [W]ork has come to dominate our existence' (2010: xv and xxi). And yet, despite Bourdieu's claim that '[w]hat is valued is activity for its own sake, regardless of its strictly economic function . . . .' (Bourdieu, 1990: 116) it is not at all obvious that work has always dominated mankind's existence. Donkin (2010) goes on to re-evaluate how much time was actually spent in work in hunter–gatherer societies and concludes that, it would generally have been a relatively small proportion of the day: in fact, he identifies one view that work should actually be defined as 'something that we would rather not be doing' (Donkin, 2010: 4).

The purpose of this brief reflection about the nature of 'work' is to remind ourselves that there is a multitude of ways in which mankind engages in work – from slavery to social enterprise and every possibility in between – and that, however ubiquitous it might be, the notion of 'employee' is neither universal nor one that we should necessarily take for granted. Indeed, as MacEwan reminds us '. . . in any society, work is carried out in a variety of ways [and different forms of work organisation] can co-exist in the same society and the same time' (MacEwan, 1999: 183). Consequently, lying just behind the discussion in this chapter sits the notion that '**employment**' itself must never be taken entirely for granted as either essentially desirable or as a singular idea.

Indeed, it is the very diversity of forms that 'employment' can take that informs much of the concern surrounding accounting for and about employees and employment. At the heart of our concerns with social accounting and accountability lies the concept of relationships (see Chapter 3) – and if the relationship of mankind with work is problematic, those of society with employment and employees with the employing organisation are more complex still. And, as you might expect, how one perceives the relationships generally reflects one's worldview (see Chapter 3). Is the relationship fundamentally and essentially one of conflict and exploitation as a socialist might see it (Thompson, 1989) and/or just one aspect of modernity's necessary exploitation of the natural environment (Curry, 2006)? Or is it, rather, the mechanism via which people find self-fulfilment through increased morale, satisfaction and pride in their organisation (Kassinis, 2012)? The answer of course will depend on a range of factors including the nature of our employment contract; our level or responsibility, authority and autonomy; gender; race; age; and the fit of our work with our abilities, aptitude, skills, experience and culture. These matters need not detain us here, but employment is a great deal more important and complex than the simple matter that human resource management (sic) would suggest to us (Mellahi *et al.*, 2010). Interestingly, whilst we might as employees consider employee satisfaction very important and recognise its importance to our goodwill towards our employer and hence our performance, such data is often not collected or reported (see Hubbard, 2011).

Once we start to address the control of, and accountability to, employees and employee-based organisations, we are still faced with a challenge. At its simplest, employees are quite probably the most complex of organisational stakeholders. As Parkinson says:

> . . . employees are often seen not merely as one of several 'outside' groups whose interests merit protection against too ruthless a pursuit of profit, but also as a special group with a claim to be regarded as 'insiders', with a right equal to that of the shareholders to demand that the company be run for their benefit
>
> (Parkinson, 1993: 397)

This question of their rights to rank at least as highly as the providers of capital is a recurrent theme to which we will return. But even here it is not as simple as it seems. Henriques (2007), for example, suggests that employees are not in any real sense 'inside' the organisation and that the statement 'employees are our greatest asset' is wrong in so many ways – not least because, slavery aside, they cannot be an 'asset' as they are not owned. And yet:

> staff are encouraged to consider themselves part of the company and identify with it . . .
> The company's values . . . are usually chosen to reflect the values which senior management would like staff to hold, so that their work is more helpful to the company's goals.
>
> (Henriques, 2007: 41)

Williams and Adams (2013) note that 'it cannot be assumed . . . that companies will have regard to their moral responsibility to take account of the interests of employees, and of their accountability both to employees and to wider society with regard to the discharge of that responsibility' (p. 483) and find an 'interesting managerial twist' (p. 482) that ultimately this lack of accountability is bad for business.

It is clear that the nature of work and employment is constantly changing and affecting different strata of different societies in continuously different ways (Donkin, 2010) and whether you, our readers, aspire to a stellar career of over-achievement and reward, yearn for a job for life, embrace the notion of a portfolio of diverse employment or would simply be grateful for any means to provide warmth, shelter and food for you and yours, you are part of this ever-changing landscape through which we as humankind organise our economic needs. Indeed:

> . . . in an age of flexibility and downsizing, the psychological contract between employee and firm has been greatly weakened. Jobs are no longer for life; conversely firms can no longer expect the same degree of loyalty from employees. Moreover, there has been an increasing divergence between managerial and employee pay; the latter has tended to stagnate, ostensibly to ensure greater competitiveness and to reduce inflation (but also reflecting the reduced bargaining power of employee collectives).
>
> (Mellahi *et al.*, 2010: 237)

At its heart, it seems difficult to shake off the notion that '. . . the pursuit of profit [is set] on a collision course with the human importance of meaningful work' (Knights and Willmott, 200: 63).

It is in this context that this chapter considers reporting to and about employees and unions. Until the early 1990s, the majority of social and environmental disclosures made by companies, predominantly within the annual financial report and accounts, related to human resource and workplace issues. In particular, reporting initiatives in Western Europe (including innovative special-purpose reports) placed an over-riding emphasis on the enterprise–employee relationship, reflecting a continuing debate at that time over the status of labour and its position within the enterprise. The advent of corporate environmental reporting in the guise of 'standalone' special-purpose reports in the early 1990s, however, signalled a major shift in reporting priorities, with labour issues, other than workplace health and safety, very much taking a back seat.

The later evolvement of the purely environmental into a more rounded 'sustainability' reporting model, with its re-introduction of a social reporting component, has led to human capital issues once again achieving some prominence. As Johansen (2008) points out, however, employees remain a somewhat neglected stakeholder group in social accounting research (see also Gray, 2002). In particular, limited attention has been paid to the consequences of possible differences in information needs between employees and other stakeholder groups.

Nevertheless, there are sound reasons for considering the particular accountability needs of employees separately from other stakeholders. Firstly, and uniquely, they are in a dual accountability position – being accountable to higher level management in the workplace whilst, at the same time, generally regarded as a beneficiary of reporting initiatives purportedly designed to ensure accountability to stakeholders (Johansen, 2008). Secondly, as Maunders (1981) and Brown (2000) suggest, employees might be regarded as being in an analogous position to that of shareholders in having 'invested' their labour and having 'capital' tied up in the organisation (significantly, capital that is by no means as mobile as financial capital).

Survey evidence indicates that corporate management regard 'employee motivation' as an important driving force behind their Corporate Social Responsibility (CSR) initiatives (KPMG, 2005), whilst seeing the development of a 'more satisfied and productive labour force' as one of its key benefits (Carroll and Buchholtz, 2006). However, whether employees themselves benefit from CSR initiatives, in terms of them actually addressing their central concerns such as employment protection, better working conditions and improved representation rights, is open to question. Royle (2005), for example, argues that voluntary social reporting initiatives are essentially employed as part of a strategy to persuade policy-makers not to introduce, or tighten up, hard forms of regulation which, at least in many western mainland European countries, have traditionally provided the most effective means for protecting workers rights. In contrast to Royle's position, proponents of social reporting, and associated reporting initiatives, of course, routinely assert that it provides an effective means of enhancing accountability to employees, and indeed stakeholders more generally (see, in particular AccountAbility, 1999, 2008).

Our key aim in this chapter is to investigate the link, if any, between social reporting and the discharging of accountability to employees. In particular, we focus on the reporting of employment information (as an integral component of corporate social/sustainability reporting) and the reporting of corporate information in general to employees. In this latter context, we also consider the institutional framework within which such information may be effectively utilised. Attention is focused here on the role of trade unions, which have traditionally provided the mechanism through which employee interests, in terms of wage levels, job security, working conditions and health and safety, amongst other issues, have been advanced.

The chapter first considers reporting of employment information from early developments to its inclusion in modern day 'sustainability' reports. Section 6.3 then considers accounting for human resources. Section 6.4 considers the issue of reporting to employees covering employee reports and accountability to trade unions. Last, but not least, in Section 6.5 we consider accountability for equality in employment whilst Section 6.6 offers a few reflections, conclusions and future possibilities.

6.2 Reporting employment information

Early developments

In the early years of social reporting, companies in both developed and developing nations (according to the limited evidence available) placed greater emphasis on disclosure about human resources than other areas of social reporting (see, for example, Guthrie and Parker, 1990; Roberts, 1990). Yet, countries vary in the types of disclosure and relative emphasis placed on disclosures about employment. Roberts (1990), for example, found that European, South African and, to a lesser extent, Australasian companies were more likely than their counterparts in other parts of the world to disclose employment policies and health and safety information, or have separate sections of their annual reports on employment or value added data (we discuss this further below). By contrast, in the USA reporting practice tended to be directed towards the interests of the general public and consumers, with employment information largely confined to issues of race and gender equality (Preston *et al.*, 1978).[1]

The most innovative early approaches to the reporting of employment information took place in Western Europe (see, for example, Lessem, 1977; Schreuder, 1979). This is perhaps not surprising given the European Union's long history of concern with the working conditions of employees and their status within the organisation, as exemplified by the adoption of the EU Social Charter and the promotion of Works Councils giving employee representatives considerable information and consultation rights. Additionally, the impact of the trade union movement has traditionally been greater in Western Europe than, for example in the USA[2] where consumerism and equal rights have historically been higher profile issues with, arguably, a consequent influence on reporting priorities.

Surveys of reporting practice across six European countries – Germany, Sweden, France, Switzerland, the Netherlands and the UK (Roberts, 1990, 1991; Adams *et al.*, 1995b) – indicated that disclosure of employment information within annual financial reports was fairly widespread, with the main areas of disclosure being pay and benefits, breakdown of employee numbers by gender, geographical location, etc., recruitment/redundancies and training. Other important issues such as employee consultation and trade union representation were much less frequently disclosed, whilst the majority of employment-related disclosures made tended to be qualitative and general, rather than specific in nature.

Leading the way in terms of both overall disclosure volume and extent of specifically quantitative disclosure were German companies, which indeed have a particularly long tradition of reporting on employee issues. Brockhoff (1979), for example, in a survey of 296 annual reports published in 1973–74 noted that 205 companies published a clearly identifiable social section (*Sozialbericht*). Even at this early stage in the evolution of reporting, 28% of the sample provided a breakdown of the workforce with respect to salaried versus hourly paid personnel; 22% gave data for German versus foreign personnel working in Germany; 17% for female versus male employees; and 14% information on the

---

[1] This rather underlines the point that rather than being developed in a theoretically coherent, logical manner, social and environmental reporting practices tend to respond to issues that arise and fall in the social and political environment. A further example of this phenomenon in the employee-related domain relates to the development of voluntary employment information disclosure in Southern Africa (arising from the issuing of a United Nations code of practice for multi-national companies) which reflected considerable popular concern over apartheid and related social conditions in the 1970s. Response by companies was, however, patchy and often partial in nature (see Patten, 1990).

[2] Although the impact of the Teamsters in the USA was clearly very significant indeed, and they still remain influential, it is rather that, as far as we are aware, this influence has had little impact on reporting, accounting and disclosure as we have discussed it here.

age distribution of the workforce. In contrast, 97% of the sample reported information on pensions and retirement benefits, 47% on apprenticeship programmes and 43% on other training programmes. Other less covered issues included employee housing and security of the workplace (21 and 20% respectively). As these figures indicate, social reporting in Germany was a voluntary phenomenon with a large discrepancy in terms of reporting sophistication between 'leading edge' companies and the rest. It was encouraged by institutions such as the Business and Society Foundation (established by business leaders to study important social developments affecting the business community), the Social and Behavioural Science Division of the Battelle Institute and the government sponsored International Institute for Environment and Society.

A more uniform approach towards social reporting developed in France following publication of the *Sudreau Report* in 1975, which suggested a wide ranging series of measures concerned with social reform of the business enterprise. Amongst issues considered were shareholder protection, relations with consumers and the environment, inflation accounting, regional development and the promotion of small businesses. The main thrust of the report's proposals, however, concerned the relationship between the enterprise and its employees, with much attention directed towards improving working conditions, together with employee consultation and information rights.[3] In the latter context, it was suggested that each enterprise produce, separately from the financial report, an annual social balance sheet (*bilan social*) based on indicators of its social and working conditions. Legislative support for this proposal followed in 1977, with a mandatory requirement initially introduced for companies employing more than 750 people to publish social balance sheets in 1979, soon extended in 1982 to those employing more than 300. The *bilan social*, publication of which continues to be mandatory, includes non-financial quantitative as well as financial information covering the seven areas itemised in Figure 6.1.

Intriguingly, in the UK the *Corporate Report* published by the (then) Accounting Standards Steering Committee (1975) was an initiative concerned with re-examining the scope and aims of published financial reports, in particular their role in promoting *public* accountability, recommended as one of six additional statements to appear in financial reports a special purpose 'employment report'. In terms of information provision, this statement had much in common with France's *bilan social*. Underpinning the recommendation was a specific concern with a perceived lack of corporate accountability to the workforce:

> Nothing illustrates more vividly the nineteenth century origin of British Company Law than the way in which employees are almost totally ignored in present Companies Acts and in corporate reports.
>
> (Accounting Standards Steering Committee, 1975: para 6.12)

Vehement opposition from influential organisations such as the Stock Exchange and the Confederation of British Industry (CBI) towards any move away from the traditional 'stewardship' concept of corporate governance, with its narrowly acknowledged obligations towards shareholders, in favour of a more public form of accountability contributed to the limited use of employment statements. The subsequent election of the Thatcher government in 1979 espousing similar views ensured they never got off the ground. A similar fate (notwithstanding a far more discernible initial uptake on the part of large companies within their published financial statements) befell the additional employee-related statement suggested in the *Corporate Report* – the **value added statement** (Bougen, 1983). This latter

---

[3]This emphasis reflected ongoing tensions at the time in French society, triggered by the events of 1968 marked particularly by widespread strikes, plant occupations and student riots.

Figure 6.1 Summary of the requirements for France's *bilan social*

1   Details of employees by physical characteristics (such as gender, age, etc.).
2   Levels of employee remuneration and other employment expenses.
3   Hygiene and security conditions (health and safety standards, etc.).
4   Other working conditions (hours of work, incidence of night work, noise levels, etc.).
5   Staff development and training.
6   Information regarding the relationship between firm and employees which provides an indication of the internal social climate.
7   Other employment-related factors (such as initiatives in participative management or employer subsidies to staff facilities).

Source: Taken from Christophe and Bebbington, (1992: 280).

statement is simply, in effect, a restatement of the profit and loss account to show employees, governments and capital providers as recipients of 'value added' (being turnover less bought-in materials and services) rather than costs to the business.[4] However, value added statements, whilst largely disappearing from financial reports, have made a re-appearance in the ongoing wave of 'standalone' sustainability reporting, a development influenced by value added forming a key economic performance indicator in sustainability reporting guidelines published by the *Global Reporting Initiative* (see GRI, 2006). One example of a value added Statement is shown in Figure 6.2.

Notwithstanding the fate of the *Corporate Report* recommendations, a longitudinal (1979–1991) study by Gray *et al.* (1995) of social and environmental reporting practices on the part of large UK companies showed workforce-related information dominating disclosure within annual financial statements. Partly this result is driven by (fairly minimal) mandatory reporting requirements on issues such as numbers employed, pension arrangements, employment of the disabled and employee share ownership schemes (ESOPs). However, alarmingly, the volume of disclosure on specific issues such as health and safety and employee consultation remained low, whilst a later study by Day and Woodward (2004) found a high degree of non-compliance with statutory requirements to disclose employee-related information.[5] Day and Woodward suggest that lack of enforcement of sanctions for non-disclosure together with inadequate monitoring of information content may provide explanatory factors here and conclude that:

> . . . even where legislation is in place the lack of monitoring of compliance suggests that the government acted symbolically by enacting the [disclosure] requirements rather than with the substantive intent of making organisations **accountable**.
>
> (Day and Woodward, 2004: 56, emphasis added).

At the time of writing there was no evidence of which we are aware that the 2006 Companies Act for companies to 'have regard to the interests of employees' and to report on their

---

[4]The employee accountability dimension of value added statements is actually somewhat open to question. On the positive side, the statements ask 'whose profit is it anyway?' and rank employees on a par with providers of finance. On the negative side, they have been used to seek to persuade employees that they already take more than their fair share out of the business (see, for example, Bougen, 1983, 1984).

[5]Specifically, Day and Woodward's study was concerned with mandatory disclosure in the Directors Report of information on matters of concern to employees; employee consultation; employee involvement in company performance (for example through participation in an employee share scheme); and achievement of employee financial and economic awareness.

Figure 6.2 A value added statement: extract from ECC Group Report for Employees 1987

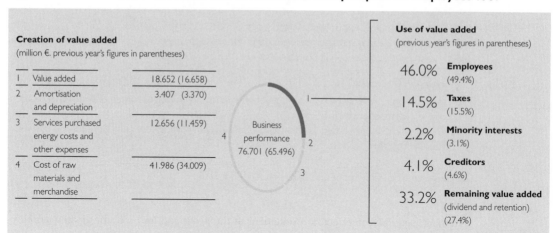

Creation of value added
(million €. previous year's figures in parentheses)

1	Value added	18.652 (16.658)
2	Amortisation and depreciation	3.407 (3.370)
3	Services purchased energy costs and other expenses	12.656 (11.459)
4	Cost of raw materials and merchandise	41.986 (34.009)

Business performance 76.701 (65.496)

Use of value added
(previous year's figures in parentheses)

46.0% **Employees**
(49.4%)

14.5% **Taxes**
(15.5%)

2.2% **Minority interests**
(3.1%)

4.1% **Creditors**
(4.6%)

33.2% **Remaining value added**
(dividend and retention)
(27.4%)

interactions with staff in their annual report had made any noticeable difference to reporting practice.

Certainly, notions of corporate accountability to employees, which would appear to have driven both the reporting developments in the 1970s in Western Europe referred to above and changes in labour law designed to enhance employee and trade union information rights (an issue we shall return to later in this chapter), disappeared from the public policy agenda, at least in Anglo-American capitalist societies, with the advent of the 1980s Thatcher/ Reagan 'greed is good' decade. Massive levels of unemployment consequent upon the decline of heavy manufacturing industry, together with the repealing of much labour law and the pulling of the teeth of the rest, led to the breaking of the power of trade unions. Additionally, a process of 'exporting jobs' to developing countries where labour legislation was so much less constraining on business grew apace. Indeed, the alacrity with which so much of corporate UK, USA, Australasia, etc. abandoned the higher wage economies of the West for the lower wage ones of developing nations was proof enough of how seriously social responsibility to employees had been taken by business.

In the light of these developments, it comes as little surprise that interest in reporting employment information waned, with the little produced being, as Day and Woodward (2004) suggest, simply symbolic in intent. Even within the European Union, where employee rights remained firmly established on the political agenda, reporting itself rather stagnated through the 1980s (Gray *et al.*, 1996), and somewhat significantly the next major reporting innovation at the beginning of the following decade centred on the production of specialist environmental reports, with employment information largely confined to health and safety issues. However, labour issues are showing some signs of returning to the reporting scene.

In an interesting development in the early years of the 21st century, the United Nations Conference on Trade and Development (UNCTAD) attempted to revitalise recognition of the importance of the disclosure of labour indicators to CSR. Indeed the UN's CSR working group saw employee indicators as forming the backbone of their voluntary reporting initiative – albeit a reporting initiative in which economic considerations clearly dominated any social or environmental ones. The recommendations followed relatively predictable lines and suggested including in a company's annual report such elements as workforce wages and benefits, employees by age and gender, trade union representation as well as training and health and safety issues. As with so many UN reporting initiatives, the reasoning and ideas are sound, but the lack of suasion behind the recommendations means that take up tends to be low (see, for example, Kamp-Roelands, 2009). A more substantial development, however, has been a shift in the production of standalone reports away from the production of purely environmental reports towards **'sustainability' reports**, a shift which gathered pace from the mid-1990s onwards.[6]

The employment dimension within sustainability reports

Pioneering the move towards a more socially rounded reporting were a number of 'values-based' organisations espousing a wide set of social and ethical objectives rather than simply being concerned with seeking profits. Prominent amongst these were Traidcraft, the Co-operative Bank and Body Shop in the UK and the Danish Sbn Bank (Sparekassen Nordjylland), whose initial reporting forays from the early to mid 1990s were increasingly taken up by more mainstream companies as the decade progressed. As KPMG's triennial surveys of international corporate responsibility reporting indicate (see particularly KPMG 2005, 2008, 2011), 'sustainability' reporting[7] has steadily displaced purely environmental reporting on an international level, at least as far as the largest companies are concerned, over recent years. Whilst environmental issues, of course, still figure prominently, and indeed most space is generally devoted to this dimension in published sustainability reports (see, for example, Larrinaga-González, 2001), some attempt is also made to address the economic and social dimensions within the same report – a process for which Elkington (1997) memorably coined the phrase **'triple bottom line reporting'** (see Chapter 9). Significantly, a study of a sample of UK reports published in 2005 by Erusalimsky *et al.* (2006) found social disclosure to be dominated by employment information, with 'the space devoted to employees . . . more than that devoted to communities, customers and suppliers (the next three biggest stakeholder groups) combined' (p. 17).

As was the case with the majority of reporting initiatives considered earlier in this chapter, and the later foray into 'standalone' environmental reporting, sustainability reporting is purely a voluntary phenomenon. However, guidelines addressing issues of both report preparation and content continue to be issued (see particularly AccountAbility, 1999, 2008, 2011; GRI, 2000, 2002, 2006, 2011). Most influential on an international scale have been the *GRI Reporting Guidelines* which specify both reporting principles (influenced by the AA1000 Standards) and required content in the form of **performance indicators**. In the latter context, employment issues are covered via indicators (both core and additional) addressing specific aspects of labour practices and human rights performance derived from internationally recognised standards of conduct such as the Conventions of the International Labour Organisation (ILO) and international instruments such as the United Nations Declaration of Human Rights. Figure 6.3 provides a summary of core performance indicators within the labour practices and decent work category. Core indicators addressing human rights issues particularly relating to the workplace dimension (GRI, 2006) are discrimination

---

[6]A further recent development worthy of brief mention here is the move on the part of some European countries, notably Denmark, France and Germany, to introduce a mandatory requirement for the disclosure of employment information in the annual financial report. Amounts disclosed are, however, in the main fairly minimal. Similarly in the UK, one can expect to find small amounts of employment information in the Business Review section, the replacement for the ill-fated Operating and Financial Review (see Chapter 11), of the annual report.

[7]It is worth reminding ourselves about this term 'sustainability' reporting. The term is widely used – (as it is used in this chapter) – as a catch-all term for selective social, environmental and possibly sustainable development data. The term has no direct link with sustainability as understood in the Brundtland report, but we have adopted the convention because most commentators seem to do so (see Chapter 9). Hereafter 'sustainability' will not appear in quotation marks – although arguably it should.

Figure 6.3 Labour practices and decent work: a summary of core indicators

> 1 Employment (including analysis of workforce by employment type, contract and region; employee turnover by age group, gender and region).
> 2 Labour/Management Relations (including percentage of employees covered by collective bargaining agreements).
> 3 Occupational Health and Safety (including management processes; formal agreements with trade unions; statistical data on injury rates, occupational diseases, fatalities, etc.).
> 4 Training and Education (including information on programmes and amount of training per year by employee category).
> 5 Diversity and Equal Opportunity (including breakdown of employees by reference to gender, age group and minority group membership; ratio of basic salary of men to women by employee category).

Source: GRI, GR3 Guidelines (2006).

(number of incidents and actions taken); freedom of association and collective bargaining; child labour; and forced and compulsory labour. In the latter three areas, disclosure should focus on identifying operations carrying significant risk that abuses may occur and measures taken which contribute to eliminating such abuses. Expanded guidance for reporting on human rights, local community impacts and gender were included in the G4 Guidelines (GRI, 2013).

From the above, it is clear that the GRI recommendations for reporting on employment issues have much in common in terms of subject matter with the French *bilan social* and UK employment statement initiatives we discussed earlier. Unfortunately, as Erusalimsky *et al.* (2006) point out, there is a dearth of systematic studies collating detailed data from standalone reports (but see Hubbard, 2011), so it is impossible to arrive at an overall global picture of the influence of the GRI indicators on employment reporting practice. Casual observation suggests that information appertaining to the core labour practices and human rights indicators is routinely included, the latter particularly by multinational companies operating in the extractive industries. There are, however, large discrepancies in both volume and rigour of disclosure (for more detail and critique see Chapter 9).

Moving on from the issue of recommended report content, one notable way in which the GRI Guidelines go further than the earlier 1970s initiatives lies in the encouraging of the reporting organisation to engage with key stakeholder groups in preparing the report, and to explain within it how the organisation has responded to their reasonable expectations and interests. Significantly, the stakeholder-centred approach to sustainability reporting is also a guiding principle of the international *AA1000 Standards* which are largely concerned with matters of reporting principles and processes rather than content. The issue of stakeholder dialogue and engagement is explored at some length in Chapter 11. We would simply note at this stage that, despite the encouragement offered to companies by the GRI and AA1000 standards to adopt a participatory and transparent approach to reporting, no rights to information are built into the process and hence power differentials between the organisation and its (in this case employee) stakeholders remain unchanged. Indeed, for Johansen (2010) there are reasons to be sceptical about the development of new reporting structures *per se* in terms of their ability to deliver accountability to employees. Rather the issue is one of bringing about institutional reform, whereby power differentials can be effectively addressed and reporting perhaps thereby made more relevant (see also Owen *et al.*, 1997). We shall return to this issue when considering the role of trade unions in the context of corporate information disclosure later in this chapter.

The need for caution when considering the many rhetorical claims companies make concerning employee accountability in the context of the reporting function is re-enforced by the findings of a study of the **human capital management (HCM)** disclosures made in a sample of 'leading edge' sustainability reports from UK companies by the Association of Chartered Certified Accountants (ACCA, 2009). Of the 40 reports analysed (all entrants to the ACCA's 2008 Sustainability Reporting Awards Scheme), whilst all but two included a section on 'people' or 'employees' in their reports, and thereby might be presumed to regard employees as key stakeholders, 43% did not mention any form of engagement with employees (via surveys, focus groups, etc.), and of those that did only 35% gave detailed feedback on the results of such dialogue. Moreover, the overall standard of reporting itself was 'lower than hoped' (p.7). Whilst being generally strong on rhetoric in the form of disclosure of broad strategies and organisational 'visions' and 'values', reporting was, by contrast, very patchy in the areas of governance structures in place to manage HCM issues and actual performance data.

A similar picture was painted by an earlier (2008) study, also conducted under the auspices of the ACCA of the HCM disclosure practices of companies comprising the Top 50 index of the Australian Securities Exchange as at 31st July 2007 (ACCA, 2008), with even high scoring companies failing to report on 'a comprehensive set of indicators that would monitor all the key Human Capital Management issues' (p.10). A particular area of weakness from an accountability perspective identified in this latter study concerned a lack of explanation of the mechanisms in place to incorporate employee feedback into decision-making processes for HCM.

Noticeable in the case of the above ACCA sponsored studies is the focus on 'human capital *management*', rather than human capital or simply employment information, disclosure. Significantly, whereas it is argued that HCM is a sustainability issue, it is also acknowledged to be a business issue:

> It is now generally considered that many business opportunities can be found in a firm's intangible assets, which include intellectual capital and a diverse, knowledgeable and skilled workforce. Conversely, it is also widely believed that poor management of human capital is one of the most significant risks facing organisations
>
> (ACCA, 2009: 3)

A similar concern with the potential business benefits arising from the management of human capital underpinned the development of another form of employee-related reporting, that of human asset or **human resource accounting (HRA)**. Originating from the late 1960s, and rather fading from view in the late 1970s, HRA has enjoyed a renewed prominence in recent years under the guise of **intellectual capital accounting (ICA)**. Regardless of the niceties of terminology, however, the over-riding objective of such initiatives is to arrive at a financial value of employees to the organisation. Clearly, ethical (as well as practical) issues arise here, as well as questions concerning the role HRA or ICA may play in promoting corporate accountability to the workforce. It is to a consideration of such issues we now turn.

6.3 Accounting for human resources

The decade from the mid-1960s to the mid-1970s witnessed a great deal of academic interest in the area of human resource accounting, defined as being:

> . . . the process of identifying and measuring data about human resources and communicating this information to interested parties.
>
> (American Accounting Association, 1973: 169)

The concept was originally popularised by the American social psychologist Rensis Likert who adopted a behavioural science approach to human resource measurement in investigating the relationship between the system of management used and productivity of the organisation (see, in particular, Likert, 1967). Likert's major concern was that traditional accounting systems with their emphasis on short-term profit maximisation tended to encourage managers to adopt authoritarian management styles and ignore factors such as the need for employee participation in decision-making and for more training for subordinates. Likert argued that short-term profit increases generated in this way were largely illusory, as the resultant increases in employee turnover and consequent additional spending on hiring and training more than offset the immediate savings. He went on to advocate incorporating human resource measures into formal organisational accounting systems in order to direct managerial attention to the value of their human resources.

The challenge set for accountants, to develop suitable measurement methods for valuing the human resource, was rapidly taken up, although Likert's broad definition of the term 'human resources', which included the value of assets such as a firm's human organisation, customer loyalty and reputation in the local community, was narrowed down to a concentration on the value of the firm's human capital – its workforce. The models developed can be conveniently classified as cost-based and value-based. In the case of the former, costs incurred in, for example, recruiting, training and developing or replacing employees, instead of being written off immediately as an expense are capitalised and then amortised over the expected useful life of the employee (or human asset). By contrast, value-based methods adopt a more forward looking economic income approach whereby an attempt is made to value the future contribution that the employee may make to the firm. Among several possible suggested approaches here are models based on discounting future salaries (Lev and Schwarz, 1971) or future forecast earnings of the firm (Brummet et al., 1968).[8]

Whilst some practical initiatives in HRA were undertaken, most notably at the R.G Barry Corporation in the USA (see Flamholtz, 1974), widespread implementation failed to materialise, with the result that interest in the area faded rapidly in the latter part of the 1970s. However, the issue never entirely left the accounting research agenda. Flamholtz, an indefatigable champion of HRA, continued to pursue practical, field-based studies (see, for example, Flamholtz, 1987) whilst occasional evaluative survey papers served to keep interest simmering (Harte, 1988; Sackman et al., 1989; Scarpello and Theeke, 1989). Indeed, this interest began to escalate in the mid-1990s as the recognition grew that in modern economies the greatest source of competitive advantage for many organisations lies in the 'intellectual capital' contributed by their workforce. A number of researchers have, in particular, drawn attention to the growing differences between the market value, or capitalisation, of a firm and net book value represented in the balance sheet. As Roslender and Fincham (2004) point out, this difference is frequently greatest in knowledge-based industries, where intellectual capital is most critical to business performance.[9]

Significantly, this surge of interest in the valuing of human resources in the new guise of ICA heralded something of a return to the broader definition of the term originally promulgated by Likert. Whilst writers have tended to differ in their definitions of intellectual capital,[10]

---

[8]For extended discussion and analysis of the various cost- and value-based measurement approaches developed see Harte (1988).

[9]It is, of course, overly simplistic to attribute the difference between market and book value of the firm as being solely attributable to the existence of intellectual capital. As Abeysekara (2008) points out, market value of a listed company (as represented by its share price) can diminish due to a wide range of factors which are beyond the control of the firm and have nothing whatsoever to do with a diminution in the stock of their intellectual capital.

[10]Particularly influential works here are those of Brooking (1996), Edvinsson and Malone (1997) and Sveiby (1997).

there is a broad consensus that it can be broadly classified into **human capital, structural capital** and **relational capital** (Tayles *et al.*, 2007). Human capital is made up of intangible factors, basically the knowledge, skill and creativity of employees, whereas structural capital comprises more tangible assets such as software, databases and patents, and relational capital refers to factors such as customer and supplier relationships and industry networks.

Whilst arriving at financial measures of key aspects of structural capital is relatively straightforward, clearly the same cannot be said of the relational and human components of intellectual capital. One notable attempt to at least make visible the individual components of intellectual capital lies in the intellectual capital statements produced as supplements to the annual financial accounts by the Scandinavian financial services company Skandia since the mid-1990s. Essentially, the intellectual capital statements are an attempt to account for (and indeed encourage) organisational value creation utilising a loosely coupled range of financial, non-financial and even pictorial representations of the firm's intellectual capital. Their nature and purpose seem somewhat abstract – as Mouritsen *et al.* put it:

> Intellectual capital statements do not attempt to form one bottom-line expression of value. Rather they attempt, through networks of sketches, stories and numbers, to form paths along which new value-creating activities can be supported. Sketches about the management of relations between employees, customers, technologies and organisational routines and procedures; stories about the effects of bundles of human capital, structural/organisational capital and customer capital; and configurations of loosely coupled numbers that accompany and make the implementation of the story-line accountable and thus serious.
>
> (Mouritsen *et al.*, 2001: 419)

Practical experimentation in intellectual capital accounting would appear to be largely confined to the Scandinavian block of countries, with the emphasis very much being on managing rather than simply accounting for human resources (see Roslender and Fincham 2001, 2004; Roslender and Stevenson, 2009). Roslender and Fincham (2004) suggest that this largely reflects the social welfare culture and underlying social democratic tradition of these countries with their associated emphasis on technological advance, the development of highly educated workforces and pursuit of extensive value added activity. External reporting is predominantly qualitative in nature, rather than an attempt simply to place a financial value on the workforce. A notable example here being the *Danish Intellectual Capital Statement*, developed out of a Danish government funded research project (Danish Agency for Trade and Industry, 2000). According to Roslender and Stevenson the presentation of such management produced 'knowledge narratives' carries significant emancipatory potential in that:

> Not only do they provide alternatives to the highly restrictive valuation emphasis implied in putting people on the balance sheet, they do so in ways that are not greatly reliant on financial numbers and reporting formats . . . The opportunity to account for people using indicators shaped by the needs of people rather than accounting is a major step forward, alongside the possibility of using this approach in association with extensive narrative content.
>
> (Roslender and Stevenson, 2009: 857)

They go on to suggest that such emancipatory potential can be yet better realised if employee 'self accounts', possibly in the form of a 'yearbook' bringing together employee reflections of how the organisation is performing in human resource development terms (Roslender and Fincham, 2001), were to replace the current management derived version.

Beyond the Scandinavian experience, intellectual capital reporting is very much a minority pursuit (Roslender and Fincham, 2004). Whilst there is considerable research interest in the area, it is somewhat significant to note that this appears to be driven by capitalist market rather than human development concerns. For example, Li *et al.* (2008) point to the role of intellectual capital in creating and maintaining competitive advantage and shareholder value, with its disclosure providing valuable information for investors in helping reduce uncertainty about future prospects and facilitating a more precise valuation of the company. Similarly, from a management control perspective, for Tayles *et al.* (2007), the key issue of interest is the value-driven transformation of human and relational capital into the structural capital of the organisation. Somewhat revealingly, in writing of the Skandia experience in developing intellectual capital, Edvinsson (1997) adopts a similar perspective here. Whilst referring to the need to nurture the roots (human resources) of the organisation and acknowledging the fact that intellectual capital is at least as important as financial capital in providing sustainable corporate earnings, he goes on to note that:

> . . . a key role of leadership is the transformation of human capital into structural capital. Furthermore, the human capital cannot be owned, it can only be rented. The structural capital can, from a shareholder's point of view, be owned and traded . . . structural capital can be used as a leverage for financing corporate growth.
>
> (Edvinsson, 1997: 369)

Edvinsson's observation gives rise to some suspicion that the claims of ICA proponents that attributing a financial valuation to the human resource (albeit somewhat more indirectly than HRA sought to do) is not of paramount importance are perhaps a trifle disingenuous. Certainly, it would appear that issues of economic efficiency and long-term financial benefit to the organisation, rather than those of social efficiency *per se*, are a key driving factor, significantly a criticism applied by a number of writers (see, for example, Glautier, 1976; Marques, 1976; Cherns, 1978) to earlier HRA research initiatives. As Abeysekera (2008) points out, the seeking of ways to convert human capital into structural capital simply amounts to a concern with maximising the market value of the organisation and indeed represents a commodification of labour in that the know-how and expertise of employees becomes embedded in systems and processes owned by the firm.

Other writers raise further ethical reservations over the thinking behind ICA and related management practices. Gowthorpe (2009), for example, points to the potential for IC management metrics to act as 'malign instruments of management control' which simply treat people as a means towards the end of economic gain, whilst ignoring the crucial issues of the distribution and appropriation of such gains. For his part, McPhail (2009) points to the whole debate over the growth of the knowledge economy, and related concerns with intellectual capital issues, being concerned with finding methods to deal with the intangible nature of contemporary capitalism, particularly reflected in the huge discrepancies between book and market values of companies referred to earlier. There is, however, 'little exploration of the new threats and opportunities it poses for greater corporate accountability and democratic progress' (p. 805).

McPhail (2009) goes on to suggest that intellectual capital reporting and corporate social disclosure possess important conceptual differences. In particular, the latter emerged largely as a response to measuring things solely in terms of their market value whilst the former seems far more concerned with communicating previously undisclosed assets in market value terms. Even more avowedly qualitative approaches towards ICA, such as the Danish Intellectual Capital Statement, as Nielsen and Madsen (2009) point out, possess limited emancipatory potential in that the voice of management is privileged with the latter's 'perception of strategy, key management challenges and key performance indicators form[ing] the basis of company disclosure' (p. 848).

ICA may, of course, have the potential to inform the development of corporate social disclosure. Certainly Roslender's long-standing championing of employee 'self reporting' has much relevance in this context. Additionally, Yongvanich and Guthrie's (2006) suggestion of developing an integrated reporting framework comprising intellectual capital, balanced scorecard and social and environmental disclosure in order to more fully address the economic and non-economic dimensions of organisational performance may be worthy of further exploration. However, it is only too clear that the overwhelming body of extant ICA research and practical experimentation implicitly presumes the same classical capitalistic objectives, essentially maximisation of shareholders returns, as did its predecessor HRA (see, for example, Cherns, 1978) with, fundamentally, people continuing to be considered a resource for the enterprise, rather than the enterprise a resource for people (Marques, 1976).

6.4 Reporting to employees

Our discussion of the reporting of employment information in the previous section strongly suggests that, whatever else it may or may not achieve, such activity has little to do with the discharging of accountability to the workforce. Its extent depends, amongst other things, on the degree to which management believes the workforce is committed to its values and whether management sees trade unions as cooperative and committed to the organisations' goals (see Peccei *et al.*, 2008) and whether or not there is joint consultation (see Peccei *et al.*, 2010). But, where it is recognised as desirable, accountability may possibly be more successfully established via the medium of reporting corporate information *to* employees. It is to this issue that we now turn our attention.

Employee reports

As we noted in the introduction to this chapter, human capital issues now figure prominently in corporate sustainability reporting practice, with many reports, at least implicitly if not explicitly, indicating that the workforce is considered a key audience for the information provided. However, there seems little attempt being made to ascertain the specific information needs of employees or to consider the consequences of these needs possibly differing from the other stakeholder groups commonly lumped together as potential report users (see Johansen, 2008). Interestingly here, an earlier reporting initiative in the form of specifically designated **employee reports** (reports to employees) which emerged in the 1970s did represent an attempt to communicate directly with the workforce.

Employee reports were (and still are) in no sense standardised and so covered a very wide range of styles and subjects. However, most attempted to convey, in an accessible form featuring a liberal use of photographs, line drawings, bar charts, pie charts and graphs, information on the company's financial performance and position together with divisional and product line information thought to be of particular interest to the workforce. For a time, employee reports represented a very popular form of communication on the part of companies, with a survey undertaken by Maunders (1982), for example, indicating that 77% of the 300 largest UK commercial and industrial companies produced such a document (sometimes in the form of simplified financial statements for both employees and shareholders) in 1981/82.

Despite the popularity of employee reporting, a number of commentators expressed severe reservations concerning the usefulness of the information provided to its intended audience. A study by Lyall (1982), for example, based on a random sample of employee reports drawn from the *Times 1,000* largest UK companies found that they generally failed to contain enough real information to satisfy employees' basic information needs over key issues such as job security, company performance and wealth sharing. Similarly, a further study by Hussey (1979) concluded that the desire on the part of employees for strategic information on future developments and plans was the need least well satisfied in the reports he studied. Further criticisms centred on a perceived degree of bias in the way information was presented. This was most apparent in the chairman's statement which Parker (1977) suggested simply tended to resort to clichés, exhortations, political dogma and management conventional wisdom to the extent that employees would be likely to reject the report in its entirety.

Lyall (1981) identified bias in the way information was presented in value added statements which featured as a commonly employed presentational device within employee reports. For example, Lyall noted that company retentions and distributions to shareholders were commonly shown net of tax whilst payments to employees were shown inclusive of tax and national insurance, thus effectively over-stating by a considerable amount the employee share of value added in terms of take home pay. A further issue here, of course, is that any increase in national insurance would suggest an apparent increase in the employees' share of value added, whereas it clearly represents an increase in the share taken by government.

Notwithstanding their deficiencies in terms of information provision, the production of employee reports might be taken as a sign of enlightened management, in that the existence, and importance, of the employee stakeholder is at least recognised. With the move towards 'harder' management styles through the 1980s, it therefore comes as little surprise that the production of such reports rapidly waned. Any mourning of their demise should, however, be tempered in that employee reports largely represented an exercise in downward communication (Parker, 1977) that fell far short of implementing any meaningful form of accountability to the workforce. The essential point here is that the importance of reporting in accountability terms doesn't just rest upon information provision *per se* but also its role in facilitating action on the part of recipients (Stewart, 1984; Bailey *et al.*, 2000). Crucial here is the institutional framework within which information transmission takes place (Johansen, 2010). A central feature of workplace institutional arrangements is that of the **collective bargaining** arena wherein employees through the medium of their trade union representatives can, rather than simply acting as passive recipients of information management chooses to disclose, actively influence the content of such disclosure and, more importantly, the decisions it goes on to inform. In particular, an avowedly pluralist, labour orientated, rationale might be brought to bear in the latter context which rejects the rationale underpinning the employee reporting initiatives we have just considered that 'what is good for capital is good for everyone else' (Brown, 2000).

Trade unions and corporate accountability

The issue of the disclosure of *financial* information to trade unions for collective bargaining purposes was the subject of much public policy debate throughout the 1960s and 1970s, with an apparent measure of consensus emerging across the political spectrum that information disclosure had a important role to play in promoting more informed and 'rational' bargaining. This culminated in the UK in the enactment of the Employment Protection Act 1975 giving unions rights to receive:

(a) information without which they would to a material extent be impeded in carrying out collective bargaining; and

(b) information it would be good industrial relations practice to disclose.

These provisions have been retained in subsequent legislation, whilst further information and consultation rights for employees have been introduced, largely emanating from European Union initiatives, particularly in potential takeover and collective redundancy situations (EU, 2002; Information and Consultation of Employees (ICE) Regulations, 2004).

Early empirical studies of the actual use being made of financial information by trade unions (Mitchell *et al.*, 1980; Moore and Levie, 1981; Reeves and McGovern, 1981; Jackson-Cox *et al.*, 1984) presented a somewhat negative picture, with union representatives showing little enthusiasm for adding a concern with the financial circumstances of the company to their more traditional emphasis on the cost of living, productivity and comparability concerns when conducting wage negotiations. These studies drew particular attention to a number of specific problems facing unions attempting to utilise corporate information in the collective bargaining situation (see McBarnet *et al.*, 1993). Key issues arising here are:

● A large difference in the degree of expert power in the use of information operates in favour of management, with union research departments being relatively small and over-worked while education services are over-stretched and under-financed.

● The potential for mobilising membership support behind demands based on detailed financial arguments appears very limited compared to support that could be mobilised behind comparability or cost of living claims.

● Management controls the communication process, thus giving them a wide discretion concerning what to disclose or not to disclose.[11]

● Management possession of the means of production enables them to command decision-making processes.

The latter two points in particular highlight a situation of managerial control over strategic decision-making issues, with the union side restricted to a reactive stance, enabling them to merely impede implementation of decisions with no real power to initiate and influence decision topics. This is a highly significant issue in that, as Ogden (1986) points out, the whole *raison d'etre* for unions to pursue the disclosure issue is for them to be in a position to *extend* collective bargaining beyond the traditional areas of concern, namely terms and conditions of employment. Additionally, merely accepting at face value routine, non-sensitive, information that management chooses to disclose whilst being excluded from key policy-making networks carries the danger of the union side simply absorbing managerial values and imperatives, or being 'sucked into management' (see, for example, Brown, 2000).

Interestingly here, empirical work by McBarnet *et al.* (1993) provides some evidence of unions beginning to use financial information strategically in order to critique and challenge management's plans. They suggest that:

> According to our pilot study, trade unions use financial information for a range of purposes, some more predictable than others: for wage bargaining, for negotiation in takeover and merger situations, in relation to closure or redundancy proposals, to argue the ability of management to improve health and safety conditions; they use it to negotiate over profit related pay and in arguing against tenders for public contracts; they use it for recruitment campaigns and in wider propaganda battle within the public or in the political arena more generally.
>
> (McBarnet *et al.*, 1993: 87)

---

[11]Significantly here, employers have frequently resorted to taking advantage of the generous exemption clauses to the main provisions of the Employment Protection Act. In particular, recourse has often been made to the 'substantial injury' clause, which exempts companies from having to disclose information that would cause substantial injury to the enterprise for reasons other than its effect on collective bargaining.

Whilst McBarnet *et al.* present a persuasive case suggesting that unions can successfully employ what they term 'adversary' accounting techniques in order to avoid being sucked into management, their analysis places the union very much in the position of reacting to management plans whilst also confining the debate to purely financial dimensions of performance. However, union concerns clearly go beyond the financial performance domain and extend into issues of corporate environmental and social impact, with the latter encompassing amongst other things equality of opportunity, health and safety at work and accountability in the supply chain.[12]

The trade union movement has indeed taken a keen interest in the environmental dimension of performance over many years. The Trades Union Congress (TUC), for example established an Environment Action Group back in 1989 which, amongst other initiatives, encouraged individual unions to negotiate 'Green Agreements' with employers and to seek active involvement in corporate environmental initiatives (see TUC, 1991). The TUC's proactive stance was enthusiastically followed by several individual unions although employer resistance proved a major stumbling block to progress in many instances (see Benn, 1992; Jackson, 1992). Signs of much more constructive engagement are, however, apparent in the success of the TUC's Green Workplaces project (TUC, 2010) which featured a number of initiatives in both private and public sector workplaces that brought together the practical engagement of both workers and management to secure energy savings and reduce organisational environmental impact. Amongst key benefits claimed flowing from the exercise (TUC, 2010: 4) were:

- mutual appreciation of the material impact such projects can have in reducing carbon emissions serving to foster improved industrial relations;
- the training of 97 environmental representatives, resulting in changes to workplace structures and the formation of environmental committees/forums;
- effectiveness in building capacity to extend the trade union consultation agenda to cover environmental issues.

The Green Workplaces project provides an example of a social partnership between employers and trade unions, which Ackers and Payne (1998) argue represents 'the institutional process of applying the spirit of business ethics and the theory of stakeholding to the employment relationship' (p. 530). Ackers and Payne draw on the writings of the Italian Marxist, Antonio Gramsci (see, particularly, Gramsci, 1978) in order to conceptualise this particular union strategy. Gramsci distinguished between a 'war of manoeuvre', or frontal assault on the prevailing economic and social system, and a 'war of position'. The latter, Ackers and Payne argue, is more appropriate to modern western societies which, whilst exhibiting vast inequalities of wealth and influence amongst their populations, are ruled more by consent than coercion. In conducting a war of position they stress that:

> . . . the union movement has to transcend the corporate pursuit of narrow economic interests, and construct a hegemonic bloc around itself linked to a programme that speaks for the needs of society in general against the interests of conservative power blocs. Moreover, such a programme is not confined to unrealizable rhetoric or propaganda, but must embed itself in the institutional practices of society.
>
> (Ackers and Payne, 1998: 545)

For Ackers and Payne, the attraction of a social partnership strategy lies, at least partly, in it representing 'a moveable feast susceptible to redefinitions in a more radical direction'

---

[12]See, for example, in this context the UNITE union's involvement in War on Want's 'Love Fashion Hate Sweatshops' campaign (www.unitetheunion.org).

(p. 546). This more radical direction is outlined in a later paper by Lee and Cassell (2008) who advocate a war of position in which unions build alliances with other civil society institutions critical of the prevailing social and economic order. Significantly, the issue of corporate social disclosure plays an important role in Lee and Cassell's analysis. Essentially, they seek to establish common ground between advocates of social and environmental reporting and critical theorists' sceptical of its role in bringing about meaningful social change. For Lee and Cassell, corporate disclosure, whilst not unimportant, is not an end in itself but the prelude to developing alternative 'accounts' capable of reflecting popular demand for change emanating from employees' challenges to employers in the workplace or from wider social movements within civil society. Interestingly, Lee and Cassell's analysis has much in common with Cooper *et al.*'s earlier (2005) stressing of the need for social and environmental accounts to be articulated to social movements, whilst their views on reporting developments largely reflect those of the seminal work of Roy Moore and his colleagues at Ruskin College (Moore *et al.*, 1979; Moore and Levie, 1981) on the issue of extending collective bargaining. The latter particularly point to the need for unions to move along an 'information scale'. Starting at the bottom of the scale, with an acceptance of company information as it comes, the union moves through points such as asking for additional information, seeking access to the management information system itself and changing the system towards the ultimate point of developing its own trade union information system. Crucial to the success of such a process, it is noted, is the ambitiousness of union demands.

There are, of course, a number of practical obstacles standing in the way of union ambitions to extend collective bargaining into the social reporting arena. First, and foremost, a similar situation to that of financial information disclosure prevails, this being management control of the communication and decision-making process. Additionally, the fear of losing power and influence is a very real one for labour, in that social reporting processes, whilst generally claimed by their proponents as being attempts to engage the workforce in issues of policy and strategy, may effectively by-pass union representatives (Preuss *et al.*, 2009).

Perhaps not surprisingly, therefore, Preuss' (2008) analysis of European unions' responses to the EU Commission's 2001 Green Paper *Promoting a European Framework for Corporate Social Responsibility* highlights a considerable degree of suspicion towards it among trade unionists. Amongst unions in mainland Western Europe in particular, there appears to be a profound reluctance to embrace the concept of stakeholderism. This can largely be explained by the systems of industrial relations prevailing in these countries. In countries such as France, Belgium, Spain, Italy, Germany and the Netherlands, a 'Romanic-Germanic' system prevails, with the state playing a key role resulting in industrial relations issues being extensively legally codified. In the Nordic countries, whilst the state plays a limited role, a network of collective agreements between unions and employer associations, which are seldom challenged, characterises the industrial relations arena. As Preuss *et al.* (2009) point out, the scope granted in these regulated environments for unions to possess rights of influence over corporate decision-making offers an uneasy fit with the corporate discretion inherent in public reporting.

By contrast in the UK, characterised by collective bargaining arrangements at company, or plant, level and with state involvement being generally minimal, union engagement in sustainability reporting initiatives might appear more conceivable. Certainly, Preuss' (2008) analysis of union responses to the EU Commission's Green Paper suggests that UK unions, at least on the surface, seem more at ease than their mainland compatriots with the whole notion of stakeholding. Evidence of UK trade unions at least beginning to explore how public reporting and the related stakeholder concept might be utilised in support of traditional union goals is provided by the union AMICUS' *Corporate Responsibility Guide* (2007). This document notes that many of the issues covered by social reporting are core trade union campaigning issues and further suggests 'there is evidence to support the premise that the benefits of positive dialogue on CSR can be beneficial for both employees and employers' (p. 3). Significantly, the Guide goes on to stress the need for rigorous reporting and auditing procedures to be established and argues that such an outcome can only be brought about through legislation, a position echoed in the TUC's response to the EU Commission's Green Paper (Preuss, 2008).

Notwithstanding the generally supportive tone adopted in the AMICUS Guide, it is made clear that CSR initiatives have to fit within prevailing institutional frameworks:

> . . . CSR is not a substitute for collective bargaining (or trade unions) and the CSR agenda should be pursued through existing workplace channels. Looking long term, AMICUS would wish to see the myriad of issues that could be defined within CSR become mainstreamed to form part of any modern progressive company's collective negotiations agenda.
>
> (AMICUS, 2007: 17)

Whilst clearly underlining the potential importance of CSR initiatives to trade unions, the above comment also re-enforces Johansen's (2010) observation concerning the fundamental importance of considering institutional arrangements when evaluating its relevance and usefulness. The main barrier to trade union engagement with CSR initiatives and reporting on them would therefore appear to lie in management's apparent reluctance to accept this point. Certainly, from a UK perspective, one sees little sign of managerial ambitions to engage trade unions centrally in CSR initiatives or the reporting process. Rather, there seems a tendency to attempt to by-pass the union constituency via spurious employee stakeholder engagement exercises which, in not offering an effective channel for the employee voice to be effective, have little to do with establishing accountability to the workforce.

6.5 Accountability for equality in employment

We consider the various intra-governmental and non-governmental organisation (NGO) initiatives to put pressure on, and give guidance to, companies in addressing human rights issues in Chapter 5 where the focus is primarily on social and community issues. But human rights also apply to employment issues, and a number of organisations incorporate statements on issues such as employee representation and gender diversity in statements about their responses to human rights guidelines, their human rights initiatives and their human rights performance.

Whilst there has been a long history of public reporting that highlights corporate and societal attitudes to (a lack of) equality in the workplace (see Chapter 5), this by no means reflects a situation worthy of attracting favourable attention. In fact, the way women and women's contribution to the workforce were portrayed reflected an acceptance of a hugely inequitable situation by the dominant forces in a patriarchal western society. Adams and Roberts (1999), for example, found a low frequency of disclosure on equal opportunities in Europe, a continent otherwise known (relatively speaking) for its historical concern for the rights of employees generally. The *lack* of reporting on the employment situation of women and ethnic minorities through the 20th century and the limited reporting of employment segregation and isolated breakthroughs, such as the appointment of the first female bank branch manager (see Figures 6.4 and 6.5), tell a story of deep-rooted attitudes to women and ethnic minorities in the workforce in the context of a patriarchal and racist society (see Adams and Harte, 1998; Adams and McPhail, 2004).

Figure 6.4 Reporting reflecting gender segregation

It seems likely that the development of electronic bookkeeping will evolve methods of dealing with some at least of the great number of entries which the Bank is called upon to handle. Looking ahead it is reasonable to visualise further changes in the staffing structure of the Bank. Almost certainly the number of ladies on our staff will increase, while the number of men may quite possibly tend to fall. If this be so it will become more important than ever before to ensure that every boy who joins the staff is of the material to become a manager and that he has ample opportunity to study banking in all its aspects.

To this end we have for some years maintained a Staff Training College through which many hundreds of men have passed. . . . We are now making such facilities available to younger men . . .

I am glad to report that young men are recognising that banking offers prospects of a happy and useful career with plenty of opportunity . . . (National Provincial, 1954 corporate annual report).

Source: Adams and Harte, (1998: 794).

Tinker and Neimark (1987) suggested that the changing nature of women's exploitation reflected changes in the crises facing capitalism, with capitalist alienation being a prime factor in the oppression of women. They used a political economy framework in their analysis of the portrayal of women in the annual reports of General Motors between 1917 and 1976. Cooper and Puxty (1996: 299) criticised their work for focusing on the economic and underplaying the social context, the role of patriarchy and for failing to allow women to 'speak for themselves'.

Adams and Harte's (1998) and Adams and McPhail's (2004) longitudinal studies (from 1935–1993 and 1935–1998, respectively) of (non) reporting on women and ethnic minority employment respectively in the UK banking and retail sectors examined disclosures, and lack of them, in the context of prevailing societal attitudes drawing from a range of historical sources. The studies provide evidence that changing notions of patriarchy and attitudes to race influence both reporting and employment practices with respect to minority groups even where this is not in the economic interests of the business. Adams and Harte (1998) concluded that a: '*his*tory of employment in banking and retail, drawing on the corporate annual reports when set in the social, political and economic contexts can be seen to be largely *his* story' *[emphasis in original]* (p. 808). Their finding that patriarchy is an important influence on disclosure of women's employment practices is consistent with studies of women's experiences in the accounting profession (see, for example, Ciancanelli *et al.*, 1990; Loft, 1992: Kirkham and Loft, 1993). With respect to the employment of ethnic minorities, Adams and McPhail contend:

> A Marxist analysis does not fully explain nondisclosure, which may be related to social attitudes and changes in the state's approach, through the CRE, to managing the evolving threats of racism.
>
> (Adams and McPhail, 2004: 431)

Adams *et al.* (1995a) studied reporting by the top 100 UK companies for reporting years ending in 1991. The study examined disclosures in the Annual Reports on: specific equal opportunities policies; other evidence of equal opportunities commitment; and reference to external pressures, initiatives and legislation. There is no UK legal requirement to disclose information on women's employment, but there is a requirement to disclose policy with respect to disabled employees in the Director's Report section of the annual report. Given interest in the relative impacts of voluntary and regulatory approaches, it is worth noting that these authors found that only 34 companies complied fully with this legislation in the

Figure 6.5 Reporting reflecting isolated successes

We have opened a number of new branches during the year, including one in the West End of London under the management of a member of our women staff, Miss E. M. Harding. This interesting experiment has been hailed in some quarters as a portent, as indeed in a sense it is, but it may also be regarded as a natural and perhaps somewhat belated recognition that the holding of responsible posts in contact with our customers is no longer necessarily an exclusively male preserve (Barclays, 1958 corporate annual report).

Source: Adams and Harte, (1998: 795–96).

corporate annual report, 52 complied partially, whilst 14 made no mention of disabled employees (Adams *et al.*, 1995a). These findings suggest that legislation requiring disclosure to the public which is not enforced is not fully effective and that other drivers, in particular social regulation, may be required to encourage public accountability. However, where a policy with respect to disabled employees was disclosed in the Director's Report, the policy generally also referred to sex, race and religion.

Adams and Harte (1999) studied the portrayal of equal opportunities performance in three British organisations from a variety of stakeholder perspectives.[13] They found that detailed performance data collected for internal purposes (e.g. to monitor compliance with equal opportunities legislation in the event of an Equal Opportunities Commission (EOC) or Commission for Racial Equality (CRE) investigation or a court case) was not reported or summarised in external company reports.

The contributions of Adams and Harte (1998, 1999, 2000) and Adams *et al.* (1995a) had identified a number of key influences on the (non) reporting of women's employment *and* the nature of women's employment, including: the second world war; unemployment levels; equal opportunities legislation; pressure from the CRE, EOC and trade unions; government rhetoric; the changing nature of work; patriarchal views; and demographic changes. A more recent increased emphasis on (reporting on) diversity issues for women found in Grosser *et al.* (2008) can be explained by: an increasing soft regulation and government (agency) intervention with regard to diversity issues; changes in society giving employees (as well as other stakeholders) a greater voice in corporate affairs; a desire to be a 'preferred employer' (in order to attract and retain the best staff, thereby reducing training and recruitment costs and increasing earnings); and responding to pressure from key stakeholders such as the media and investors.

6.6 Some conclusions, reflections and possibilities

It seems quite apparent that the nature of employment and the relationship(s) between employers and employees are most complex and continue to change in previously unanticipated ways. Employment is certainly not homogeneous – if ever it was – but seems to be becoming even less so. It is no longer enough – if ever it was – to imagine the principal form of employment as some idealised European partnership of workforce and management, or as a sort of 1970s Anglo union–management conflict model. Within the developed countries, the gaps between those in successful employment and those in casual or no employment

---

[13]These included: trade unions representing their workers; the Equal Opportunities Commission (EOC) and the Commission for Racial Equality (CRE); two organisations which monitor the ethical and social performance of organisations, the Ethical Consumer Research Association (ECRA) and the Ethical Investment Research Service (EIRIS); academic literature; a database search for legal cases; and contact with the Industrial Tribunal Offices in Scotland and England.

grows ever wider. We are seeing the return of slavery and sweatshops all across the globe whilst new forms of employment through social enterprises and employee-ownership schemes and cooperatives offer such promising futures (see Chapter 12). Attempts to break down old distinctions between (say) management and (say) employees are successful to a degree – but with unforeseen consequences as senior managers becoming more seemingly psychopathic (Hines, 2007) and white collar employees more ambitious and anxious.

It is unfortunate then that the accounting literature seems to have backed off from the area just as the issues became more complex and demanding. As we have seen, the main thrust of employee-related research in (social) accounting seems to have moved away from an interest in unions and the well-being of employees to a concern with the far more managerialist and technist concerns of intellectual capital – albeit researchers such as Roslender can still see emancipatory potential here.

But there are signs that researchers have not abandoned the area altogether. As we saw above, issues of gender, race and equality remain of vibrant concern (Adams and McPhail, 2004, Haynes, 2008) and there is a stirring of interest in the conditions of and accounting for supply chain labour in sweatshops and factories in the developing world (Islam and McPhail, 2011) as well as a wider concern with human rights (Gray and Gray, 2011). But scholarship now appears to be a long way behind the state it was 40-odd years ago when academics and research offered complex and detailed narratives about the nature and conditions of employees and employment. It will take a substantial effort by new researchers to open this area up again, but this remains, in our view, one of the many potentially fruitful areas for further research in social accounting.

References

Abeysekera, I. (2008) Intellectual capital practices of firms and the commodification of labour, *Accounting, Auditing and Accountability Journal,* 21(1): 36–48.

ACCA (2008) *Disclosure on Human Capital Management.* Sydney: Association of Chartered Certified Accountants Australia and New Zealand.

ACCA (2009) *Human Capital Management: An Analysis of Disclosure in UK Reports.* London: Association of Chartered Certified Accountants.

AccountAbility (1999) *AA 1000 Framework: Standard, Guidelines and Professional Qualification.* London: AccountAbility.

AccountAbility (2008) *The AA 1000 AccountAbility Principles Standard.* London: AccountAbility.

AccountAbility (2011) *AA1000 Stakeholder Engagement Standard.* London: AccountAbility.

Accounting Standards Steering Committee (1975) *The Corporate Report.* London: ICAEW.

Ackers, P. and Payne, J. (1998) British trade unions and social partnership: rhetoric, reality and strategy, *The International Journal of Human Resource Management,* 9(3): 529–50.

Adams, C. A. and Harte, G. (1998) The changing portrayal of the employment of women in British banks' and retail companies' corporate annual reports, *Accounting, Organizations and Society,* 23: 781–812.

Adams, C. A. and Harte, G. (1999) *Towards Corporate Accountability for Equal Opportunities Performance,* Occasional Paper 26. London: ACCA.

Adams, C. A. and Harte, G. F. (2000) Making discrimination visible: the potential for social accounting, *Accounting Forum,* 24(1): 56–79.

Adams, C. A. and McPhail, K. (2004) Reporting and the politics of difference: (non)disclosure on ethnic minorities, *Abacus,* 40(3): 405–35.

Adams, C. A. and Roberts, C. B. (1999) Corporate ethics: an issue worthy of report?, *Accounting Forum,* 19(2/3): 128–42.

Adams, C. A., Coutts, A. and Harte, G. F. (1995a) Corporate equal opportunities (non) disclosure, *British Accounting Review,* 27(2): 87–108.

Adams, C. A., Hill, W. Y. and Roberts, C. B. (1995b) *Environmental, Employee and Ethical Reporting in Europe*, Research Report 41. London: ACCA.

American Accounting Association (1973) Report of the committee on human resource accounting, *Accounting Review Supplement*, 169–85.

AMICUS (2007) *Amicus Corporate Social Responsibility Guide*. London: AMICUS.

Bailey, D., Harte, G. and Sugden, R. (2000) Corporate disclosure and the deregulation of international investment, *Accounting, Auditing and Accountability Journal*, 13(2): 197–218.

Benn, H. (1992) Green negotiating: the MSF approach, in Owen, D. (ed.), *Green Reporting: Accountancy and the challenge of the nineties*. London: Chapman and Hall.

Bougen, P. (1983) Value added, in Tonkin, D. J. and Skerratt, L. C. L. (eds), *Financial Reporting 1983/84*. London: ICAEW.

Bougen, P. (1984) Review of *Linking Pay to Company Performance* by Vernon Harcourt T, *British Accounting Review*, 16(1): 96.

Bourdieu, P. (1990) *The Logic of Practice*. Trans Richard Nice. Cambridge: Polity.

Brockhoff, K. (1979) A note on external social reporting by German companies: A survey of 1973 company reports, *Accounting, Organizations and Society*, 4: 77–85.

Brooking, A. (1996) *Intellectual Capital: Core Asset for the Third Millennium*. London: International Thompson Business Press.

Brown, J. A. (2000) Competing ideologies in the accounting and industrial relations environment, *British Accounting Review*, 32(1): 43–75.

Brummett, R. L., Flamholtz, E. G. and Pyle, W. C. (1968) Human resource measurement – a challenge for accountants, *The Accounting Review*, 43(2): 217–24.

Carroll, A. B. and Buchholtz, A. K. (2006) *Business and Society: Ethics and Stakeholder Management*. Mason, OH: Thomson Southwestern.

Cherns, A. B. (1978) Alienation and accountancy, *Accounting, Organizations and Society*, 3: 105–14.

Christophe, B. and Bebbington, K. J. (1992) The French Bilan Social: A pragmatic model for the development of accounting for the environment? A research note, *The British Accounting Review*, 24(3): 281–90.

Ciancanelli, P., Gallhofer, S., Humphrey, C. and Kirkham, L. (1990) Gender and accountancy: some evidence from the UK, *Critical Perspectives on Accounting*, 1: 117–44.

Commission of the European Communities (2001) *Green Paper: Promoting a European Framework for Corporate Social Responsibility*. Brussels: Commission of the European Communities.

Cooper, C. and Puxty, A. (1996) On the proliferation of accounting (his)tories, *Critical Perspectives on Accounting*, 7: 285–313.

Cooper, C., Taylor, P., Smith, N. and Catchpole, L. (2005) A discussion of the political potential of social accounting, *Critical Perspectives on Accounting*, 16(7): 951–74.

Curry, P. (2006) *Ecological Ethics: an introduction*. Cambridge: Polity.

Danish Agency for Trade and Industry (2000) *A Guideline for Intellectual Capital Statements: A Key to Knowledge Management*. Copenhagen: Danish Agency for Trade and Industry.

Day, R. and Woodward, T. (2004) Disclosure of information about employees in the Directors' Report of UK published financial statements: substantive or symbolic?, *Accounting Forum*, 28(1): 43–59.

Donkin, R. (2010) *The History of Work*: London: Palgrave Macmillan.

Edvinsson, L. (1997) Developing intellectual capital at Skandia, *Long Range Planning*, 30(3): 366–73.

Edvinsson, L. and Malone, M. (1997) *Intellectual Capital: Realising your company's true value by finding its hidden brain power*. New York: Harper Collins.

Elkington, J. (1997) *Cannibals with Forks: The triple bottom line of 21st century business*. London: Capstone.

Erusalimsky, A., Gray, R. and Spence, C. (2006) Towards a more systematic study of stand-alone corporate social and environmental reporting: an exploratory pilot study of UK reporting, *Social and Environmental Accounting Journal*, 26(1): 12–19.

EU (2002) *EU Directive 2002/14/EC of the European Parliament and of the Council. Establishing a General Framework for informing and Consulting Employees in the European Community*, 11 March.

Flamholtz, E. G. (1974) *Human Resource Accounting*. California: Dickenson.

Flamholtz, E. G. (1987) Valuation of human assets in a securities brokerage firm: an empirical study, *Accounting, Organizations and Society*, 12: 309–18.

Glautier, M. W. E. (1976) Human resource accounting – a critique of research objectives for the development of human resource accounting models, *Journal of Business Finance and Accounting*, 3(2): 3–21.

Gowthorpe, C. (2009) Wider still and wider? A critical discussion of intellectual capital recognition, measurement and control in a boundary theoretical context, *Critical Perspectives on Accounting*, 20(7): 823–34.

Gramsci, A. (1978) *Selections from the Prison Notebooks*. London: Lawrence and Wishart.

Gray, R. (2002) The social accounting project and accounting, organizations and society: privileging engagement, imaginings, new accountings and pragmatism, *Accounting, Organizations and Society*, 27: 687–708.

Gray, R. and Gray, S. (2011) Accountability and human rights: a tentative exploration and a commentary, *Critical Perspectives on Accounting*, 22(8): 781–9.

Gray, R. H., Kouhy, R. and Lavers, S. (1995) Corporate social and environmental reporting: a review of the literature and a longitudinal study of UK disclosure, *Accounting, Auditing and Accountability Journal*, 8(2): 47–77.

Gray, R. Owen, D. and Adams, C. (1996) *Accounting and Accountability: Changes and challenges in corporate social and environmental reporting*. Harlow: Prentice Hall Europe.

GRI (2000, 2002, 2006, 2011, 2013) *Sustainability Reporting Guidelines*. Amsterdam: Global Reporting Initiative.

Grosser, K., Adams, C. A. and Moon, J. (2008) *Equal Opportunity for Women in the Workplace: A study of corporate disclosure*. London: ACCA.

Guthrie, J. and Parker, L. D. (1990) Corporate social disclosure practice: a comparative international analysis, *Advances in Public Interest Accounting*, 3: 159–76.

Harte, G. (1988) Human resource accounting: a review of some of the literature, in *Making Corporate Reports Valuable - The Literature Surveys*, pp. 217–19. Glasgow: ICAS.

Haynes, K. (2008) Moving the gender agenda or stirring chicken's entrails: Where next for feminist methodologies in accounting?, *Accounting, Auditing and Accountability Journal*, 21(4): 539–55.

Henriques, A. (2007) *Corporate Truth: The limits to transparency*. London: Earthscan.

Hines, J. A. H. S. (2007) The shadow of MacIntyre's manager in the Kingdom of Conscience constrained, *Business Ethics: A European Review*, 16(4): 358–71.

Hubbard, G. (2011) The quality of the sustainability reports of large international companies: an analysis, *International Journal of Management*, 28 (3 Part 2 Sept), 824–47.

Hussey, R. (1979) *Who Reads Employee Reports?* Oxford: Touche Ross.

Information and Consultation of Employees Regulations (ICE Regulations) (2004) *Statutory Instrument No. 3426*. London: HMSO.

Islam, M. and McPhail, K. (2011) Regulating for corporate human rights abuses: the emergence of corporate reporting on the ILO's human rights standards within the global garment manufacturing and retail industry, *Critical Perspective on Accounting*, 22(8): 790–810.

Jackson, A. (1992) The trade union as environmental campaigner; the case of water privatisation, in Owen, D. (ed.), *Green Reporting: Accountancy and the challenge of the nineties*. London: Chapman and Hall.

Jackson-Cox, J., McQueeney, J. and Thirkell, J. E. M. (1984) The disclosure of company information to trade unions – the relevance of the ACAS Code of Practice on Disclosure, *Accounting, Organizations and Society*, 9: 253–73.

Johansen, T. R. (2008) Blaming oneself: examining the dual accountability role of employees, *Critical Perspectives on Accounting*, 19(4): 544–71.

Johansen, T. R. (2010) Employees, non-financial reports and institutional arrangements: a study of accounts in the workplace, *European Accounting Review*, 19(1): 97–130.

Kamp-Roelands, N. (2009) Corporate responsibility reporting, in UNCTAD (eds), *Promoting Transparency in Corporate Reporting: A Quarter Century of ISAR*, pp. 99–112. New York: United Nations.

Kassinis, G. (2012) The value of managing stakeholders, in Bansal, P. and Hoffman, A. J. (eds), *The Oxford Handbook of Business and The Natural Environment*, pp. 83–100. Oxford: Oxford University Press.

Kirkham, L. M. and Loft, A. (1993) Gender and the construction of the professional accountant, *Accounting, Organizations and Society*, 18: 507–58.

Knights, D. and Willmott, H. (2007) *Introducing Organizational Behaviour and Management.* London: Thomson Learning.

KPMG (2005) *KPMG International Survey of Corporate Responsibility Reporting 2005.* Amsterdam: KPMG International.

KPMG (2008) *KPMG International Survey of Corporate Responsibility Reporting 2008.* Amsterdam: KPMG International.

KPMG (2011) *KPMG International Survey of Corporate Responsibility Reporting 2011* (kpmg.com).

Larrinaga-González, C. (2001) The GRI Sustainability Reporting Guidelines: a review of current practice, *Social and Environmental Accounting Journal,* **25**(1): 1–4.

Lee, B. and Cassell, C. (2008) Employee and social reporting as a war of position and the union learning representative initiative in the UK, *Accounting Forum,* **32**(4): 276–87.

Lessem, R. (1977) Corporate social reporting in action: an evaluation of British, European and American practice, *Accounting, Organizations and Society,* **2**: 279–94.

Lev, B. and Schwarz, A. (1971) On the use of the economic concept of human capital in financial statements, *Accounting Review,* **46**(1): 103–11.

Li, J., Pike, R. and Haniffa, R. (2008) Intellectual capital disclosure and corporate governance structure in UK firms, *Accounting and Business Research,* **38**(2): 137–59.

Likert, R. (1967) *The Human Organisation.* New York: McGraw Hill.

Loft, A. (1992) Accountancy and the gendered division of labour: a review essay, *Accountancy, Organizations and Society,* **17**: 367–78.

Lyall, D. (1981) Financial reporting for employees, *Management Decision,* **19**(3): 33–8.

Lyall, D. (1982) Disclosure practices in employee reports, *Accountants Magazine,* July: 246–8.

MacEwan, A. (1999) *Neo-liberalism or Democracy? Economic strategy, markets and alternatives for the 21st Century.* London: Zed Books.

Marques, E. (1976) Human resource accounting: some questions and reflections, *Accounting, Organizations and Society,* **1**: 175–8.

Maunders, K. T. (1981) Disclosure of company financial information to employees and unions: the state of the art, *AUTA Review,* **13**(2): 5–19.

Maunders, K. T. (1982) Simplified and employee reports, in Tonkin, D. J. and Skerratt, L. C. L. (eds), *Financial Reporting 1982–1983,* pp. 173–7. London: ICAEW.

McBarnet, D., Weston, S. and Whelan, C. J. (1993) Adversary accounting: strategic uses of financial information by capital and labour, *Accounting, Organizations and Society,* **18**: 81–100.

McPhail, K. (2009) Where is the ethical knowledge in the knowledge economy? Power and potential in the emergence of ethical knowledge as a component of intellectual capital, *Critical Perspectives on Accounting,* **20**(7): 804–22.

Mellahi, K., Morrell, K. and Wood, G. (2010) *The Ethical Business: Challenges and controversies.* London: Palgrave Macmillan.

Mitchell, F., Sams, K. T., Tweedie, D. P. and White, P. I. (1980) Disclosure of information - some evidence from case studies, *Industrial Relations Journal,* **11**(5): 53–62.

Moore, R. and Levie, H. (1981) *Constraints upon the Acquisition and Use of Company Information by Trade Unions,* Occasional Paper No. 67. Trade Union Research Unit, Ruskin College.

Moore, R., Gold, M. and Levie, H. (1979) *The Shop Stewards Guide to the Use of Company Information.* Nottingham: Spokesman.

Mouritsen, J., Larsen, H. T. and Bukh, P. N. (2001) Valuing the future: intellectual capital supplements at Skandia, *Accounting, Auditing and Accountability Journal,* **14** (4): 399–422.

Nielsen, C, and Madsen, M. T. (2009) Discourses of transparency in the intellectual capital reporting debate: moving from generic reporting models to management defined information, *Critical Perspectives on Accounting,* **20**(7): 847–54.

Ogden, S. G. (1986) *Trade Unions and the Disclosure of Information,* working paper 86/4. The University of Leeds, School of Economic Studies.

Owen, D. L, Gray, R. H. and Bebbington, J. (1997) Green accounting: cosmetic irrelevance or radical agenda for change?, *Asia Pacific Journal of Accounting,* **4**(2): 175–98.

Parker, L. D. (1977) *The Reporting of Company Financial Results to Employees.* London: ICAEW.

Parkinson, J. E. (1993) *Corporate Power and Responsibility.* Oxford: Oxford University Press.

Patten, D. M. (1990) The market reaction to social responsibility disclosures: the case of the Sullivan Principles signings, *Accounting, Organizations and Society*, **15**: 575–87.

Peccei, R., Bewley, H., Gospel, H. and Willman, P. (2008) Look who's talking: sources of variation in information disclosure in the UK, *British Journal of Industrial Relations*, **46** (2): 340–66.

Peccei, R., Bewley, H., Gospel, H. and Willman, P. (2010) Antecedents and outcomes of information disclosure to employees in the UK, 1990–2004: The role of employee voice, *Human Relations*, **63**(3): 419–38.

Preston, L. E., Rey, F. and Dierkes, M. (1978) Comparing corporate social performance: Germany, France, Canada and the USA, *California Management Review*, **20**(4): 40–49.

Preuss, L. (2008) A reluctant stakeholder? On the perception of corporate social responsibility among European trade unions, *Business Ethics: A European Review*, **17**(2): 149–60.

Preuss, L., Haunschild, A. and Matten, D. (2009) The rise of CSR: implications for HRM and employee representation, *The International Journal of Human Resource Management*, **20**(4): 953–73.

Reeves, T. K. and McGovern, T. (1981) *How Shop Stewards Use Company Information – Ten Case Studies of Information Disclosure*. London: Anglian Regional Management Centre.

Roberts, C. B. (1990) *International Trends in Social and Employee Reporting*, Occasional Research Paper 6. London: ACCA.

Roberts, C. B. (1991) Environmental disclosures: a note on reporting practices in Europe, *Accounting, Auditing and Accountability Journal*, **4**(3): 62–71.

Roslender, R. and Fincham, R. (2001) Thinking critically about intellectual capital accounting, *Accounting, Auditing and Accountability Journal*, **14**(4): 383–98.

Roslender, R. and Fincham, R. (2004) Intellectual capital accounting in the UK: a field study perspective, *Accounting, Auditing and Accountability Journal*, **17**(2): 178–209.

Roslender, R. and Stevenson, J. (2009) *Accounting for people*: a real step forward or more a case of wishing and hoping?, *Critical Perspectives on Accounting*, **20**(7): 855–69.

Royle, T. (2005) Realism or idealism? Corporate social responsibility and the employee stakeholder in the global fast food industry, *Business Ethics: A European Review*, **14**(1): 42–55.

Sackman, S. A., Flamholtz, E. G. and Bullen, M. L. (1989) Human resource accounting: a state of the art review, *Journal of Accounting Literature*, **8**: 235–64.

Scapello, V. and Theeke, H. A (1989) Human resource accounting: a measured critique, *Journal of Accounting Literature*, **8**: 265–80.

Schreuder, H. (1979) Corporate social reporting in the Federal Republic of Germany: an overview, *Accounting, Organizations and Society*, **4**: 109–22.

Stewart, J. D. (1984) The role of information in public accountability, in Hopwood, A. and Tomkins, C. (eds), *Issues in Public Sector Accounting*. Oxford: Philip Allen.

Sveiby, K. E. (1997) *The New Organisational Wealth: Managing and Measuring Knowledge Based Assets*. San Francisco, CA: Berret-Koehler.

Tayles, M., Pike, R. H. and Sofian, S. (2007) Intellectual capital, management accounting practices and corporate performance: perceptions of managers, *Accounting, Auditing and Accountability Journal*, **20**(4): 522–48.

Thompson, P. (1989) *The Nature of Work: An introduction to debates on the labour process*. London: Macmillan.

Tinker, T. and Neimark, M. (1987) The role of annual reports in gender and class contradictions at General Motors: 1917–1976, *Accounting, Organizations and Society*, **12**: 71–88.

TUC (1991) *Greening the Workplace: A TUC guide to environmental policies and issues at work*. London: Trades Union Congress.

TUC (2010) *Greenworks: The TUC Green Workplaces Report 2008–2010*. London: Trades Union Congress.

Williams, S. J. and Adams, C. A. (2013) Moral accounting? Employee disclosures from a stakeholder accountability perspective, *Accounting, Auditing and Accountability Journal*, **26**(3): 449–95.

Yongvanich, K. and Guthrie, J. (2006) An extended performance reporting framework for social and environmental accounting, *Business Strategy and the Environment*, **15**(50): 309–21.

CHAPTER 7

Environmental issues

7.1 Introduction

Environmental issues have already featured earlier in this text. As we now know, the data relating to the global environment is far from encouraging: environmental sustainability is looking less and less possible under current forms of organisation and this, in turn, threatens social sustainability and social justice (see also **Chapter 9**). And yet one could not but be aware of the vast increase in initiatives about 'the environment' in all walks of life. Business, accounting and organisational life are no exceptions and have, indeed, often been the leader in such initiatives. To clarify this considerable organisational response to environmental issues, it is first necessary to establish how 'the environment' is being conceived. This might seem obvious, but it significantly affects what follows.

Chapter 3 introduced a range of worldviews including that of deep ecology: a point of view which sees the global ecology as a holistic system. But more pertinently, a deep ecological view would see humanity as an inseparable component of that system and would reason that any violation of the natural environment is morally wrong in and of itself and is additionally morally wrong as a direct violation of humanity. One can see quite quickly how such a view would sit uncomfortably with modernity and capitalism (York *et al.*, 2003). The essential point is that the environment *is* a holistic system, but reacting to it as a whole is not something that organisations (or often individuals) are very good at (Whiteman *et al.*, 2013). Indeed, as Whiteman *et al.* (2013) show, the typical business and political reaction to environmental issues has been piecemeal – a focus on 'carbon', a concern for waste, measurement of certain pollutants, etc., but with little examination of the likely systemic causes of these environmental issues or an explicit recognition of the inter-connected nature of these environmental manifestations.

But more importantly still, organisations – and especially business organisations – only recognise the natural environment in very limited ways. Predominantly 'organisations' *can only* recognise 'the environment' through the lenses that they have to the outside world: lenses such as price and costs, powerful stakeholders, external pressure and so on. In a sense this is quite obvious; the loss of habitat, extinction of a species, pollution of a river, drought, etc. are not recognised by organisations unless they affect the prices they have to pay or unless one of their salient stakeholders responds to the event in a manner which requires acknowledgement from the organisation. This is crucial because it generally means that most organisations do not – and cannot – respond to the natural environment *per se* but are reacting to manifestations from other systems and groups that are themselves responding to environmental issues. Furthermore, even when an issue is brought to the attention of the organisation, it still has to be filtered through the organisation's values and mission – typically it has to be worth responding to. So, when we talk of organisational response to the environment, we rarely mean this: we typically mean responding to cost or price changes, changes in tastes, changes in stakeholder views or changes in risk or other matters which fit within the entity's **business case**.

We see this at its most vivid when we consider how conventional accounting responds to the environment: at its most basic, financial accounting only responds to environmental issues in so far as they affect the financial numbers and risk; and management accounting broadly is concerned with categories of costs and potential costs and risk. When we look at environmental management systems the lens is a little wider and, as we shall see, the potential ambit of environmental reporting is considerable. It is these matters that we examine in this chapter.

We have chosen to take a slightly more managerial view in this chapter and approach the issues less through the exigencies of accountability and more through the lens of the organisation itself. As a result, the chapter is organised to reflect this. The next section revisits stakeholder engagement in order to emphasise that this chapter is not about the environment but about the 'perceptions of the environmental' as they appear to have impact upon entities. This leads directly into environmental reporting (in Section 7.3) as the most visible manifestation of the organisational response to environmental issues. Then we move on to explore, in Section 7.4, environmental management systems as the basis of organisational response to environmental issues and the basis for environmental accounting and reporting. Environmental management accounting and capital budgeting are then considered in Section 7.5 before we briefly look at the way environmental issues are affecting financial accounting in Section 7.6. The final section (Section 7.7) considers how this debate can be opened out and, in particular, how the accounting profession has (and has not) responded to the issues.

7.2 Background and stakeholders

Over the last three to four decades organisations have become increasingly aware of the importance of measuring and managing their environmental impacts for a variety of reasons, including building stakeholder trust, enhancing their reputation, legitimising their ongoing business activities, responding to stakeholders and/or international developments, decreasing risk and reducing costs.[1] Organisations use various approaches to control their environmental interactions and embed environmental management within organisational processes, practices and thinking. These include environmental reporting, managing environmental risk within financial reporting, environmental management accounting, environmental management systems and capital investment appraisal. We consider how integration of these various processes, often operating largely separately from one another, might improve organisational environmental performance (see Adams, et al., 2008a, b).

Measuring environmental impact is an important initial step to making decisions which minimise harmful impacts – assuming an organisation is clear about its reasons for doing so. The nascent literature on organisational environmental impact measurement, although

---

[1] The rationale for a greater focus on environmental issues in corporate activities has a wide variety of stimuli and has been related to a range of theoretical arguments (see Chapters 1, 2 and 4).

still relatively limited perhaps, has sought to explore, explain and stimulate this corporate practice in recognition of the occurrence of what is increasingly referred to as 'sustainability performance measurement and reporting' – a practice which has been increasing over the last two decades (see Deegan and Gordon, 1996; Lamberton, 2000; Schaltegger and Burritt, 2000; Tregidga and Milne, 2006; Adams and McNicholas, 2007; Unerman *et al.*, 2007). Practice has been spurred on by a wide variety of stimuli, including: formal legislative requirements (including emergent national emissions trading schemes); social pressures from various realms; voluntary initiatives such as the Global Reporting Initiative (GRI); and the principles of corporate social responsibility (CSR) and citizenship (see Adams, 2004; Owen, 2008).

In the dialogue around the development of integrated reporting (despite its primary concern with the providers of capital), this focus is termed 'natural capital'[2] (also sometimes referred to as environmental or ecological capital) which Forum for the Future have defined as:

> the natural resources (energy and matter) and processes needed by organisations to produce their products and deliver their services. This includes sinks that absorb, neutralise or recycle wastes (e.g. forests, oceans); resources, some of which are renewable (timber, grain, fish and water), whilst others are not (fossil fuels); and processes, such as climate regulation and the carbon cycle, that enable life to continue in a balanced way.
>
> (The Five Capitals Model: a framework for sustainability)[3]

and an undated document posted on the IIRC website notes:

> The measurement and management of non-financial factors, of which natural capital forms a part, must be embedded in business strategies and become part of mainstream decision-making and reporting. We support efforts to reach a global agreement to develop methodologies that will account for and value natural capital. This would be an important step towards integrating natural capital into mainstream decision making.
>
> (Rio + 20 Policy)[4]

This would be a significant change in the way businesses make decisions. The extent to which they can change from their profit maximisation mode remains to be seen (see also Burritt, 2012 and **Chapter 11**). Nevertheless, this emphasis in such initiatives as Integrated Reporting points to an increased prominence of environmental accounting and the need for the development of new approaches informed by theory and developed by multi-disciplinary teams (Burritt, 2012). As we have emphasised throughout (see especially **Chapters 3, 5 and 11**), one key element in both the drivers for this change and the mechanism through which change can be developed is stakeholder engagement.

The importance of engaging stakeholders to improve the organisation's awareness of environmental impacts is now widely supported. Stakeholders obviously must include not just the traditional pool of shareholders, customers, employees and suppliers, but also wider interest groups such as local communities, pressure groups, regulators, non-governmental organisations (NGOs) and the environment (and even future generations). Such engaging with stakeholders allows identification of a wide range of views and perceptions of environmental

---

[2]The idea of natural or environmental capital has a long history in both economics and accounting. For more detail see, for example, Daly (1980), Turner (1987, 1988/1993), Pearce *et al.* (1989); Gray (1992); Bebbington and Gray (2001).

[3]See www.forumforthefuture.org/project/five-capitals/overview.

[4]For more detail see www.theiirc.org/rio20 and see also the IIRC's briefing paper on capitals (IIRC, 2013).

impacts that supplements other, more physical measures such as **eco-balance** (see Section 7.4 and **Chapter 13**). When undertaken through sound processes, stakeholder consultation accentuates the notion that corporations are viewed as operating at the centre of a network of inter-related stakeholders that, it is sometimes argued, enhance value creating capacity (see Jamali, 2008).[5]

Although we have extensively considered the key elements of a stakeholder approach elsewhere in this book, the words of Benn *et al.* are worth consideration. The authors argue that from stakeholder engagement:

> Resulting shifts in practice entail a replacement of short-termism by long-termism and a balancing of organizational competition with interdependence and mutuality. Yet we have argued that this ideal cannot be achieved without innovative practices fostered by a governance system which enables debate, challenges the established order and uses embeddedness and interconnectedness to foster the exchange of ideas and values. We also reviewed several practical approaches to problem definition, decision-making and action planning that represent tested strategies to be integrated into an emergent set of participative processes for decision-making in the sub-political arena. We hope that, as new horizontal governance systems for dealing with environmental impact and risk of industry and business develop, the production of intractable wastes will cease and there will be more equitable distribution of any remaining risks arising from the activities of companies.
>
> (Benn *et al.*, 2009: 1574)

Those horizontal processes are much needed. Embedding environmental management requires a cross-functional, integrated approach, but evidence suggests we are some way off seeing that realised (see, for example, Adams *et al.*, 2008b).

O'Riordan and Fairbrass (2008) construct a framework (see also **Chapter 5**) that conceives of stakeholder dialogue practices as made up of four inter-related but analytically distinct domains – context, events, stakeholders and management responses – which would seem to us to be critical steps in moving from the current reactive stance of many organisations towards a pro-active consideration of environmental issues.

In an era dominated by global communications through internet enabling technologies, engaging with stakeholders around the world has become much easier at a time when environmental issues and means of addressing them are also increasingly seen as global in nature (see also **Chapter 5**). One of the consequences of a global perspective is the ease with which comparisons may be drawn between regional governance and behaviours. Such comparisons provide justification for the improvement of environmental management in developing economies and the need for developed economies to lead (where appropriate) by demonstrating improved environmental outcomes. Investment is increasingly flowing from the developed world, where social and environmental regulations are relatively comprehensive and well enforced (at least to a degree), to the developing world where there can be thought to be (for the western modern mind) inadequate resources to enforce such laws and regulations as do exist.[6] This situation has added weight to calls for mandatory legislation for multi-national corporations, and some governments have attempted to influence behaviour through foreign policy, aid and trade (Aaronson, 2005).

---

[5]Chapters 12 and 13 briefly explore how the entity concept might itself be treated as a more permeable and flexible notion if stakeholders were more fully integrated with organisations.

[6]Levels of voluntary CSR allegedly are higher in western countries, particularly the US and Europe, than in developing economies (Chapple and Moon, 2005; Welford, 2005).

In addition to the influence of western governments upon international environmental practices, another sphere of influence comes from NGOs. Poor environmental practices (at least as currently defined) are usually brought to the attention of the wider media by NGOs (see Adams, 2004), and the response of multi-national corporations to this attention may take several forms.

Nijhof *et al.* (2008) identify three corporate orientations that have an impact upon behaviour towards NGOs. The first, a business case, maintains NGO relations to mitigate risks from particular business activities and usually relates to minimising bad publicity and reputational risk from polarising NGOs. The second, an identity orientation, investigates the meaning of CSR (by which we may be able to infer also environmental management) to the organisation and then informs stakeholders, a one-sided approach which leaves little opportunity for influence from NGOs. The third, a stewardship orientation, takes a broader production and consumption view and engages all stakeholders in dialogue towards improving CSR outcomes across the supply chain. In this case, NGO involvement is more likely to take the form of a discerning integrator.

The decision to engage in partnership with NGOs or other companies in formulating and delivering (what are increasingly referred to as) 'sustainable' outcomes for an organisation is a strategic position that recognises the need for greater inclusivity of stakeholders in business. In their literature review, Dahlmann *et al.* (2008) identified three types of responses to environmental issues: cost reduction and economic efficiency; reaction to pressure from stakeholders; and a proactive, enlightened approach which can improve reputation leading to competitive advantage opportunities not available to organisations adopting an economic efficiency approach. The first, addressing environmental change management from an economic viewpoint, focuses on shorter term financial incentives for corporate profitability such as cost reduction and improved efficiency as well as economic incentives such as tax breaks, interest-free loans and revenue opportunities such as emissions trading schemes (Dahlmann *et al.*, 2008). As discussed in Chapter 5 and above, engaging with stakeholders may be able to build trust and other intangible assets and competencies that are important in gaining and maintaining competitive advantage. Proactive firms go further, integrating environmental issues into their strategic positioning and all aspects of operations (Adams *et al.*, 2008b; Dahlmann *et al.* 2008).

7.3 Environmental reporting

There is little question that it is environmental reporting (and its later development into 'sustainability' reporting) that has been the most prominent aspect of accounting and the environment as well as the most prominent aspect of organisational response to environmental issues – especially for the last two decades or so. Environmental reporting has a long history (see, for example, Lessem, 1977; Preston *et al.*, 1978; Brockhoff, 1979; Schreuder, 1979), particularly in Western Europe. However, it was only during the 1980s and 1990s that it became widespread (e.g. Harte and Owen, 1991; Freedman and Stagliano, 1992; Owen, 1992; KPMG, 1993, 1994, 2008; Adams *et al.*, 1995, 1998). The reasons for this increase in reporting are complex and the subject of a vast volume of research (see, for example, Gray, 1990; Owen, 1992; Adams, 2002; Adams and Larrinaga-González, 2007). However, important early drivers seem to include action by European legislators in the early 1990s (resulting in the Agreement on the European Economic Area 1992, the European Union's Fifth Action Programme and the Environmental Management and Audit Scheme, EMAS), as well as the development of a range of voluntary reporting guidelines and the increasing attention given by professional accounting bodies to environmental reporting issues. In more recent years, initiatives from diverse entities including SustainAbility/ UNEP,[7] UNCTAD,[8] AccountAbility,[9] the GRI,[10] the United Nations Global Compact[11] and the Prince of Wales' Accounting for Sustainability project[12] have appeared to be increasingly influential.

The early global history of environmental reporting is a relatively elusive one (see also **Chapter 4**) and there are many good histories that can be consulted (see, for example, Bloom and Heymann, 1986; Gray *et al.*, 1996; Buhr, 2007; Deegan, 2007). Figure 7.1 provides one overview of the historical context we are exploring here.

Initially environmental (and indeed social) reporting was mostly seen as a part of an organisation's annual report – a part of the wider, typically non-financial disclosures. This was not the case in the USA where financially-related disclosure within the financial statements was probably the principal way in which companies manifested any environmental response. (We deal with this briefly in Section 7.6.) It seems, however, to have been first the Brundtland Report (UNWCED, 1987) and then the run up to the Rio de Janeiro Earth Summit in 1992 (see, for example, Dresner, 2002, Chapter 4) which transformed the global awareness and prominence of environmental issues, and it is within that changed atmosphere that the 'standalone' environmental reports emerged. Following the initial reports from Canada (Noranda Minerals and Noranda Forestry) and the UK (Norsk Hydro and British Airways), the practice seemed to gain a life of its own with much of the initiative – at least in the early years – seemingly coming from Europe. In this connection, it is probably worth examining the developments in Europe in a little more detail.

Despite the early development of environmental policy for the European Economic Area in 1992, there were no mandatory rules concerning individual companies' policy and disclosure. However as Hibbitt and Collison (2004) note, some European countries, such as Denmark and the Netherlands, interpreted EU initiatives in such a way as to develop national requirements for reporting. More relevantly the EU Commission's Fifth Action Programme on the Environment (European Commission, 1992), entitled *Towards Sustainability,* called for enterprises to: disclose in their annual reports details of their environmental policy and activities; detail in their accounts the expenses on environmental programmes; and make provision in their accounts for environmental risks and future environmental expenses (EC, 1992, Vol. II: 67).[13]

In contrast, the European Union's EMAS, adopted in 1993, whilst primarily concerned to encourage enterprises to develop their environmental management systems (see Section 7.4), contained an explicit requirement that organisations should, as regards their environmental performance, make information available to the public in a '. . . concise, comprehensible form . . .' [Article 5(2)]. EMAS required that environmental statements of the company's activities should be made at the site level and should cover a range of matters, including the company's environmental policy, a summary of the figures on pollutant emissions, waste generation, consumption of raw material, energy, water, noise and other significant environmental aspects. All of this information had to be externally verified (see **Chapter 11**).

---

[7]See, for example, Sustainability/UNEP (1998).

[8]See, for example, Moore (2009).

[9]www.accountability.org.

[10]www.globalreporting.org.

[11]http://www.unglobalcompact.org/.

[12]www.acountingforsustainability.org.

[13]The 5th Action Programme also called for product pricing based on the 'full cost of a product' including the use and consumption of environmental resources.

Figure 7.1 Influential initiatives in environmental accounting and reporting

The period 1970–1989

1968 The Club of Rome first met to consider the interactions of economic, social, natural and political factors.

1970 The US Environmental Protection Agency (EPA) established to protect human health and to safeguard the natural environment.

1971 The US Securities and Exchange Commission adopted environmental regulations to be taken into account when assessing a company's financial position.

1972 The United Nations Environment Programme (UNEP) established to monitor the world environmental situation.

1976–1986 Many US federal statutes enacted covering water, air, resource conservation and hazardous waste clean-up.

The period 1990–the 2000s

This period has seen the growth in international organisations, agreements and protocols aimed at fostering global environmental health and safety (EHS) excellence, economic success and corporate social responsibility (CSR) reporting standards. For example:

1990 The Global Environmental Management Initiative (GEMI) created tools and provided strategies for EHS management and sustainable development. In the same year, the European Environment Agency and the International Institute for Sustainable Development (IISN) were established.

1993 The Confederation of British Industry (CBI) issued *Introducing Environmental Reporting Guidelines for Business* and The European Federation of Accountants (FEE) established an Environmental Working Party.

1994 The World Industry Council for the Environment issued *Environmental Reporting – A Managers Guide*.

1996 The Institute of Social and Ethical AccountAbility (ISEA) (now named AccountAbility) formed.

1997 The Global Reporting Initiative (GRI) formed. In this same year, the Board of EHS Auditor Certifications was established in Florida (US) to provide certification programmes for the professional practice of EHS auditing. The professional designation is Certified Professional Environmental Auditor.

1999 The ISEA published the AA1000 standard, establishing the principles of reporting.

2001 The International Federation of Accountants (IFAC, www.ifac.org) International Auditing Practices Committee produced an exposure draft on environmental reporting.

2002 The European Commission issued the White Paper, *Promoting a European Framework for CSR*.

2005 IFAC published exposure draft on sustainability assurance engagements.

2006 GRI's G3 sustainability reporting guidelines published.

2006 The International Auditing and Assurance Standards Board (IAASB) published a consultation paper of assurance aspects of the GRI's G3 sustainability reporting guidelines.

2008 AccountAbility's AA1000 Assurance Standard published.

2009 Climate Disclosure Standards Board (CDSB) issued a reporting framework exposure draft dealing with carbon measurement and disclosure.

Figure 7.1 (*continued*)

> **The period from 2010 onwards**
>
> 2010 The European Commission held stakeholder focused workshops to explore the desirability and the feasibility of stakeholders moving towards an agreed set of key performance indicators for environmental, social and governance (ESG) performance.
>
> 2010 UN Global Compact and GRI signed a Memorandum of Understanding.
>
> 2010 The Prince of Wales' Accounting for Sustainability project established an International Connected Reporting Committee.
>
> 2013 The International Integrated Reporting Council (IIRC) released the Consultation Draft of the Integrated Reporting Framework.
>
> 2013 GRI released its G4 sustainability reporting guidelines.

Sources: Özbirecikli (2007) with adaptations by the authors and drawing on Adams (2010) and Adams and Petrella (2010).

While individual accountants tended to remain outside discussions on the environmental agenda within their organisations (Adams *et al.*, 2008a, b), their professional bodies have been keen to promote their environmental and sustainability credentials. By the early 1990s, professional accounting bodies in Belgium, Denmark, Germany and the UK had set up working parties concerned with environmental issues.[14] The UK professional bodies were probably the most active in Europe with four of the major bodies[15] having sponsored research projects concerned with environmental accounting. In addition, the ACCA developed its immensely influential Environmental Reporting Award Scheme[16] and for many years remained at the forefront of professional accounting, addressing sustainability accounting, reporting and assurance issues – including taking a leading role in both GRI and Integrated Reporting.

The early response of UK companies was not at all promising (Touche Ross, 1990: 22) and this was likely one of the stimuli for a growing concern in business about possible regulatory responses and/or risk minimisation. In the early 1990s, we see the emergence of an array of voluntary guidelines for reporting. The Hundred Group of Finance Directors (1992) provided guidance on what a 'meaningful' statement of environmental policy should contain (including realistic and measurable targets); the London Stock Exchange faced pressure to adopt standards of environmental disclosure in its listing requirements (for which the accounting professions could provide guidelines on accounting for the environment); and the Confederation of British Industry (CBI) launched its *Agenda for Voluntary Action* in 1992 which required companies to publish their environmental policy and report progress towards meeting the targets and objectives for meeting the policy. Much of this really came to nought in the end.

Owen (2008) is not alone in seeing these early initiatives (and, indeed, many since) as failing to deal with inadequacies in target setting as well as presenting over-aggregated or

---

[14]Lest there be a mistaken view that this was a novel initiative; we should note that the American Accounting Association was considering such matters in the 1970s (AAA, 1973).

[15]The Institute of Chartered Accountants of England and Wales (ICAEW), Institute of Chartered Accountants of Scotland (ICAS), Association of Chartered Certified Accountants (ACCA) and Certified Institute of Management Accountants (CIMA).

[16]Which later morphed into a Sustainability Reporting Awards scheme operating in many countries.

incomplete data. It is often impossible to obtain a clear view of the company's strategic thinking from these reports. Report users need to know what the key environmental impacts are and how the company is going about tackling these issues. This remained a concerning omission throughout the early decades of the 21st century and was oft commented upon in the judges reports of the ACCA sustainability reporting awards. Adams (2004), in her comparison of a company's own portrayal of its environmental performance with that obtained from external sources, such as the media and NGOs, concluded that the reports were verging on the worthless and displayed a lack of accountability which would not be tolerated in financial reporting.[17] And yet, what is at stake here, the well-being of our planet and future generations, is enormous and infinitely more significant than the short-term wealth of any shareholder.

Despite the growth of voluntary environmental (and, increasingly, 'sustainability') reporting since those early days, it is still practised by only a minority of (typically larger) organisations (KPMG, 2011) and it remains significantly incomplete as either a picture of organisational environmental performance or as a mechanism for the discharge of accountability (Livesey and Kearins, 2002; Whiteman et al., 2013). The patterns in such reporting are relatively easy to establish through surveys (see, for example, KPMG, 2011) and databases (such as those maintained by GRI and Corporate Watch) and so attention has turned more to why organisations do (or more usually do not) voluntarily produce public environmental information (**Chapter 4** examines such matters in some detail). Researchers continue to offer partial explanations about such things as an organisation's search for legitimacy (Deegan, 2002), its need to manage its stakeholders (Mitchell et al., 1997) or its need to follow the herd (Larrinaga-González, 2007). But such examples are attempts to explain behaviour from outside the organisation (Adams, 2002; Adams and Whelan, 2009) whereas we are seeking to adopt something closer to a manager's perspective in this chapter.

To see this through the eyes of management, we need to recognise that reporting on environmental (and broader sustainability) issues might well be an important part of the process of managing performance and minimising negative environmental impacts. Ideally, the environmental report would be an outcome of a stakeholder engagement and performance management process (see, for example, Zambon and Del Bello, 2005). This is just normal, sensible management: defining the organisation's environmental values; establishing a governance and management process; identifying the key social and environmental impacts for the organisation's industry and any specific to the organisation; setting targets; identifying responsibilities and accountabilities; developing plans to achieve targets; measuring performance against targets; reviewing trends in data; benchmarking performance against similar organisations; and involving stakeholders at each stage (see AccountAbility, 1999). The report then reflects these processes and, again ideally, would ensure that material issues are reported and that internal stakeholders are committed to environmental targets and would signal to external stakeholders the quality of the environmental management systems (see Section 7.4). Sound environmental management is increasingly linked to the quality of management itself and particularly to risk management. Indeed, research suggests that companies issuing quality 'sustainability' reports exhibit significantly more positive market reactions than companies with low-quality reports (Guidry and Patten, 2010).

Undoubtedly the single biggest influence in environmental (and what is often called 'sustainability' (see **Chapter 9**)) reporting is the GRI which has increasingly (following the influence of the AA1000 standards) sought to provide support for and encourage

---

[17]Examples like the Adams performance-portrayal gap are explored in more detail in Chapter 10.

organisations to adopt the soundest processes in order to firmly underpin their reporting.[18] The GRI is a multi-stakeholder organisation with members from the broader business and public sectors (organisational stakeholders) who elect a stakeholder council which, in turn, elects the Board of Directors. It receives feedback from stakeholders on their priorities which feeds into a plan which is posted on its website for public comment. The final set of priorities for implementation for the next fiscal year is approved by the Board following feedback from the Technical Advisory Committee and the Stakeholder Council. Working groups are established to develop and review proposals which are reviewed by the TAC and forwarded to the SC for advice before the Board makes a final decision.

But the GRI is by no means alone as an international influence on environmental (and 'sustainability') reporting. Influence can be seen in a range of bodies including the International Standards Organisation (ISO), the World Business Council for Sustainable Development (WBCSD), AccountAbility, the Global Compact, the Sustainability Integrated Guidelines for Management project (SIGMA) and, in a more specific sense, the Carbon Disclosure Project. Of these, AccountAbility[19] has perhaps had the most impact in recent years, particularly with respect to stakeholder engagement processes and assurance (see **Chapter 11**). The initial AA1000 framework was designed to serve as a standalone framework for the processes of reporting, with a particular emphasis on stakeholder engagement, and also to link together other specialised standards covering aspects of sustainability reporting such as the GRI guidelines, Social Accountability International (SA 8000) standards and the ISO standards through a 'common currency of principles and processes' (AccountAbility, 1999: 1). Indeed, the 1999 Framework, with its focus on processes rather than performance indicators, provided organisations with guidance on management approaches to improving performance which are still relevant and which appear to be still lacking in many large companies' reports. In addition to developing its process framework, AccountAbility has now derived a set of principles[20] to support organisational response and which have been influential in developments in the GRI reporting guidelines.

Although broader than just the environment, the early years of the 21st century saw the initiation of the Prince of Wales' Accounting for Sustainability project (A4S)[21] which sought to influence reporting and bring environmental issues to the attention of investors. Through direct involvement with leading global professional accounting bodies and the International Federation of Accountants (IFAC), A4S has been a key factor in the collaboration with GRI that has led to the Integrated Reporting[22] initiative (see **Chapter 9** and Hopwood *et al.*, 2010).

These initiatives are, of course, all voluntary. And voluntary initiatives have significant limitations, hence our repeated calls for mandatory requirements (see, for example, Adams, 2004; Gray, 2006; Owen, 2008). For example, the KPMG *International Survey of Corporate Responsibility Reporting* 2008 finds that a significant proportion of even the very largest companies are not producing standalone reports – indeed, in 2008 only 37% of Australia's largest 100 companies did so. Further, as Gray and Herremans (2102) for example show, probably far less than 5% of the world's multi-national companies actually produce anything that might be considered as substantial reporting. Yet, governments are reluctant to mandate reporting on social and environmental issues (Deegan,1999; Cooper and Owen, 2007).

---

[18]See www.globalreporting.org where the latest guidelines and data can be accessed.

[19]www.accountability.org.

[20]See for example the *AA1000 AccountAbility Principles Standard 2008* at http://accountability.org/publications.aspx?id=3040.

[21]http://www.accountingforsustainability.org.

[22]See http://www.integratedreporting.org/.

There are a number of barriers to the more widespread acceptance of reporting on environmental impacts. Özbirecikli (2007) argues that a key issue impeding the further reduction of the environmental impacts of the corporate sector is the lack of participation in environmental accounting and reporting by small and medium sized enterprises (SMEs). The lack of infiltration of environmental and broader sustainability reporting into mainstream reporting is another issue which A4S has sought to address (see Adams, 2010). Another key issue limiting the ability of environmental reporting to improve environmental performance is the lack of integration of reporting with environmental management systems and stakeholder engagement processes (Adams and Frost, 2008; Adams *et al.*, 2008a, b).

So the picture around reporting on environmental issues is at best mixed and looks likely to remain so for the foreseeable future. The barriers to further progress would appear to derive from issues of a political and organisational nature rather than from any reasons of practicability. Although academe (and the more imaginative of practice) may have not managed to encourage the profession, organisations and governments to adopt environmental reporting as seriously as accountability and the environment need, there has been no shortage of suggestions, experiments and careful developments. Whether we look to the occasional examples worthy of considering as best practice in environmental reporting,[23] the innovative suggestions that some governments have adopted (the Danes come to mind here), the best of the guidelines on how to report for and about the environment,[24] the development of carbon disclosure, the outstanding notions of accounting for biodiversity (as represented by, for example, Jones, 2003) or the possibilities for the future we consider in Chapter 13, reporting on the environment is not constrained by practicable considerations.

One of the most surprising insights is that after more than two decades of guidance and experimentation, most organisations still do not have the environmental information systems in place to support sensible management or sensible reporting. It is to that which we now turn.

7.4 Environmental management systems

Environmental management systems (EMS) have already featured in the foregoing. They are clearly a pre-requisite for good environmental reporting and, as we shall see, they are also the foundation for any substantive environmental accounting.

Environmental management systems (EMS) can be defined as *'the organizational structure, responsibilities, practices, procedures, processes and resources, for determining and implementing environmental policy'* (Netherwood, 1996). They are one part of the organisation's wider management control systems and, whilst they may develop at the whim of the organisation, it is much more likely that they will follow the guidance provided by the EMAS or by the International Organization for Standardization ISO14001 series (Buhr and Gray, 2012: 428). EMS are complex matters in their own right and increasingly seen as an essential component

---

[23]There are many such examples. Some examples that you will find on the CSEAR website include Eastern Gas and Fuel Associates (1972), BSO/Origin (1991), Danish Steel Works (1991), Novo Nordisk (2003), CFS (2005) (see http://www.st-andrews.ac.uk/csear/sa-exemplars/reporting-practice/). Other exemplars can be found through a number of sources, but reporting awards schemes are amongst the best. More detail along these lines can be found in Gray and Bebbington (2001, Chapter 12).

[24]The dominant source of guidance would certainly be the GRI current guidelines. Some further suggestions can be found in the work of SustainAbility/UNEP (1996 *et seq.*) and in reports from reporting awards schemes such as that from ACCA (for more detail see Gray and Bebbington, 2001).

of a well-run organisation (Brady *et al.*, 2011). They are also expected to play an important role in improving environmental performance with the potential to effect real change (Larrinaga-González *et al.*, 2001). Albelda-Pérez *et al.* (2007), for example, find EMAS to be a catalyst for change, with the EMAS sites they had studied having developed valuable intangible assets for improving environmental performance – the key to this being the extent to which EMAS was fully embedded into the organisation.

Unfortunately, given their potential impact on environmental performance, EMS are not well used. For example, Dahlmann *et al.* (2008) found that only about half of the UK companies they studied had formal environmental management systems. Economic considerations such as cost, risk reduction and compliance with environmental legislation dominated the firms' environmental behaviour, with small and medium-sized firms in particular much less likely to have an EMS and appearing to rely on relatively short-term planning horizons. EMS, even in the larger companies, were under-developed. These limitations are undoubtedly related to the fact that EMS development has largely been driven by regulation rather than being seen as adding strategic value (Adams *et al.*, 2008a, b).

Adams and Frost (2008) argue that the process of developing key performance indicators (KPIs) for the purposes of sustainability reporting has focused attention on social and environmental performance leading to developments in data collection systems and the integration of social and environmental performance data into decision-making, risk management and performance measurement. Much remains to be done, and Adams *et al.* (2008a, b) in particular have called for greater integration of reporting and environmental management systems, noting the importance of leadership in those organisations which had had the most success in embedding environmental concerns.

Of course, EMS do not and cannot operate in a vacuum. They are part of – and interact with – the other control systems of the organisation. Accounting is one of (if not *the*) most important of those control systems. Modifications to traditional management accounting systems in such areas as accounting for energy and waste and activity-based and life-cycle costing, when used as support mechanisms for corporate environmental management systems, can promote environmental efficiency by focusing management attention on environmental protection at the design stage of products, processes and systems rather than 'end of the pipe' preoccupation with liability for past environmental transgressions (Stone, 1995). Also, in signalling a move away from short-term financially driven decision-making they are very much in harmony with evolving total quality management systems (TQM) in their emphasis on continuous improvement in performance (see Bennett and James, 1998; Gray and Bebbington, 2001).

It is rarely this simple, however. As we shall see, it is very easy for the EMS and the accounting systems to act, not in concert, but in opposition. Were it not the case, then every environmental initiative would be financially beneficial to the organisation and have the full support of that organisation's accounting systems – and environmental degradation from organisations would be a thing of the past. This is clearly not so. There are two crucial and central notions in this potential for harmony and conflict between traditional accounting and the environmental designs of the organisation. These concepts are the deceptively simple ideas of the **business case** and **win–win** scenarios.

The *business case* refers to the notion that an organisation that is tightly run will only undertake activities that are in its interest. Whether the organisation is considering hiring staff, opening a new market, closing a hospital or investing in environmental technology, the decision would be expected to be formally evaluated as to the extent to which it contributed to the organisation's goals. These might be expansion, risk reduction, improved reputation, costs savings or profit enhancement, for example. For most organisations – and virtually all businesses – the input of accounting is crucial to that decision. The EMS might suggest an environmental initiative to (say) reduce the use of single trip packaging or invest in some more efficient processes, but the business case has to be made: if it cannot be shown that this initiative will save money, increase income or reduce risk, for example, it is unlikely to fly – no matter how environmentally desirable. That is not to say that the business case is always cut and dried or that accounting will always have a negative influence on environmental progress. Business cases vary considerably across organisations and can often be very subtle indeed (see especially Gray and Bebbington, 2001; Spence and Gray, 2007; Schaltegger *et al.*, 2008). This is one of the major reasons why the management accounting system needs to be sensitively adjusted for environmental matters. The other principal reason arises from the so-called win–win scenarios (Walley and Whitehead, 1994).

Win–win scenarios arise when an activity or opportunity is simultaneously environmentally *and* financially beneficial. Saving energy, saving water, reducing waste or packaging are all examples where there is likely to be a clear business case that is also environmentally-desirable. The term coined for such situations is **eco-efficiency** which is sometimes (if rather crassly) defined as 'doing more with less'. Efficiency is always a ratio of input to output. Eco-efficiency is, broadly, the ratio of resource input (energy, materials, transport) for any given amount of output (typically units of goods or services). One can easily see that as long as the costs of reducing the inputs do not exceed the benefits (i.e. that the initiative is indeed a win–win) then an organisation will normally be expected to be in favour of eco-efficiency and be able to pursue it most seriously. Indeed, the levels of eco-efficiency achieved by business globally have been astonishing (see, for example, Schaltegger, 1998; Gray and Bebbington, 2001; Porritt, 2005) and, as commentators such as Schaltegger and colleagues continue to show, accounting and accounting systems have a crucial role to play in their development.

This, then, is the background to our consideration of the potentials of environmental management accounting and environmental capital budgeting.

7.5 Environmental management accounting and capital budgeting

Environmental management accounting refers to environmental accounting information used internally and it requires the adaptation of existing management accounting systems to incorporate financial and non-financial information to enable managers to improve their organisation's environmental performance (see also Yakhou and Dorweiler, 2004). It typically involves such things as life-cycle costing, full-cost accounting, benefits assessment and strategic planning for environmental management (see Deegan, 2008). **Environmental cost accounting** is the use of accounting records to directly assess environmental costs (immediate costs, taxation implications and costs of preventing external failure) to products and processes (Yakhou and Dorweiler, 2004). At its heart, environmental management accounting is the process of both identifying costs associated with environmentally-related activities (such as waste disposal, transport, energy, fines, resource use, etc.) and teasing out financial cases to support EMS initiatives.

The field – both as a practice and as an area of study and research – has grown remarkably over recent decades (Schaltegger *et al.*, 2013), but the challenge posed to management accounting systems design by the ever-developing environmental agenda is significant. Initially, systems are in need of modification so that environmentally related areas of expenditure (and revenue) may be indentified separately. Beyond that, even more fundamental change is called for in ameliorating the environmentally negative elements of existing systems, in particular the restrictive, short-term 'bottom line' perspective, and introducing a more forward-looking focus whereby potential environmental threats and opportunities can be taken into account. New systems need to be developed, probably employing both physical and financial measures. There is still considerable work to be done in developing such systems to facilitate organisational decision-making.

But progress is being made (Schaltegger *et al.*, 2008). It would now be unusual for an organisation not to recognise and capture their costs of energy, water, waste and other areas of obvious environmental impacts as essential components of the management accounting system. And newer areas of importance to the firm are emerging all the time. Probably the most important issue of recent years is that of climate change and its associations with greenhouse gas emmissions and 'carbon'[25] (Bebbington and Larrinaga-González, 2008/2010). Despite the impact of the Carbon Disclosure Project, carbon taxation and Carbon Trading Schemes (see, for example, Kolk *et al.*, 2008; Ratnatunga and Balachandran, 2009), it is not at all obvious that businesses generally and accounting in particular have responded in the ways expected. Dahlmann *et al.* (2008) actually found that climate change, the issue of the greatest public and policy concern, was considered of very low significance among their accounting respondents despite a high awareness and involvement in energy and waste issues – largely as a function of price and taxation changes (p. 278). But even here the situation is not clear cut. Whilst reductions in costs may arise through changes in the nature of the business or production process, or perhaps through switching between energy sources, such changes do not measure adequately changes in energy efficiency. Furthermore, a focus on greenhouse gas emissions, rather than energy useage/waste creation, may mask inefficiencies where carbon credits are used. One major way around this has been to encourage accounting in physical units (as well as in financial units) as this facilitates the setting of energy/waste targets and subsequent assessments of volume variances. In accounting for energy, waste and the other direct environmental costs, the essential aim is to identify where, and how, the resource is used/waste arises and to identify inefficiency and wastage, thus focusing attention on the areas where savings may be made and particular managers held accountable (see, for example, Gray and Bebbington, 2001).

In essence, management accounting for the environment is concerned to identify and then charge all identifiable costs to the cause of their creation and ultimately to allocate them to products and services. Activity-based costing (ABC) has been used as one approach to achieving this by allocating costs to products on the basis of the individual product's demand for particular activities. Whereas traditional costing systems allocated overheads to products on the basis of simplistic volume-related bases such as direct labour hours, ABC systems recognise that different products make different demands on organisational resources for reasons that are not necessarily related simply to production volume. Kreuze and Newell (1994) point out that product-specific environmental costs may require a particularly sophisticated use of ABC, in that, in addition to operating costs such as energy usage and waste disposal, and regulatory compliance cost, being important components of a full environmental costing analysis, future costs also have to be considered. Their call for the use of **life-cycle costing**[26] recognises the growing importance of both contingent

---

[25]'Carbon' in this connection is shorthand and only indirectly related to the element carbon whilst more generally related to the carbon cycle. When used in situations like 'carbon-trading', it is a collective noun referring to the greenhouse gases which comprise (principally) carbon dioxide, methane, ozone and nitrous oxide.

[26]While life-cycle analysis focuses on the physical environmental impacts over a product's life, life-cycle costing attempts to internalise some of the costs associated with the impacts, particularly those traceable to particular activities and measurable with a particular degree of reliability. Life-cycle cost has been defined as the amortised annual cost of a product, including capital costs, installation costs, operating costs, maintenance costs and disposal costs discounted over the lifetime of a product (Deegan, 2008). It is sometimes called 'cradle-to-grave' costing. Life-cycle costing may also include societal costs or externalities, such as adverse health effects caused by generated emissions, and capital costs allocated for the prevention of global warming or to prevent ozone depletion.

liability costs and the intangible cost and benefits connected with higher environmental performance standards.

An environmental costing framework employing a combination of ABC and life-cycle concepts was developed by the US EPA (Tellus Institute, 1992a, b). The framework consists of four ascending tiers of costs, each with increasing sophistication and subjectivity, in terms of information supplied:

- **Tier 0** Includes only direct environmental costs associated with a particular product or process.
- **Tier 1** Includes indirect costs, or overheads such as would be captured in an ABC system, in addition to direct environmental costs.
- **Tier 2** Encompasses an estimate of future legal liability costs in addition to the actual costs currently incurred addressed by the two lower tiers.
- **Tier 3** Goes further still in taking into account the intangible benefits (including costs saved) arising from environmentally responsible business practice.

(We will return to this shortly as the framework is important when considering investment appraisal. See also, Bennett and James, 1998; Gray and Bebbington, 2001).

Researchers have explored life-cycle costing in order to identify the sheer range of social and environmental issues that need to be interpreted. Deegan (2008) in particular identifies a number of 'costs' which are difficult to quantify in financial terms, but nevertheless require consideration in life-cycle costing. These are equally relevant to capital investment decision-making (as we shall see shortly).[27] The broad issue, as Milne (1991) identified, is that, once accounting for environment starts to become more pro-active and inclusive, it must start to include items which are not immediately and directly costed to the organisation concerned – and this takes the accountants out of their comfort zone into areas such as cost benefit analysis (Milne, 1991).

It is quite apparent that as with the focus of the EMS on win–win situations, the environmental management accounting system has precisely the same concerns. At its heart lies the questions as to whether improving environmental performance is costly or whether pollution is a form of economic inefficiency and thus economic and environmental efficiency go hand in hand. The question is far from clear cut (see Burnett and Hansen, 2008; **Chapter 8**).

These tensions between the costs and the benefits of environmental performance and between the identification of traditional and more elusive costs are at their most acute in the **environmental aspects of capital investment appraisal**. Investment appraisal (or capital budgeting – the two terms are almost interchangeable) involves organisations in making long-term decisions about such matters as new products, new processes, investment in infrastructure, new plant or buildings and so on. Such decisions are typically undertaken on a purely economic and strategic basis in which the long-term costs and benefits associated with the range of choices are considered carefully (Emmanuel *et al.*, 2010). But this is also the place where the long-term environmental (and social) impacts of the organisation are largely determined (Epstein and Roy, 1998; Sloan, 2011).

It can be argued that for many decisions it is no longer in the interests of organisations or their stakeholders to base capital investment decisions purely on expected short-term economic returns to the neglect of long-term environmental impacts and associated social

---

[27]These included costs associated with breaching community expectations; costs associated with climate change mitigation efforts of alternate investment decisions; and supply chain costs – the social and environmental impact of the processes employed by their suppliers.

impacts. The development of a formal process for incorporating social, environmental and economic impacts into the assessment of alternative capital investment options, including those that cannot be monetised or quantified, is needed to address the limitations of existing informal (or non-existent) processes which omit consideration of material impacts. The final decision in many contexts may not be based on an all-encompassing 'bottom line', but a more complex and nuanced outcome that reflects an implicit rather than explicit balance between competing elements and priorities. The quest for a single and all-encompassing assessment of a project, which is typical of current practice, is limiting.

Multi-stakeholder perspectives and communication processes for assessing sustainability impacts have been advocated (Adams, 2004; Adams and McNicholas, 2007; Bebbington *et al.*, 2007; Frame and Brown, 2008; Brown, 2009). This emerging emphasis focuses on the importance of dialogue and participation in organisational processes, recognising the potential of accounting technologies to facilitate the foregrounding of significant issues (creating visibilities) (Boyce, 2000). However, specific accounting and related technologies for assessment of interrelated social, environmental and economic effects still have a long way to go (but see Durden, 2008).

We might expect non-profit organisations (see **Chapter 12**) to have developed processes for considering social and environmental impacts in capital investment appraisal decisions. As Ball (2004, 2005) and Ball and Seal (2005) argue, accountability to multiple stakeholders and the inherent need to balance social, environmental and economic impacts are clearly recognised in local government (Kloot and Martin, 2000). Indeed, the public sector as a whole is significantly involved in investing significant funds to capital projects and making the appraisal and selection of capital investment proposals on a local, regional and national basis. But still too little is known about the detail: so, for example, only one of the public sector organisations studied by Adams *et al.* (2008a, b) – a water authority – had developed a formal process for assessing a range of social, environmental and economic impacts. This involved a highly sophisticated method and process taking in a wide range of stakeholders in the identification and assessment of various impacts.

Examples of bodies succeeding in balancing social, economic and environmental criteria in their strategic decisions do of course exist in practice. For example, a local government decision to pedestrianise a street raises concerns about negative economic impacts to traders and negative social impacts to shoppers. At the same time, it may bring positive environmental and social impacts in terms of reduced air, noise and visual pollution and improved aesthetics. Particular difficulties are presented by the need to prioritise competing project proposals in the face of funding constraints: for example, alternative proposals for the establishment of a sporting field versus a community support education programme or a passive recreation reserve. When hierarchies of project proposals are drawn up, social and environmental aspects are often marginalised in the short-term, but they may be the most important over the long-term. A case in point is provided by Ball (2005) whose study of UK local government capital investment found that the main role of environmental accounting was costing environmentally sensitive schemes (i.e. waste management and a school bus scheme) in connection with the public–private finance initiative. It has long been her argument that it is the embedding of quality of life indicators into the decision-making practices that constituted the principal contribution that environmental management accounting could play.

Experiences of organisations with capital investment and environmental issues are beginning to emerge (see, especially, Schaltegger *et al.*, 2008) and, whilst the public sector is beginning to learn from for-profit organisations (see Hoque and Adams, 2008), it is not obvious that the converse is yet happening. Economic imperatives and the profit motive remain dominant within most corporations.

There is considerable potential for the future of environmental management accounting (Schaltegger *et al.*, 2008). We can see both an increase in a number of potential techniques that management accounting can embrace as well as a number of initiatives which suggest that progress may be possible. In essence, environmental costs for the purposes of capital investment decision-making or life-cycle costing need to be defined broadly to include all costs (financial, technical, social), in relation to organisational activities, that affect the environment. Techniques such as eco-balance (see **Chapter 13**) and (despite the scepticism of Deegan, 2008) life-cycle assessment and costing may help us along the way towards a more comprehensive approach to accounting for environmental costs that will potentially move organisations and the economic system towards sustainable development.

The pricing of carbon emissions is one of those areas having a discernible impact on accounting and accountants. The potential of a carbon pollution reduction scheme in Australia, for example, brought large numbers of new corporate players into the debate about business impacts on the environment as evidenced by increased numbers at workshops, seminars and conferences. And with the introduction of the National Greenhouse and Energy Reporting Act (2007), Chief Financial Officers started to talk about the importance of environmental performance to the business as environmental reporting was moved to their area of responsibility, now being a compliance issue. Not everybody is convinced by these changes (see Adams, 2010: 87) and developments and initiatives certainly vary by country. So, for example, Burritt *et al.*'s (2011) study of carbon accounting in German companies found that physical information dominates the financial and that information was collected across various parts of the organisation. And while Burritt (2012) and Burritt *et al.* (2002, 2011) note the progress that has been made in carbon management accounting by corporations, others, such as Young (2010), highlight the many unresolved issues associated with greenhouse gas accounting around, especially, the determination of boundaries (see also Schaltegger and Csutora, 2012). By way of contrast though, Kirschbaum and Cowie (2004) examined an accounting scheme in Canada, the USA, New Zealand and Australia which would resolve many of the biospheric carbon accounting anomalies identified in the four nations studied and lead to better decisions concerning land use. (For more detail see, for example, Kolk *et al.*, 2008; Bennett *et al.*, 2013.)

The analysis presented in this section of the chapter has indicated some of the immense difficulties faced in developing management accounting systems that are capable of incorporating the environmental performance dimension (let alone beginning to address the concept of sustainability – see **Chapter 9**). It is debatable how much real progress has been made in advancing techniques of addressing eco-efficiency issues (certainly there is little which addresses the eco-justice dimension) in the last few decades. Opportunities clearly exist and the techniques are potentially available, but accounting, accountants *and* organisations still seem to be insufficiently committed to embracing these possibilities. Perhaps more surprising still, we discern a somewhat similar message when we examine financial accounting and environmental issues.

7.6 Financial accounting and the environment

Section 7.3 looked briefly at the phenomena of (what is usually) the voluntary disclosure of environmental information by organisations. Such disclosure could take place in a variety of media but would typically be found in either the annual report or the standalone report. That review did not touch upon the equally significant matter of accounting for environmental issues within the financial statements. This is what we attempt in this neccesarily brief section (for more detail see Schaltegger and Burritt, 2000; Gray and Bebbington, 2001; Unerman *et al.*, 2007).

Recall that conventional accounting, under normal circumstances, only responds to environmental issues when they are reflected in prices and costs. This is especially true of financial accounting. Whilst we might like to see evidence of more speculative attempts at valuing environmental impacts within the financial statements, this is a very rare occurence (see **Chapter 13**, Bebbington and Gray, 2001; Herbohn, 2005). So we can only normally expect to see a profit and loss account or balance sheet reflecting environmental issues when either the numbers become sufficiently large to warrant separate identification ('material' in other words) or when a new category of disclosure is required by law or standard. For most of those costs typically thought of as 'environmental', despite the best efforts of, *inter alia*, the UN (see, for example, Moore, 2009) neither of these conditions is likely to hold (see, for example, CICA, 1993; Gray *et al.*, 1998)[28] and only major oil spills or environmental disasters will generally produce numbers which are both environmental and material in their own right. Consequently, the financial accounting issues associated with environmental accounting tend to be clustered around the categories of environmental liabilities, contingencies and, increasingly, carbon disclosure as the following quotation (albeit from the UK) illustrates:

> Depending on the nature of a business, certain accounting standards and interpretations will be relevant to the treatment and disclosure of environmental issues in financial statements. For instance, the valuation and reporting of tangible and intangible assets, including the measurement of inventories, can be affected by environmental impairment. Businesses should account for their allowances and transactions associated with the EU emissions trading scheme. Financial provisions could be required for liabilities arising from costs of waste disposal, pollution, decommissioning and environmental contamination, and wildlife habitat restoration. Where environmental issues have a material impact, specific disclosures may be necessary. Some environmental items may require special treatment due to their harmful impact.
>
> Irrespective of the size and value of an environmental item, its nature, societal importance, and impact on a company's reputation might be sufficient to be regarded as financially material. Where supply and disposal chain risks and impacts are material to the business they should be taken into account. If different reporting boundaries are used they should be stated.
>
> (ICAEW/Environment Agency, 2009)

Such considerations are likely to apply increasingly in most jurisdictions where the impact of standards and legislation is growing steadily.

There are some jurisdictions, however, where more specific and longer-standing requirements are in place and where (at least part of) the impact on the financial statements is much more apparent. The USA is the most obvious candidate here having had stringent financial reporting requirements governing environmental issues and particularly liabilities – especially those relating to contaminated land (Cho *et al.*, 2012). Cho *et al.*, however, argue that, despite this raft of regulation and guidance, reporting by US companies of the financial impact of the environmental on their operations is irregular and limited. They go on to show that similar situations obtain in other countries in Europe and conclude that whether or not the investors are much exercised by environmental issues they are not likely to be especially well-informed on the subject.

---

[28]For instance, the UK has a requirement that environmental issues be disclosed in the financial statements if material. Such disclosure is fairly rare – except (more recently) in the case of carbon disclosure (Environment Agency, 2007). The 1998 Gray *et al.* study examined UK company reporting in order to try and explain how it was that virtually no mention of environmental issues appeared in UK company accounts. It transpired that the issues were not considered material and the impact of changing legislation so slow that costs were absorbed piecemeal and did not need to be reflected in the financial statements.

So, no matter the importance of environmental degradation to society and life on the planet, environmental issues remain matters on which the financial community is largely indifferent and largely uninformed (but see **Chapter 8**). Furthermore, whilst the statutory auditor (who is required by law to express an independent opinion on the financial statements of organisations) might also be expected to be concerned by the environmental issues as they affect the organisation and its reporting, results here are also fairly mixed and not especially promising (Collison *et al.*, 1996; Beets and Souther, 1999; Owen, 2007; and see **Chapter 11**).

Perhaps this will change as Integrated Reporting gathers momentum and a more nuanced understanding of risk and opportunity (for the organisation if not for nature) is increasingly reflected in corporate reporting and their supporting notes (Percy, 2013). Perhaps so, but if the decades of environmental reporting have taught us anything it is that until there is either a substantial legislative framework and/or the issues have substantial financial implications, financial accounting, financial markets and businesses are, on the whole, likely to remain largely indifferent.

7.7 Conclusions and concluding comments

The accounting profession worldwide has acknowledged that present accounting systems are inadequate in terms of their ability to incorporate the full effects of corporate environmental impacts (see, for example, ICAEW, 1992; CMA, 1992; FEE, 1993; Stone, 1995; Gray and Bebbington, 2001). Disappointingly, despite this recognition of the need for change, accountants and the accounting profession have been slow to lead further change, with organisational participants engaged in managing environmental performance reporting little involvement of accountants in their work (see Bebbington *et al.*, 1994; Adams *et al.*, 2008a, b). This is highlighted by Adams *et al.*'s (2008a) finding of the relatively low influence of accounting systems on the environmental data collected by the top 200 Australian companies. The most important influences on such data collection were risk assessment procedures (influencing 93% of respondents) followed by activity-based management and then performance management using a balanced score card approach (influencing 50% and 43% of respondents, respectively). The financial reporting system, cost accounting system and management accounting system were somewhat less important.

Where accountants do get involved, their approach tends to be reactive, rather than proactive, i.e. they are not seeking out opportunities to use their data collection and systems development skills to assist their colleagues with responsibilities for environmental performance management. Indeed, as an outcome of their detailed case study work in six organisations examining the relationships between environmental management systems, external environmental reporting, stakeholder engagement and drivers of organisational change leading to improved environmental performance, Adams *et al.* (2008b) made the following recommendations to accountants and accounting professions (pp. 2–3):

- Greater understanding is needed of how sustainability issues impact on organisational performance.
- The financial implications of non-financial quantified and qualitative performance measures should be highlighted.
- A team approach to sustainability data collection, measurement and reporting systems is required.
- Introduction of robust sustainability assurance processes is critical to the improvement of sustainability performance and stakeholder confidence.

Adams *et al.*'s (2008a, b) findings would seem to indicate little progress in accounting's involvement in environmental performance management since Bebbington *et al.*'s (1994) survey of the UK's top 1000 companies. This indicated that only in the areas of accounting for energy and waste and investment appraisal could *any* discernible accounting input be observed. Initial returns from a later international survey (Gray and Bebbington, 2000) broadly confirmed the impression that accounting is one of the least developed areas of the corporate response to the environmental agenda.

These studies also draw attention to the fact that whilst accounting was, and still is, of marginal relevance in the context of corporate environmental management, the accounting system still clearly dominates traditional areas of decision-making such as medium- and long-term planning, capital expenditure and divisional performance evaluation. There is therefore a clear danger that the short-term, restrictive focus adopted by traditional accounting systems, which ignore the environmental dimension, may be transmitting signals that encourage environmentally malign behaviour and offer resistance to initiatives designed to encourage more environmentally sensitive behaviour. The capital budgeting system offers one example here where the conventional approach to discounting handles uncertainty by employing short-term payback criteria and inflated discount rates, thus discriminating against giving a fair weighting to long-term environmental factors. Even more fundamentally, as Gray and Bebbington (2001) point out, while systems which lie at the heart of the organisation – notably budgeting and performance appraisal systems – emphasise conventional financial factors and remain largely untouched by the changing environmental agenda, the former are always likely to dominate the latter.

References

Aaronson, S. A. (2005) 'Minding our business': what the United States Government has done and can do to ensure that U.S. multinationals act responsibly in foreign markets, *Journal of Business Ethics*, **59**: 175–98.

AccountAbility (1999) *Accountability 1000: A foundation standard for quality in social and ethical accounting, auditing and reporting*. London: AccountAbility.

AccountAbility (2008) *AccountAbility Principles Standard*. London: AccountAbility.

Adams, C. A. (2002) Internal organisational factors influencing corporate social and ethical reporting, *Accounting, Auditing and Accountability Journal*, **15**(2): 223–50.

Adams, C. A. (2004) The ethical, social and environmental reporting-performance portrayal gap, *Accounting, Auditing and Accountability Journal*, **17**(5): 731–57.

Adams, C. A. and Frost, G. (2008) Integrating sustainability reporting into management practices, *Accounting Forum*, **32**(4): 288–302.

Adams, C. A. and Larrinaga-González, C. (2007) Engaging with organisations in pursuit of improved sustainability accountability and performance, *Accounting, Auditing and Accountability Journal*, **20**(3): 333–55.

Adams, C. A. and McNicholas, P. (2007) Making a difference: sustainability reporting, accountability and organisational change, *Accounting, Auditing and Accountability Journal*, **20**(3): 382–402.

Adams, C. A. and Petrella, L. (2010) Collaboration, connections and change: The UN Global Compact, the Global Reporting Initiative, Principles for Responsible Management Education and the Globally Responsible Leadership Initiative, *Sustainability Accounting, Management and Policy Journal*, **1**(2): 292–6.

Adams, C. A. and Whelan, G. (2009) Conceptualising future changes in corporate sustainability reporting, *Accounting, Auditing and Accountability Journal*, **22**(1): 118–43.

Adams, C. A., Hill, W. Y. and Roberts, C. B. (1995) *Environmental, Employee and Ethical Reporting in Europe*. London: ACCA.

Adams, C. A., Hill, W. Y. and Roberts, C. B. (1998) Corporate social reporting practices in Western Europe: legitimating corporate behaviour?, *British Accounting Review*, **30**(1): 1–21.

Adams, C. A., Burritt, R. and Frost, G. (2008a) Environmental management, *Financial Management*, September: 43–44.

Adams, C. A., Burritt, R. and Frost, G. (2008b) *Integrating Environmental Management Systems, Environmental Performance and Stakeholder Engagement.* London: ACCA.

Adams, R. (2010) It's (already) beginning to look a bit like Christmas, *Sustainability Accounting, Management and Policy Journal*, 1(1): 85–8.

Albelda-Pérez, E., Correa-Ruiz, C. and Carrasco-Fenech, F. (2007) Environmental management systems as an embedding mechanism: a research note, *Accounting, Auditing and Accountability Journal*, 20(3): 403–22.

American Accounting Association (1973) Report of the committee on environmental effects of organisational behaviour, *The Accounting Review* (Supplement to Vol.XLVIII): 75–119.

Ball, A. (2004) A sustainability accounting project for the UK local government sector? Testing the social theory mapping process and locating a frame of reference, *Critical Perspectives on Accounting*, 15(8): 1009–35.

Ball, A. (2005) Environmental accounting and change in UK local government, *Accounting, Auditing and Accountability Journal*, 18(3): 346–73.

Ball, A. and Seal, W. (2005) Social justice in a cold climate: could social accounting make a difference?, *Accounting Forum*, 29(4): 455–73.

Bebbington K. J. and Gray, R. H. (2001) An account of sustainability: failure, success and a reconception, *Critical Perspectives on Accounting*, 12(5): 557–87.

Bebbington, J. and C. Larrinaga-González, C. (2008) Carbon trading: accounting and reporting issues, *European Accounting Review*, 17(4): 697–717, reprinted in Gray, R. H., Bebbington, K. J. and Gray, S. (eds) (2010) *Social and Environmental Accounting: Sage Library in Accounting and Finance*, Volume 1, pp. 311–32. London: Sage.

Bebbington, K. J., Gray, R. H., Thomson, I. and Walters, D. (1994) Accountants attitudes and environmentally sensitive accounting, *Accounting and Business Research*, 94: 51–75.

Bebbington, J., Brown, J. and Frame, B. (2007) Accounting technologies and sustainability assessment models, *Ecological Economics*, 61(2/3): 224–36.

Beets, S. D. and Souther, C. C. (1999) Corporate environmental reports: the need for standards and an environmental assurance service, *Accounting Horizons*, 13(2): 129–45.

Benn, S., Dunphy, D. and Martin, A. (2009) Governance of environmental risk: new approaches to managing stakeholder involvement, *Journal of Environmental Management*, 90: 1567–75.

Bennett, M. and James, P. (1998) *Environment Under the Spotlight: Current practices and future trends in environment-related performance measurement for business*, Research Report 55. London: ACCA.

Bennett M., Schaltegger, S. and Zvezdov, D. (2013) *Exploring Corporate Practices in Management Accounting for Sustainability.* London: ICAEW.

Bloom, R. and Heymann, H. (1986) The concept of social accountability in accounting literature, *Journal of Accounting Literature*, 5: 167–82.

Boyce, G. (2000) Public discourse and decision making: Exploring possibilities for financial, social and environmental accounting, *Accounting, Auditing and Accountability Journal*, 13(1): 27–64.

Brady, J., Ebbage, A. and Lunn, R. (eds) (2011) *Environmental Management in Organizations: The IEMA Handbook*, 2nd Edition. London: Earthscan.

Brockhoff, K. (1979) A note on external social reporting by German companies: A survey of 1973 company reports, *Accounting, Organizations and Society*, 4(1/2): 77–85.

Brown, J. (2009) Democracy, sustainability and dialogic technologies: taking pluralism seriously, *Critical Perspectives on Accounting*, 20(3): 313–42.

Buhr, N. (2007) Histories of and rationales for sustainability reporting, in Unerman, J., Bebbington, J. and O'Dwyer, B. (eds), *Sustainability Accounting and Accountability*, pp. 57–69. London: Routledge.

Buhr, N. and Gray, R. (2012) Environmental management, measurement and accounting: information for decision and control?, in Bansal, T. and Hoffman, A. (eds), *Oxford Handbook of Business and the Environment*, pp. 425–43. Oxford: Oxford University Press.

Burnett, R. D. and Hansen, D. R. (2008) Ecoefficiency: defining a role for environmental cost management, *Accounting, Organizations and Society*, 33: 551–81.

Burritt, R. (2012) Environmental performance accountability: planet, people, profits, *Accounting, Auditing and Accountability Journal*, 25(2): 370–405.

Burritt, R., Hahn, T. and Schaltegger, S. (2002) Towards a comprehensive framework for environmental management accounting, links between business actors and environmental management accounting tools, *Australian Accounting Review*, **12**(2): 39–50.

Burritt, R., Schaltegger, S. and Zvezdov, D. (2011) Carbon management accounting: explaining practice in leading German companies, *Australian Accounting Review*, **21**(1): 80–98.

Chapple, W. and Moon, J. (2005) Corporate Social Responsibility (CSR) in Asia, *Business and Society*, **44**(4): 415–41.

Cho, C., Patten, D. M. and Roberts, R. W. (2012) Corporate environmental financial reporting and financial markets, Bansal, T. and Hoffman, A. (eds), *The Oxford Handbook of Business and the Natural Environment*, pp. 444–61. Oxford: Oxford University Press.

CICA (1993) *Environmental Costs and Liabilities: Accounting and financial reporting issues.* Toronto: Canadian Institute of Chartered Accountants.

CMA (1992) *Accounting for the Environment.* Hamilton: Society of Management Accountants of Canada.

Collison, D. J., Gray, R. H. and Innes, J. (1996) *The Financial Auditor and the Environment.* London: ICAEW.

Cooper, S. and Owen, D. (2007) Corporate social reporting and stakeholder accountability: the missing link, *Accounting, Organizations and Society*, **32**(7-8): 649–67.

Dahlmann, F., Brammer, S. and Millington, A. (2008) Environmental management in the United Kingdom: new survey evidence, *Management Decision*, **46**(2): 264–83.

Daly, H. E. (ed.) (1980) *Economy, Ecology, Ethics: Essays toward a steady state economy.* San Francisco, CA: W. H.Freeman.

Deegan, C. (1999) Mandatory public environmental reporting in Australia: here today, gone tomorrow, *Environmental and Planning Law Journal*, **16**(6): 473–81.

Deegan, C. (2002) The legitimising effect of social and environmental disclosures: a theoretical foundation, *Accounting, Auditing and Accountability Journal*, **15**(3): 282–311.

Deegan, C. (2007) Organizational legitimacy as a motive for sustainability reporting, in Unerman J., Bebbington, J. and O'Dwyer, B. (eds), *Sustainability Accounting and Accountability*, pp. 127–49. London: Routledge.

Deegan, C. (2008) Environmental costing in capital investment decisions: electricity distributors and the choice of power poles, *Australian Accounting Review*, **18**(44): 2–14.

Deegan, C. and Gordon, B. (1996) A study of the environmental disclosure practices of Australian corporations, *Accounting and Business Research*, **26**(3): 187–99.

Dresner, S. (2002) *The Principles of Sustainability*. London: Earthscan.

Durden, C. (2008) Towards a socially responsible management control system, *Accounting, Auditing and Accountability Journal*, **21**(5): 671–94.

Emmanuel, C., Harris, E. and Komakech, S. (2010) Towards a better understanding of capital investment decisions, *Journal of Accounting and Organizational Change*, **6**(4): 477–504.

Environment Agency (2007) *Environmental Disclosures: The Second Major Review of Environmental Reporting in the Annual Report and Accounts of the FTSE All-Share*. Bristol: Environment Agency.

Epstein, M. J. and Roy, M.-J. (1998) Integrating environmental impacts into capital investment decisions, in Bennett, M. and James, P. (eds), *The Green Bottom Line: Environmental Accounting for Management, Current Practice and Future Trends*, pp. 100–114. Sheffield: Greenleaf Publishing Ltd.

European Commission (1992) *Towards Sustainability (The Fifth Action Plan)*, Com(92) 23 final, Vol. I-III, 27 March. Brussels: EC.

Fédération des Experts Comptables Européens (1993) *Environmental Accounting and Auditing: Survey of current activities and developments*. Brussels: FEE.

Frame, B. and Brown, J. (2008) Developing post-normal technologies for sustainability, *Ecological Economics*, **65**(2): 225–41.

Freedman, M. and Stagliano, A. J. (1992) European unification, accounting harmonization and social disclosures, *International Journal of Accounting*, **27**(2): 112–22.

Gray, R. H. (1990) *The Greening of Accountancy: The profession after Pearce*. London: ACCA.

Gray, R. H. (1992) Accounting and environmentalism: an exploration of the challenge of gently accounting for accountability, transparency and sustainability, *Accounting, Organizations and Society*, **17**(5): 399–426.

Gray, R. (2006) Social, environmental and sustainability reporting and organisational value creation? Whose value? Whose creation?, *Accounting, Auditing and Accountability Journal,* **19**(6): 793–819.

Gray, R. H. and Bebbington, K. J. (2000) Environmental accounting, managerialism and sustainability: Is the planet safe in the hands of business and accounting?, *Advances in Environmental Accounting and Management,* **1**: 1–44.

Gray, R. H. and Bebbington, K. J. (2001) *Accounting for the Environment,* 2nd Edition. London: Sage.

Gray, R. and Herremans, I. (2012) Sustainability and social responsibility reporting and the emergence of the external social audits: the struggle for accountability?, in Bansal, T. and Hoffman A. (eds), *Oxford Handbook of Business and the Environment,* pp. 405–24. Oxford: Oxford University Press.

Gray, R., Owen, D. and Adams, C. (1996) *Accounting and Accountability. Changes and challenges in corporate social and environmental reporting.* London: Prentice Hall.

Gray, R. H., Bebbington, K. J., Collison, D. J., Kouhy, R., Lyon, B., Reid, C., Russell, A. and Stevenson, L. (1998) *The Valuation of Assets and Liabilities: Environmental law and the impact of the environmental agenda for business.* Edinburgh: ICAS.

Guidry, R. P. and Patten, D. M. (2010) Market reactions to the first-time issuance of corporate sustainability reports: evidence that quality matters, *Sustainability Accounting, Management and Policy Journal,* **1**(1): 33–50.

Harte, G. and Owen, D. L. (1991) Environmental disclosure in the annual reports of British Companies: A research note, *Accounting, Auditing and Accountability Journal,* **4**(3): 51–61.

Herbohn, K. (2005) A full cost environmental accounting experiment, *Accounting, Organizations and Society,* **30**(6): 519–36.

Hibbitt, C. and Collison, D. J. (2004) Corporate environmental disclosure and reporting: Developments in Europe, *Social and Environmental Accounting Journal,* **24**(1): 1–11.

Hopwood, A., Unerman, J. and Fries, J. (eds) (2010) *Accounting for Sustainability: Practical Insights.* London: Earthscan.

Hoque, Z. and Adams, C. A. (2008) *Measuring Public Sector Performance: A study of government departments in Australia.* Melbourne: CPA Australia.

Hundred Group of Finance Directors (1992) *Statement of Good Practice: Environmental reporting in annual reports.* London: 100 Group.

ICAEW (1992*) Business, Accountancy and the Environment: A policy and research agenda* (eds), Macve, R. and Carey, A. London: Institute of Chartered Accountants in England and Wales.

ICAEW/Environment Agency (2009) *Sustainable Business: Turning questions into answers – Environmental issues and annual financial reporting.* London: Institute of Chartered Accountants in England and Wales.

Integrated Reporting (2012) *Rio+20* (www.theiirc.org/rio20).

IIRC (2013) *Capitals Background Paper for IR.* London: Integrated Reporting Council.

Jamali, D. (2008) A stakeholder approach to corporate social responsibility: a fresh perspective into theory and practice, *Journal of Business Ethics,* **82**: 213–31.

Jones, M. J. (2003) Accounting for biodiversity: operationalising environmental accounting, *Accounting, Auditing and Accountability Journal,* **16**(5): 762–89.

Kirschbaum, M. U. F. and Cowie, A. L. (2004) Giving credit where credit is due. A practical method to distinguish between human and natural factors in carbon accounting, *Climatic Change,* **67**: 417–36.

Kloot, L. and Martin, J. (2000) Strategic performance management: a balanced approach to performance management issues in local government, *Management Accounting Research,* **11**(2): 231–51.

Kolk, A., Levy, D. and Pinkse, J. (2008) Corporate responses in an emerging climate regime: the institutionalization and commensuration of carbon disclosure, *European Accounting Review,* **17**(4): 719–45.

KPMG (1993) *International Survey of Environmental Reporting.* London: KPMG.

KPMG (1994) *Environmental Risks and Opportunities: Advice for Boards of Directors.* Toronto: KPMG.

KPMG (1996) *UK Environmental Reporting Survey 1996.* London: KPMG.

KPMG (1997) *Survey of Environmental Reporting 1997.* London: KPMG.

KPMG International Environmental Network (1997) *Environmental Reporting.* Copenhagen: KPMG.

KPMG (1999) *KPMG International Survey of Environmental Reporting 1999.* Amsterdam: KPMG/WIMM.

KPMG (2002) *KPMG 4th International Survey of Corporate Sustainability Reporting.* Amsterdam: KPMG/WIMM.

KPMG (2005) *KPMG International Survey of Corporate Responsibility 2005.* Amsterdam: KPMG International.

KPMG (2008) *KPMG International Survey of Corporate Responsibility Reporting 2008.* Amsterdam: KPMG International.

KPMG (2011) *KPMG International Survey of Corporate Responsibility Reporting 2011.* Amsterdam: KPMG.

Kreuze, J. G. and Newell, G. E. (1994) ABC and life-cycle costing for environmental expenditures, *Management Accounting (US),* **75**(8): 38–42.

Lamberton, G. (2000) Accounting for sustainable development – a case study of city farm, *Critical Perspectives on Accounting,* **11**(5): 583–605.

Larrinaga-González, C. (2007) Sustainability reporting: insights from neoinstitutional theory, in Unerman, J., Bebbington, J. and O'Dwyer, B. (eds), *Sustainability Accounting and Accountability,* pp. 150–67. London: Routledge.

Larrinaga-González, C., Carrasco-Fenech, F., Javier Caro-González, F., Correa-Ruíz, C. and María Páez-Sandubete, J. (2001) The role of environmental accounting in organizational change - an exploration of Spanish companies, *Accounting, Auditing and Accountability Journal,* **14**(2): 213–39.

Lessem, R. (1977) Corporate social reporting in action: an evaluation of British, European and American practice, *Accounting, Organizations and Society,* **2**(4): 279–94.

Livesey, S. M. and Kearins, K. (2002) Transparent and caring corporations? A study of sustainability reports by The Body Shop and Royal Dutch/Shell, *Organization and Environment,* **15**(3): 233–58.

Milne, M. J. (1991) Accounting, environmental resource values and non-market valuation techniques for environmental resources: a review, *Accounting, Auditing and Accountability Journal,* **4**(3): 81–109.

Mitchell, R., Agle, B. R. and Wood, D. J. (1997) Toward a theory of stakeholder identification and salience: defining the principle of who and what really counts, *Academy of Management Review,* **22**(4): 853–86.

Moore, D. (2009) Reporting on environmental performance, in UNCTAD (ed.), *Promoting Transparency in Corporate reporting: A quarter century of ISAR,* pp. 21–52. Geneva: United Nations.

Netherwood, A. (1996) Environmental management systems, in Welford, R. (ed.), *Corporate Environmental Management: systems and strategies,* pp. 35–58. London: Earthscan.

Nijhof, A., de Bruijn, T. and Honders, H. (2008) Partnerships for corporate social responsibility: a review of concepts and strategic options, *Management Decision,* **46**(1): 152–67.

Netherwood, A. (1996) Environmental management systems, in Welford, R. (ed.), *Corporate Environmental Management: systems and strategies,* pp. 35–58. London: Earthscan.

O'Riordan, L. and Fairbrass, J. (2008) Corporate social responsibility (CSR): models and theories in stakeholder dialogue, *Journal of Business Ethics,* **83**: 745–58.

Owen, D. L. (1992) *Green Reporting: The challenge of the nineties.* London: Chapman and Hall.

Owen, D. L. (2007) Assurance practice in sustainability reporting, in Unerman J., Bebbington, J. and O'Dwyer, B. (eds), *Sustainability Accounting and Accountability,* pp. 168–83. London: Routledge.

Owen, D. (2008) Chronicles of wasted time? A personal reflection on the current state of, and future prospects for, social and environmental accounting research, *Accounting, Auditing and Accountability Journal,* **21**(2): 240–67.

Özbirecikli, M. (2007) A review on how CPAs should be involved in environmental auditing and reporting for the core aim of it, *Problems and Perspectives in Management,* **5**(2): 113–26.

Pearce, D., Markandya, A. and Barbier, E. N. (1989) *Blueprint for a Green Economy.* London: Earthscan.

Percy, S. (2013) The bigger picture, *economia,* May: 80–81.

Porritt, J. (2005) *Capitalism: as if the world matters.* London: Earthscan.

Preston, L. E., Rey, F. and Dierkes, M. (1978) Comparing corporate social performance: Germany, France, Canada and the US, *California Management Review,* **20**(4): 40–9.

Ratnatunga, J. T. D. and Balachandran, K. R. (2009) Carbon business accounting: the impact of global warming on the cost and management accounting profession, *Journal of Accounting Auditing and Finance,* **24**(2): 333–55.

Schaltegger, S. (1998) Accounting for eco-efficiency, in Nath, B., Hens, L., Compton, P. and Devuyst, D. (eds), *Environmental Management in Practice. Volume I - instruments for environmental management,* pp. 272–87. London: Routledge.

Schaltegger, S. and Burritt, R. (2000) *Contemporary Environmental Accounting. Issues, concepts and practice*. Sheffield: Greenleaf.

Schaltegger, S. and Csutora, M. (2012) Carbon accounting for sustainability and management. Status quo and challenges, *Journal of Cleaner Production*, **36**: 1–16.

Schaltegger, S., Bennett, M., Burritt, R. and Jasch, C. (eds) (2008) *Environmental Management Accounting for Cleaner Production*. Dordrecht: Springer.

Schaltegger, S., Gibassier, D. and Zvezdov, Z. (2013) Is environmental management accounting a discipline? A bibliometric literature review, *Medatari Accountancy Research* 21(3) eprint.

Schreuder, H. (1979) Corporate social reporting in the Federal Republic of Germany: an overview, *Accounting, Organizations and Society*, **4**(1/2): 109–22.

Sloan, T. W. (2011) Green renewal: incorporating environmental factors in equipment replacement decisions under technological change, *Journal of Cleaner Production*, **19**(2–3): 173–86.

Spence, C. and Gray, R. (2007) *Social and Environmental Reporting and the Business Case*. London: ACCA.

Stone, D. (1995) No longer at the end of the pipe but still a long way from sustainability: a look at management accounting for the environment and sustainable development in the United States, *Accounting Forum*, **19**(2/3): 95–110.

SustainAbility/UNEP (1996) *Engaging Stakeholders: The benchmark survey*. London/Paris: SustainAbility/UNEP.

SustainAbility/UNEP (1997) *Engaging Stakeholders: The 1997 Benchmark survey*. London/Paris: SustainAbility/UNEP.

SustainAbility/UNEP (1998) *Engaging Stakeholders: The non-reporting report*. London/Paris: SustainAbility/UNEP.

SustainAbility/UNEP (1999) *Engaging Stakeholders: The internet reporting report*. London/Paris: SustainAbility/UNEP.

Tellus Institute (1992a) *Total Cost Assessment – Accelerating industrial pollution prevention through innovative project financial analysis*. Washington, DC: US EPA.

Tellus Institute (1992b) *Alternative approaches to the financial evaluation of industrial pollution prevention investments*. Boston, MA: Tellus Institute.

Touche Ross (1990) *Head in the clouds or head in the sands? UK managers' attitudes to environmental issues – a survey*. London: Touche Ross.

Tregidga, H. and Milne, M. J. (2006) From sustainable management to sustainable development: a longitudinal analysis of a leading New Zealand environmental reporter, *Business Strategy and the Environment*, **15**: 219–41.

Turner, R. K. (1987) Sustainable global futures: common interest, interdependence, complexity and global possibilities, *Futures*, **10**: 574–82.

Turner, R. K. (ed.) (1988/1993) *Sustainable Environmental Management: Principles and practice*. London: Belhaven Press.

Unerman, J., Bebbington, J. and O'Dwyer, B. (eds) (2007) *Sustainability Accounting and Accountability*. Oxford: Routledge.

United Nations World Commission on Environment and Development (1987) *Our Common Future (The Brundtland Report)*. Oxford: Oxford University Press.

Walley, N. and Whitehead, B. (1994) It's not easy being green, *Harvard Business Review*, **72**(3): 46–52.

Welford, R. (2005) Corporate social responsibility in Europe, North America and Asia: 2004 survey results, *Journal of Corporate Citizenship*, **17**: 33–52.

Whiteman G., Walker, B. and Perego, P. (2013) Planetary boundaries: ecological foundations for corporate sustainability, *Journal of Management Studies*, **50**(2): 307–36.

Yakhou, M. and Dorweiler, V. P. (2004) Environmental accounting: an essential component of business strategy, *Business Strategy and the Environment*, **13**(2): 65–77.

York, R., Rosa, E. A. and Dietz, T. (2003) Footprints on the Earth: the environmental consequences of modernity, *American Sociological Review*, **68**(2): 279–300.

Young, A. (2010) Greenhouse gas accounting: global problem, national policy, local fugitives, *Sustainability Accounting, Management and Policy Journal*, **1**(1): 89–95.

Zambon, S. and Del Bello, A. (2005) Towards a stakeholder responsible approach: the constructive role of reporting, *Corporate Governance*, **5**(2): 130–41.

CHAPTER 8

Finance and financial issues[1]

8.1 Introduction

> Are the world's financial markets . . . a force for sustainable human progress, or are they
> an impediment against it? [T]he . . . role played by financial markets in the way we
> organize our commercial, industrial and personal [is critical].
>
> (Schmidheiny and Zorraquin, 1996: xxi)[2]

This chapter attempts to provide an overview (albeit a brief overview) of some of the major
intersections between social, environmental and sustainability accounting and accountability
on the one hand and the world of finance and financial institutions on the other. It is perhaps
worth recognising, before we embark on this exploration, that it might seem redundant to
have a chapter on 'finance' in a text such as this. After all, accounting as conventionally understood arose largely because of the needs of those who had finance available to invest or lend –
and so the two things are inextricably intertwined. Equally, it is through the development of
accounting that organisations have been able to grow to the point were the owners (those with
the finance) might have no knowledge of or involvement with the activities and effects that
their finance is funding – thus leading to the accepted need for both management and financial accounting systems.[3] But, more subtly still, there has been a steadily growing separation
between (on the one hand) the acts of owning and investing finance and (on the other) the
activities in which one is investing or to which one is lending. This has slowly but irrevocably
led to a world in which a great deal of finance is actually unrelated to anything other than
finance itself (what McGoun, 1997, calls hyper-reality). It brings the world economy to the
point where (what is called) the 'financial economy' is creating, buying and selling 'financial
products' which make and lose money with little (or sometimes no) direct relationship with
goods and services or the activities of people and organisations in (what is called) the 'real
economy'.

Periodically, the hyper-reality of the financial world is exposed by crashes, crises or
exposed frauds: the Wall Street Crash of 1929; the US Savings and Loans Scandal of 1985;
the 1987 world stock market crash; Leeson's breaking of Barings Bank in 1995; the 1997 Asian
financial crisis and the collapse of Long-Term Capital Market in 1998; the collapse of the dot.
com bubble in 2000; the world wide financial crisis of 2007 including the collapse of Lehman

---

[1]We are pleased to acknowledge the advice and comments of John Wilson and Andreas Hoepner in the preparation
of this chapter.

[2]This reference comes from an important text produced by the World Business Council for Sustainable Development (WBCSD), some of whose analysis and virtually all of whose conclusions we strongly disagree with.

[3]That size problem (see Chapters 1 and 3) is one of the key elements in our need for new accountings and
accountability.

Brothers and AIG in 2008; and so on and so on.[4] How well these crises are really understood is unclear, but what seems to happen is that after much wringing of hands and an apparent desire for increased regulation of financial institutions, all seems to go back to (what passes for) normal and we await the next, bigger and more damaging crisis, crash or fraud.

Whilst we are not dealing specifically with finance (see, for example, Mishkin and Eakins, 2008; Berger *et al.*, 2010) and we cannot therefore offer to deal comprehensively with the financial world, we do need to establish a bit of context before we move on to focus on the more specific elements of the financial world that more directly affect our concerns for accounting and accountability. Consequently, this chapter is structured as follows. In the next section, we will briefly review some of the principal themes within the global financial world in order to set the backdrop for the rest of the chapter. Section 8.3 offers a word of caution about the subtleties of market and financial prices before Sections 8.4 and 8.5 turn to look, respectively, at the nature of shareholding and investing and the crucial question of whether the pursuit of social and environmental performance is in conflict or is in harmony with financial performance. Section 8.6 then explores the emergence of the socially responsible investor whilst Section 8.7 examines the question of whether investing ethically is in conflict or in harmony with investing for financial returns. In Section 8.8 we return to our major theme of accountability and examine what socially responsible investment (SRI) can mean for its development. Section 8.9 offers a few words on the narrowness of our normal assumptions about 'investment' and offers some alternative examples of good social and environmental investments which may not be seen that way in conventional finance and touches, very briefly, on the emerging issue of micro-finance. Section 8.10 contains the summary and conclusions.

8.2 A brief glimpse into the world of finance

> Practices of the unscrupulous money changers stand indicted in the court of public opinion, rejected by the hearts and minds of men. . . . . Faced by the failure of credit, they have proposed only lending more money. Stripped of the lure of profit by which to induce our people to follow their false leadership, they have resorted to exhortations, pleading tearfully for restored confidence. They know only the rules of a generation of self-seekers. They have no vision, and when there is no vision the people perish. . . . . Yes, the money changers have fled from their high seats in the temple of our civilisation. We may now restore that temple to the ancient truths. The measure of the restoration lies in the extent to which we apply social values more noble than mere monetary profit.
>
> (Roosevelt, Franklin D., 1933)[5]

Finance is typically defined as the 'management of money' or 'pecuniary resources'.[6] Bannock *et al.* (1987) refer to it as 'the provision of money where and when required'

---

[4]For illustration, the World Bank has a section which examines (what it calls) the 'number of systemic banking crises in progress worldwide'. A graph available at http://go.worldbank.org/OQS5FMQ8B0 shows the number (if not the intensity) of such crises in recent years: rising from two or three per year in the 1980s to about 10 per year currently. To illustrate the extent of the concern, the credits for the feature film *The Other Guys* contain a brief tutorial on Ponzi schemes and some of the issues in the early 20th century financial crisis. This can be viewed on YouTube (http://www.youtube.com/watch?v=ueUPvPk0Q00).

[5]Roosevelt, Franklin D. (1933/2007) '*The only thing we have to fear is fear itself*', Inauguration Speech in Washington, DC March 4, 1933 reprinted (London: Guardian News and Media).

[6]*The Pocket Oxford Dictionary* (London: Oxford) 1967.

(p. 156). Economists typically refer to it as a store of wealth (how much potential access to resources you have is measured in money) or as the medium of equivalence and exchange (what things are worth and their exchange values – costs and prices in other words). We use it in accounting to measure the 'success' or 'failure' of economic enterprises (profit or loss) and, increasingly, it is used 'to keep the score' of the 'winners and losers' in corporate and financial life (I'm/we're doing better/worse than you because I/we have more/less money than you do).

You'd have thought it was quite easy to work out how much money there was, but defining money is actually quite difficult. Once one gets beyond coins and notes one is struggling with what actually is 'money'. It clearly includes all sorts of things like deposits with banks, bonds, inter-bank lending and so on. It is the 'so on' that causes the real problems. It is probable that nobody really knows, but a range of estimates put the total at about $40 trillion ($40,000,000,000,000) which is close to estimates of the world's GDP. Figure 8.1 suggest some of the ways it which this money moves around (see Allen *et al.*, 2004, 2010 and the World Bank website Open Data for much more detail).

At about this point, it all begins to get a little strange. If you spend any time with sources like the World Bank and the International Monetary Fund you will learn that something like $15 trillion is invested in stock markets (an important figure to which we will return) and there is something like $16 trillion on the world bond market (publicly issued debt in effect). There is something like $2 trillion in foreign direct investment (FDI). Caulkin (2008) reports that $1 trillion is traded *daily* on currency markets and, of this, only 20% relates to any tourism or trade requirement – the rest is gambling. MacWhirter (2008) reports data from the Bank of International Settlement in Basel that $516 trillion is invested in the global market for derivatives and reports this as 10 times the value of all the world's stock markets put together.[7]

The point of these numbers – accurate or not – is not only that they are very, very large and not only that they seem elusive to anyone not steeped in macro-economics but that they are indicative of a system which has a life of its own (McGoun, 1997; Collison and Frankfurter, 2000). Our instincts are to think of money as (say) so many hours of work, or so many apples or a new pair of shoes or (as we will see shortly) a part ownership of an asset, but we begin to get a glimpse here of money as something which refers only to itself. It has become something which can be traded – bought and sold – and a range of financial instruments and financial products can be created on the basis of it ('securitised' as it was widely known in the bank credit crisis on the early 21st century) and these, in turn, are bought and sold and profits (and losses) made on each transaction. This 'financial economy' is enormous – much bigger than the 'real economy' of goods and services – and very complex indeed. It is this self-referential complexity and the considerable difficulty that many people (business people, politicians and probably even bankers) have in understanding it that is in large part responsible for the patchy control of the financial economy. This, in turn, seems to lead inevitably to the financial system's periodic lurches into crisis. The financial economy has an increasingly important influence on the real economy when, ironically, the financial system was originally devised in order to assist and help develop the organs of the real economy (see, for example, Corporate

---

[7]These data are derived from trying to make sense of a range of sources – the specific references are: Caulkin, S. (2008) 'There's life yet beyond the British super-casino', *Observer*, Business and Media 3 February (p. 8) and MacWhirter (2008) 'The Red Menace', *Sunday Herald*, 23 March (p. 10). Please note the inconsistency in the data here but also note that it doesn't really matter – the figures are so big they overwhelm any sense of how it all fits together and what is being controlled – or not controlled. If you need help or counselling speak with a kindly macro-economist – if you can find one. It is not without interest to note that Rudyard Kipling wrote a poem published in 1919 called 'The Gods of the Copybook' which essentially foretold of the inevitability of financial meltdown.

Figure 8.1 An overview of the financial system

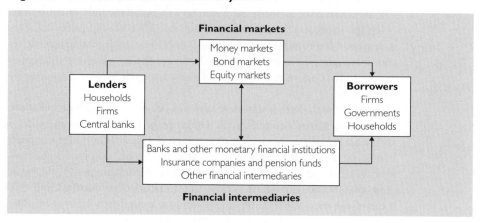

Source: Taken from Allen, *et al.,* (2004: 491).

Watch, 2012). It may very well now be the other way around: the tail of the financial economy may well be wagging the dog of the real economy.[8]

More importantly, we can (relatively) easily see how actions in the real economy of consumption, production, manufacturing and waste disposal (for example) directly affect people, societies and the planet itself and we can (as we are doing) seek to exercise some control over that relationship. If, as is increasingly the case, the bizarre and disjointed financial economy is actually driving the organs of the real economy, how then might society intervene and seek to regain any kind of control or accountability of its corporations? Organisations are, in principle at least, the creations of human society and intended for the benefit of that society. Our attempts at control and accountability – the core of our discourse – may actually be in vain if we cannot exercise similar levels of accountability and control over the organs of the financial economy.

8.3 A cautionary note about numbers, measurement and remoteness

At the heart of our concerns over the relative size of the financial economy and its dislocation from normal 'real' activities there lies a more subtle and important range of issues to do with the reification of financial numbers – or 'economisation' as it is often known. The potential dangers of this run throughout accounting, finance and economics. As long as money is only used as a way of keeping score and a reflection of command over assets, it may be that its intrinsic influence need not be baleful. (Of course, the existence of those who have access to more, rather than less, financial resources; the concerns we may have about the way in which power accrues because of ownership; and the way in which there is so much acceptance of the apparent objectivity of prices: all crucially and critically reflect the inequalities and potential abuses of embedded power relationships.) However, not only do finance and financial numbers potentially take on a life of their own, as we saw above, but they also (like many numbers) have the potential to suggest, quite inappropriately, a sense of objectivity and

---

[8]Indeed, the financial crisis of 2007–8 and the resultant financial 'bail out' using public money to save banks from the consequences of their own actions – i.e. failure – contained a distinct irony. Capitalism is supposed to be good at weeding out weakness through collapse – but the use of government funds to secure it is a lot more like communism – but without the advantages!

accuracy. This is not just the apparent accuracy and apparent objectivity of (say) a profit figure or a cost saving (neither of which is usually either accurate or objective in this simple sense). It goes so much further to the point where a share price, for example, might look objective – might look as though it can be relied upon to show some 'real' worth (as opposed to a synthesis of a few people's expectations). In turn, this leads to simplistic assumptions that the rise and fall of asset (especially share) prices might reflect real success or failure in the underlying assets – when this *need* not be the case at all.[9] The final step is when the less subtle of financial markets participants begin to treat the numbers themselves, not as one reflection of reality but as the reality itself (the hyper-reality perhaps).

As measurement and numbers become more ubiquitous – often for fairly reasonable and useful reasons – so the potential for unexpected consequences increases. More measurement starts to look better than less measurement to the point where we start to first measure, and then make valuations of things which really do not lend themselves to such treatment (see Figure 8.2).[10]

The advantage of measurement – and especially accounting measurement – is that it helps simplify complex situations, but the danger is that the number comes to be seen as the whole of the story. Very rarely is this the case: profit can never be 'right'; a director's salary can never precisely capture his or her whole worth; rising or falling prices may – or may not – be a sensible reflection of real economy activities.

Matters become really serious when we eventually start to think that financial numbers are the ideal representation of the world and, consequently, the answer to everything. Then we start to value air, the natural environment, social well-being and even pollution. Once we have a price for something, then we can start to buy it and sell it – we can create a market; and once we have a market then we have market prices and these can be treated as objective, can't they?

Two examples can be used to illustrate this point: the economisation of the public and not-for-profit sectors in the late 20th and early 21st centuries and the bewildering emergence and growth of **carbon markets**. The public sector and other non-profit organisations, under the influence of the growing neo-liberal agenda, were increasingly subject to financial representation and the demands of a financial accountability – despite the fact that this is, to a significant degree, highly inappropriate. Through political demands for a spurious 'accountability' more and more financial accounting is imposed on these organisations and 'best' becomes 'cheapest'; output not measured becomes discounted; major issues in society are reduced to the economic – life is 'economised' (see Broadbent and Laughlin, 1998; Broadbent *et al.*, 2001).

In the case of the carbon market (which, actually has nothing to do with carbon – see Chapter 7), rather than privatise air and ask us to buy and sell clean air to each other, the decision was made to allow licences which permit organisations to emit greenhouse gases up to a certain level. If an organisation was likely to exceed its permitted level then it had to buy unused permits from organisations which had them for sale – hence a market in carbon credits and carbon permits (for more detail see, for example, Bayon *et al.*, 2009; Brohe *et al.*, 2009). The only reason an organisation would do this is because the government would charge it for exceeding its quota. The charge for exceeding the quota became a 'price for carbon' and a market was born (Bebbington and Larrinaga-González, 2008), and actually, it became rather more hyper-real than this – but that is another story (Callon, 2009; Lohmann, 2009).

---

[9]The *Independent* newspaper reported that a false rumour of impending bankruptcy of the bank HBOS had resulted in a major fall in share prices: those who manufactured the rumour bought the very cheap shares and watched the price rally as millions are made on a false rumour (*Independent*, 23 March 2008: 9).

[10]The corollary can be valuable: there may be value in trying to ascribe numbers to things that do not have prices, but this will always be political and arbitrary and only really helpful if we never believe the outcomes.

Figure 8.2 The 'MacNamara fallacy'

The first step is to measure whatever can be easily measured. This is OK as far as it goes. The second step is to disregard that which can't be easily measured or give it an arbitrary quantitative value. This is artificial and misleading. The third step is to presume that what can't be measured easily isn't important. This is blindness. The fourth step is to say that what can't be easily measured really doesn't exist. This is suicide.

Source: Taken from Yankelovich (1972).

The key thing upon which to keep an eye is that there is a dangerous point at which all the important things – love, sunshine, water, family, music, friendship, nature, chocolate – are defined as being important only in so far as they are measured and reflected in financial numbers. And once something has a financial number, it has a price and it can be bought, sold, evaluated in a way which drives out (what Thielemann, 2000; calls) the market alien values – i.e. humanity (see, for example, Craig *et al.*, 1999; McPhail, 1999; Gray, 2002).[11]

The dangers and advantages of measurement and numbers can go even further than this, and the whole issue becomes really quite subtle. Perhaps the most important illustration of this is the way in which **risk** has become such an important part of the debate around environmental, social and governance (ESG) issues. Risk, strictly speaking, refers to the probability distribution of future events or states occurring – although its use in areas such as **risk management** generally focuses on undesirable future states. That might seem a good approach to organising and quantifying future undesirable states and, of course, it is – to a point. The absolutely key question from our point of view is *whose* risk is being considered. Environmental risk might well seem to refer to matters such as species extinction, habitat destruction or water course damage. These are risks to polar bears, spiders or flora. However, what is nearly always meant when *risk* appears is the risk to an organisation – i.e. ultimately an *economic* risk. Polar bear extinction is *not* a risk to (say) most mining companies. Drought is not a risk to an electronics company unless its supply chain or customers are affected. That is, 'risk' means risk to the organisation, not risk to the ecology or society. These two only converge in the win–win situations we discussed earlier when a social or environmental risk to a society or an ecology may have substantive potential impact on the economic functioning of the organisation concerned. This merging of two very different notions of risk is widespread even amongst entities who might know better: so there are occasions when WWF talks of *water risk* but is predominantly concerned about the impact on utility companies and the financing of those (Nattero *et al.*, 2009); and occasions when the United Nations Principles of Responsible Investment (UNPRI) talks about *risk management* and it predominantly means the risk to earnings and (financial) returns.[12]

Of course, it would be an exaggeration – perhaps even downright wrong – to say that everything about finance is, in some sense, 'bad', but this remoteness that the financial economy involves is both its strength and its weakness. It allows exchange and control at a distance which allows, consequently, those who own and potentially have the power over resources to lose sight of those resources. People buying and selling shares are no longer exchanging ownership and responsibility in a (say) mining company but are exchanging financial instruments which make or lose money: the connection with the real economy is

[11]It is more than likely that you are experiencing just this process as either a student (with fees, evaluations and grading, etc.) or as a faculty member (with rankings and journal impacts, and output orientation and cost-centres, etc.). It is not a very healthy development (Tinker and Puxty, 1995).

[12]See, for example, http://academic.unpri.org/.

again broken. A major function of a really well-developed accountability would be to re-connect the investor (owner or lender) with the things owned or financed and bring back the Rawlsian notion of **closeness** (Chapter 13) – a more direct involvement between the financial transaction and the consequences and implications of that transacting. It is this that sits at the heart of much that follows.[13]

8.4 Shareholders, investors and investment

> The business case for CSR . . . has two dimensions. First, there are clear risks to shareholder value from poor management of supply chain issues, inadequate environmental management, human rights abuses and poor treatment of workers, suppliers or customers. Shareholders want to see that companies are managing such risks. Concerned investors will apply pressure to those which are not and reward those which are. Ultimately, companies which do not engage in this process will incur a higher cost of capital. The second element of the business case is the potential for competitive advantage from CSR.
>
> (Cowe, 2001: 4)

It remains the case that purely profit-seeking investors are likely to be almost entirely uninterested in corporate social responsibility (CSR), sustainability and other non-financial matters except in so far as these influence their financial position. This observation has led to two major strands in the wider literature. First, there is the tantalising question of whether or not there is an essential conflict between a concern for responsibility, accountability and sustainability on the one hand and financial performance on the other. Does it, indeed, 'pay to be good but not too good' as Mintzberg (1983) has it? The second and still developing strand argues the necessity to distinguish investors aiming simply for short-term, speculative profit from those who seek to practise more responsible share ownership involving long-term commitment (see, for example, Church of Scotland, 1988; Moore, 1988; Charkham, 1990). It has been suggested that the latter group may have a significant role to play in the development of social responsibility and (perhaps) sustainability and an associated widening of corporate accountability (Owen, 1990; and see also Solomon and Darby, 2005; Mallin, 2007).

We will concentrate upon the 'conventional' investor and their 'conventional investments' in this and the subsequent section of this chapter. In later sections of the chapter, we will move onto the more promising territory of considering both the reform of investing and what possibilities are available to the more committed investor.

We have already seen that the financial world involves a very wide range of activities. Not all of these activities attract the same attention within accounting and social accounting because our principal concern tends to be with that part of the financial world that involves the corporate investor; that is, we tend to be concerned with those who hold shares in (i.e. own) the companies and, especially, those who hold shares in the large (typically multi-national) corporations (but see O'Sullivan and O'Dwyer, 2009; Corporate Watch, 2012). The reason for this focus is two-fold. First (as argued through the rest of the book, and notably in Chapter 1), corporations and especially Multi-National Corporations (MNCs) are a major force – perhaps the major force – economically, environmentally and socially on the planet today. Studying corporations inevitably demands that we also study the ownership and governance of these entities. The second reason is a little more pragmatic. Our analysis

---

[13]A helpful initiative in this area is that by Corporate Watch (2012) and which can be warmly recommended (http://www.corporatewatch.org/?lid=4171).

emerges from the discipline of accounting and the consequential emergence of social accounting. Accounting as a practice has grown up and developed its worldwide influence principally through (i) its ability to provide a mechanism for control at a distance (what we tend to call management accounting) so that organisations can grow and expand internationally; and (ii) its ability to provide accountability information about organisations' management and performance to their owners (what we know as financial accounting). Rightly or not, our predominant traditional focus in both accounting and social accounting tends to be on these large organisations where our discipline most notably evolved (although see, Chapter 12, for a widening of this focus).[14]

Shareholders have typically bought shares in a company (or companies) in order to give themselves a financial return at some point in the future – through a combination of the dividends paid by the company plus any capital gain that the investor hopes to achieve when they sell the shares at some time in the future. Their funds (at least in the primary financial markets – see below) are effectively handed over to the corporate management who put them to work in the corporation's interests and, in return, that management (and particularly the directors of the company) provide an annual account of what they have done with the shareholders' funds and on the shareholders' behalf; that is, they report on their financial performance through the annual (or more frequent) financial statements.

Where it gets interesting is that, assuming that the shareholders have not chosen to invest for some social or environmental concern (a matter we will examine in detail below), the shareholders are effectively demanding that the directors make as much money for the investors as possible (and the shareholders will have the full weight of law behind them in this demand in most jurisdictions). Should the directors want to undertake some social or environmental act that they consider to be genuinely important but which looks less likely to make money (or at least fit the business case) for the shareholders then that act is (in many jurisdictions) illegal. Of course there are exceptions[15] to this simple view and, as we have said, should the matter be expressed within a **business case** or factored into **social and/or environmental risk** then there will be convergence in the interest of the finance community and the society on that issue. However, in the majority of cases one can reasonably expect the shareholders to rapidly punish the directors, and they may well sack them for stepping outside their fiduciary duties (see especially Friedman, 1970; Bakan, 2004).

The most important point is that if you put any saint or environmental hero into any company boardroom, they will be forcibly asked to leave the moment that their social or environmental concerns conflict with the organisation's business case, assessment of risk and/or its pursuit of profit (although this proviso may well be less draconian in private companies).[16] The point is that a company owned by shareholders must – ultimately – satisfy these shareholders *and cannot do otherwise* (Sunstein, 1997; Bakan, 2004).

---

[14]The avoidance of examination and exploration of social and environmental accounting in the public and NGO sectors is a major weakness in the literature (Gray *et al.*, 2009) and one which we reflect to a degree throughout this text. There is, however, a growing substantive literature – some of which we have touched upon elsewhere (see Chapter 12) – but which is interestingly synthesised and stimulated in Osborne and Ball (2011).

[15]Corporate giving in the USA after the New Orleans hurricane or in Japan and Australia after the Japanese tsunami are examples here, although both were far from uncontroversial. Both could be seen as possible exceptions where concern for human values was sufficiently strong that unusually it transcended the economic concerns of the financial public.

[16]A private company is one whose shares are held by a relatively few people who can only buy and sell shares privately. They are not listed on a stock exchange – as public companies – where the shares are publicly bought and sold. Evidence shows that the management of a private company may have much more room for discretion – with, of course, the approval of the shareholders.

Now, if we were to assume that all shareholders were individuals who owned shares, who turned up to vote at the company's Annual General Meetings and who took the time to consider and express their views, the world might be a lot simpler. The world is not like this. Most shareholders, most importantly in public companies, buy their shares through 'the market' from other shareholders who are selling them. It is this act of buying and selling which produces the 'share price' that has become such an important indicator of corporate success and (thus) of how well the directors are thought to be performing. Therefore, if those who are buying and selling shares in a company (typically a relatively small number of people at any one time) react negatively to some new information concerning the company, these buyers and sellers will cause the price of the share to fall. The share price drop is a signal sent both to the rest of the market and to the company itself that the market does not approve of some action, omission or decision. This disapproval could be the result of a perceived bad investment, a failure to deal with a new area of risk or it could be a reaction to an altruistic act by the directors but which has no potential financial pay-off. The directors (whether they like it or not) are governed by these daily – even hourly – swings in their share price as the company shares are bought and sold by people that the directors may actually never meet.[17]

Indeed, the reality is slightly different still. Whilst the traditional language of finance and the traditional notions of financial accountability to shareholders seem to implicitly assume that shareholders are diverse and individual, the reality is a long way from this. Figure 8.3 shows the range of persons and institutions which own shares in UK companies.[18] As shown, it is quite clear that only about 10% of shares are held by individuals: the rest are held by organisations of one sort or another – many of which are actually not in the UK at all. This means that to understand what is happening in reporting to shareholders one needs to be explicitly clear that one is speaking to the trustees of (for example) pension funds and investments trusts and the brokers who buy and sell their holdings for them; only at the margins are we seeking to speak to individuals who are able to act freely as individuals. It complicates matters immensely.

8.5 The profit-seeking investor, CSR and performance

So, to what extent are typical investors interested in social and environmental issues and do they actually respond to social and environmental (and 'sustainability') disclosures? Such a question leads us, inevitably, to ask what does the evidence tell us about the relationship between, crudely, making money, social responsibility and environmental stewardship? We examine these questions in this section.

There has been a steady stream of work over many years which has sought to discover whether investors actually find disclosure about a corporation's social and environmental activities of any use.[19] On the face of it, the results seem quite encouraging. Early work (see, for example, Benjamin and Stanga, 1977; Firth, 1984) suggested that shareholders find social and environmental information of some interest and of some value in their decision-making. Work such as Epstein and Freedman (1994) continues to find that investors demand and are interested in social and environmental disclosure. Chan and Milne (1999) and Milne and Chan (1999) revisit the work of Belkaoui (1980) and find that, whilst narrative disclosures were important in the investors' decisions, they were marginal and any 'ethical' effect

---

[17]We emphasise again that this is less likely in private companies.

[18]It proved very difficult to find the same information for other countries.

[19]This work tends to be associated with 'decision usefulness' as the theoretical lens – we touch on this in Chapter 4 and in Gray et al. (2010).

Figure 8.3 Beneficial ownership of UK shares: end-2010

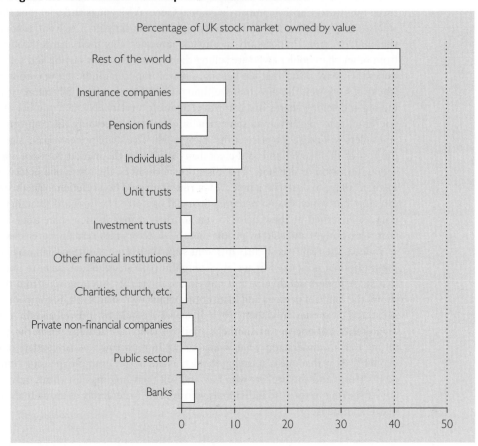

Source: Taken from ONS, (2010: 20).

Notes: 'Ordinary shares' are the most common type of share in the ownership of a corporation. Holders of ordinary shares receive dividends.

Rest of world investors are equivalent to foreign investors.

that such disclosure might have was more than swamped by the financial concerns in the decision-making. Both Solomon and Solomon (2006) and Hunt and Grinnell (2004) find in very different contexts that the enthusiasm for social and environmental information is muted. The interest seems to be influenced by both the extent to which the investors (and their representatives) are familiar with and understand social and environmental matters as they affect corporations as well as the extent to which these social data seem to be relevant to future earnings. So, although at the margins shareholders appear to have a generalised interest in such matters, it tends to be driven out by more pressing financial exigencies.

Of course, there are going to be situations when the financial and the social/environmental information is potentially mutually reinforcing (one area of the win–win situations that are introduced in Chapter 7). This will typically occur when the disclosure is itself mandated and has clear, direct financial implications. The most striking example would probably be the 'Superfund' legislation in the USA which requires (to put it simply) that polluted land has to be cleaned up – a task involving considerable costs and, consequently, a significant potential liability (see, for example, Gray and Bebbington, 2001). The requirement to disclose such liability is bound to have – at least in principle – significant impact on investors' perceptions of corporate worth (see, for example, Freedman and Stagliano, 1995; but see also Patten, 2002b). As we mentioned earlier, a similar situation increasingly obtains with carbon/greenhouse gas emission accounting and (ultimately) disclosure, linked, as it is, to initiatives with real financial consequences such as emissions trading and carbon reduction commitments. Indeed, it would appear that the cost of carbon now increasingly features as a standard part of company valuation on the part of many analysts, most notably in the electricity sector, whilst anticipation of further regulation in the area has also led the investment community to encourage companies to take action to reduce greenhouse gas emissions (Sullivan, 2011).

This is an elusive, rapidly changing and diverse field of increasing importance, but there is little doubt that the accounting/disclosure and the financial consequences have an increasingly direct relationship (Bebbington and Larrinaga-González, 2008; De Aguiar and Fearfull, 2010). But to what extent are the disclosures, the social and environmental issues and the economics always in line? There will inevitably be situations where social and environmental risks attract unwelcome attention from the media and/or non-governmental organisations (NGOs), there will be occasions when social and environmental incompetence reflects poor management of the company itself. In these circumstances, the interests of the conventional investor and the needs of those more exercised by social and environmental issues will be (or appear to be) broadly in harmony. But how extensive is this harmony? There can be few questions in the broad area of CSR which have exercised the research community as much as this one (see, for example, De Bakker *et al.*, 2005).

An examination of whether or not financial performance, social and environmental disclosure and social and environmental performance are in harmony or related comprises three simple elements: what companies reveal about themselves (**social and environmental disclosure**), what companies actually do economically (**financial performance**) and what companies actually do – or are thought to do – socially and environmentally (**social and environmental performance**). These three elements are shown in Figure 8.4.[20]

Each of these elements is of continuing concern in the management and accounting literature: not least because the implications of these relationships are potentially very important for understanding corporate behaviour in general and CSR in particular. However, these relationships are also important because the literature illustrates a really important but frequently ignored point; that is, researching important relationships is often very difficult indeed and we always have to be very careful to ensure that the inferences we draw from such research actually relate to the issues of concern.[21] Drawing the wrong inferences on matters as important as this can (and we will argue does) have dire and baleful effects. We will briefly explore each of the relationships in turn.

Social and environmental disclosure and social and environmental performance

On the face of it, it must seem very odd that we might query this relationship at all. After all, surely what a company *says* about its social and environmental performance and its *actual* social and environmental performance must be closely related? Well . . . no, actually. Ever

---

[20]This diagram and much of this section of the chapter is taken from Gray (2006b). Ullmann (1985) provides an excellent early introduction to this literature.

[21]Theories are typically thought of as relationships between *concepts* – for example, between social responsibility and financial performance. However, as is often the case, neither of these *concepts* can be uniquely measured and so most theories examine *constructs* which are empirical approximations of the *concepts*. To the extent that there is a difference between the construct examined and the concept about which we wish to speak, we must always tread very carefully indeed so as not to mislead ourselves (or others) or rush to overly simple (and wrong) answers.

Figure 8.4 Are social responsibility + profitability compatible?

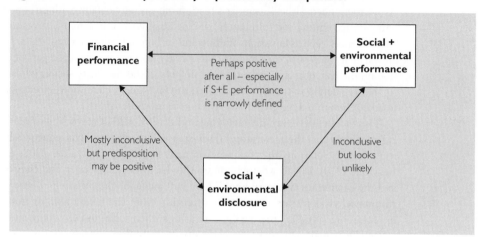

since Wiseman (1982) and Rockness (1985) identified that there was no relationship – or possibly even a *negative* relationship – between what an organisation says that its does and what it actually does, there has remained considerable scepticism about this relationship. Indeed, a study by Deegan (2002) suggests that unfavourable media attention acts as a catalyst for positive corporate disclosure, whilst, even more damningly, Deegan and Rankin (1996) provide evidence of companies significantly increasing the reporting of favourable environmental information when being prosecuted for environmental misdemeanours and, when one stops and thinks about it, there is no particular surprise here. The reasons for this are twofold: the disclosures are unreliable and the performance is difficult to measure. We will return to the question of measuring social and environmental performance below, but for the moment let us consider the disclosures.

The disclosures by companies tend to be voluntary in most cases, and such disclosures are generally thin, trivial and general (Freedman and Jaggi, 2006) as well as being partial and selective (Milne *et al.*, 2009). Consequently, corporate disclosure is unlikely to tell you the complete story of an organisation's social and environmental performance but rather a selective view that corporations – understandably – wish to convey (Adams, 2004; Adams and McPhail, 2004; Milne *et al.*, 2009). Such a story is bound to mislead to some degree at least. The situation is not necessarily much improved when the disclosure is either required by statute or is influenced directly by state requirements (see Freedman and Walsey, 1990; Patten, 2002a).

Of course there are examples where specific disclosures are an accurate reflection of specific areas of performance: philanthropic giving, volume of greenhouse gases emitted or level of accidents and deaths, for example. Such disclosures can probably be relied upon, but they do not give a full picture of performance. It *is* possible to construct accounts which would give a fairly full story of an organisation's social and environmental performance (see, Chapters 9 and 13), but virtually no organisation does so. Therefore, researchers (and others) cannot rely upon disclosure to provide a reliable indicator of social and environmental performance organisation. And so they look elsewhere and end up using a wide range of proxies which may range from the accurate but narrow through the accurate but obliquely relevant (e.g. pollution expenditures) to the accessible but probably irrelevant (e.g. reputation ratings) (see, for example, Patten, 2002b, 2005; Gray, 2006a).

So we have the peculiar situation that – to a significant degree at least – corporate social and environmental disclosure cannot be relied upon to tell us about organisational performance and, until there is a major change in the regulation of social and environmental disclosure, this is likely to remain the case. Equally important – perhaps even more so – is the notion that if disclosure was substantial (i.e. it was sufficient to start to discharge a substantial accountability) then it is very possible that the quality of social and environmental *performance* could itself potentially rise (but see Adams and Larrinaga-González, 2007). The notion of information inductance (Prakash and Rappaport, 1977) tells us that we tend to perform better in those areas for which we are held accountable – not least because nobody likes reporting what might be thought of as a poor performance. It leads to the old cliché that 'you get what you measure', and accounting research has known this for decades and recognised that if you measure the wrong things, then you get the performance you didn't want. Good social and environmental accountability has considerable potential.

Social and environmental disclosure and financial performance

There has been relatively little direct study of the relationship between disclosure and performance for two (fairly obvious) reasons. The first reason is that major studies require data and data on social disclosure has not been easily available in database format since 1977.[22] The more substantive reason is that, on the face of it, there is no major reason why one should expect this relationship to be interesting; that is, although it is well established that voluntary social disclosure is related to both size and industry (see, for example, Gray *et al.*, 2001, but see also Bouten *et al.*, 2012), it is not immediately obvious why a decision to disclose social information might be related to financial performance. In fact, the relationship proves to be potentially more interesting than it looks. The original speculation in this field was that richer companies disclosed because they could – having made money they could afford to 'feel good' (Gambling, 1985). More persuasive arguments have emerged subsequently and, although none of them has been fully substantiated, they have a broad plausibility. Remembering that the interest in this area concerns a correlation, a statistical relationship – not a causality – it is quite plausible that shareholders are concerned that the company's management is covering potential social and environmental risks and exposures. Consequently, the management might use social and environmental disclosures to signal their awareness of these issues. Social and environmental disclosure might then be taken as an indicator of a better quality of management – one which is not exclusively focused on short-term financial gain but takes a wider and longer focus (Buhr and Reiter, 2006). And, of course, it could be that management are simply responding to a desire expressed by shareholders for this data – however unlikely that might seem (Cormier *et al.*, 2004, 2005). Whatever the reasoning behind the putative relationship, there have been relatively few studies, and these have not been convincing. So, for example, Ingram (1978) finds no relationship although Anderson and Frankle (1980) find some marginal evidence of a positive relationship. A more substantive and longitudinal study (albeit in the UK) by Murray *et al.* (2006) finds that the *predisposition* to undertake voluntary disclosure seems to be related to longer term profitability, and that paper concludes that it seems unlikely that they are directly influencing each other but, rather, are probably both functions of the style and quality of the company's management: good companies make profits, manage risks and report voluntarily (or so we might be led to believe).

---

[22]Ernst and Ernst maintained a database of annual report disclosures by US companies during the 1970s. This facility no longer exists.

Social and environmental performance and financial performance

Of the three relationships, the one that inevitably excites the most interest is that which concerns the harmony or conflict between behaving 'properly' and making money – or 'doing well by doing good' as it is sometimes tritely referred to. This is, in many ways, the current holy grail of management research, and there is no shortage of (largely unsubstantiated) claims made by business and business organisations in this area. Indeed, these claims extend beyond just social and environmental performance and even (however bewilderingly) embrace sustainability as the route to profitability (see, for example, Oberndorfer, 2004; White and Kiernan, 2004; and see Milne *et al.*, 2009; and, Chapter 9 of this title).

Let us say from the outset that the research on this topic is seriously inconclusive: whilst some studies find a relationship others find no relationship at all. There are many useful reviews of the field (Griffin and Mahon, 1997; Edwards, 1998; Richardson *et al.*, 1999; Margolis and Walsh, 2003), and a particularly useful review by Wagner (2001) captures the inconclusiveness – or confusion – that has beset this literature:

> Although there is ample anecdotal evidence on the considerable economic benefits of individual firms from environmental performance improvement . . . systematic evidence for larger samples of firms across several industries is much more inconclusive. . . . The variability of the results based on different methodological approaches raises the question whether the variability encountered . . . represents more an artefact of the methodology or the research design or more is due to the intrinsically wide variance in the relationship between environmental and economic performance.
>
> (Wagner 2001: 44 and 46)

One attempt to overcome these issues is offered by Orlitsky *et al.* (2003) who undertook a thorough meta-analysis[23] of prior studies and concluded that there is, indeed, a positive relationship between these variables. The results are potentially persuasive – especially as the authors infer that the relationship between financial and social performance is reflexive (not in a single direction as is usually implicitly assumed) and that there are intervening variables, the most likely candidates being management quality and corporate reputation. That is, they infer, organisations *both* do well and do good.[24]

This would be a crucial and very important finding if we could rely upon it. It would suggest a range of implications such as: financially successful companies can generally be assumed to do 'good' things. We might even consider the potential inference that companies that do not undertake CSR and/or embrace sustainable development are more likely to be run by less able and/or more stupid managers. The final inference must be treated with considerable care as it is unreliable – despite its influence underlying a lot of the CSR field – that if the good in the world is being done by (*inter alia*) financially successful companies, the plethora of bad things in the world are probably being committed by the less wealthy (maybe even the poor): which gets the rich nicely off the hook and is a conclusion every rich westerner would want to be true!

This highly implausible conclusion is, indeed, highly implausible. It transpires that the problem of measuring social and/or environmental performance (see Patten, 2002b) hides an important and misleading tautology. That is, there are, inevitably, many acts that an organisation undertakes which are clearly in its financial interest and clearly in society's and

---

[23] A meta-analysis is a systematic collation of the results from prior studies. It is more than a review as it tries to combine the details from the original studies in order to create a mega-study whose results should be so much more rigorous. It is an important method in medicine, but its application in social science is difficult.

[24] Murray *et al.* (2006) make a similar inference regarding social and environmental disclosure.

the environment's interest. We touched on these economic win–wins earlier: protecting the health of employees is clearly good for the company and for society; selling products which obviously cause damage to consumers is good for neither the company nor the society; polluting water upon which the corporation's employees and/or processes depend, similarly. Many of the measures of social or environmental performance which are employed in these studies are of this nature and are, broadly, in the nature of a tautology that socially and environmentally responsible acts are those which benefit stakeholders and the company simultaneously. Such acts must, *by definition,* be correlated with some aspects of economic performance. The only surprise is then that it has taken research this long to spot what is increasingly recognised as a truism. Thus, we begin to see that what the corporate world increasingly might mean by 'social responsibility' relates to matters that are, when not considered in terms of the economic costs and benefits of corporate life, likely to be trivial at best. Levy and Egan (2003) raise an exceptionally persuasive issue in this context. The importance and the persuasiveness of the win–win situation in, what they call, 'eco-modernist rhetoric' is that it is being used to establish a new consensus about society, the environment and corporations. After all, as they argue, the win–win scenarios are not designed to ameliorate the dire ecological situation or reverse trends in social justice, although this may be how they appear – they are there to justify and bolster corporate and market primacy and autonomy.

So we are led to conclude that whilst at the margins there are shareholders who ask for and even use social and environmental information, we need to be fairly sceptical about the likelihood of investors *qua* investors finding that the pursuit of CSR (and even sustainability) is in their financial interests: however attractive such a conclusion might at first appear.

The really sad thing about all of this is that if there was substantive social, environmental and sustainability disclosure then these matters would be perfectly clear: we would know what social and financial benefits and detriments flowed from the corporations' pursuit of economic well-being. But, because the quality of the social and environmental information disclosed is so very poor, we continue to work in ignorance: an ignorance which business and business leaders exploit with their implausible, dangerous and unsupported claims. (We return to this again in later chapters.)

So we now move away from the conventional investor and look at the rather more promising developments in the field of (what is generally known as) SRI (or socially responsible investment).

8.6 The emergence of the socially responsible investor?

> Any individual or group which truly cares about ethical, moral, religious or political principles should in theory at least want to invest their money in accordance with their principles
>
> (Miller, 1992: 248)

The whole ethos of ethical, or socially responsible, investment might be argued to have its roots in the notion of individual responsibility.[25] Indeed, the social investment movement, with its emphasis on personal commitment, may be viewed, at least in part, as a reaction to the growing impersonality – even the sense of personal distance – of most savings and investment (Cadman, 1986; Sullivan and Mackenzie, 2006; Coulson, 2007).

---

[25] Always assuming, for example, that one is not part of a society which prohibits notions of investment or investment for one's own benefit and/or rewards investment that resonates with other non-economic or non-market values.

In seeking to use the investment function to pursue social goals, the social investment movement is choosing to keep broadly within modern capitalism's 'rules of the game' in order to achieve evolutionary change in society (see Owen, 1990).[26] In this, the social investment movement has much in common with the consumer movement (Elkington, 1987; Elkington and Hailes, 1989) in that consumer (or investor) pressure attempts to impose new social controls on the business enterprise under the essential principle of consumer (investor) sovereignty (Craig-Smith, 1990).

Miller (1992) traces the origins of social investment back to Victorian England and to social reformers who sought to provide good standard housing for the poor whilst offering acceptable financial returns to those providing funds for this purpose. A prominent early role was also played by religious groups, notably the Quaker movement with its traditional antipathy towards investing in areas such as armaments and alcohol production. The movement reached new levels of prominence in the late 1960s and early 1970s as the social proxy movement in the USA spawned an increasing number of shareholder resolutions on various social issues.[27] Few large companies anywhere in the world are now immune from the organised efforts (often orchestrated by NGOs and/or religious groups) of social responsibility resolutions proposed at the companies' annual general meeting on matters as diverse as weaponry, human rights, anti-union activities and oil spills (Domini and Kinder, 1984).

A more structured and institutional response to the concerns of a socially-minded investor is (what is typically termed) **socially responsible investment (SRI)** and the successful launching of a rapidly growing number of ethical mutual funds/unit trusts, which provide an important potential as a mechanism for providing individuals with access to investments that were more in line with their preferences. In addition, the funds offered the possibility of both exercising some ethical control over investment and imposing a wider social accountability on business enterprise.

Kreander (2001) identified 1965 as the date at which the first ethical funds available to private investors were launched. Since then they have grown considerably and, although estimates vary (see below), something between 2% and 10% of all funds under management are invested with some degree of socially responsible consideration. Indeed the UNPRI claims that those who have signed up to comply with the principles have approximately $18 trillion under management in 36 countries.[28] The percentages and figure may seem relatively small, but the overall impact is in excess of its immediate size as we shall see.

Defining SRI is not simple. Mallin (2007) suggests that it involves 'considering the ethical, social, and environmental performance of companies selected for investment as well as their financial performance' (p. xix) and EuroSiF says 'SRI (socially responsible investment) is investing that is mindful of the impact those investments have on society. SRI traditionally combines investors' financial objectives with their concerns about environmental, social and governance (ESG) issues' (http://www.eurosif.org/sri-resources/sri-faq). So, at its most basic, it involves the addition of non-financial criteria (increasingly referred to as ESG – environmental, social and governance criteria) to the more immediate financial criteria on which securities are normally selected for purchase or sale.

---

[26]Recognising that there are different forms of capitalism each with different issues and with different potentials for change. For an un-reconstituted socialist, however, it should be noted that this statement is nonsensical and that all capitalism has the same root problems.

[27]See Domini and Kinder (1984 Chapter 13) and Bruyn (1987, Chapters 2 and 3) for further discussion of the social proxy movement and other US initiatives from the late 1960s onwards.

[28]*Principles for Responsible Investment, Annual Report of the PRI Initiative 2009*, UNEP Finance Initiative/Global Compact.

Figure 8.5 Socially responsible investment: selected countries by investing criteria

In Billions		2007	Total SRI	Total SRI in Euros
United States (2007)	Social screening	US$2098		
	Shareholder advocacy	US$739		
	Screening and shareholder*	(US$151)	US$2710	€1917.3
	Community investing	US$26		
Canada (2006)	Core SRI	Cnd$57.4		
	Broad SR	Cnd$446	Cnd$503.6	€333.6
Australia/NZ (2007)	Core SRI	Au$19.4		
		Au$52.8	Au$72.2	€41.4
Japan (30/09/2007)	Broad SRI	¥840	¥ 840	€5.5
Europe (2007)	Core SRI	€511.7		
	Broad SRI	€2153.7	€2665.4	€2665.4
TOTAL WORLD				€4963.2

Notes: Exchange rate as of 10/09/2008.

NB: core = (thematic + positive screen); broad = (simple screen, standards of engagement).

*Negative number to avoid double counting.

Sources: Social Investment Forum, RIAA, SIO, EuroSIF, SIF Japan.

The ethical funds set about this selection with (predominantly) one or more of three broad approaches to choosing the stocks in which they are willing to invest – their 'investment universe' (Kreander, 2001; Coulson, 2007). These three are: **positive screening, negative screening** and **engagement** (each of these appears in some form in Figure 8.5).

Positive screening relates to when an investor or fund seeks out companies which meet certain criteria: a good safety record, innovation in pollution control, a leadership in human rights and so on. This approach itself comes in three broad forms: thematic (when the fund is built around a specific theme such as, for example, innovative low-impact technologies or direct assistance to poverty); seeking out leading organisations more generally; or picking companies, across all industries, who are considered to be the 'best in class' (the best oil company, the best car manufacturer, etc.).

Negative screening tends to refer to either the avoidance of specific issues and/or activities (armaments, tobacco, gambling, pornography, child labour and so on) and/or the avoidance of companies with unusually bad records in areas of ethical concern.

The third approach – engagement – refers to the process by which investors and investment funds seek to spend time with companies trying understand their issues and encouraging them through dialogue and cajolery (as well by threat) to raise their game. Each approach has its strengths and weaknesses, its advocates and detractors (see, for example, Kreander, 2001; Sparkes and Cowton, 2004).

As the interest in and use of SRI has grown so has the industry that supports it: most notably information intermediaries have emerged which help the investment funds make their SRI decisions.[29] Equally important has been the panoply of indices and standards and guidelines

---

[29]Such information intermediaries would include bodies such as EIRIS (which stands for both Ethical Investment Research Services and, more recently, Experts in Responsible Investment Solutions), Kinder, Lyndenberg, Domini (KLD) and Franklin Research and Development Corporation (FRDC).

Figure 8.6 United Nations Principles for Responsible Investment

1 We will incorporate ESG issues into investment analysis and decision-making processes.
2 We will be active owners and incorporate ESG issues into our ownership policies and practices.
3 We will seek appropriate disclosure on ESG issues by the entities in which we invest.
4 We will promote acceptance and implementation of the Principles within the investment industry.
5 We will work together to enhance our effectiveness in implementing the Principles.
6 We will each report on our activities and progress towards implementing the Principles.

which has grown up around SRI. These, to an increasing degree, have become the criteria by which it is decided whether or not an investment might be thought of as 'ethical'. The better known of these metrics include indices such as the UK's FTSE4Good, the FTSE KLD 400 Social Index in the USA and the international Dow Jones Sustainability Index. All provide an inclusive listings and rankings of the 'best' companies: rankings based, to a degree at least, on the organisation's social, environmental and sustainability performance. Picking companies listed in these indices can offer some funds the appearance of a straightforward positive screen for investment (see, for example, Cobb *et al.*, 2005; Zeigler and Shroder, 2010).[30]

These indices are complemented by a range of standards and guidelines that also assist, amongst others, SRI practitioners (Leipziger, 2010). There is a plethora of such standards, and they are issued by a diverse range of organisations from time to time (SA8000 on Child Labour and the Universal Declaration of Human Rights are examples – see Chapters 5 and 6). However, the most universal of such standards is almost certainly the United Nations *Global Compact* (UNGC)[31] which was issued in 1999 and which outlines 10 principles in the areas of human rights, labour standards, anti-corruption and environmental protection. Companies are invited to sign up to the Compact and thereafter required to report on their progress in compliance. Related to the UNGC is the UNPRI (UNPRI)[32] which asks investors and investment funds to sign up to six broad principles of responsible financing (see Figure 8.6). SRI practitioners can be expected to comply with the UNPRI and, broadly at least, to invest in companies which are registered with the UNGC.

The apparent effect of these initiatives and the many that are emerging all the time is that SRI looks a lot as though it is becoming part of the mainstream. In this latter context, one could particularly point to examples such as the British Association of Insurers lobbying for increased disclosure of ESG matters featuring the publication of a demanding set of social and environmental disclosure guidelines (see Solomon, 2007) together with the move on the part of many major banks to sign up to the Equator Principles, an international voluntary code designed to encourage the consideration of environmental and social issues in project financing.[33] The question therefore arises as to whether we can envisage a time when all investment is 'ethical'?[34] We will return to this question shortly, but it seems potentially

---

[30]In fact, the rankings are based on a complex scoring which includes both a range of financial performance characteristics, reputation and social and environmental issues. The impact of social and environmental issues is relatively minor. As we have seen (and see also Chapter 9), there is no 'sustainability performance' on which to rank the companies.

[31]http://www.unglobalcompact.org/.

[32]http://www.unpri.org/principles/. Related principles are included in *The Equator Principles* which defines itself as 'A financial industry benchmark for determining, assessing and managing social & environmental risk in project financing' (http://www.equator-principles.com/principles.shtml). For more detail, see O'Sullivan and O'Dwyer (2009).

[33]http://www.equator-principles.com/principles.shtml.

[34]One crucial matter is that all investment is 'ethical' in some sense or other (Prindl and Prodhan, 1994: 21). Friedmanite self-interest is a form of utilitarianism which is an ethical position. The decision to ignore (what Thielemann, 2000, calls) 'market alien values' – i.e. humanity – is an ethical decision.

unlikely that SRI would have moved so strikingly towards the mainstream if such investment did not, in actuality, produce the financial returns that investors typically appear to seek. This is a matter we examine briefly in the next section.

8.7 Financial performance of the funds: how socially responsible is SRI?

Conventionally, if we follow the normal economically rationalist arguments, investing in socially responsible stocks should result in the investor receiving lower financial returns – or, at least, experiencing higher levels of risk. The simple argument is that an investor focusing on SRI is reducing the investment universe from which to select stocks and doing so not on financial but on 'ethical' grounds (Kreander *et al.*, 2002). The research work in this field is still emerging, but the dominant conclusion appears to be, from studies done in the USA, UK and mainland Europe, that investment in SRI does not necessarily involve financial penalties to the investor compared to the returns from conventional investment (Hamilton *et al.*, 1993; Mallin *et al.*, 1995; Gregory *et al.*, 1997; Reyes and Grieb, 1998; Kreander *et al.*, 2005). Why this seemingly counter-intuitive result might be the case is a matter we will return to but, at least in part, the reasons seem to have some connection with the types of companies towards which SRI may converge and consequential issues of size and variability in returns (risk as typically measured in finance). That is, *a priori*, an SRI portfolio of stocks is likely to contain more companies which have embraced (for example) environmental management and human rights (for example), more companies that are engaged in (for example) innovative alternatives to fossil fuels or methods for treating wastes; and fewer companies involved in the 'sin' industries or subject to the detrimental attentions of pressure groups or environmental activists. The larger of such companies might be thought to have lower risk and more enlightened management. At the same time, the SRI portfolio may well contain more smaller, innovative companies which are seeking new products or processes which could in theory increase both potential returns and potential volatility. Taken across the piece, these factors might well wash out in the research methods used and produce the results we find reported here.[35]

On the face of it, this brings us back to the 'it pays to be good' arguments we saw earlier. That is, the evidence suggested here is that by using the surplus funds that one has the good fortune to own in a way dictated by one's beliefs and conscience, one will continue to gain financially from the operation: one can, it seems, have one's moral cake and eat it.

Rather as we saw with the financial and social/environmental performance literature earlier, it all depends upon how one defines one's terms: that is, an ethical fund with an environmental concern might be expected to insist on not investing in (for example) any fossil fuel that was not used as capital or investing in any company that disrupted eco-systems. An SRI fund with a social focus might be expected to insist on avoiding any company with any human rights abuses, with any exploitation of labour or with any advertising to children. A fund claiming to be directed towards sustainability might be expected to refuse to have anything to do with any corporation that was increasing its material impact, encouraging consumers to increase their material consumption or distributing money from the poor to the rich. One can see how quickly an investor would be left with very few companies in which to invest. If one wished to invest wholly ethically and combined all the criteria, it seems fairly likely that no conventional investment would be available.

So, ethical investment is a question of degree, clearly. It is not about environmental protection but limiting environmental damage; it is not about social justice but about limiting social

---

[35]The research in this field is becoming increasing sophisticated and embracing the increased availability of data. See Statman (2006) and Lopez *et al.* (2007) for examples.

injustice. Equally, therefore, SRI is about allowing some of one's moral concerns into consideration: but only some. One's 'ethics' in this case are not requiring one to sacrifice very much of one's material prosperity and comfort. The 'ethics' are also, clearly, a question of degree.

This is important because it leads to the realisation that SRI is, very appropriately, a marginal, gradualist movement that seeks partial reform slowly. The more partial and the more gradual the reform, the less the economic 'penalty' investors are being asked to accept and, consequently, the more popular SRI becomes with investors. Indeed, a considerable amount of the academic, policy and professional interest in the field relates to the impact on the investor rather than the impact on society, the environment or on ethics as such (see, for example, Sullivan *et al.*, 2006; Godfrey *et al.*, 2009; Lee and Faff, 2009).

So, although it remains feasible, as Owen (1990) suggests, that financial capitalism and investment in particular is capable of substantive reform, we do well to consider the arguments of Cooper (1988) and Cooper and Sherer (1984). International financial capitalism is constructed on the basis that a relatively few rich people are encouraged to pursue their own – notably financial – interests, and only at the margins does the law prevent this. If this is the heart of a society which is primarily interested in whether a few rich people are likely to lose a little financial return then we are asking the wrong questions. The right questions are: how do we change the power structure to allow wider democracy? how do we alleviate oppression and poverty? how do we seek to ameliorate the appalling desecration of the planet? Only the most rabid of neo-liberals would believe we achieve it through endorsing and encouraging financial pursuit in unbridled international financial markets. That SRI offers a partial brake on such conventional expansion may be a cause for celebration, but it may also be window dressing and a cause for some alarm that marginalist, mainstream investment is now claiming to be socially responsible. Once again, if we allow ourselves to infer that 'making money' (i.e. gathering investment return) is synonymous with 'doing good', and therefore that it is the rich people who are doing the good in the world, we are only one small (if ill-considered) step from inferring that it must be those poor people again who are doing all the harm. This is not very plausible is it?

This brings us back to the implications that SRI may (or may not) have for accountability which we will explore in the next section before finishing with a brief look at what an ethical investment might really look like!

8.8 SRI, disclosure and accountability?

A key factor throughout our examination of finance in this chapter has been that investors, researchers and citizens actually know very little about what is actually happening inside the companies. The principle of accountability asks that, rather than us having to root around to try and find if (say) social and financial performance are correlated, we should be able to assess this information directly from corporate reports of one sort or another. Such suggestions are, of course, resisted to varying degrees by companies and, indeed, by politicians. However, one of the things that makes the whole SRI area of especial interest to social accountants (as Owen, 1990, pointed out) is that in advanced capitalist societies, it has long been established that the investor has primacy. This, in turn, means that investors can expect to be given the information that they need for whatever purposes they deem fit. Whether or not one accepts the moral reasoning behind such a simple view of capitalism this suggests, really quite strongly, that, even if civil society cannot manifest its rights to information, investors should be able to do so and that includes SRI investors. But, generally speaking, SRI investors *cannot* obtain the information they require to make informed investment decisions in line with the values that they espouse.

This problem has been recognised for some time. For example, Rockness and Williams' (1988) study of US ethical funds suggested that the lack of availability of appropriate information on social performance, rather than frequency of negative criteria, was the main reason for excluding firms from ethical portfolios. In the UK, deficiencies in information provision are particularly highlighted by Harte *et al.* (1991), who find that ethical investors cannot acquire the information they need, and in a study by Perks *et al.* (1992) who conclude that:

> Although company annual reports may be presented as meeting the information needs of investors in general terms, they are clearly lacking in relation to the specific needs of ethical investors who are concerned with environmental issues.
>
> (Perks *et al.*, 1992: 55)

The emergence of standalone reporting from the mid-1990s – together with the increased sophistication of the various information intermediaries – held the promise of beginning to overcome the serious information asymmetry experienced by social investors. However, the quality and partiality of the information is such that it remains unlikely that investors are having their information needs met by voluntary disclosure (Solomon and Solomon, 2006).

One unexpected development has been charted by Solomon and Darby (2005) and Solomon and Solomon (2006) who explored what might be meant by the increasing use of engagement in SRI and discovered that one element of the engagement was a significant increase in the exchange of *private* social and environmental information; that is, fund managers and other financial professionals, in the process of engaging with companies on ESG issues, could become the recipient of information which (on occasions) was not available to the market in general. There are several ironies here: apart from the generally held view that the exchange of private information is perilously close to insider trading and thereby both illegal and unethical, the expectation that Bruyn (1987) expresses that the increase in SRI will have the potential effect of increasing wider and more *public* social and environmental disclosure is actually having the opposite effect. Nevertheless, many of the SRI movers and shakers continue to lobby, argue and press for increase in formal, mandatory and substantive social and environmental accountability, and such efforts probably deserve our support, albeit that they have, as yet, had only marginal success.

8.9 Extending the nature of social investment

SRI is only one – albeit a potentially powerful – theme in the development of wider notions of what comprises an investment. We have noted briefly above that other areas of the financial world have sought to include social and environmental information within (at least some of) their activities. We saw the UNPRI and the banks' involvement in the *Equator Principles* as examples of this (see, for example, Coulson, 2007; O'Sullivan and O'Dwyer, 2009; Bhimani and Soonwalla, 2010; Macve and Chen, 2012). Individual financial institutions have, over several decades, chosen to try and take matters much further, often on an individual basis. Examples here are legion and growing. The UK's Co-operative Bank has sought to bring elements of ethical financing into the mainstream and is well-known for its ethical stance on a range of social and environmental issues. In Europe, the International Association for Investors in the Social Economy has a long-standing role in coordinating 'alternative' social investment practice through promoting social banking initiatives across Europe by connecting investors directly with social projects (Campanale *et al.*, 1993). A key member of the Association has been the Triodos bank which has long been a leader in what community-based banking should look like. Similarly, Mercury Provident, a small UK licensed deposit-taking institution founded in 1974, allows investors to direct their funds to individual projects of their own choice and choose the interest rate they wish to receive.

The more one investigates the more one finds that this principle of sacrificing financial returns in order to achieve greater social returns is a fundamental feature of the 'alternative' investment scene, and it stands in clear contrast to the notion promoted by the larger, more conventional, institutional social investors that both values may be maximised together. As Sullivan and Mackenzie (2006) state:

> It is hard to see why investment analysts would attempt to incorporate companies' environmental externalities in their valuations because, in the absence of regulation to internalise these externalities, they are, by definition, external to the company's value. In such cases, government or other social interventions will be necessary to address these market failures.
>
> (Sullivan and Mackenzie 2006: 348)

One important mechanism to overcome this *impasse is* **mutuality**. Mutuality is a very old notion and refers to the idea that the lenders and borrowers were also the owners of the financial institution and, in fact, the institution was established predominantly to help a group of people help each other. The notion of mutuality finds one of its most profound manifestations in the financial sector in British building societies and mutual funds. Most of the UK's building societies have been de-mutualised on the altar of private efficiency and profit, but some remain: perhaps most notably the Ecology Building Society which restricts its interest rates and lends only for the purchase and restoration of ecologically sensible buildings. Another UK example of mutuality is represented by Shared Earth who 'are an ethical investment co-operative . . . [which lends] money to fair trade businesses in the developing world'.[36]

Such examples – and there are many others – demonstrate that financial stewardship and prudent financial management are perfectly compatible with higher social and environmental principles. Indeed, there are examples all over the world of micro-credit institutions (perhaps the Grameen Bank of Bangladesh specifically set up to provide access to small but vital amounts of loan capital for the very poor is the most famous), credit unions and community financial cooperatives that all offer both the less financially blessed(?) and the developed West examples of a sensible way to organise and manage financial transactions and needs (see, for example, Fisher and Sriram, 2002; McKillop *et al.*, 2007). There is considerable potential – and probably a great deal less risk – in financial institutions built upon closeness, mutuality and membership.

Once we escape the simplistic notions of what is 'finance', the very notion of investment becomes problematic. So, for example, two well-known examples of companies whose shares are held with no real expectation of regular dividends are Traidcraft plc (Dey, 2007) and the Centre for Alternative Technology (there are many others). One might think of such organisations as offering the investor the enormous privilege of placing their funds where they can do excellent work that the investor themselves would be unable to undertake. It is really only a question of how one expresses these things.

---

[36]For more detail on mutuality and a wider global perspective see, for example, Cuevas and Fischer, (2006).

8.10 Summary and conclusions

It is probably a simple truism that one cannot really begin to understand capitalism and the complex systems with which modern humanity organises itself without a good grasp of the financial world. As we have seen in even this brief outline, the financial world is a complex and diverse one and, in all probability, it is little understood by significant proportions of society. Given its immense influence on both the real economy and the lives of society – and thereby ultimately on the planet itself – this should concern us. So, perhaps the first thing to take away from this chapter is the realisation of the enormity and illusory nature of the financial economy and the burning question (which regular financial crises do not seem to address) of whether this is a situation which humanity is content to let become yet more complex? It seems very unlikely that more sustainable and responsible economic organisation will be possible unless the financial world is much better understood and (at least) better controlled. The lack of a socially and environmentally sympathetic financial world is almost certainly one of the major impediments to any major moves towards a more sustainable future (Schmidheiny and Zorraquin, 1996).

That said, we have also seen that there is an enormous number of very positive initiatives and possibilities within the financial sector from which one can draw various levels of optimism. The most substantial impetus is certainly that which comes from the considerable and exciting developments in SRI along with the increasing integration of ESG into the mainstream of investment practice. This sector continues to make great strides and to achieve great things but, as Sullivan and Mackenzie have argued, it seems very unlikely that even the best efforts form this part of the financial world will make the breakthroughs necessary without global government changing of the rules.

Although much less significant in economic terms, the new growth in interest in cooperative finance, credit unions and truly alternative investments offers important new ways of looking at how we, as societies, organise ourselves financially. Being relatively small, they also offer individuals of varying levels of wealth, opportunities for involvement and engagement – we can all have some influence, however small it might be.

The challenges for social accounting in this arena are legion. Those challenges are really no different from those we have covered through the text. It is just that when we consider the financial sector, the challenges seem more elusive, more complex, larger and more vivid (Coulson, 2007). Can the financial sector support the urgent and essential necessity of organisational accountability? Can the financial sector be held accountable itself? How can new accountings be devised to help the sector redirect? Is size a real impediment? How can social accounting support the third sector and the cooperative and mutual potential of other organisations? Any attempted solution to social accounting problems that ignores the vast influence of the financial sector has little hope of success.

References

Adams, C. A. (2004) The ethical, social and environmental reporting-performance portrayal gap, *Accounting, Auditing and Accountability Journal*, **17**(5): 731–57.

Adams, C. A. and Larrinaga-González, C. (2007) Engaging with organisations in pursuit of improved sustainability accountability and performance, *Accounting, Auditing and Accountability Journal*, **20**(3): 333–55.

Adams, C. A. and McPhail, K. (2004) Reporting and the politics of difference: (non)disclosure on ethnic minorities, *Abacus*, **40**(3): 405–35.

Allen, F., Chui, M. and Maddaloni, A. (2004) Financial systems in Europe, the USA, and Asia, *Oxford Review of Economic Policy*, **20**: 490–508.

Anderson, J. and Frankle, A. (1980) Voluntary social reporting: an Iso-beta portfolio analysis, *The Accounting Review*, **55**(3): 467–79.

Bakan, J. (2004) *The Corporation: The pathological pursuit of profit and power*. London: Constable and Robinson.

Bannock, G., Baxter, R. E. and David, E. (1987) *Dictionary of Economics*. London: Penguin Books.

Bayon R., Hawn, A. and Hamilton, K. (2009) *Voluntary carbon markets*, 2nd Edition. London: Earthscan.

Bebbington, J. and Larrinaga-González, C. (2008) Carbon trading: accounting and reporting issues, *European Accounting Review*, **17**(4): 697–717.

Belkaoui, A. (1980) The impact of socio-economic accounting statements on the investment decision: an empirical study, *Accounting, Organizations and Society*, **5**(3): 263–83.

Benjamin, J. J. and Stanga, K. G. (1977) Difference in disclosure needs of major users of financial statements, *Accounting and Business Research*, **27**: 187–92.

Berger, A. N., Molyneux, P. and Wilson, J. O. S. (2010) *The Oxford Handbook of Banking (Oxford Handbooks in Finance)*. Oxford: Oxford University Press.

Bhimani, A. and Soonwalla, K. (2010) Sustainability and organizational connectivity at HSBC, in Hopwood, A., Unerman, J. and Fries, J. (eds), *Accounting, for Sustainability: Practical Insights*, pp. 173–90. London: Earthscan.

Bouten, L., Everaert, P. and Roberts, R. W. (2012) How a two-step approach discloses different determinants of voluntary social and environmental reporting, *Journal of Business Finance and Accounting*, **39**(5-6): 567–605.

Broadbent, J. and Laughlin, R. (1998) Resisting the 'new public management': absorption and absorbing groups in schools and GP practices in the UK, *Accounting, Auditing and Accountability Journal*, **11**(4): 408–35.

Broadbent, J., Jacobs, K. and Laughlin, R. (2001) Organisational resistance strategies to unwanted accounting and finance changes: The case of general medical practice in the UK, *Accounting, Auditing and Accountability Journal*, **14**(5): 565–86.

Brohe, A., Eyre, N. and Howarth, N. (2009) *Carbon Markets: an international business guide*. London: Earthscan.

Bruyn, S. T. (1987) *The Field of Social Investment*. Cambridge: Cambridge University Press.

Buhr, N. and Reiter, S. (2006) Ideology, the environment and one world view: a discourse analysis of Noranda's environmental and sustainable development reports, in Freedman, M. and Jaggi, B. (eds), *Advances in Environmental Accounting and Management*, Vol.3, pp. 1–48. Oxford: Elsevier.

Cadman, D. (1986) Money as if people mattered, in Ekins, P. (ed.), *The Living Economy: A new economics in the making*. London: Routledge.

Callon, M. (2009) Civilizing markets: carbon trading between *in vitro* and *in vivo* experiments, *Accounting, Organizations and Society*, **34**(3-4): 535–48.

Campanale, M., Willenbacher, A. and Wilks, A. (1993) *Survey of Ethical and Environmental Funds in Continental Europe*. London: Merlin Research Unit.

Chan, C. C. C. and Milne, M. J. (1999) Investor reactions to corporate environmental saints and sinners: an experimental analysis, *Accounting and Business Research*, **29**(4): 265–79.

Charkham, J. A. (1990) Are shares just commodities?, in NAPF, *Creative Tension*, pp. 34–42. London: National Association of Pension Funds.

Church of Scotland (1988) *Report of the Special Commission on the Ethics of Investment and Banking*. Edinburgh: Church of Scotland.

Cobb, G., Collison, D., Power, D. and Stevenson, L. (2005) *FTSE4Good: Perceptions and performance*, ACCA Research Report 88. London: ACCA.

Collison, D. and Frankfurter, G. (2000) Are we really maximizing shareholder wealth? Or: What investors must know if we do, *The Journal of Investing*, **9**(3): 55–63.

Cooper, D. J. (1988) A social analysis of corporate pollution disclosures: a comment, *Advances in Public Interest Accounting*, **2**: 179–86.

Cooper, D. J. and Sherer, M. J. (1984) The value of corporate accounting reports: arguments for a political economy of accounting, *Accounting, Organizations and Society*, **9**(3-4): 207–32.

Cormier, D., Gordon, I. M. and Magnan, M. (2004) Corporate environmental disclosure: contrasting management's perceptions with reality, *Journal of Business Ethics*, 49(2):143–65.

Cormier, D., Magnan, M. and van Velthoven, B. (2005) Environmental disclosure quality in large German companies: economic incentives, public pressures or institutional conditions?, *European Accounting Review*, 14(1): 3–39.

Corporate Watch (2012) *Demystifying the Financial Sector: Part 1 of Corporate Watch's Guide to Banking and Finance* (http://www.corporatewatch.org/?lid=4171).

Coulson, A. B. (2007) Environmental and social assessment in sustainable finance, in Unerman, J., Bebbington, J. and O'Dwyer, B. (eds), *Sustainability Accounting and Accountability*, pp. 266–82. London: Routledge.

Cowe, R. (2001) *Investing in Social Responsibility: Risks and opportunities*. London: Association of British Insurers.

Craig, R. J., Clarke, F. L. and Amernic, J. H. (1999) Scholarship in university business schools: Cardinal Newman, creeping corporatism and farewell to the 'disturber of the peace'?, *Accounting, Auditing and Accountability Journal*, 12(5): 510–24.

Craig-Smith, N. (1990) *Morality and the Market: Consumer pressure for corporate accountability*. London: Routledge.

Cuevas, C. E. and Fischer, K. P. (2006) *Cooperative Financial Institutions: Issues in governance, regulation and supervision*. Washington, DC: World Bank.

De Aguiar, T and Fearfull, A. (2010) Global climate change and corporate disclosure: Pedagogical tools for critical accounting?, *Social and Environmental Accountability Journal*, 30(2): 64–79.

De Bakker, F. G. A., Groenewegen, P. and Den Hood, F. (2005) A bibliometric analysis of 30 years of research and theory on corporate social responsibility and corporate social performance, *Business and Society*, 44(3): 283–317.

Deegan, C. (2002) The legitimising effect of social and environmental disclosures: a theoretical foundation, *Accounting, Auditing and Accountability Journal*, 15(3): 282–311.

Deegan, C. and Rankin, M. (1996) Do Australian companies report environmental news objectively? - an analysis of environmental disclosures by firms prosecuted successfully by the Environmental Protection Authority, *Accounting, Auditing and Accountability Journal*, 9(2): 50–67.

Dey, C. (2007) Social accounting at Traidcraft plc: a struggle for the meaning of fair trade, *Accounting, Auditing and Accountability Journal*, 20(3): 423–45.

Domini, A. L. and Kinder, P. D. (1984) *Ethical Investing*. Reading, MA: Addison Wesley.

Edwards, D. (1998) *The Link Between Company Environmental and Financial Performance*. London: Earthscan.

Elkington, J. (with Tom Burke) (1987) *The Green Capitalists: Industry's search for environmental excellence*. London: Victor Gollancz.

Elkington, J. and Hailes, J. (1989) *The Green Consumer's Supermarket Shopping Guide*. London: Victor Gollancz.

Epstein, M. J. and Freedman, M. (1994) Social disclosure and the individual investor, *Accounting, Auditing and Accountability Journal*, 7(4): 94–109.

Firth, M. (1984) The extent of voluntary disclosure in corporate annual reports and its association with security risk measures, *Applied Economics*, 16: 269–77.

Fisher, T. and Sriram, M. S. (2002) *Beyond Micro-credit: Putting development back into microfinance*. London: New Economics.

Freedman, M. and Jaggi, B. (2006) Editorial: environmental accounting: commitment or propaganda, in Freedman, M. and Jaggi, B. (eds), *Advances in Environmental Accounting and Management*, Iss.3, pp. xiii–xv

Freedman, M and Stagliano, A. J. (1995) Disclosure of environmental clean-up costs: the impact of the Superfund Act information, *Advances in Public Interest Accounting*, 6: 163–78.

Freedman, M. and Walsey, C. (1990) The association between environmental performance and environmental disclosure in annual reports and 10Ks, *Advances in Public Interest Accounting*, 3: 183–93.

Friedman, M. (1970) The social responsibility of business is to increase its profits, *The New York Times Magazine*, September 13: 122–26.

Gambling, T. (1985) The accountants' guide to the galaxy, including the profession at the end of the universe, *Accounting, Organizations and Society,* **10**(4): 415–25.

Godfrey, P. C., Merrill, C. B. and Hansen, J. M. (2009) The relationship between corporate social responsibility and shareholder value: an empirical test of the risk management hypothesis, *Strategic Management Journal,* **30**(4): 425–45.

Gray, R. H. (2002) Of messiness, systems and sustainability: towards a more social and environmental finance and accounting, *British Accounting Review,* **34**(4): 357–86.

Gray, R. (2006a) Social, environmental, and sustainability reporting and organisational value creation? Whose value? Whose creation?, *Accounting, Auditing and Accountability Journal,* **19**(3): 319–48.

Gray, R. H. (2006b) Does sustainability reporting improve corporate behaviour? Wrong question? Right time?, *Accounting and Business Research (International Policy Forum),* **36**: 65–88.

Gray, R. H. and Bebbington, K. J. (2001) *Accounting for the Environment,* 2nd Edition. London: Sage.

Gray, R. H., Javad, M., Power, D. M. and Sinclair, C. D. (2001) Social and environmental disclosure and corporate characteristics: a research note and extension, *Journal of Business Finance and Accounting,* **28**(3–4): 327–56.

Gray, R. H., Dillard, J. and Spence, C. (2009) Social accounting as if the world matters: An essay in Postalgia and a new absurdism, *Public Management Review,* **11**(5): 545–73.

Gray, R. H., Owen, D. L. and Adam, C. (2010) Some theories for social accounting?: a review essay and tentative pedagogic categorisation of theorisations around social accounting, *Advances in Environmental Accounting and Management,* **4**: 1–54.

Gregory, A., Matatko, J. and Luther, R. (1997) Ethical unit trust financial performance: small company effects and fund size effects, *Journal of Business Finance and Accounting,* **24**(5): 705–25.

Griffin, J. G. and Mahon, J. F. (1997) The corporate social performance and corporate financial performance debate: twenty-five years of incomparable research, *Business and Society,* **36**(1): 5–31.

Hamilton, S., Jo, H. and Statman, M. (1993) Doing well while doing good? The investment performance of socially responsible mutual funds, *Financial Analysts Journal,* **49**(6): 62–6.

Harte, G., Lewis, L. and Owen, D. L. (1991) Ethical investment and the corporate reporting function, *Critical Perspectives on Accounting,* **2**(3): 227–54.

Hunt III, H. G. and Grinnell, D. J. (2004) Financial analysts' views of the value of environmental information, in Freedman, M. and Jaggi, B. (eds), *Advances in Environmental Accounting and Management,* vol. 2, pp. 101–20. Oxford: Elsevier.

Ingram, R. W. (1978) An investigation of the information content of (certain) social responsibility disclosure, *Journal of Accounting Research,* **16**(2): 270–85.

Kreander, N. (2001) *An Analysis of European Ethical Funds,* Research Paper 33. London: ACCA.

Kreander, N., Gray, R. H., Power, D. M. and Sinclair, C. D. (2002) The financial performance of European ethical funds 1996–1998, *Journal of Accounting and Finance,* **1**: 3–22.

Kreander, N., Gray, R. H., Power, D. M. and Sinclair, C. D. (2005) Evaluating the performance of ethical and non-ethical funds: a matched pair analysis, *Journal of Business Finance and Accounting,* **32**(7-8): 1465–93.

Lee, D. D. and Faff, R. W. (2009) Corporate sustainability performance and idiosyncratic risk: a global perspective, *Financial Review,* **44**(2): 213–37.

Leipziger, D. (2010) *The Corporate Responsibility Code Book,* Revised 2nd Edition. Sheffield: Greenleaf.

Levy, D. L. and Egan, D. (2003) A neo-Gramscian approach to corporate political strategy: conflict and accommodation in the climate change negotiations, *Journal of Management Studies,* **49**(4): 803–29.

Lohmann, L. (2009) Toward a different debate in environmental accounting: the cases of carbon and cost–benefit, *Accounting, Organizations and Society,* **34**(3–4): 499–534.

Lopez, M. V., Garcia, A. and Rodríguez, L. (2007) Sustainable development and corporate performance: a study based on the Dow Jones Sustainability Index, *Journal of Business Ethics,* **75**: 285–300.

Macve, R. and Chen, X. (2010) The 'equator principles': a success for voluntary codes?, *Accounting, Auditing and Accountability Journal,* **23**(7): 890–919.

Mallin, C. (2007) *Corporate Governance.* Oxford: Oxford University Press.

Mallin, C, Saadouni, B. and Briston, R. (1995) The financial performance of ethical investment trusts, *Journal of Business Finance and Accounting*, 22(4): 483–96.

Margolis, J. D. and Walsh, J. P. (2003) Misery loves companies: rethinking social initiatives by business, *Administrative Science Quarterly*, 48: 268–305.

McGoun, E. G. (1997) Hyperreal finance, *Critical Perspectives on Accounting*, 8(1-2): 97–122.

McKillop, D. G., Ward, A.-M. and Wilson, J. O. S. (2007) The development of credit unions and their role in tackling financial exclusion, *Public Money and Management*, 27: 37–44.

McPhail, K. (1999) The threat of ethical accountants: an application of Foucault's concept of ethics to accounting education and some thoughts on ethically educating for the other, *Critical Perspectives on Accounting*, 10(6): 833–66.

Miller, A. (1992) Green investment, in Owen, D. (ed.), *Green Reporting: Accountancy and the challenge of the nineties*, pp. 242–55. London: Chapman Hall.

Milne, M. J. and Chan, C. C. (1999) Narrative corporate social disclosures: how much difference do they make to investment decision-making?, *British Accounting Review*, 31(4): 439–57.

Milne, M. J., Tregidga, H. M. and Walton, S. (2009) Words not actions! The ideological role of sustainable development reporting, *Accounting, Auditing and Accountability Journal*, 22(8): 1211–57.

Mintzberg, H. (1983) The case for corporate social responsibility, *The Journal of Business Strategy*, 4(2): 3–15.

Mishkin, F. S. and Eakins, S. G. (2008) *Financial Markets and Institutions*. London: Pearson.

Moore, G. (1988) *Towards Ethical Investment*. Gateshead: Traidcraft Exchange.

Murray, A., Sinclair, D., Power, D. and Gray, R. (2006) Do financial markets care about social and environmental disclosure? Further evidence and exploration from the UK, *Accounting, Auditing and Accountability Journal*, 19(2): 228–55.

Nattero, M., Orr, S. and Farrington, R. (2009) 21st Century Water: Views from the finance sector on water risk and opportunity. Goldaming: WWF-UK.

Oberndorfer, M. (2004) *Sustainability Pays off: An analysis about the stock exchange performance of members of the World Business Council for Sustainable Development (WBCSD)*. Vienna: Kommunalkredit Dexia Asset Management.

ONS (2010) *Ownership of UK Quoted Shares, 2010*. London: Office for National Statistics.

Orlitsky, M, Schmidt, F. L. and Rynes, S. L. (2003) Corporate social and financial performance: a meta analysis, *Organization Studies*, 24(3): 403–41.

Osborne, S. and Ball, A. (eds) (2011) *Social Accounting and Public Management: Accountability for the Common Good*. London: Routledge.

O'Sullivan, N. and O'Dwyer, B. (2009) Stakeholder perceptions on a financial sector legitimation process: the case of NGOs and the Equator Principles, *Accounting, Auditing and Accountability Journal*, 22(4): 553–87.

Owen, D. L. (1990) Towards a theory of social investment: a review essay, *Accounting, Organizations and Society*, 15(3): 249–66.

Patten, D. M. (2002a) Media exposure, public policy pressure and environmental disclosure: an examination of the impact of TRI data availability, *Accounting Forum*, 26(2): 152–71.

Patten, D. M. (2002b) The relation between environmental performance and environmental disclosure: A research note, *Accounting, Organizations and Society*, 27(8): 763–73.

Patten, D. M. (2005) The accuracy of financial report projections of future environmental capital expenditures: a research note, *Accounting, Organizations and Society*, 30(5): 457–68.

Perks, R. W., Rawlinson, D. and Ingram, L. (1992) An exploration of ethical investment in the UK, *British Accounting Review*, 24(1): 43–66.

Prakash, P. and Rappaport, A. (1977) Information inductance and its significance for accounting, *Accounting, Organizations and Society*, 2(1): 29–38.

Prindl, A. R. and Prodhan, B. (1994) *Ethical Conflicts in Finance*. Oxford: Blackwell.

Reyes, M. and Grieb, T. (1998) The external performance of socially-responsible mutual funds, *American Business Review*, 16: 1–7.

Richardson, A. J., Welker, M. and Hutchinson, I. R. (1999) Managing capital market reactions to social responsibility, *International Journal of Management Reviews*, 1(1): 17–43.

Rockness, J. W. (1985) An assessment of the relationship between US corporate environmental performance and disclosure, *Journal of Business Finance and Accounting*, **12**(3): 339–54.

Rockness, J. and Williams, P. F. (1988) A descriptive study of social responsibility mutual funds, *Accounting, Organizations and Society*, **13**(4): 397–411.

Schmidheiny, S. and Zorraquin, F. J. (1996) *Financing Change: The financial community, eco-efficiency and sustainable development*. Cambridge, MA: MIT Press.

Solomon, J. (2007) *Corporate Governance and Accountability*, 2nd Edition. Chichester: Wiley.

Solomon, J. and Darby, L. (2005) Is private social, ethical and environmental reporting mythicizing or demythologizing reality?, *Accounting Forum*, **29**(1): 27–47.

Solomon, J. and Solomon, A. (2004) *Corporate Governance and Accountability*. London: Wiley.

Solomon, J. F. and Solomon, A. (2006) Private social, ethical and environmental disclosure, *Accounting, Auditing and Accountability Journal*, **19**(4): 564–91.

Sparkes, R. and Cowton, C. J. (2004) The maturing of socially responsible investment: A review of the developing link with corporate social responsibility, *Journal of Business Ethics*, **52**(1): 45–57.

Statman, M. (2006) Socially responsible indexes. Composition, performance, and tracking error, *Journal of Portfolio Management*, **32**(3): 100–109.

Sullivan, R. (2011) *Valuing Corporate Responsibility: How do investors really use corporate responsibility information?* Sheffield: Greenleaf.

Sullivan, R. and Mackenzie, C. (eds) (2006) *Responsible Investment*. Sheffield: Greenleaf.

Sullivan, R., Mackenzie, C. and Waygood, S. (2006) Does a focus on social, ethical and environmental issues enhance investor performance?, in Sullivan, R. and Mackenzie, C. (eds), *Responsible Investment*, pp. 56–61. Sheffield: Greenleaf.

Sunstein, C. R. (1997) *Free Markets and Social Justice*. New York: Oxford University Press.

Thielemann, U. (2000) A brief theory of the market – ethically focused, *International Journal of Social Economics*, **27**(1): 6–31.

Tinker, T. and Puxty, T. (1995) *Policing Accounting Knowledge*. London: Paul Chapman.

Ullmann, A. E. (1985) Data in search of a theory: a critical examination of the relationships among social performance, social disclosure and economic performance of US firms, *Academy of Management Review*, **10**(3): 540–57.

Wagner, M. (2001) *A Review of Empirical Studies Concerning the Relationship Between Environmental and Economic Performance*. Lüneburg: Center for Sustainability Management.

White, A. and Kiernan, M. (2004) *Corporate Environmental Governance: A study into the influence of environmental governance and financial performance*. Bristol: Environment Agency/Innovest.

Wiseman, J. (1982) An evaluation of environmental disclosure made in corporate annual reports, *Accounting, Organizations and Society*, **7**(1): 53–63.

Yankelovich, D. (1972) *Corporate Priorities: A continuing study of the new demands on business*. Stamford, CT: Daniel Yankovich, Inc.

Zeigler, A. and Schroder, M. (2010) What determines the inclusion in a sustainability stock index? A panel data analysis for European firms, *Ecological Economics*, **69**: 848–56.

CHAPTER 9

Seeking the Holy Grail: towards the triple bottom line and/or sustainability?

9.1 Introduction

A core element of our text concentrates upon specific aspects of the social and environmental accounting agenda and explores some of the issues involved and how they are developing. This chapter, at its simplest, tries to bring this all together and examines what our social and environmental accounts might comprise when they attempt to capture the wider interactions of an organisation in a more holistic fashion.

If you have read this far, you will, hopefully, appreciate that this seemingly simple ambition – to provide a complete account of an organisation – is actually an enormous – and probably impossible – task. This won't stop us trying – nor should it: accountability and responsible management demand the best accounts we can construct even if they can never be 'perfect'. The earlier chapters identified how all accounts simplify the world by including some aspects of some things and excluding others. We also show there how different ways of looking at the world (the theories and worldviews[1]) led to different conceptions of what accounts might be trying to achieve and different beliefs about the underlying issues which accounts might be seeking to represent. In this chapter, we will concentrate on the attempts – both practical and speculative – to pull this all together and produce a complete – or at least a fuller – account for and of the organisation.

So, how would one go about this? Very broadly there seem to be four (overlapping) ways of approaching the problem. We might think of these as follows.

- *Fully monetised accounts*: First, we might try and take the economist's approach and express all interactions with society and the natural environment in financial terms. This would then allow us to add these numbers to those behind the profit and loss account and balance sheet in order to produce a total, financial, account of the organisation. Examples of this approach are provided by, for example, Abt (1972 *et seq.*), Linowes (1972), BSO/Origin's 1991 'green accounts' (Gray and Bebbington, 2001) and Mathews' Total Impact Accounting (Mathews, 1984).

- *Integrated accounts*: Second, we might try for other ways in which to integrate all the data that encompasses our interactions in some composite form of communication. This might use financial expression, but it might also use other means – some of which might be additive and some which might not. Examples of this approach might include Schaltegger and Burritt's (2000) application of Kaplan and Norton's (1996) 'balanced scorecard' approach, Guthrie's suggestions for 'extended performance reporting' (see, for example, Yongvanich and Guthrie, 2005) and the (so-called) 'integrated reporting

---

[1] See also Gray *et al.* (2010b) where these matters are more fully developed.

framework' which was developed in the UK following the Prince of Wales' initiatives in this area (Hopwood *et al.*, 2010) and has subsequently led to the formation of the International Integrated Reporting Committee.

● *Multiple accounts*: Next, we might recognise that not everything can be added together and we might look, instead, for a range of accounts covering the various aspects of the organisation. This is, in effect, what we do in the core chapters, looking at accounts relating to communities, human rights, the natural environment and employees for example. The accountability model that we outlined there would suggest a form of accounting that involved multiple accounts – at its crudest one complex account for each relationship of the organisation (Gray *et al.*, 1997; Gray, 2000). We re-examine this suggestion in later chapters. But perhaps the most widely known and widely discussed approach to multiple accounts is, what is called the 'triple bottom line' (TBL) in which accounts are produced for the economic, the social and the environmental interactions of the organisation (Elkington, 1997; Henriques and Richardson, 2004).[2] And, as we shall see, the TBL is (loosely, anyway) the basis on which the very successful Global Reporting Initiative (GRI) is based.

● *Sustainability accounts*: Emphasising that these ways of approaching the accounting are not discrete, the fourth approach we see is that of trying to capture a holistic notion within the account and constructing the account accordingly. The dominant notion for some time (and likely to remain so) is that of sustainability. So, an increasingly important thrust is to seek out ways in which to capture notions like 'the sustainability of the organisation' or 'the relationship of the organisation to sustainable development'. This takes many forms from versions of fully monetised accounting (known as 'full cost accounting' or 'sustainable cost accounts') through to a range of powerful mechanisms like the measurement of an organisation's ecological footprint (Wackernagel and Rees, 1996; Bebbington *et al.*, 2001a; Bebbington and Gray, 2001; Unerman *et al.*, 2007). Whilst significant progress has been made on the environmental component of sustainability, the challenge of how (if at all) to account for social justice remains elusive.

This diversity of approaches produces, you will not be surprised to hear, a plethora of potential social, environmental (and, possibly, sustainability) accounts. None is 'correct', all have strengths and all have weaknesses. In order to make any sense of this stuttering progress towards a more complete accounting, we need to carefully analyse each of these approaches, explore the lessons to be learnt from each and see how they have got on attempting to change practice.

Consequently, this chapter is structured so that it (broadly and initially) follows the four approaches we have identified above. The next section briefly reviews a number of the key aspects of the fully monetised approach to social and environmental accounting. Section 9.3 considers integrated accounts and Section 9.4 looks at multiple accounts with an especial focus on TBL and GRI. Section 9.5 confronts the notion of sustainability again and explores how, if at all, we might account for it. This forms the platform for Section 9.6 which briefly assesses the attempts to produce accounts of (un)sustainability. Section 9.7 picks up a theme often ignored in sustainability accounting – the matter of social justice – before the final section reviews our progress and tries to tease out a few key points.

---

[2]The economic accounts of an organisation would probably not be identical to the financial accounts of that organisation. Although there would be some considerable overlap, the economic account would draw its context more widely than does the financial.

9.2 Fully monetised accounts

Although approaches to social and environmental accounting have been around for many years – and in exceptionally diverse ways (Maltby, 2005; Buhr, 2007) – the first substantial attempts to provide a coherent, focused account of all of an organisation's interactions with society and its natural environment were probably those which tried to put together a full financial account of the organisation. The reasons for such an approach are fairly obvious. In the first place, accounting (in financial terms) was the dominant information medium for both managers and external participants. Managers and external participants were used to financial information, and social and environmental accounting could be thought of as a development of a successful, existing process (Solomons, 1974). Secondly, the historically dominant intellectual paradigm for both business and accounting teachers and researchers was economics. Thus it must have made obvious (if un-questioned) sense that any approach to account for completeness should start with 'economisation'. Now, as we know, traditional economics is pretty much only interested in things which have a price attached and it more or less ignores anything which doesn't have a price (see the MacNamara fallacy in Chapter 8). Consequently, an economically trained accountant could be expected to approach the problem by assigning financial numbers to the social and environmental matters that were 'missing' from the basic (financial) account (Churchman, 1971).

The first such attempt of which we are aware was made by a USA practitioner, David Linowes (Linowes,1972). Linowes argued that a 'good' company – a company which embraced the highest standards of social responsibility – should not be penalised in its financial statements. He argued that social responsibility will, at least in the short run, cost a company money.[3] This will, in turn, reduce profit and, as a result, the 'responsible' company will appear to be less successful. He went on to propose an additional accounting statement that could be published in the Annual Report which would show how well the organisation had performed in the social domain. The *Socio-Economic Operating Statement* captured selected 'improvements' and 'detriments' relating to 'people', 'environment' and 'product'. For each category, the costs incurred by the company (e.g. costs of installing pollution prevention equipment) were included as improvements whilst costs that a truly responsible organisation *would* have undertaken but our company did not (e.g. what it would have cost to install some potential safety devices) were identified as detriments. The total was summed to produce the 'Grand total socio-economic actions for the year'.

> **NOTE**: Examples of Reporting Practice – including the Linowes, Abt and BSO/Origin Social and Environmental Accounts – are available on the CSEAR website under **'Approaches to Practice'**.
>
> **www.csear.co.uk**

Linowes' proposal was especially important for a number of reasons: not least because it explicitly attempted to link the 'social' and the 'economic' in one statement. However, the proposed statement was not without considerable problems: (a) it was highly subjective; (b) it used different valuation and cost bases in different parts of the statement and then added and subtracted the resultant oranges, apples and pears; and (c) Linowes was unclear as to whether he was taking the corporation's view looking out to society, or the society's view looking into the corporation.[4] Linowes' work was, and remains many decades later,

---

[3]Chapter 8 identifies how the discourse on this issue has changed substantially and now the more widely held view is that 'doing good' makes money – however implausible that might seem.

[4]This issue of which perspective to take when constructing a social account is very important – although this may not be immediately obvious. It is a problem which has bedevilled attempts at CSR and, for an accountant, might be most easily thought of as the CSR analogue of the 'entity' versus 'proprietorship' concept in conventional financial accounting.

seminal and an important starting point for any discussion about the practicalities of social accounting. (For more detail, see, for example, Burton, 1972; Estes, 1976; Jensen, 1976; and, especially, Gray *et al.*, 1987; Mathews, 1997b.)

The second seminal attempt at a holistic, monetised social accounting was the Abt model (Abt, 1972 *et seq.*). Clark C. Abt was a consultancy firm which initiated and developed a set of 'social accounts' that were intended to show the total impact of the company in financial terms. The accounts were conceived, in part, as a public relations device but were published (and refined) by Abt throughout the 1970s and were accompanied by very detailed and thoughtful notes on how items like pollution or employee remuneration were treated. The accounts comprised both a *Social and Financial Income Statement* and a *Social and Financial Balance Sheet*. The income statement identified income related to each of the company's/ stockholders, staff, clients and community as the financial benefit each group gained from the company's activities. Some of these were actual financial numbers (such as sales revenue or staff salaries) and others were imputed (such as career advancement or contribution to knowledge). Costs incurred were also a combination of actual and imputed costs covering matters like energy use and the opportunity cost of staff time worked. The balance sheet comprised traditional financial categories and these were summed together with imputed categories related to (for example) staff as an asset or environmental pollution as a liability. It was a major and impressive attempt at a holistic financial social account and one which was refined for a number of years. Equally importantly, the method was applied by The Cement Corporation of India in their social accounts for 1981 and, more obliquely, the same thinking can be seen in the environmental accounts of BSO/Origin, a Dutch company, in the early 1990s (see also Huizing and Dekker, 1992; Gray and Bebbington, 2001; and the CSEAR website). These examples are the tip of an iceberg of perennial interest in continuing attempts to offer partial financial accounts of an organisation's social and environmental interactions. There continue to be many examples of these as the years go by, and particularly notable examples include such things as the Baxter Health Care's continuing *Environmental Financial Statement* (see, for example, Bennet and James, 1997) and PUMA's *Environmental Profit and Loss Account* published for the first time in 2011. These approaches resonate with the examples like BSO/Origin, Abt and Linowes while offering some attempt at measuring an approximation of *sustainable cost*: a matter we return to later in this chapter.

The Abt accounts demonstrated (rather more clearly than was intended) that it is very difficult (if not impossible) to capture all social and environmental interactions in financial terms *on the same valuation basis*. Even more acutely than the Linowes model, the Abt accounts end up comparing, adding and subtracting figures calculated on fundamentally different bases. What, if anything, the resultant bottom-line purports to represent is therefore anybody's guess.[5]

These attempts to produce holistic accounts in financial terms raise some really interesting issues, for example, questions about the nature of a price. Crudely speaking, a price arises when the ownership claim in something changes hand in return for money. Conventional accounting recognises any priced transaction that crosses the boundary of the organisation (Bebbington *et al.*, 2001b). If a transaction is not priced – e.g. volunteer labour, fresh air, good intentions, rainwater, pristine habitat, national social stability and so on – then

---

[5]For more information on the Clarke C. Abt model and examples of the final published social accounts, see Mueller and Smith (1976); Epstein *et al.* (1977); Belkaoui (1984); Gray *et al.* (1996) and the CSEAR website. For other examples of early attempts to integrate the social, environmental and financial see American Accounting Association (AAA, 1973a, b), Eastern Gas and Fuel Associates report of 1972 (see Gray *et al.*, 1996 and the CSEAR website) and Deutsche Shell Report of 1975 (Schreuder, 1979). Several of these seminal articles are also reprinted in Gray *et al.* (2010a).

conventional accounting cannot do anything with it. One way around that is to draw from economics and assign or impute a value; and that is when the problems start because there are just so many ways in which a value can be assigned: what is clean air worth? how clean? to whom? etc. . . . (Milne, 1996).

A second conceptual and practical conundrum that these accounts raise is: whose value are we interested in? The conventional commercial organisation is almost entirely uninterested in (say) nature or justice as such – it is a predominantly economic creature that will only recognise economic things (Bakan, 2004). So consider the example of the last pair of breeding golden eagles: they would be worthless to a corporation unless there was some way to buy or sell them; on the other hand, bird watchers and nature lovers might pay a fortune to prevent their extermination; and, if we could consider a non-anthropocentric view, to the natural world (of which humans may or may not be a part?) they are (literally) beyond price.

Prices and valuations are political things and reflect the political power obtaining at the time: they are not and cannot be objective things. So we should always tread carefully when prices are being considered; but we can usefully take this a little further still and ask: why would we want to express everything in financial terms in the first place? The increasing dominance of a financial/economic mindset may well be an important part of the social, environmental and sustainability problems that act as the stimulus for social accounting (see Chapters 1, 2 and 8). If they are part of the problem, we should hesitate before assuming that more of the thing that caused the problem may cure the problem.[6] Now, obviously, we live in a predominantly market economy in which major movements occur through markets in which prices operate. The more efficient these markets, the more the prices are ideal and markets clear (as economists would say). Furthermore, money is a useful means of keeping the score and, if we want to seek comparability, then money is our obvious first candidate. Such a line of reasoning has power, is widely applied (Pearce *et al.*, 1989; Lomborg, 2001) and certainly does not lack usefulness. But to follow this reasoning wholly requires an *a priori* faith in the ubiquity and benign nature of economic thinking. Given that economic thinking is amongst the principal likely suspects in our anxieties about responsibility and sustainability then such faith must be given cautiously, if at all (Maunders and Burritt, 1991).

More fundamentally still, though, there is a wider moral concern that the reduction of important aspects of life to financial numbers is simply wrong[7] – whether as an aesthetic, religious or intellectual concern it offends the very foundations of what it is to be human. That is, there must be a very real anxiety that any attempt to reduce the natural environment, human life, spirituality, love or simply the joys of existence to a financial description is to destroy them (Hines, 1991) (although it might be worth just mentioning that a fundamentalist intolerance of pricing may equally well not be a constructive attitude to adopt in most rational circumstances).

Potential difficulties and objections notwithstanding, it seems inevitable that as long as liberal economics and financial markets dominate the social scene then attempts (both well-meaning and otherwise) will be made to develop the current models of accounting to the point where they can increasingly incorporate that which was previously excluded. Mathews (1984, 1997b), for example, long pushed for the idea that these difficulties could be overcome and that it is possible – and desirable – to derive something he called **Total Impact Accounting**. Exploring all avenues seems a useful aim, but the holistic financial approach to social accounting has not yet come to dominate social accounting research or practice.

---

[6]This notion of curing a thing with the thing that caused the problem has an analogue in social science called 'juridification' – derived (as you might guess) from legal studies.

[7]One could echo this with a more practical or pragmatic suggestion that to do so is actually also just silly.

Equally, however, there is no immediate likelihood of financial measurement being usurped either. Not only is profit the ubiquitous measure of organisational success and failure, but nations are equally wedded to measuring economic well-being (national income) through the financial metric of gross national (or domestic) product. Despite the considerable drawbacks of such a measure of well-being (Daly and Cobb, 1990), nations seem more likely to seek to adapt this old and doubtful measure (as 'green GDP' or similar, for example) than to abandon it altogether.[8]

This takes us neatly into one final factor that we must consider with regard to monetisation and that is the matter of the **internalisation of externalities**. Bakan (2004) refers to corporations as 'externalising machines'. That is, corporations seek to minimise their costs wherever possible by trying to 'externalise' them and, wherever possible, ensure that society (or parts of society) bear that cost: whether it be the costs of pollution, health, training of the workforce and so on. Over time, however, changes in (say) law or customer preferences or the expectations of civil society have effectively forced organisations to embrace (internalise) some of these costs. So, for example, many societies have required corporations to reduce polluting emissions, protect the health of employees and safeguard customers' rights (to a degree at least). Each of these steps forces a company to adopt what were previously externalities as an internalised cost of doing business. This process is a continuing one. An economist would see this process as the only realistic – and certainly as the most efficient – way in which to recognise the increasing social and environmental issues within a society. So, for example, as society becomes more anxious about global climate change, those organisations which produce greenhouse gases (and all do one way or another) have been slowly forced to incorporate the 'costs of carbon' into their ways of doing business (Bebbington and Larrinaga-González, 2008). It is possible to imagine that, as society becomes more and more anxious about a wider range of social and environmental concerns, the process of law, regulation, taxation and pricing will ensure that these are incorporated into the cost structures of all organisations. Ultimately, it is possible to imagine that *all* externalities will be internalised – although of course this seems profoundly unlikely to happen any time soon. But it does mean that economists might argue against the need for additional (social and environmental) accounts because the market will always ensure that all matters that are of import are already incorporated into prices. You can believe this if you wish!

So, on the whole, there is still much work to be done with the attempts to express social and environmental accounting in financial numbers. The problems are legion, but it is not a pursuit which we would necessarily see as offering the best chance for the fulfilment of organisational social, environmental and sustainability accountability (although see later when we look at some of the attempts to account for (un)sustainability).

So, if we cannot capture everything in financial numbers, can we find some other way of integrating the social and the environmental with the economic? This is what is attempted in the next section.

9.3 Integrated accounts

For some time there has been a dissatisfaction with accounting, and especially management accounting, as the dominant (sometimes only) source of information on which managers base their decisions. Basically, it can be thought of as backward looking (it is mainly

---

[8]The early decades of the 21st century saw a welcome increase in attention to these issues, including such initiatives as seeking to measure happiness or encourage root and branch examinations of the thinking behind growth and well-being. See, for example, Jackson (2009).

historic data) and too narrow (it excludes any wider ambitions that the organisation might have) and, consequently, management accounting has little value in supporting an organisation's strategic vision (Norris and O'Dwyer, 2004). This is the problem that Kaplan and Norton sought to remedy with their **balanced scorecard** (BSC) (Kaplan and Norton, 1996). The idea of the BSC 'provides a selected set of performance measures that, when taken together, show whether a company, its sub-units and its individual managers have improved their (past) performance across a range of activities and outcomes' (Schaltegger and Burritt, 2000: 151). Schaltegger and Burritt pick this idea up and suggest it as a means of integrating 'the environment into decision making, planning and control' (p. 155) by offering both financial and non-financial strategic targets against which management performance will be measured and rewarded. Indeed, they take this further and suggest that the environmental (and presumably social) aspect of the GRI (see later) could be integrated to give a nested set of embedded key performance indicators reflecting the organisation's view about the strategically important social and environmental issues. The notion of the BSC continues to be of interest to management accounting scholars and, whilst it is a far from settled technology, the suggestions that it be employed in the integration of environmental (and possibly social) matters into the organisation continue to engage researchers and practitioners (Figge *et al.*, 2002; Wagner and Schaltegger, 2006; Dey and Burns, 2010).

Broadly similar motivations to seeking integration also appear to underlie both Guthrie's attempt to develop (what he calls) 'extended performance reporting' (Yongvanich and Guthrie, 2005) and the increasingly influential *Integrated Reporting Initiative* which emerged from the Prince of Wales's 'Accounting for Sustainability' project (Hopwood *et al.*, 2010). The Yongvanich and Guthrie framework recommends that organisations compile data under the three broad headings: external capital, internal structure and human capital. The framework explicitly includes both social and environmental indicators but also includes work conditions and governance matters and suggests that this range of information systems is likely to enhance both internal and external accountability and decision-making (Yongvanich and Guthrie, 2005). In this regard, the Yongvanich and Guthrie framework seems to have the same roots and progenitors as the Integrated Reporting Initiative. This latter started life as part of the Prince of Wales' recognition of the importance of accounting if businesses were ever to succeed in embracing sustainability. The project which emerged drew from the large accountancy firms, UK and overseas accountancy institutes, private and public sector businesses and even academe to produce what became known as the *Connected Reporting Framework* (Fries *et al.*, 2010). The framework was intended to serve the decision-making needs of both internal management and major external participants by explaining 'the connection between delivery of the business's strategy and its financial and non-financial performance' (p. 37). This link to strategy echoes the BSC but, in its attempt to integrate sustainability (sic) and to link directly to external reporting, it goes beyond that initiative. The internal and external information systems are developed to systematically capture additional data on (depending upon the organisation concerned) such things as wastes, welfare and fair trade, which are then reported alongside the traditional financial metrics. Each of the additional data elements is tracked formally through the organisational systems, across supply chains, customers' needs, investors needs and so on, back to the strategy of the organisation. This, it is argued, produces an approach to management and information which connects up the traditional financial with the emerging social and environmental exigencies. The model has been applied in practice (Hopwood *et al.*, 2010), and the success of that application led to the formation of the *International Integrated*

Reporting Committee (IIRC)[9] which was able to draw from a very wide constituency including the GRI (of which more in a moment).

These are not the only attempts to offer a new integration of the economic, the social and the environmental (see, especially Mathews, 1997a, Jones, 2010) but they provide a flavour of the admirable intentions that motivate integration: even if the results remain largely unsatisfactory. What is important, as we stressed earlier, is that these categories of approaches to combining the economic, the social and the environmental are by no means discrete. Indeed, it is probably a very fine line, more of intention than fact, which separates these attempts at integrated accounting for social, financial and environmental issues from the (to our mind more realistic and transparent) attempts at producing a multiple of accounts.

9.4 Multiple accounts – TBL, GRI + the UN Global Compact

What the preceding section outlined was a series of attempts to integrate what are, in effect, a series of multiple accounts. Such attempts at integration meet with varying levels of success – not least because one is trying to meld different sorts of things from different sorts of perspectives – and when there is conflict between the (say) social criteria and the corporate strategy/business case, then no amount of 'embedding' is going to persuade an organisation to go against its own business case if it does not want to. In our view, any failure to recognise that there will (eventually) be conflict between social, and environmental and economic desiderata is little more than a dangerous myopia (Walley and Whitehead, 1994). Some (although not all) of this problem is obviated by the pursuit of multiple accounts which do not need to be summed, added or integrated. As we mentioned in the opening section, this is effectively what we have been looking at with accounts relating to communities, human rights, the natural environment and employees, for example, and effectively what is implied by the accountability model (as outlined in Chapter 3).

But perhaps the most influential approach to multiple accounts is, what is known as, the **'Triple Bottom Line' (TBL)** in which accounts are produced for the economic/financial, the social and the environmental interactions of the organisation (Elkington, 1997; Henriques and Richardson, 2004). The TBL has proved to be a powerful metaphor and its influence has been greatly enhanced by its adoption (in broad terms at least) within the GRI.

The term TBL was coined by John Elkington in the mid-1990s (Elkington, 1997, 2004). Its intention was to focus 'corporations not just on the economic value they add, but also on the environmental and social value that they add – or destroy' (2004: 3). It was a metaphor – sometimes articulated as 'people, planet and profits'[10] – intended to encourage a profound change in the way in which business was thought of and the way in which businesses were managed. In that regard, the conception was targeted at the organisations themselves and was part of a larger movement to persuade businesses that making money and being (say) a responsible citizen were not just compatible but mutually re-enforcing. The TBL was, therefore, initially an exclusively managerialist idea and the extent to which one liked or disliked it depended, largely, on how one felt about issues of conflict. Basically did one fear that the notion of the TBL was (deliberately) failing to address the central issues of conflict between

---

[9]The IIRC is a continuing project about which there are a widespread range of views and about which readers are encouraged to form their own opinion. See http://www.integratedreporting.org/ for current information and watch for special issues of major accounting journals for updated research on the field.

[10]This phrase was adopted by Shell when, following the disasters of both the Brent Spar and the Ogani people, they produced their first substantial standalone report in 1998.

the economic, the social and the environmental concerns in organisations? Alternatively, did one, rather, hope that the TBL might help to tease out those conflicts which were currently buried? The real fear was that its supporters preferred to consider that there were no matters of conflict there in the first place (Henriques and Richardson, 2004).

As a managerial metaphor, TBL was widely adopted and the term entered the language of business relatively painlessly. It is this lack of pain that concerns us, actually.[11] To change business so that it fundamentally re-directs attention towards social and environmental matters requires fundamental re-engineering: fundamental change of organisations is not simple (Norman and MacDonald, 2004). Equally, of course, the TBL fails to help managers actively and sensibly balance the three 'bottom lines', and it is this, more than anything else, that probably leads to the search for integrated social, environmental and financial accounts that we touched upon above.

Where the metaphor has been more powerful, however, is in the field of external reporting. The provenance of the idea remains a little hazy but, at its heart, there is a notion that if each organisation produced *three* sets of accounts – one financial, one social and one environmental – we might be on our way towards a more interesting accountability. For this to work, each set of accounts would have to be more or less equivalent in terms of reliability, rigour and space given to them, but one can easily imagine that a more obviously balanced annual report might change the way in which stakeholders, policy-makers and management viewed the organisation itself (Gray, 2000).

On the face of it, the growth of standalone reports in the late 1990s and into the 21st century seemed to be a response to this call for a TBL approach to reporting (see Figure 9.1). Indeed, the reference to TBL was not uncommon in these reports as they slowly morphed their way from their initial manifestation as environmental reports, through the emergence of corporate social responsibility (CSR) reports to the increasingly common presentation of sustainable development (sic) reports (of which more later). But, regardless of the language or even the intentions, no organisation was producing reports that gave equal billing to each of the components of the TBL. The waters remained distinctly muddy as to what corporate self-reporting was actually seeking, what was really driving the apparent changes in voluntary reporting practice and what one could (or could not) learn from the publication of such reports.

Into this muddiness stepped the GRI. The GRI was initiated by the Coalition for Environmentally Responsible Economies (CERES) in collaboration with the Tellus Institute and the United Nations Environment Programme (UNEP) in 1997 (Leipziger, 2010). It became an independent body in 2002. GRI defined itself as a multi-stakeholder organisation drawing members from business, the public sector, NGOs, the accountancy profession, academe, etc. (Adams and Narayanan, 2007). The 'Global Reporting Initiative (GRI) is a network-based organization that has pioneered the development of the world's most widely used sustainability reporting framework and is committed to its continuous improvement and application worldwide.'[12] Those who chose to adopt the GRI guidelines are invited to announce this fact on the GRI's website and the number of organisations (not just companies) doing so passed the 2,000 mark by the second decade of the century (see Figure 9.1). GRI became an important player in the reporting world: a role given momentum with its involvement in the International Integrated Reporting Initiative.

Figure 9.1 provides a snapshot of reporting in the late 20th and early 21st centuries. The data is derived from the GRI website Reports List as well as GRI (2012) and the KPMG

---

[11]A very similar concern lies at the arguments about the easy way in which 'sustainability' similarly seemed to enter business parlance in the opening decades of the 21st century (Milne *et al.,* 2009).

[12]http://www.globalreporting.org/AboutGRI/WhatIsGRI/.

Figure 9.1 Development of standalone reporting

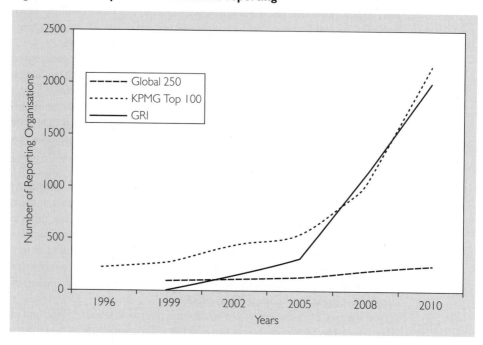

Reporting Surveys (1997, 1999, 2002, 2005, 2008, 2011). The GRI data reports the number of organisations that have self-declared on the GRI website as reporting around the GRI guidelines. The KPMG data reports the number of companies in the Global Fortune 250 (the world's largest companies) producing voluntary standalone reports and the total number of large companies reporting from a sample of the 100 largest in a range of countries. The numbers are indicative only but show both the very significant rise in reporting as well as how small the number of global reporters there are compared to the total number of potentially reporting organisations.[13]

The GRI is 'built on TBL foundations' (Elkington, 2004: 4) and the website states that a GRI report will 'reflect significant economic, environmental, and social impacts and enable stakeholders to assess the reporting organization's performance in the reporting period'.[14] The data, the key performance indicators as recommended by the GRI, broadly followed this three-part structure of economic, social and environmental as well.

GRI rapidly established itself as the global standard for voluntary reporting by all organisations (Willis, 2003; Leipziger, 2010) despite its significant limitations (Doane, 2004; Henriques, 2004; Levy *et al.*, 2010). These limitations seem to be inherent in any voluntary reporting initiative – or even in a mandatory reporting which requires the prior approval of business and markets (Laufer, 2003). First, the GRI remains unbalanced in that, whilst the development of its environmental indicators was relatively coherent and thorough, the development of the economic and social indicators has been beset with difficulties. A GRI report gives much less information about the social and economic dimensions of

---

[13]The increase in the KPMG Top 100 is partly explained by the greater number of countries that KPMG survey. They included 10 countries in 1996 and 23 in 2008. For comparison, it is usually assumed that there are about 60,000 MNCs. There are, of course, many millions of SMEs and non-profit organisations worldwide.

[14]GRI website accessed 24th August 2010.

organisational performance than it does about the environmental performance. Second, the GRI (as is almost any other initiative which has such broad support) is a compromise between what the different parties want and/or are willing to disclose. Consequently, it comes as little surprise to discover that a GRI report rarely, if ever, offers complete, reliable and rigorous accounts of the social and environmental activities of an organisation (Moneva *et al.*, 2006). It seems uncontroversial to suggest that if financial statements were constructed to the same level of quality as most social and environmental accounts, auditor qualification would be the least of their worries (Gray, 2006b, c).

But by far the biggest problem with the GRI (and most attempts at TBL) is not that they fail as TBL accounts – after all this is a potentially fixable problem. The problem is rather that they sport the title 'sustainability guidelines' and reporting organisations consequently claim that their reports are related to sustainability. The GRI guidelines actually have nothing at all to do with sustainability – at least with any form of sustainability that relates to the Brundtland definition (addressed in Chapter 3).

This is a line of argument that has been developed for a number of years by a New Zealand academic, Markus Milne, working with a range of colleagues (Milne and Gray, 2002, 2007; Milne *et al.*, 2003, 2006, 2009; Gray and Milne, 2004; Tregidga and Milne, 2006; Laine, 2010, and see also Hawken, 2002; Henriques and Richardson, 2004; Gray, 2006a, 2010a; Moneva *et al.*, 2006). The essence of the argument has three equally important themes. First, the claims that are made by reporting organisations, by business associations and, by implication, the GRI itself take many forms. So, for example, you will read phrases such as the 'path towards sustainability', or 'reporting on sustainability' or 'our sustainability principles' but – and this is the key point – these statements will not be supported by any explanation or evidence as to why they might be true. They are vacuous assertions in the main. This extends into the research arena where 'sustainability' will be used when something entirely different is intended (see, for example, Lopez *et al.*, 2007). The second theme of Milne's argument is that 'sustainability' is an important concept related to nothing less that the species' interaction with its planet and with each other. To the extent that any meaning *is* ascribed to 'sustainability', it is taken to mean something like 'continuing for the immediate future' which appropriates and diminishes this important concept. Indeed, a detailed study of the GRI reporting and assurance guidelines and the accompanying performance indicators came to the arresting (although not entirely surprising) conclusion that the nature of sustainability implied in GRI has few connections with the TBL, is clearly managerialist and seems to exhibit little or no regard for the planet.[15] The final thread in Milne's argument is perhaps the most telling of them all: sustainability simply is *not* an organisational concept and to apply it at the organisational level is either very complex or very stupid. This, coupled with an exploration of why we might prefer to consider all organisations *un-sustainable*, is something we will address in the next section.

Before leaving this section looking at multiple accounts, we should be remiss if we ignored one of the most interesting developments of the last few decades – the Millennium Development Goals (MDG) and the related United Nations Global Compact (UNGC). The MDG were formulated at the turn of the century under the impetus of the then Secretary General of the United Nations, Kofi Annan. The Goals[16] were an aspirational statement by all the peoples of the world that we as a species must ensure that we collectively provide certain minimum conditions for both people and planet. (These conditions included such matters as poverty, education, gender equality and environmental sustainability.) That we currently do not meet these

---

[15]Later revisions of GRI sought to introduce some notion of context and capacity which were at least initial steps towards dealing with this problem.

[16]http://www.un.org/millenniumgoals/.

Figure 9.2 United Nations Global Compact – 10 principles

Human Rights

- Principle 1: Businesses should support and respect the protection of internationally proclaimed human rights; and
- Principle 2: make sure that they are not complicit in human rights abuses.

Labour

- Principle 3: Businesses should uphold the freedom of association and the effective recognition of the right to collective bargaining;
- Principle 4: the elimination of all forms of forced and compulsory labour;
- Principle 5: the effective abolition of child labour; and
- Principle 6: the elimination of discrimination in respect of employment and occupation.

Environment

- Principle 7: Businesses should support a precautionary approach to environmental challenges;
- Principle 8: undertake initiatives to promote greater environmental responsibility; and
- Principle 9: encourage the development and diffusion of environmentally friendly technologies.

Anti-Corruption

- Principle 10: Businesses should work against corruption in all its forms, including extortion and bribery.

minimum standards is not contentious: what is more bothersome is that the world remains *off*-track to meet these basic standards. These Goals are for all people and especially for nations. In an attempt to enlist corporate support for progress towards these goals, the UN formulated the UNGC[17] with the support and encouragement of business leaders. The Compact covers the areas of human rights, labour rights, environment and anti-corruption, and it comprises (at the time of writing) 10 principles as shown in Figure 9.2.

The UNGC is not principally a reporting guideline, but it is the most influential of international frameworks employed by companies to manage their CSR (KPMG, 2008). Importantly, the UNGC asks 'participants to communicate annually to all stakeholders their progress in implementing the ten principles' and to communicate that progress to the UNGC website. Interestingly, participants who do not communicate in this way can de-listed and 'named and shamed'. As a result, more companies are relating their voluntary reporting to the UNGC and reporting their compliance to the Compact through their standalone reports (KPMG, 2008). By 2010, the UNGC had claimed that over 7,000 companies in 130 countries were signed up to and were following and reporting on these principles.

There is much more that can be done in both theory and practice with respect to varieties of multiple accounts, including the TBL as a mechanism for developing organisational accountability. But without wishing to offer any substantive criticism of the TBL or the UNGC as such, the claims to sustainability *are* a matter of concern, so much so that if these TBLs are not reflecting sustainability what are they doing? And what would sustainability

---

[17]http://www.unglobalcompact.org/AboutTheGC/TheTenPrinciples/index.html.

accounting actually look like? Some answers to the first question are offered in Milne and Gray (2012). Now we will try and answer the second part.

9.5 So what is this sustainability we wish to account for?

There are two central problems with 'accounting for sustainability'. The first problem is deciding what we mean by sustainability: how demanding are its conditions for example? The second problem is that (to follow Milne's arguments) sustainability is a concept which only makes any sense at a planetary, ecological or social system level. As such, it simply cannot map to an organisation. These problems do not prevent the construction of something we might legitimately consider forms of accounting for sustainability, but they are essential pre-considerations to any such endeavour (Gray, 2010b).

Perhaps the simplest way to explore what we mean by sustainability is to consider the notions of 'weak' and 'strong' sustainability. This is what Bebbington *et al.* (2001a) do in Figure 9.3. Figure 9.3 outlines two distinct views about what sustainability entails. What is crucial is that both views are entirely clear that humanity's current ways of organising itself are not sustainable. They are unsustainable in the sense that whole species are critically damaged and even wiped out; large proportions of the human species live under the most inconceivably brutal (inhumane?) conditions; and whole areas of the planet and the biosphere are suffering irreparable damage (Dresner, 2002). The only real questions – which Figure 9.3 attempts to address – are (i) how far are we from some potentially sustainable position? and (ii) how drastic are the changes that are needed to bring us to this position?

There is a potentially infinite number of sustainable positions that the planet (with or without humanity) could settle on. To try and make sense of that infinite number, it is convenient to consider the notions of **weak sustainability** and **strong sustainability**. Under assumptions of weak sustainability, on the one hand, the consequences of human action are not seen to be either critical or irreversible and we assume that linear 'progress' of the kind we have experienced in the last century or so (especially in the West) will bring us to a sustainable position (often associated with *eco-modernisation*). The assumptions of strong sustainability, on the other hand, suggest: that the gap between where we currently are and any sustainable position is enormous and growing; that linear progress is what got us into the mess and will not get us out; and that the very assumptions on which our ways of organising are based are faulty and this requires the most fundamental of re-thinks (more usually associated with *deep ecology*). The distinction is, of course, crucial. We share the view of most commentators who have spent much time with the data that the need for strong sustainability is irrefutable (Gladwin *et al.*, 1995; Kilbourne *et al.*, 2002). Indeed, York *et al.* (2003) go so far as to argue that there simply isn't any evidence or argument of substance behind the weak sustainability view. Consequently, they argue that attachment to such a view owes more to preference and convenience than to reasoning.

The second of the major problems we identified above was whether it was possible to translate a broad and global concept like sustainability to the organisational level. Well, Milne argues that it cannot: sustainability 'implies the need to consider the scale of development relative to the available resource base; the fairness with which access is provided to those resources and the outputs from them, both among current generations and between current and future generations; and the efficiency with which resources are used' (Gray and Milne, 2004: 76). Not only is this a virtually impossible calculation[18] but it is not something

---

[18]Although, as we mentioned, GRI have recognised this problem in principle and some experiments are seeking to address it (see, for example, McElroy and van Engelen, 2012).

Figure 9.3 'Strong' and 'weak' sustainability

Aspect	'Strong' Sustainability	'Weak' Sustainability
Focus of the pursuit of sustainability and the impetus for change	Fundamental examination of the relationship between humans and their environment and with each other.	Concerned to prevent an environmental catastrophe that would threaten human society.
View of nature–human interaction	Humans and nature are not separate from each other and harmony between the two is sought.	The natural environment is a resource, humans need to better master the environment to solve present problems.
What do we wish to sustain?	Other species, not just the human species, are to be maintained.	The human species is what we are seeking to sustain.
The gap between the present and a sustainable future	The present situation is a long way from a sustainable one, it is so far away it is almost impossible to imagine what sustainability looks like. The time span of change may take 150–200 years.	Present situation is near to a sustainable one, over next 30–50 years it should be reached.
Extent of change required	Fundamental, structural change is likely to be required.	Sustainability is achievable with incremental adjustment of the current system.
Nature of the process of getting to a sustainable path.	Likely to require a participatory, transparent and democratic process. Technical fixes may generate more side effects than they solve.	Authoritative and coercive structures can be utilised (for example, market forces). Greater technological development will allow problems to be solved.
Relevance of eco-justice concerns – who is to be sustained?	Intragenerational equity is an integral and essential part of sustainability. Focus on third world conditions and aspirations cannot be avoided.	Intragenerational equity is a separate issue, sustainability focus is primarily on ecological issues, equity issues will follow from them. Primary focus is on sustaining Western populations.
Sustainable in what way?	The nature of economic growth may need to be redefined or abandoned as a dominant goal. This raises questions about how we currently measure and view development.	Sustainability of the Western civilisation at, at least, the current level of economic development. There is a belief that economic development is actually essential for the pursuit of sustainability.

Source: Extracted from: Bebbington and Thomson (1996) and adapted from Redclift (1987), Gray *et al.* (1993) and Turner (1993).

to which (predominantly) private sector profit-seeking organisations are capable of adapting. But it is slightly more complex still in that it is systems or parts of systems that may or may not be sustainable – not individual units within the system. So it is perfectly possible to imagine an organisation which we might consider as detracting from sustainability (e.g. a mining company) to be doing so within a social and environmental system that was overall considered to be sustainable. Equally, as we shall see, if the system is overall not sustainable it is probably almost impossible for a single entity within that system to be acting in a manner which does not contribute to that non-sustainability.

Figure 9.4 The IPAT/I = PCT equation

Impact = (population) × (affluence) × (technology)

Also specified as:

impact = (population) × (per capita consumption) × (environmental impact of productive technology)

Source: Dresner, (2002: 24).

So, as far as we can tell, the relationship between any organisation and the state of sustainability will be complex and, to a degree at least, elusive. But, we can make some progress here. The first step is to stop assuming (as most corporate voluntary reporting does) that the organisation is sustainable. Such an unexamined assumption tends to lead to asking seriously fallacious questions such as 'how well are we contributing to sustainability?' when, of course, one is probably detracting from sustainability. Rather, the presumption of argument must actually be in the opposite direction. That is, the evidence suggests that we must start from the assumptions that most (if not all) large and medium sized western organisations are significantly *un*-sustainable – and then our task becomes trying to assess how un-sustainable the organisation is.

One of simplest way in which to illustrate this is through the **IPAT** equation (Figure 9.4) (see Chapter 1; Ehrlich, and Holdren, 1971; Commoner, 1972). Impact – meaning environmental impact in this case – relates to the extent to which humans are treating the planet within or without its carrying capacity. Ehrlich postulated that this impact arose from a combination of how many of us there are (population) multiplied by how much each of us consumes. That seemed plausible: the same number of people in Europe or the USA, for example, are likely to have 5–10 times more impact on the planet than an equivalent number of folk from Africa, for example, because they consume enormously more per person. However, technological development means that the volume we consume might be less per unit of the thing consumed (because we have become more efficient in the use of technology) or higher (because the processes of production and shipping, for example, might be more harmful).[19] So each mile driven uses less fuel, each ream of paper involves less water and so on, but there might be more miles driven and the technology involved in the production and disposal of the cars and paper may be more (or less) harmful to the finite environment.

Overall, the total impact of humanity on the planet is clearly rising – and one major way in which this can be measured is by the *ecological footprint* (see Chapter 1; Wackernagel and Rees, 1996). So, despite (or possibly because of) technological advances, human impact is rising. In part this is due to population – which we do not discuss here – but a considerable part of that rise in impact derives from the affluence and technology components. The source of those components is, to a considerable extent and most obviously, markets, corporations and business (Dresner, 2002, Meadows *et al.*, 2005).

Consequently, our line of reasoning says that: impact is rising; affluence and technology are amongst the principal causes; the very success of international financial capitalism is what allows that affluence and encourages that technology. Therefore, it seems self-evident that we might start from the assumptions that most organisations are actually *un-sustainable*. If we then notice that, whilst public sector and non-profit organisations often have the counter-purpose, private sector organisations are assumed to have the principal purpose of increasing

---

[19]The issues of eco-efficiency (which this is) are explored in Chapter 7.

profits for shareholders (making rich people richer in effect) then we can also see that the relationship between corporations and social sustainability is a contentious matter as well (Kovel, 2002). So whether we (for example) monitor the key elements of environmental sustainability (such as, for example, water, emissions, habitat, resource use, etc.) and determine whether an organisation is adding more to or detracting more from the stock of that thing; or whether we measure the ecological footprint of the organisation and see that it continues to grow; or whether we try and assess whether an organisation contributes to a positive or negative redistribution of wealth and access to environmental resources; it is probable that we *can* produce accounts of *un*-sustainability (Gray, 2010a).

9.6 Accounting for sustainability?[20]

To begin to explore what an account of (*un*-)sustainability might look like, we need to turn to the range of experiments that we find in the business and academic literature. We have conveniently categorised these around three broad approaches to constructing a narrative of sustainability: those which focus upon **indicators**; those which use a financial narrative (**financial quantification**); and those which employ other non-financial quantitative narratives (**non-financial quantification**) (Gray, 2010a).

Indicators

Attempts to employ indicators that might capture moves towards/away from sustainability are inevitably partial and they rarely provide a full narrative of the sustainability interactions of the organisation. Ranganathan (1998) of the World Resources Institute synthesised over 50 studies of the use of indicators with a sustainability leaning. She found that, whilst some attempts managed to derive clever and helpful indicators (of such things as indicator species or key social variables), it was the users' insistence that the results should be both simple and presented as a coherent story that fundamentally undermined the value of any indicator approach to such accounts (see, for example, Glatzer, 1981; Parke and Peterson, 1981; Cave *et al.*, 1988). Such concerns for simplicity and coherence may remain a major problem as sustainability – and its interactions with economics – is not something that lends itself to simple coherent pictures.[21]

Financial quantification

The range of attempts to offer financial accounts of organisations and their (un-)sustainability appears to be motivated by a recognition that financial representation speaks to business in a language that it recognises. One such approach to the construction of a financial account of an organisation's un-sustainability is that of identifying the 'sustainable costs' of organisation activity. This approach employs the concept of the maintenance of capital as an analogue for environmental sustainability and identifies man-made, renewable/substitutable and critical natural capital at the level of the organisation. The 'sustainable organisation' would be one which maintained these three capitals over an 'accounting' period. The 'sustainable cost' is the amount that the organisation *would have had to spend* if it *had been*

---

[20]This section draws heavily from Gray (2010a).

[21]It is appropriate to remind ourselves here that the GRI is largely an indicator-based initiative and in its attempts to recognise issues of context would seem to be both seeking to embrace the first steps towards an account for un-sustainability as well as heading for an area of inevitable conflict (McElroy and van Engelen, 2012).

sustainable. The figures that result from this tend to be enormous (see, for example, Gray, 1992). Few, if any, corporations are sustainable by these calculations and, perhaps most significantly, the calculation would wipe out almost any company's profit – and that for a considerable period going back into history. The calculations thus offer significant potential challenges to business, but the approach inevitably has significant practical problems. Two major such problems are (i) that organisations are largely unwilling to entertain an accounting system which produces such a 'wrong answer' and (ii) the non-availability of 'sustainable options' within current markets means that the *amount an organisation would have to spend* is unknowable in any realistic sense – although clearly very large indeed (Bebbington and Gray, 2001, and see also Herbohn, 2005, for a variation on this experience and, especially, Bebbington *et al.*, 2001a for a review of – what is normally called – **full cost accounting**).[22]

Ekins *et al.* (2003) take a variant on this and identify 'sustainability gaps': the distance between current activity and activity which might be considered 'sustainable' (e.g. current versus some defined level of energy use per capita). They then estimate the costs necessary to close the sustainability gap: which might be initially specified as the gap between current and current best practice.[23]

By way of a contrast, BP's 'Sustainability Assessment Model' (SAM – see, for example, Baxter *et al.*, 2004) is a project based system of analysis which produces a *signature* of the project's economic, social, environmental and resource impacts over its life. The resultant signature seeks to inform managers on where the 'positives' and 'negatives' of the project may lie, how that signature compares with other competing possible projects and, indeed, in an ideal world, how that signature conforms to a standard set by the company concerned (Bebbington, 2007).

Non-financial quantification

There is also a range of experiments which have sought to express aspects of the sustainability of an organisation through non-financial quantification. Jones (1996, 2003) and Jones and Matthews (2000) place bio-diversity at the heart of an account and provide for the monitoring of and changes in (especially) 'critical' habitat and flora in a particular nature reserve as a means of monitoring stewardship of natural assets. Lamberton (2000) builds a complete account of sustainability by monitoring all physical inputs and outputs and assessing progress in these against specified 'sustainability' targets. His work illustrates that even a values-based organisation looks more likely to be contributing to un-sustainability.

Probably the most powerful discourse around sustainability is that of the **ecological footprint** (Wackernagel and Rees, 1996). As we have already seen, the ecological footprint, derived initially from debates around environmental space, is linked directly to the carrying capacity of the planet and seeks to measure the amount of land usage that any activity requires for its support (see, for example, Dresner, 2002; WWF 2004; Meadows *et al.*, 2005; and for a critique, Fiala, 2008). Despite its difficulties, it retains a very powerful potential as

---

[22]Howes (2004) and Taplin *et al.* (2006) take a more mellow approach to the same basic principle and considered **remediation** – what would it cost to repair the damage caused by organisational activity? (The most obvious example of this is the sequestration of carbon dioxide, over which there is much controversy – see, for example, Lohmann, 2009.) The approach still shows that businesses are not environmentally sustainable despite adopting a weaker form of sustainability.

[23]By contrast Figge and Hahn's (2004) use of incrementalism involves the calculation of 'sustainable value' to provide comparative data on companies within industries on the relative performance on selected sustainability criteria. The approach relies on an active enlisting of the media through which the dramatic approximations are reported and which, consequently, suggest the levels of financial and economic work that most organisations need to do in order to meet current benchmarks standards.

the primary measure of environmental (un-)sustainability at the organisation level (Chambers and Lewis, 2001; McElroy and van Engelen, 2012).

The synthesis of the foregoing is that few (if any) organisations are making substantial attempts to account even for environmental sustainability, despite the many claims to the contrary (see, especially, Milne *et al.*, 2009). Equally, it is clearly now possible to produce, albeit approximate, accounts of *un*-sustainability for the organisational level (for more detail, see Bebbington, 2007; Unerman *et al.*, 2007). That no organisation does so speaks volumes and permits the inference that the brighter organisations know that they are un-sustainable but would just as soon not admit it publicly. So, researchers continue to make significant progress on accounting for the environmental component of sustainability but the challenge of how (if at all) to account for social justice remains elusive.

9.7 Accounting for social justice?

Sustainability is explicitly both an environmental and a social concept (indeed, one of its principal weaknesses is said to be its anthropomorphism). However, as one can see here, most of the substantial attempts to engage with sustainability have focused on environmental sustainability, not social sustainability. This is almost certainly because the environmental problems might *appear* to be (a) the more pressing and (b) the more tractable. Put more simply: (for example) poverty has always been with us but (for example) floodwaters on the Thames, Rhine, Murray or Hudson rivers are more immediate – and they affect the policy-makers, educators and media professionals more directly. Also, there is a sense in modernity that mankind can fix almost anything: but solving social problems involves far more difficult and more explicit ethical and political issues. Now, of course, this distinction is trite and so many social and environmental issues are co-determined: drought, salination and famine most obviously, but the links between development, social well-being and environmental health are clearly very important indeed – even if they are complex and sometimes elusive.

Consequently, attempts to provide full/holistic accounts of any organisation's contributions to social well-being, social justice and access to environmental resources have proved difficult. Indeed, in the holistic sense we have been talking here we are unaware of any such attempts. There are, however, a number of partial attempts that deserve mention. Perhaps the most established is the use of **economic multipliers** to estimate the economic benefit flowing to/losses flowing from a region, a country or a community through the organisation's activities. The injection or removal of economic resources – wages paid, payments to suppliers, purchase of capital and consultants' services and so on – have knock on effects as that money flows through the economy. This is known as the 'multiplier' effect and $100 paid in wages might have $200 of economic benefit eventually. The withdrawal of economic resources may also be thought to have the equivalent negative effects (an effect that Harte and Owen, 1987, use in their exploration of plant closure social audits, see Chapter 10). In so far as economic well-being is conventionally thought to be a good proxy for social well-being, this method offers some insight into the issues. However, the relationship between economic and social well-being is far from straightforward – especially in the so-called developed world (Daly and Cobb, 1990; Collison *et al.*, 2007; Jackson, 2009). There is a widespread view that injection of economic resources into the so-called lesser developed countries helps to 'lift them out of poverty', but that is also a far from simple relationship (Dresner, 2002).

If the economic approach has its problems, we are somewhat short of serious alternatives. The usual approach is to see the social justice elements as captured within the notion of CSR. If we were seeking a TBL then this approach would work but, as we have discussed, a holistic pursuit of sustainability and the TBL are not the same thing. How we would unpick what is and what is not social justice and sustainability remains unclear. Alternatively, we could explore the attempt from McElroy and van Engelen (2012) to capture a **social footprint** which seeks to mirror the ecological footprint. It also runs into political problems about defining our political assumptions about who has what rights and responsibilities.

This is the essence of the problem for social sustainability, even more than for environmental sustainability: it has an infinite number of potential forms and all of them suggest wealth distributions, ownership and human rights issues that are bound to favour some people and groups over others. The problem is made worse because, as we have suggested from time to time, it is not obvious that corporations especially and financial markets in particular are either designed for or are capable of delivering a flatter more equitable income distribution and/or lifting (whatever that means) peoples out of poverty. Indeed, any such effect may well be an unintended bi-product of advanced financial capitalism – a system which, basically, is designed to make rich people richer – and this may often be at the expense of the poorer people and the planet.[24]

In this sense, the MDG and the associated UNGC look like brave initiatives and the best we seem likely to manage in the near future. Of course, this is just the sort of challenge that new researchers need to address and develop.

9.8 Summary and conclusions

The significant growth in organisational voluntary self-reporting (Kolk, 2008; KPMG, 2008) over recent decades has been remarkable. The reporting developed over many years to the point where social responsibility and sustainability (sic) reporting, influenced by *inter alia,* the GRI and the MDG, seemed to be trying to provide a wider, more complete view of the organisation and its social and environmental interactions. Against this apparent good news of more expansive voluntary reporting, we have set some not so good news: most organisations do not report voluntarily and those that do report do so selectively and thereby fail significantly to provide any substantive accountability. If, as we believe, reporting should seek to be inclusive and encompassing, what options exist? This chapter has sought to outline a number of those options.

It is obvious that the purpose one has for social accounting determines how one goes about it and how one interprets the attempts of others. We have seen in this chapter a series of practical and speculative approaches which we have (artificially) categorised into monetised, integrated, multiple and sustainability accounting. Each of them is trying to produce a more holistic or complete social accounting and each has its own strengths and weakness depending upon the purpose in mind and for whom the accounts are intended – managers, employees, investors or civil society generally. But perhaps the key point we have seen is that there is precious little practice concerned with a genuine attempt to provide a narrative around organisational activity and planetary sustainability – and, perhaps even more striking, little progress (even in research) has been made in attempting to understand how social justice plays through the analysis.

---

[24]This opens up an enormous area which might question, for example, what do we mean by wealth creation, whether profit is created or is it simply appropriated from others and from the environment. There is also the whole area concerning foreign direct investment (FDI) and the benefits to the resource-rich, but socially poor nations – especially related to oil and mineral extraction. These are beyond our scope here.

We will return to these issues elsewhere where we outline what we see as the leading edge of practice to illustrate that it is possible to satisfy the most demanding forms of accountability and that (to an albeit lesser extent) useful accounts of (un-)sustainability can be constructed – and they look little like those so labelled in current practice.

References

AAA (1973a) Report of the committee on environmental effects of organisational behaviour, *The Accounting Review,* Supplement to Vol. XLVIII: 75–119.

AAA (1973b) Report of the committee on human resource accounting, *The Accounting Review,* Supplement to Vol. XLVIII: 169–85.

Abt C. C. and Associates. (1972 *et seq.) Annual Report and Social Audit.*

Adams, C. A. and Narayanan, V. (2007) The standardisation of sustainability reporting, in Unerman, J., Bebbington, J. and O'Dwyer, B. (eds), *Sustainability Accounting and Accountability,* pp. 71–85. London: Routledge.

Bakan, J. (2004) *The Corporation: The pathological pursuit of profit and power.* London: Constable and Robinson.

Baxter, T., Bebbington, J. and Cutteridge, D. (2004) Sustainability assessment model: modelling economic, resource, environmental and social flows of a project, in Henriques, A. and Richardson, J. (eds), *The Triple Bottom Line: Does it all add up?* pp. 113–20. London: Earthscan.

Bebbington, J. (2007) *Accounting for Sustainable Development Performance.* London: CIMA.

Bebbington, J. and Larrinaga-González, C. (2008) Carbon trading: accounting and reporting issues, *European Accounting Review,* **17**(4): 697–717.

Bebbington K. J. and Gray, R. H. (2001) An account of sustainability: failure, success and a reconception, *Critical Perspectives on Accounting,* **12**(5): 557–87.

Bebbington, K. J., Gray, R. H., Hibbitt, C. and Kirk, E. (2001a) *Full Cost Accounting: An Agenda for Action.* London: ACCA.

Bebbington, K. J, Gray, R. H. and Laughlin, R. (2001b) *Financial Accounting: Practice and Principles,* 3rd Edition. London: Thomson Learning.

Belkaoui, A. (1984) *Socio-Economic Accounting.* Westpont, CJ: Quorum Books.

Bennet, M. and James, P. (1997) Making environmental management count: Baxter International's environmental financial statement, *Greener Management International,* **17**: 114–27.

Buhr, N. (2007) Histories of and rationales for sustainability reporting, in Unerman, J., Bebbington, J. and O'Dwyer, B. (eds), *Sustainability Accounting and Accountability,* pp. 57–69. London: Routledge.

Burton, J. C. (1972) Commentary on 'Let's get on with the social audit', *Business and Society Review,* Winter: 42–3.

Cave, M., Hanney, S., Kogan, M. and Trevett, G. (1988) *The Use of Performance Indicators in Higher Education: A critical analysis of developing practice.* London: Jessica Kingsley.

Chambers N. and Lewis, K. (2001) *Ecological Footprint Analysis: Towards a sustainability indicator for business,* ACCA Report 65. London: ACCA.

Churchman, C. West (1971) On the facility, felicity and morality of measuring social change, *The Accounting Review,* **XLVI**: 30–35.

Collison, D., Dey, C., Hannah, G. and Stevenson, L. (2007) Income inequality and child mortality in wealthy nations, *Journal of Public Health,* **29**(2): 114–17.

Commoner, B. (1972) The environmental cost of economic growth, in *Population, Resources and the Environment,* pp. 339–63. Washington, DC: Government Printing Office.

Daly, H. E. and Cobb, J. B. Jr (1990) *For the Common Good: Redirecting the economy towards the community, the environment and a sustainable future.* London: Greenprint.

Dey, C. and Burns, J. (2010) Integrated reporting at Novo Nordisk, in Hopwood A., Unerman, J. and Fries, J. (eds), *Accounting for Sustainability: Practical Insights,* pp. 215–32. London: Earthscan.

Doane, D. (2004) Good intentions – bad outcomes? The broken promises of CSR reporting, in Henriques, A. and Richardson, J. (eds), *The Triple Bottom Line: does it all add up?* pp. 81–8. London: Earthscan.

Dresner, S. (2002) *The Principles of Sustainability*. London: Earthscan.

Ehrlich, P. R. and Holdren, J. P. (1971) Impact of population growth, *Science,* **171**: 1212–17.

Ekins, P., Simon, S., Deutsch, L., Folke, C. and De Groot, R. (2003) A framework for the practical application of the concepts of critical natural capital and strong sustainability, *Ecological Economics,* **44**(2–3): 165–85.

Elkington, J. (1997) *Cannibals with Forks: the triple bottom line of 21st century business.* Oxford: Capstone Publishing.

Elkington, J. (2004) Enter the triple bottom line, in Henriques, A. and Richardson, J. (eds), *The Triple Bottom Line: Does it add up?,* pp. 1–6. London: Earthscan.

Epstein, M. J., Epstein, L. B. and Weiss, E. J. (1977) *Introduction to Social Accounting.* New York: National Association of Accountants.

Estes, R.W. (1976) *Corporate Social Accounting.* New York: Wiley.

Fiala, N. (2008) Measuring sustainability: why the ecological footprint is bad economics and bad environmental science, *Ecological Economics,* **67**: 519–25.

Figge, F. and Hahn, T. (2004) Sustainable value added – measuring corporate contributions to sustainability beyond eco-efficiency, *Ecological Economics,* **48**: 173–87.

Figge, F., Hahn, T., Schaltegger, S. and Wagner, M. (2002) The sustainability balanced scorecard – linking sustainability management to business strategy. *Business Strategy and the Environment,* **11**: 269–84.

Fries J., McCulloch, K. and Webster, W. (2010) The Prince's Accounting for Sustainability Project: creating 21st century decision-making and reporting systems to respond to 21st century challenges and opportunities, in Hopwood, A., Unerman, J. and Fries, J. (eds), *Accounting for Sustainability: Practical Insights,* pp. 29–45. London: Earthscan.

Gladwin, T. N., Kennelly, J. J. and Krause, T.-S. (1995) Shifting paradigms for sustainable development: implications for management theory and research, *Academy of Management Review,* **20**(4): 874–907.

Glatzer, W. (1981) An overview of the international development in macro social indicators, *Accounting, Organizations and Society,* **6**(3): 219–34.

Gray, R. H. (1992) Accounting and environmentalism: an exploration of the challenge of gently accounting for accountability, transparency and sustainability, *Accounting, Organizations and Society,* **17**(5): 399–426.

Gray, R. H. (2000) Current developments and trends in social and environmental auditing, reporting and attestation: a review and comment, *International Journal of Auditing,* **4**(3): 247–68.

Gray, R. (2006a) Social, environmental, and sustainability reporting and organisational value creation? Whose value? Whose creation?, *Accounting, Auditing and Accountability Journal,* **19**(3): 319–48.

Gray, R. H. (2006b) Does sustainability reporting improve corporate behaviour? Wrong question? Right time?, *Accounting and Business Research (International Policy Forum),* **36**: 65–88.

Gray, R. H. (2006c) Trustworthy plc, *Green Futures,* March/April: 45.

Gray, R. (2010a) Is accounting for sustainability actually accounting for sustainability . . . and how would we know? An exploration of narratives of organisations and the planet, *Accounting, Organizations and Society,* **35**(1): 47–62.

Gray, R. (2010b) A re-evaluation of social, environmental and sustainability accounting: an exploration of an emerging trans-disciplinary field?, *Sustainability Accounting and Management Policy Journal,* **1**(1): 11–32.

Gray, R. H. and Bebbington, K. J. (2001) *Accounting for the Environment,* 2nd Edition. London: Sage.

Gray, R. H. and Milne, M. (2004) Towards reporting on the triple bottom line: mirages, methods and myths, in Henriques, A. and Richardson, J. S. (eds), *The Triple Bottom Line: does it all add up?* pp. 70–80. London: Earthscan.

Gray, R. H., Owen, D. L. and Maunders, K. T. (1987) *Corporate Social Reporting: Accounting and accountability.* Hemel Hempstead: Prentice Hall.

Gray, R. H., Owen, D. L. and Adams, C. (1996) *Accounting and Accountability: Changes and challenges in corporate social and environmental reporting.* London: Prentice Hall.

Gray, R. H., Dey, C., Owen, D., Evans, R. and Zadek, S. (1997) Struggling with the praxis of social accounting: stakeholders, accountability, audits and procedures, *Accounting, Auditing and Accountability Journal,* **10**(3): 325–64.

Gray, R. H., Bebbington, K. J. and Gray, S. (eds) (2010a) *Social and Environmental Accounting: Sage Library in Accounting and Finance,* Volumes 1–4. London: Sage.

Gray, R. H., Owen, D. L. and Adams, C. (2010b) Some theories for social accounting?: A review essay and tentative pedagogic categorisation of theorisations around social accounting, *Advances in Environmental Accounting and Management,* 4: 1–54.

GRI (2012) *GRI Sustainability Reporting Statistics: Publication Year 2011.* Amsterdam: Global Reporting Initiative (https://www.globalreporting.org/resourcelibrary/GRI-Reporting-Trends-2011.pdf).

Harte, G. and Owen, D. L. (1987) Fighting de-industrialisation: the role of local government social audits, *Accounting, Organizations and Society,* 12(2): 123–42.

Hawken, P. (2002), 'McDonald's and Corporate Social Responsibility?', Press Release from Food First, April 27 (http://www.mcspotlight.org/media/press/mcds/distributedbyfo270402.html).

Henriques, A. (2004) CSR, sustainability and the triple bottom line, in Henriques, A. and Richardson, J. (eds), *The Triple Bottom Line: Does it add up?* pp. 26–33. London: Earthscan.

Henriques, A. and Richardson, J. (2004) *The Triple Bottom Line: does it add up?* London: Earthscan.

Herbohn, K. (2005) A full cost accounting experiment, *Accounting, Organizations and Society,* 30(6): 519–36.

Hines, R. D. (1991) Accounting for nature, *Accounting, Auditing and Accountability Journal,* 4(3): 27–9.

Hopwood, A., Unerman, J. and Fries, J. (2010) *Accounting for Sustainability: Practical Insights.* London: Earthscan.

Howes, R. (2004) Environmental cost accounting: Coming of age? Tracking organisation performance towards environmental sustainability, in Henriques, A. and Richardson, J. (eds), *The Triple Bottom Line: Does it all add up?* pp. 99–112. London: Earthscan.

Huizing, A. and Dekker, H. C. (1992) Helping to pull our planet out of the red: an environmental report of BSO/Origin, *Accounting Organizations and Society,* 17(5): 449–58.

Jackson, T. (2009) *Prosperity Without Growth? The transition to a sustainable economy.* London: Sustainable Development Commission.

Jensen, R. E. (1976) *Phantasmagoric Accounting: Studies in Accounting Research No.14.* Sarasota, FL: AAA.

Jones, M. J. (1996) Accounting for biodiversity: a pilot study, *British Accounting Review,* 28(4): 281–303.

Jones, M. J. (2003) Accounting for biodiversity: operationalising environmental accounting, *Accounting, Auditing and Accountability Journal,* 16(5): 762–89.

Jones, M. J. (2010) Accounting for the environment: towards a theoretical perspectives for environmental accounting and reporting, *Accounting Forum,* 34: 123–38.

Jones, M. J. and Matthews, J. (2000) *Accounting for Biodiversity: A natural inventory of the Eden Valley Nature Reserve,* Occasional Paper 29. London: ACCA.

Kaplan, R. and Norton, D. (1996) *The Balanced Scorecard: Translating strategy into action.* Boston, MA: Harvard Business School Press.

Kilbourne, W. E., Beckmann, S. C. and Thelen, E. (2002) The role of the dominant social paradigm in environmental attitudes: a multinational examination, *Journal of Business Research,* 55: 193–204.

Kolk, A. (2008) Sustainability, accountability and corporate governance: exploring multinationals' reporting practices, *Business Strategy and the Environment,* 17(1): 1–15.

Kovel, J. (2002) *The Enemy of Nature: The end of capitalism or the end of the world?* London: Zed Books.

KPMG International Environmental Network (1997) *Environmental Reporting.* Copenhagen: KPMG.

KPMG (1999) *KPMG International Survey of Environmental Reporting 1999.* Amsterdam: KPMG/WIMM.

KPMG (2002) *KPMG 4th International Survey of Corporate Sustainability Reporting.* Amsterdam: KPMG/WIMM.

KPMG (2005) *KPMG International Survey of Corporate Responsibility 2005.* Amsterdam: KPMG International.

KPMG (2008) *KPMG International Survey of Corporate Responsibility Reporting 2008.* Amsterdam: KPMG International.

KPMG (2011) *KPMG International Survey of Corporate Responsibility Reporting 2011* (kpmg.com).

Laine, M. (2010) Towards sustaining the status quo: business talk of sustainability in Finnish corporate disclosures 1987-2005, *European Accounting Review*, **19**(2): 247–74.

Lamberton, G. (2000) Accounting for sustainable development – a case study of City Farm, *Critical Perspectives on Accounting*, **11**(5): 583–605.

Laufer, W. S. (2003) Social accountability and corporate greenwashing, *Journal of Business Ethics*, **43**(3): 253–61.

Leipziger, D. (2010) *The Corporate Responsibility Code Book*, Revised 2nd Edition. Sheffield: Greenleaf.

Levy, D. L., Brown, H. S. and de Jong, M. (2010) The contested politics of corporate governance: the case of the Global Reporting Initiative, *Business and Society*, **49**(1): 88–115.

Linowes, D. (1972) Let's get on with the social audit: a Specific Proposal, *Business and Society Review*, Winter: 39–42.

Lohmann, L. (2009) Toward a different debate in environmental accounting: The cases of carbon and cost–benefit, *Accounting, Organizations and Society*, **34**(3–4): 499–534.

Lomborg, B. (2001) *The Skeptical Environmentalist: Measuring the real state of the world.* Cambridge: Cambridge University Press.

Lopez, M. V., Garcia, A. and Rodríguez, L. (2007) Sustainable development and corporate performance: a study based on the Dow Jones Sustainability Index, *Journal of Business Ethics*, **75**: 285–300.

Maltby, J. (2005) Showing a strong front: corporate social reporting and the 'Business Case' in Britain, 1914-1919, *Accounting Historians Journal*, **32**:145–67.

Mathews, M. R. (1984) A suggested classification for social accounting research, *Journal of Accounting and Public Policy*, **3**: 199–221.

Mathews, M. R. (1997a) Towards a mega-theory of accounting, *Asian-Pacific Journal of Accounting*, **4**(2): 273–89.

Mathews, M. R. (1997b) Twenty-five years of social and environmental accounting research: is there a silver jubilee to celebrate?, *Accounting, Auditing and Accountability Journal*, **10**(4): 481–531.

Maunders, K. T. and Burritt, R. (1991) Accounting and ecological crisis, *Accounting, Auditing and Accountability Journal*, **4**(3): 9–26.

McElroy, M. W. and van Engelen, J. M. L. (2012) *Corporate Sustainability Management: The art and science of managing non-financial performance.* London: Earthscan.

Meadows, D. H., Randers, J. and Meadows, D. L. (2005) *The Limits to Growth: The 30-year Update.* London: Earthscan.

Milne, M. (1996) On sustainability, the environment and management accounting, *Management Accounting Research*, **7**: 135–61.

Milne, M. and Gray, R. (2002) Sustainability reporting: who's kidding whom?, *Chartered Accountants Journal of New Zealand*, **81**(6): 66–70.

Milne, M. and Gray, R. H. (2007) Future prospects for corporate sustainability reporting, in Unerman, J., Bebbington, J. and O'Dwyer, B. (eds), *Sustainability Accounting and Accountability*, pp. 184–208. London: Routledge.

Milne, M. and Gray, R. (2012) W(h)ither ecology? The triple bottom line, the global reporting initiative, and corporate sustainability reporting, *Journal of Business Ethics*, November, SSN 0167-4544 DOI 10.1007/s10551-012-1543-8.

Milne, M. J., Tregigda, H. M. and Walton, S. (2003) The triple bottom line: benchmarking New Zealand's early reporters, *University of Auckland Business Review*, **5**(2): 36–50.

Milne, M. J., Kearins, K. N. and Walton, S. (2006) Creating adventures in wonderland? The journey metaphor and environmental sustainability, *Organization*, **13**(6): 801–39.

Milne, M. J., Tregigda, H. M. and Walton, S. (2009) Words not actions! The ideological role of sustainable development reporting, *Accounting, Auditing and Accountability Journal*, **22**(8): 1211–57.

Moneva, J. M., Archel, P. and Correa, C. (2006) GRI and the camouflaging of corporate unsustainability, *Accounting Forum*, **30**(2): 121–37.

Mueller, G. G. and Smith, C. H. (eds) (1976) *Accounting: A book of readings*, pp. 225–244. Hillsdale, IL: Dryden.

Norman, W. and MacDonald, C. (2004) Getting to the bottom of the 'Triple Bottom Line', *Business Ethics Quarterly*, **14**(2): 243–62.

Norris, G. and O'Dwyer, B. (2004) Motivating socially responsive decision making: the operation of management controls in a socially responsive organization, *The British Accounting Review*, **36**(2): 173–96.

Parke, R. and Peterson, J. L. (1981) Indicators of social change: developments in the USA, *Accounting, Organizations and Society*, **6**(3): 235–46.

Pearce, D., Markandya, A. and Barbier, E. B. (1989) *Blueprint for a Green Economy*. London: Earthscan.

Ranganathan, J. (1998) *Sustainability Rulers: Measuring Corporate Environmental and Social Performance: Sustainable Enterprises Perspectives Series*. Washington, DC: World Resources Institute.

Schaltegger, S. and Burritt, R. (2000) *Contemporary Environmental Accounting: Issues, concepts and practices*. Sheffield: Greenleaf.

Schreuder, H. (1979) Corporate social reporting in the Federal Republic of Germany: an overview, *Accounting, Organizations and Society*, **4**(1–2): 109–22.

Solomons, D. (1974) Corporate social performance – a new dimension in accounting reports?, in Edey, H. and Yamey, B. S. (eds), *Debits, Credits, Finance and Profits*, pp. 131–41. London: Sweet and Maxwell.

Taplin, J. R. D., Bent, D. and Aeron-Thomas, D. (2006) Developing a sustainability accountability framework to inform strategic business decisions: a case study from the chemicals industry, *Business Strategy and the Environment* (Special issue on sustainability accounting), **15**(5): 347–60.

Tregidga, H. M. and Milne, M. J. (2006) From sustainable management to sustainable development: a longitudinal analysis of a leading New Zealand environmental reporter, *Business Strategy and the Environment*, **15**(4): 219–41.

Unerman J., Bebbington, J. and O'Dwyer, B. (eds) (2007) *Sustainability Accounting and Accountability*. London: Routledge.

Wackernagel, M. and Rees, W. (1996) *Our Ecological Footprint: Reducing Human Impact on the Earth*. Gabriola Island, BC: New Society Publishers.

Wagner, M. and Schaltegger, S. (2006) Mapping the links of corporate sustainability, in Schaltegger, S. and Wagner, M. (eds), *The Business Case for Sustainability*, pp. 108–26. Sheffield: Greenleaf.

Walley, N. and Whitehead, B. (1994) It's not easy being green, *Harvard Business Review*, May/June: 46–52.

Willis, A. (2003) The role of the global reporting initiative's sustainability reporting guidelines in the social screening of investments, *Journal of Business Ethics*, **43**(3): 233–7.

WWF (2004) *Living Planet Report 2004*. Gland: World Wide Fund for Nature.

Yongvanich, K. and Guthrie, J. (2005) Extended performance reporting: an examination of the Australian mining industry, *Accounting Forum*, **29**(1): 103–19.

York, R., Rosa, E. A. and Dietz, T. (2003) Footprints on the Earth: the environmental consequences of modernity, *American Sociological Review*, **68**(2): 279–300.

CHAPTER 10

The social audit movement

10.1 Introduction

In this chapter, we turn our attention away from reporting initiatives emanating solely from within the organisation itself, with which we have hitherto been largely concerned, and examine in some depth the phenomenon known as the **external social audits,** where external bodies or individuals assume at least some responsibility for producing the report. As we shall see, a particularly significant feature of the development of the social audit movement lies in the diversity of groups and individuals undertaking such exercises and, relatedly, the range of issues addressed. However, what the various initiatives do have in common is a desire to increase the public accountability of powerful economic organisations and, to varying degrees, question the desirability of unfettered market capitalism. As Geddes (1992) puts it:

> . . . conventional accountancy attempts to reduce the social to the economic and the economic to the cash nexus. The importance of the social audit movement lies in its commitment to the restoration of social and political control over the economy.
>
> (Geddes 1992: 237)

Initially, compilation of the reports was solely the province of external participants (see Chapter 4, Figure 4.1). Indeed, most of the examples we will consider in this chapter are 'audits' of organisations prepared by external stakeholders without any cooperation from the organisation being held accountable. In this sense, external social audits are an explicitly conflictual attempt by civil society to hold to account organisations which would apparently rather not be held accountable. However, this distinction is not always so clear cut and the issue of stakeholder engagement raises the possibility of accounts and narratives being produced cooperatively by organisations and their stakeholders.[1] However, as we saw in Chapter 5 (and will further explore in Chapter 11), the whole area of stakeholder engagement is a contentious and difficult one which must be approached with care – not least in recognition that its considerable potential for cooperative development is matched by its equally considerable potential for manipulation and abuse of power.

Our purpose in this chapter is to trace the development of the social audit movement in the UK from its beginnings in the early 1970s, to critically evaluate the progress it has made and to provide pointers towards future developments. Our reason for adopting a largely UK perspective lies in a desire to paint a coherent picture of how the concerns of the movement and methodologies employed have evolved over time. Furthermore, whilst social audit is

---

[1] It is also worth reminding ourselves that in the early years the term 'social audit' was used interchangeably with other terms like social accounting and was indeed considered by some to be primarily a tool for the *internal* monitoring of social performance (see, for example, Humble 1973).

clearly a world-wide movement, it is in the UK where much of the impetus for it appears to have arisen and from where we shall draw a lot of our examples.

Nevertheless, it is necessary at the outset to acknowledge developments in the USA which have had an undoubted influence on events in the UK. Worthy of particular mention here are the activities of the Council on Economic Priorities (CEP), a non-governmental organisation (NGO) which, since the early 1970s, has, amongst other things, published a series of authoritative environmental performance rankings of major US companies (see Gray *et al.*, 1996: 152). Evidence of American influence on the UK social audit movement is indeed further apparent in the first example of early social audit activity within the UK to which we now turn our attention, that of the organisation Social Audit Ltd, one of whose leading figures, Charles Medawar, had previously gained considerable campaigning experience with Ralph Nader, the doyen of the US consumer movement.

10.2 Early developments in external social audit

Social Audit Ltd

Social Audit Ltd described themselves as 'an independent non-profit making body concerned with improving government and corporate responsiveness to the public generally'. Their fundamental ideals are neatly summarised by Medawar in the following terms:

> We have, in fact, a democratic bias. We believe that corporate power should be exercised to the greatest possible extent with the consent and understanding of ordinary people. We believe that people should be encouraged and allowed to share responsibility in society, but that at present they are not and are imposed upon instead. This question of secrecy and accountability is fundamental here.
>
> (Medawar, 1976: 390)

The work of Social Audit Ltd conveniently splits into two categories. First, we have the *Social Audit Quarterlies* published between 1973 and 1976 which, as well as containing general articles on issues such as the armaments industry, the social cost of advertising and company law reform, also featured their most influential work, this being 'social audits' conducted on four major companies and one government body (the no longer extant Alkali Inspectorate). Second, following the demise of the quarterlies, the organisation published a plethora of books and pamphlets ranging from handbooks on pollution and consumer audits and a guide to chemical hazards to in-depth investigations of the pharmaceuticals industry and advertising of food and drugs in the third world (see Gray *et al.*, 1996: 267). Notwithstanding the importance of this latter work, it is the published social audits that we shall focus upon in this chapter.

The reports on specific organisations[2] produced by Social Audit Ltd are highly detailed lengthy documents which are mainly narrative in form, although photographs, cartoons, statistical summaries, compliance with standard and financial data are also employed. The most substantial report was that on Avon Rubber (some 90 pages long) which was also unique in that, initially at least, it was carried out with the full cooperation of the company. The report commences with a review and analysis of Avon's business. This is followed by a substantial section addressing a wide range of employment issues, such as pay and job security, industrial relations, training, equal opportunities and health and safety. Next there

---

[2] These being Tube Investments Ltd, Cable and Wireless Ltd, Coalite and Chemical Products Ltd and Avon Rubber Co. Ltd.

comes a section on consumer products and services, which notably features strong criticism of industry-wide practices concerning advertising and customer information. Finally, a section on the environment and local community addresses key issues such as air, water and noise pollution, waste disposal and energy use and conservation employing statistical summaries and compliance with standard approaches to particularly good effect.[3]

One particularly intriguing feature of the Avon report lies in Social Audit's cataloguing of the difficulties they faced in gaining access to information held by public bodies, along with a disarmingly frank attempt to communicate any biases they perceived as informing their work. Significantly, these perceived biases were instrumental in the company withdrawing cooperation during the 'audit' process and expressing 'acute disappointment and concern at the number of inaccuracies and misrepresentations' that the report contained. It is, of course, highly unlikely that any outside body without the full cooperation of the organisation concerned and the appropriate government bodies can realistically hope to produce a wholly accurate and complete report. In addition, every 'social auditor' is likely to bring her or his personal value judgements to bear on matters of selection and description of 'relevant' material. It is likely, therefore, that an objective such as 'freedom from bias' is little more than a pious hope. Certainly, as we shall see throughout this chapter, it is apparently not an objective pursued by many of the campaigning groups which have engaged in widely varying forms of external social auditing activity over the years.

Notwithstanding concerns over completeness of information provision and bias in presentation, the work of Social Audit Ltd remains even to this day among the most thorough and important examples of independently produced social reporting in the UK. Unfortunately, the very independence of Social Audit Ltd, and consequent lack of wider economic or institutional involvement with the organisations concerned, both limited their power and made articulation of a framework of accountability involving them impossible. The inevitable result was that the reports produced were apparently ignored by company management, particularly where (as was mostly the case) the message conveyed was an unpalatable one. Indeed, similar problems in making a lasting impact, beyond providing a 'toolbox' of information and ideas, have generally bedevilled other early social audit initiatives conducted by bodies external to the 'audited' organisation.

Other early social audit initiatives

Another prominent pioneer of the external social movement, one which was particularly concerned with promoting the interests of labour, Counter Information Services (CIS), certainly didn't subscribe to the notion of freedom from bias as being a desirable objective for social audit activity. On the contrary, this collective of Marxist journalists was overtly dedicated to seeking radical changes in society as Ridgers (1979) makes clear when outlining the goals of the organisation as 'providing information resources for workers engaged in specific struggles and exposing the nature of the social and economic system which is the cause and content of these struggles' (p. 326). Throughout the 1970s, CIS issued a series of *Anti-Reports* focusing on, among other prominent organisations, Lucas, Ford, Unilever, Consolidated Gold Fields, GEC, Rio Tinto Zinc and the NHS.[4]

The CIS *Anti-Reports,* which are deliberately undated, are largely of a narrative format with tables of financial and statistical data provided where available. The effects

---

[3]The Avon Rubber report can be accessed on the CSEAR website (www.st.andrews.ac.uk/csear/approaches/external-social-audits).

[4]The Ford and Rio Tinto Zinc reports can be accessed on the CSEAR website (www.st-andrews.ac.uk/csear/approaches/external-social-audits).

of redundancies, strikes and working conditions are given high priority, although other dimensions of social performance are frequently touched upon.[5] The real power of the reports, however, lies in the use of vivid emotive phraseology. For example, in one report we are told that 'The history of Consolidated Gold Fields is one of brutal and inhuman exploitation which still continues. It is a case history of our current economic system operating in its purest form' (*Anti-Report* no. 3, p. 35, *c*.1973). Photographs are also frequently used in direct appeal to the emotions, with that of Hitler, for example, gracing the Lucas and Unilever *Anti-Reports* (p. 11, *c*.1976 and p. 81, *c*.1976), whilst a caption reading 'the effects of bombing in Vietnam' appears over the picture of a burnt-out hospital in the GEC *Anti-Report* (p. 31, *c*. 1973).

The *Anti-Reports* clearly cannot be considered as examples of 'objective', 'balanced' and 'unbiased' communication. They are, however, important as a (at the time) somewhat rare example of a 'radical' approach to reporting produced on a regular basis. Furthermore, much of the overtly anti-capitalist flavour of their approach has been taken on board by a number of campaigning groups producing what are commonly termed **counter accounts** over recent years whose work we will turn our attention to later in this chapter.

A further example of social audits conducted from a labour perspective and which offered a somewhat more balanced and systematic approach were isolated attempts at auditing the economic and social effects of plant closures conducted by trade union and worker groups (Institute for Workers Control (IWC) 1971, 1975) and academic economists (Rowthorn and Ward, 1979). These reports adopted a macro-economic, as opposed to simply an organisational, perspective towards the issue of plant closures in analysing GDP and balance of payments consequences of the closure decision in addition to employment effects. They are particularly significant in being a forerunner to more sustained local authority activity in this area in the 1980s, a consideration of which we turn to in the next section of this chapter. The 1971 IWC report (on Upper Clyde Shipbuilders) is also of some historic interest being one of the earliest published works in the UK to actually use the term 'social audit' (Zadek and Evans, 1993).

Whilst much early social audit activity adopted a largely labour perspective, the consumer and environmental constituencies were not entirely neglected. An early example of systematic social auditing from a consumerist perspective is, for example, provided by the Consumers Association. Best known for their magazine *Which*, this private sector body initially concerned itself with quality and value for money concerns together with ethical issues relating to the supply of goods and services to the public. Indeed, Gray *et al.* (1993) consider the work of the Consumers Association to be 'an important early example of the social audit as a mechanism for challenging the passivity of the individual in the face of the growing power of organisations and their capacity to exploit advertising and the nature of choice' (p. 263). Gray *et al.* do, however, go on to point out that commercial pressures in the late 1980s reduced this campaigning dimension, leading to a much narrower focus upon questions of product cost and efficiency (see also Geddes, 1992). Despite this development, the issue of social accountability to consumers didn't go away as new journals launched in the late 1980s such as *New Consumer* and *Ethical Consumer* took up the baton apparently abandoned by the Consumers Association. Indeed, the latter has retained a highly influential campaigning influence today, with its consumer orientated social auditing approach which, in addition to employing the *Ethical Consumer* journal as a means of communication, has also developed a highly sophisticated web-based ethical rating system for both companies and products.

---

[5] The first *Anti-Report*, that on Rio Tinto Zinc, for example, made a number of observations concerning environmental performance culled from *The Ecologist* magazine.

Examples of somewhat less systematic, *ad hoc* social auditing initiatives from an environmental perspective are offered by the leading environmental groups Greenpeace and Friends of the Earth, which from their earliest days have mounted investigative campaigns concentrating on the environmental ramifications of the activities of companies operating in particularly sensitive industrial sectors. Indeed, the early initiatives of Greenpeace and Friends of the Earth have been the precursor to much social audit activity, albeit somewhat sporadic in nature, carried out today by a wide range of NGOs and charitable organisations which again we shall return to later in the chapter.

The few illustrations touched upon above give only a flavour of the whole range of social audits that have been conducted from an environmental, consumer and labour perspective. Other initiatives have been yet more narrowly defined. To give but one example, the End Loans to Southern Africa Organisation produced a series of **shadow reports** during the 1980s (see Wells, 1985); interesting in particular for being modelled on the bank's own annual report and accounts. The shadow reports aimed to 'provide a substantial dossier exposing the bank's support for apartheid', with the contents varying from the simply provocative to the apparently factual and having the sole aim of attempting to embarrass Barclays into withdrawing from South Africa and Namibia.[6]

As may be seen, the term 'social audit' has from the outset embraced a wide range of aims and perspectives which encompass investigative journalism, commissioned research, special interest group campaigns and the efforts of self-appointed 'watchdog' organisations. The common thread is that all are part of the broad, if usually partisan and *ad hoc*, process of opening up, exposing, explaining, developing and attempting to control the myriad aspects of organisational activity in modern society.

10.3 Local authority social audits

Plant closure audits

Rapid large-scale de-industrialisation within the UK economy and the attendant rise in unemployment levels, particularly in regions dependent upon traditionally labour-intensive manufacturing activity, gave rise to a new manifestation of social audit activity in the early to mid-1980s. The prime movers this time were local authorities in the hardest hit areas (notably Merseyside) who, facing both declining income and a deteriorating industrial infrastructure, responded by initiating a series of plant closure audits (generally termed **social cost analyses**).

The plant closure audits, drawing upon work done at the macro-economic level which sought to quantify the exchequer, or public, costs of unemployment (House of Lords, 1982; CAITS, 1984a), exhibit two major features. First, an estimate of the total impact on unemployment in the locality of the plant closure is made. In addition to highlighting direct job losses consequent upon the closure decision, this entails making an estimate of the knock-on effect in terms of the number of indirect redundancies occurring in local firms supplying materials and services to the affected plant together with induced job losses in retailing and other businesses serving the local market due to decreases in purchasing power. The latter impacts are generally arrived at by utilising employment 'multipliers' which may be standard industrial or regional multipliers or, alternatively, specific figures derived from specially commissioned surveys. Additionally, a number of the reports made an attempt to estimate

---

[6]The 1982 shadow report can be accessed on the CSEAR website (www.st-andrews.ac.uk/csear/approaches/external-social-audits).

the likely duration of unemployment, which largely hinges on factors such as the age and skill composition of the workforce together with the pattern of local job opportunities.

The above information is then utilised in deriving the second major feature of the reports, an estimate of the public costs of unemployment imposed on local government and the national exchequer by companies implementing plant closure decisions. The public costs focused upon are predominantly those arising from redundancy payments made from public funds, income tax and national insurance contributions foregone, social benefits paid (funded both nationally and locally) and loss of local (Council) tax revenue.[7]

There is, of course, much subjectivity inherent in the plant closure social audit exercises. In addition to the problem of estimating the full employment impact (particularly the indirect and induced effects), and its likely duration, there is an issue concerning what costs to include. For example, certain costs considered in the macro-economic studies referred to earlier (House of Lords, 1982; CAITS, 1984a), such as loss of indirect tax revenue and costs of ill health consequent upon significantly increased unemployment levels, are ignored in the reports. Also generally omitted is any consideration of balance of payment implications (Rowthorn and Ward, 1979) and loss of output (Glyn and Harrison, 1980). However, although clearly no 'uniquely right' figure can be produced to represent the cost of unemployment at least an indication of the *order* of the cost to the public purse – for example, whether it be £3 million, £30 million or £300 million (CAITS, 1984b) – is conveyed

Significantly, none of the plant closure audits was successful in bringing about a reversal of the closure decision. One problem here was that the initiatives were purely reactive in nature, being hastily undertaken after the closure announcement. More fundamentally, the impotence of local authorities in the face of corporate self-interest and national government indifference (at best) to the plight of affected communities meant that little pressure could be brought to bear in forcing the companies concerned to accept some degree of accountability towards the communities in which they operated. However, what they did succeed in doing was offering a fundamentally different conceptualisation of social welfare to that of private wealth maximisation, to which traditional accounting techniques address themselves, in measuring organisational performance. The alternative employed, that of employment and consequent spending power, is, it must be acknowledged, undoubtedly a crude one. For example, little attention is paid to qualitative factors such as quality of employment or social usefulness of production. In essence, the reports tend to consider productive activity *per se* as beneficial, which in itself poses an interesting contrast to much social and environmental accounting theory and practice which focuses on costs imposed on parties external to the organisation arising from such activity.

A further issue arises in the reports' focus on financial quantification. This leads inevitably to the use of market data, such as level of redundancy payments made together with state benefit and taxation levels prevailing. Such data is, of course, open to government manipulation. More fundamentally, it fails to capture the full effects of the various social stresses, for example strains placed on family life or increased incidence of ill health emanating from unacceptably high unemployment levels in particular localities.[8] Interestingly, a small number of subsequent local authority social auditing initiatives which moved beyond the reactive, single entity focus of the closure audits did signal something of a shift away from crude financial quantification in seeking to address more qualitative, albeit less tractable, issues.

---

[7]For more detailed descriptive analysis of the content and methodology of plant closure social audits see Harte and Owen (1987).

[8]A number of non-quantified social factors are, in fact, briefly referred to in most of the reports. However, the overriding emphasis is on financial quantification and generation of a 'bottom line' figure for the 'cost' of unemployment (see Harte and Owen, 1987).

Later local authority social audit initiatives

A move away from a reactive, single entity focus was, for example, signalled by two exercises designed to highlight the importance of specific industries to particular regions – these being steel (County of Cleveland, 1983) and coal (Barnsley Metropolitan Borough Council, 1984), although both were still largely concerned with financial quantification issues. A more marked step forward from the limited methodology of the plant closure audits was, however, apparent in the Newcastle upon Tyne Social Audit (Newcastle City Council, 1985) and the Sheffield Jobs Audit (Sheffield City Council, 1985).

The Newcastle upon Tyne Social Audit outlined the results of an enquiry into the impact of government policy on the welfare of residents of the city over the time period 1979–84. Whilst indicating that households were, on average, £700 per annum worse off due to government policies, the report also drew attention to the fact that the impact was not evenly spread and went on to identify the major losers – these being the unemployed, pensioners and those on below average earnings. Detailed financial analysis highlighted the relative importance of increased unemployment, changes in the system of welfare benefits and social security provision together with cost increases for fuel, rates and rent in contributing to overall income reduction. However, the latter part of the report went beyond simply the financial figures in order to highlight less quantifiable reductions in the overall quality of life, represented by diminutions in sundry public service provision and higher incidence of ill health, crime and family stress, with case studies drawn from welfare rights work being effectively utilised to illustrate the human effects of policy change. The Sheffield Jobs Audit, for its part, adopted a similar regional perspective in attempting to assess the volume and, equally significantly, the quality of direct and indirect employment created by Council expenditure as well as studying the impact of rates and other charges on local jobs. One particularly fascinating aspect of the 260 page report produced was a section analysing the social impact of Dudley Council's policies of spending cuts, privatisation and establishment of an Enterprise Zone in order to provide a contrast with Sheffield's very different policies of public provision and job preservation.

For Geddes (1992) these latter two initiatives are particularly important, not only in that they are positive and forward looking, rather than being simply reactive, but also in the community perspective adopted which goes far beyond the confines of a single economic entity. Essentially, they are concerned with defending the principle of public sector provision at the level of the community and pose an alternative to the ubiquitous 'value for money' approach towards state services with its emphasis on efficiency and cost savings rather than the satisfaction of human needs. In the previous edition of this text (Gray *et al.*, 1996), we predicted that community-based initiatives would continue to be at the forefront of developments in social auditing practice. In this we have been sadly proved wrong. However, the promotion of policies by national government, notably in the UK but also elsewhere throughout Europe, in recent years under the guise of 'austerity', which have sought to decimate support for the weakest members of the community reliant on public provision for meeting their essential needs, suggests that the notion of a community-based social audit deserves better than being simply consigned to the dustbin of history.

10.4 Involving internal participants in the social audit exercise: a false dawn?

Zadek and Evans (1993) suggest that because the social audit initiatives considered thus far have been undertaken by outsiders, and not validated by 'neutral' auditors, they tend to be confrontational in nature. Hence they have not been accepted by the organisation concerned and have, as a result, largely been marginalised or ignored. In Puxty's (1991) Habermasian-informed analysis, it is further argued that such initiatives cannot be regarded as attempts to develop a discursive dialogue.[9] That is, they are not designed to reach an understanding through working with the organisation concerned. A further problem with audits solely conducted by external participants is that they tend to be restricted to either a limited range of social issues or stakeholder groups. A new approach to social auditing pioneered by the third world trading organisation Traidcraft in the early 1990s, which drew upon the involvement of internal as well as external participants and addressed itself to a wide range of constituencies and issues, appeared initially to offer a means of avoiding the above pitfalls.

Traidcraft's approach to social auditing has been described as:

> a process of defining, observing and reporting measures of an organisation's ethical behaviour and social impacts against its objectives, with the participation of its stakeholders and the wider community.
>
> (Zadek and Evans 1993: 7).

Particular stress is laid upon consulting all stakeholders in formulating performance indicators and, in the interest of process transparency, explaining decisions to use specific indicators. The aim is to make the social audit report reflect the views of all stakeholders as well as the company and the external auditors (who are considered to play a crucial role in the whole reporting process). Thus it is not intended to present a 'universal' view of performance. Rather, the audit is regarded as a social document, 'one that reflects the reality of diversity that is intrinsic to any living community' (Zadek and Evans, 1993: 29).

The participatory nature of the audit exercise, involving both internal and external participants, is highlighted in the development of the methodology employed.[10] This drew on the work of:

- An internal reference group comprising Traidcraft staff with expertise in marketing, personnel, product selection and producer support; an audit consultant from the New Economics Foundation (NEF), a charitable body responsible for external validation of the report; the director of a consumer organisation; and an academic having previous experience of working with the organisation on social responsibility issues. This group had responsibility for assessing, commenting on and offering direction on all aspects of method and process.

- An external advisory group made up of people with expertise in social audit and related areas.

- A NEF audit group who, in addition to having final responsibility for approving (or otherwise) the social accounts produced by Traidcraft in conjunction with researchers from NEF, had the duty of establishing the adequacy of treatment of information produced in the audit process.

Whilst Traidcraft's initiative falls somewhat short of being a fully independent external social audit in the sense of the examples we considered earlier in the chapter, it does appear to offer a significant degree of external *check* over the reporting process. The notion of involving stakeholders centrally in the reporting process was quickly and enthusiastically embraced by other 'values-based' organisations, most notably Body Shop. Furthermore, it has also been exceptionally influential in the work undertaken by a range of non-profit

---

[9]Puxty's analysis specifically refers to the local authority plant closure audits. It would, however, appear to apply more generally to the range of social audits considered so far.

[10]Traidcraft's social reporting in recent times, as far as can be ascertained by scrutinising the reports, has jettisoned the initial participatory structure described here.

organisations. Caught between the need to meet their social goals whilst also having to persuade potential funds providers concerning their commercial viability, these latter such organisations both adopted and, most importantly, further developed this approach to social audit with vigour. The resultant Social Audit Network (SAN) provided a vibrant supportive network of entities working at this intersection between the true external social audits and the later NEF driven 'inside–outside' audits. SAN's work remains important and is examined in a little more depth in Chapter 12 (see also Pearce and Kay, 2005; Gibbon, 2012).

Intriguingly, more commercially-orientated organisations, with the oil giants Shell and BP in the vanguard, also appeared to endorse the concept of stakeholder consultation in the reporting process. However, at this stage it rapidly became clear that a subtle change had occurred concerning the whole nature of the reporting and 'auditing' process. Rather than acting as a vehicle for holding the organisation accountable to its stakeholders, the process became a 'learning' exercise for the organisation itself. Furthermore, a managerialist slant became apparent with social audit sold as a vehicle for strengthening and enhancing an organisation's management procedures. Prominent benefits promised were the identification of weaknesses in management control of high-risk activities, and enhanced stability which might enable an organisation to militate against unexpected shocks (see, for example, SustainAbility, 1999). Ironically, even for Traidcraft it would appear that the development of a social bookkeeping system largely based on quantified social indicators, which underpinned the social audit process, rather than augmenting accountability relationships with key stakeholders simply moved the organisation towards adopting a more commercial interpretation of its fundamental religious principles (Dey, 2007a).

Essentially, if (external) social audit means anything at all it is an increasingly necessary component of a well-functioning democracy as various groups, with varying levels of power and access, offer alternative voices and alternative views in an attempt to hold organisations to account and, importantly, to counter claims made by politicians, businesses or whatever. At heart, social audit is about highlighting the tensions between maximising return on investment and not violating social trust. Moreover, social audit attempts to provide a mechanism for decision-makers to evaluate economic and social planning, facilitate popular involvement in economic decisions and identify social need as a primary criterion for resource allocation (Owen *et al.*, 2000). Significantly, for Geddes (1992) these latter values promoted by the movement represent nothing less than a fundamental assault on prevailing market-based economic orthodoxy. As we shall see in Chapter 11, the stakeholder 'engagement' approaches to social reporting, arguably spawned by the (undoubtedly well-intentioned) Traidcraft initiative, which many leading companies claim to adopt have very little in common with the ideals outlined above. However, a number of contemporary approaches to external social audit do apparently hold such ideals dear, and it is to a consideration of these that we now turn our attention.

10.5 Contemporary approaches to external social audit

Consumer group social audits

Since the late 1980s with the advent of journals such as *New Consumer* and *Ethical Consumer,* which we briefly referred to earlier in this chapter, the consumer movement has been consistently very much to the fore in social audit activity. Particularly prominent today in this regard have been the Ethical Company Organisation and Ethical Consumer. The aim of the former is stated as being 'to encourage the world's companies to treat humans, animals and the environment with the highest possible levels of respect' and to report accordingly on those that fail to do so (www.ethical-company-organisation.org). This desire to promote corporate transparency and accountability has, of course, much in common with the earlier work of Social Audit Ltd. Amongst other activities, the organisation publishes an annual *Good Shopping Guide* whilst also offering ethical accreditation to those companies that score highly in an overall analysis of their CSR record. The *Good Shopping Guide* provides ethical rankings for over 700 companies and brands across 60 product sectors. A scoring system based on 15 ethical criteria covering the core issues of people, animals and environment is employed in order to arrive at an overall ethical score. Further analysis outlines the key ethical issues to consider in individual purchasing decisions together with in-depth tabular coverage for each ethical criterion evaluated in order to explain how the score awarded has been arrived at.

Ethical Consumer, whose stated mission is to 'make global businesses more sustainable though consumer pressure' (www.ethicalconsumer.org), adopts a similar, although arguably somewhat more sophisticated, approach towards rating companies and products. In their case, what they call an Ethiscore is computed via evaluating performance against 23 ethical criteria covering the categories of environment, people, animals, politics and product sustainability (see Chapter 1, Figure 1.1). The ethiscore is designed to assist users in differentiating companies attracting significant levels of criticism from those with cleaner records and, in particular, to enable benchmarking of companies within product or market sectors. In computing the ethiscore, a wide range of information, primarily available in the public domain, is utilised including:

- publications by environmental, animal rights and Third World campaigning NGOs (e.g. Greenpeace, Friends of the Earth and Amnesty);
- corporate publications (such as annual reports and company websites, codes of conduct and animal testing policies);
- commercial defence and nuclear industry directories;
- pollution and health and safety prosecution records;
- a wide range of other international sources and daily media reports.

The Ethical Consumer database holds detailed information going back over 20 years on the social, ethical and environmental records of over 50,000 companies, although the ethiscores themselves (which are updated on a daily basis) are based solely on information published in the last five years. This information is accessible online via the Corporate Critic database which, amongst other facilities, offers the user the opportunity to customise ratings by varying the weights attached to the evaluative criteria employed. In addition to their rating activities, Ethical Consumer undertakes research and consultancy work for campaign groups and NGOs whilst also engaging in campaign work themselves, for example encouraging the boycott of irresponsible firms, highlighting tax avoidance issues and pressing for the introduction of ethical purchasing procedures on the part of local and national government.

The social audit activities of bodies such as the Ethical Company Organisation and Ethical Consumer form an integral part of their attempts to impose some degree of social and ethical control over corporate operations. Utilising consumer pressure in this way is perhaps particularly apposite in that, as Craig-Smith (1990) points out, consumer sovereignty provides the essential rationale for capitalism itself. However, arguably, only the relatively wealthy consumer enjoys the luxury of 'shopping with a conscience' and thereby paying more for this privilege. Additionally, of course, in conducting the audits, overwhelming reliance has to be placed on information available in the public domain, much of which originally emanates from companies themselves and which may not therefore be wholly reliable. Nevertheless, such initiatives do succeed in conveying information to the user in a clear and succinct fashion, whilst the ongoing and standardised nature of the work produced is a strength not shared by the more *ad hoc*, one-off nature of many other contemporary manifestations of social audit activity to which we now turn our attention.[11]

Silent and shadow accounting and the reporting–performance portrayal gap

Gray (1997), whilst noting that although many companies (at the time of writing) did not publish standalone reports, also went on to point out that various bits and pieces of social and environmental information were nevertheless scattered throughout their annual report and accounts. He suggested collating this information together in the form of a single report with sections covering mission and policy; directors and employees; community; environment; and customers. Such a document, which Gray termed a '**silent report**', could, he argued, be taken to represent the 'voice of the company' on environmental and social issues considered pertinent to their commercial activities.

Later work carried out under the auspices of CSEAR went on to experiment with exercises in shadow accounting undertaken in conjunction with the production of silent reports which, whilst resulting in reports similar in content to the latter, entailed researchers gathering information from non-company sources (mainly newspaper and journal articles) in order to provide an alternative independent, although of course not necessarily objective, perspective on corporate performance. Two initial silent and shadow reports were compiled for HSBC (Gibson *et al.*, 2001a) and Tesco plc (Gibson *et al.*, 2001b).

Figure 10.1 featuring an extract from a later report on RyanAir Ltd (undertaken as a student project) gives a flavour of the reporting style employed in silent and shadow reporting exercises. Particularly noteworthy is that, whilst the very existence of the parallel shadow report serves to problematise the company's version of events and provide alternative insights into their social and environmental impacts, there is an absence of any editorial comment and analysis, with the reader left to draw their own conclusions as to the adequacy and reliability of the company's version of events.[12]

Further academic work by Adams (2004) adopted a similar approach to the above silent and shadow reporting exercises in investigating the extent to which an unnamed company's reporting on social and environmental and ethical issues (for the years 1993 and 1999) adequately reflected actual corporate performance. However, in this instance the researcher was able to draw upon standalone environmental reports published by the company rather than having to construct a silent account.[13]

Additionally, the approach adopted was somewhat more rigorous in nature in that a wider range of external sources was consulted (including databases, reference books and internet sites as well as newspapers and business journals) whilst the analysis itself utilises a specific accountability framework informed by the requirements of AA1000 and the

---

[11]The impact of these consumer initiatives is also reflected in such developments as food labelling – especially the increasing use of 'traffic lights' for health content and 'food miles' to capture the distances food has had to travel to reach the consumer. Other forms of 'audit' which emerge from this sort of initiative might further include accreditation of products and processes through bodies such as the Marine Stewardship Council, the Soil Association and the Forestry Stewardship Council.

[12]The three silent and shadow reports referred to are accessible at www.st-andrews.ac.uk/csear/approaches/silent-and-shadow.

[13]The overwhelming majority of contemporary shadow reporting exercises focus on target companies who produce substantial social and environmental reports. This is, of course, largely due to the sensitive nature of the industries these companies operate in (for example, mining, oil and gas, pharmaceuticals and chemicals).

Figure 10.1 Extract from RyanAir Ltd: The Silent Report and Shadow Report [Customers and Products Report], Hamling, A., Kalolian, C., Lloyd, Z. and Yuill, Z. (pp. 7–9)

Silent and Shadow Report: RyanAir

Customers and Products Report:

SILENT ACCOUNT

Business Policy

RyanAir will conform to all competition and antitrust laws enacted to prevent interference with a competitive market system. Under these laws, no company / individual may enter into any formal or informal agreement with another company / individual, or engage in certain other activities, that unreasonably restrict competition. Employees are required to report any instance in which a competitor has suggested collaboration to their department head.

It is essential that RyanAir understand its competitors and be able to collect legitimate intelligence about them. RyanAir employees must not obtain, process, use or disclose confidential information of any third parties without appropriate authorisation from the applicable third party. Employees must not use any illegal or unethical means of gathering data about competitors.

RyanAir does not seek competitive advantage through illegal or unethical business practices. All employees / directors should endeavour to deal fairly with customers, competitors and employees. No employee / director should take unfair advantage of anyone through manipulation, concealment, abuse of privileged information, misrepresentation of material facts, or any unfair dealing practice.

Customers

Customer care

RyanAir is committed to fulfilling customers' needs in an honest and fair manner. The Company is committed to generating sales through price, quality and the ability to fulfil commitments. In September 2006, less than one complaint (0.36) was received per 1000 passengers. Of these, RyanAir responded to 99% within 7days.

RyanAir has "the lowest fares, the most on-time flights and the best customer service" (www.ryanair.com, 2006)

SHADOW ACCOUNT

Business Policy

In 2004, the European Commission ruled that a deal between RyanAir and Charleroi airport, 46 kilometres from Brussels, was illegal and amounted to state aid. The agreement under scrutiny included the airport paying RyanAir €160,000 (£111,000) for each of the first 12 routes opened by the airline to Charleroi, the pledging of €768,000 to fund crew training, the provision of free offices to RyanAir and a total of €250,000 to cover hotel and subsistence costs while the new facilities were set up (Clark, 2003). A preferential rate of EUR 1 per passenger was charged for ground handling services, with normal rates charged to other airlines ranging between EUR 8–13. The airport also offered landing charges about 50% lower than the standard rate set in a decree of the Walloon Region (Bird & Bird, 2004). While local authorities claimed they were trying to revive deprived areas, the commission ruled it unlawful for Belgian taxpayers' money to be used in the deal. They concluded that "the State aid involved could not be qualified as regional development aid taking the form of investment in the airport infrastructure, but constituted illegal operational aid" (Bird et al., 2004).

The Commission considered some of the aid granted to RyanAir compatible with the common market in the context of the transport policy. This includes the aid granted to the start-up of new routes. "Concerning the other aid measures, in particular the rebates on airport charges and ground handling fees, which amount to 25–30% of the EUR 15 million initially given, the Commission asked RyanAir to reimburse them" (Bird et al., 2004).

Customers

Customer care

By the top 20 airlines by number of complaints to Airtransport Users Council (AUC) (1 April 2001–31 March 2002): RyanAir came third after British Airways and AirFrance, with the AUC receiving 77 written complaints (BBC News, 2002). With RyanAir carrying 11 million passengers in this time, complaints averaged at one per 225 000 passengers. BA, who carried 40 million passengers, averaged at one complaint per 350 000 passengers (BBC NEWS, 2002).

In 2006, a poll of 4,000 travellers around the world voted RyanAir the world's least favourite airline. Unfriendly staff was cited as the main contributing factor, followed by delays and poor leg room (Milmo, 2006).

SILENT ACCOUNT

Delayed/cancelled Flights

RyanAir updated passenger treatment in the event of a long day. Passengers are entitled to meal vouchers and hotel accommodation, which are delayed for reasons within RyanAir's control. If your flight is cancelled or before the date of travel, is rescheduled so as to depart more than three hours before or after the original departure time then you will be entitled to a travel credit or full refund of all monies paid if the alternative flight/s offered are not suitable to you. In September 2006, 79% of RyanAir's flights were on time.

Baggage Handling

RyanAir's liability for loss, delay or damage to baggage is limited unless a higher value is declared in advance and additional charges paid. RyanAir do not accept liability for fragile, valuable, perishable articles or baggage, which is packed in damaged or unsuitable containers (RyanAir's Delayed or Damaged baggage). In September 2006, RyanAir received less than one mislaid bag (0.63) per 1000 passengers (RyanAir customer service).

Many items of sporting equipment, large musical instruments and infant equipment are considered by RyanAir to be inherently unsuitable for carriage by air. Upon payment of an additional charge of £17/€25 (or local currency equivalent) per item, per sector (flight) irrespective of weight, RyanAir is prepared to carry such items on a 'limited release' (i.e. entirely 'at your own risk' for damage or delay) basis. Passengers should ensure that you have suitable private insurance cover in force for such items (RyanAir terms and conditions).

SHADOW ACCOUNT

Delayed/cancelled Flights

In February 2006, the Norwegian consumer protection agency, backed by consumer groups in Denmark, Finland and Sweden, petitioned Norway's commercial court, the Markedsraadet, "over RyanAir's policy of charging high handling fees when refunding unused tickets or transferring tickets between passengers". RyanAir, failing to improve on customer care, were subsequently fined 500,000 Norwegian kroner (76,000 usd, 63,000 eur) for the unfair treatment of passengers (NTB news agency, as quoted in Nachrichten, 2006). RyanAir, under threat of court action, were given until mid-March, 2006, to change its policy (Nachrichten, 2006). The carrier was also criticized for "being unclear towards travellers about liability rights if planes are delayed or baggage is destroyed, lost or damaged" (Nachrichten, 2006).

Baggage Handling

A British consumer watchdog ordered RyanAir to change its terms of contract "after deciding the airline was unfairly turning down claims for lost or damaged luggage" (Brignall, 2006). The Office of Fair Trading (OFT) has also told the company to increase awareness in passengers of their rights in the event of delay or cancellation (Brignall, 2006). The OFT, upon finding that RyanAir's stance on baggage liabilities, had contravened the Montreal Convention. RyanAir's terms and conditions "excluded damage to non-standard objects such as pushchairs, sporting equipment and musical instruments carried in the holds of its planes. It effectively said they were carried at the customer's own risk" (Brignall, 2006). RyanAir has subsequently been ordered to change its terms and conditions and to pay future claims. Furthermore the airline was told to "make it easier for passengers to bring these claims" (Brignall, 2006).

Figure 10.1 (continued)

SILENT ACCOUNT

Reduced Mobility Passengers and Blind/Partially sighted Passengers

For safety reasons RyanAir can only carry a maximum of four disabled/reduced mobility passengers on any flight. This limit was originally agreed with the UK's Disability Rights Commission for safety reasons (BBC, 2005).

Requests for assistance must be made through RyanAir Direct on the same day as your original booking. Failure to advice RyanAir of your requirements on the day of booking will result in the service being unavailable on your arrival to the airport and your being refused carriage (RyanAir terms and conditions of travel).

Suppliers

For suppliers, RyanAir is committed to obtaining the best value on the basis of open and truthful communication.

RyanAir serves fair trade on the basis of lowering costs. RyanAir considers the fact that the new tea and coffee supplier is a Fairtrade brand a welcome bonus (Coyle, 2006).

SHADOW ACCOUNT

Reduced Mobility Passengers and Blind/Partially sighted Passengers

In 2002, Mr Ross, who has cerebral palsy and arthritis causing stamina problems, was charged £18 for a wheelchair when he flew from Stansted Airport to Perpignan, France. Supported by the Disability Rights Commission (DRC), Mr Ross brought a case against RyanAir and Stansted, under Part 3 of the Disability Discrimination Act 1995 and was awarded more than £1,300 (BBC News, 2005). The Appeal Court overturned the ruling that RyanAir was unlawful in charging Mr Ross for the use of a wheelchair, however found that both RyanAir and Stansted were responsible for providing a free wheelchair service for disabled travellers ensuring "disabled people receive the same standard of service as non disabled travellers". Thirty similar cases were reported by the DRC, for whom Stansted Airport and RyanAir are to compensate the passengers for the wheelchair charge (DRC, 2004).

In 2005, RyanAirs nomination for the prestigious Deaf-blind Friendly Corporate Award was withdrawn by the charities, Sense and Deafblind UK. RyanAir had made it to the last three in the travel and transport category of the awards. The judges decided to withdraw the company's nomination after RyanAir's blind people incident (AEBC, 2005).

In 2005, a group of six blind and three partially sighted passengers were ordered off a flight bound for Italy from Stansted Airport. After checking in as normal and given priority boarding, a member of the cabin crew informed the group they had to leave (BBC News, 2005). This was due to the flight being over its quota for disabled people. Scope, a leading UK disability charity, has subsequently urged a boycott of RyanAir (AEBC, 2005).

Suppliers

RyanAir's Michael O'Leary recently ordered "his entire fleet to serve only Fairtrade tea and coffee, which are generally regarded as more expensive than rival brands" (Coyle, 2006). The fairtrade stamp guarantees "farmers and producers in the developing world are paid a fair price for their crops (Coyle, 2006).

Global Reporting Initiative (GRI) reporting guidelines. Much more is also offered in the way of analytical and editorial comment, with particular attention focused on instances where the company's account conflicts with information gleaned from shadow sources together with those where the company omits to provide information on issues picked up by shadow sources as being of material interest to its stakeholders. Put simply there would appear to be a distinct 'reporting-performance portrayal gap' with Adams concluding in somewhat damning terms that:

> There is little coverage of negative impacts, insufficient evidence that Alpha accepts its ethical, social and environmental responsibilities, an arguably one-sided view of sustainability issues facing the company and a lack of completeness. The different coverage in external sources also raises questions as to the inclusivity of stakeholders in the reporting process. The report itself provides insufficient information on the reporting process and governance structures in place with respect to ethical, social and environmental reporting
>
> (Adams 2004: 749).

In addition to the academic initiatives outlined above, NGOs and sundry campaign groups have long employed shadow reporting as part of their armoury in confronting, and attempting to influence, the behaviour of powerful corporations. The reports produced do, however, vary considerably in terms of breadth and depth of analysis. The most rigorous are undoubtedly those produced by the anti-smoking pressure group Action on Smoking and Health (ASH) as a riposte to the social reporting efforts of the tobacco giant British American Tobacco (BAT). Their first report, *British American Tobacco: The Other Report to Society* (ASH, 2002), was particularly noteworthy in that it adopted a similar approach to that employed by the company itself in utilising both the provisions of AA1000 and the GRI reporting guidelines, with the aim here of providing evidence of shortcomings in disclosures made by the latter. A further shadow report, *BAT in its own words* (ASH, 2005), produced in association with Christian Aid and Friends of the Earth, again closely followed the company's own reporting format, although it went a little further in calling for regulation of corporate governance and disclosure practice in order to address reporting shortcomings rather than simply highlighting them (Thomson *et al.*, 2010).

Shadow reports produced by other prominent NGOs, most notably Christian Aid, Friends of the Earth and Greenpeace, tend to be far more patchy and selective in terms of issues addressed and depth of analysis offered, with emphasis often placed on specific aspects of the target company's activities rather than a full shadow reporting exercise being essayed. Thus, for example, Greenpeace (2005a, b) have produced two short (five page) 'climate crime files' on Land Rover and Esso[14] which focus heavily on the lobbying activities of these companies designed to play down fears over climate change and to forestall government intervention. For their part, Christian Aid in their (2003) report, *Behind the Mask: The real face of corporate social responsibility*, adopt a geographical-based case study approach in critiquing the activities of Shell in the Niger Delta, BAT in Kenya and Coca-Cola in India which relies heavily on narrative accounts of harm suffered by affected individuals and communities. A similar approach underpins a series of 'other Shell reports' produced in recent years by Friends of the Earth (available at www.foe.co.uk/campaigns/corporates/case_studies/index.html). Dey (2007b) draws particular attention to the partial nature of the analysis offered (again relying heavily on somewhat emotive case study material, not all of which appears to relate to events in the

---

[14]These reports are accessible at www.st-andrews.ac.uk/csear/approaches/external-social-audits.

reporting year) in the 2002 report entitled *Failing the challenge: The other Shell report*[15] in noting that:

> Only two pages out of the 28 page report dealt with claims made by Shell in its own disclosures, and while the document explicitly acknowledged the existence of wider sources of third-party evidence on Shell's behaviour, it does not make use of most of this evidence.
>
> (Dey, 2007b: 318)

Counter accounting and the role of the internet

As Dey (2007b) points out, reports such as that of Friends of the Earth (FoE) fall far short as a piece of systematic shadow accounting. Indeed he notes that:

> Whilst FoE are in many ways right to draw attention to the selective bias and unreliability of the Shell report, they counter this with what is arguably an even more selective and unreliable report of their own.
>
> (Dey, 2007b: 318)

A somewhat more accurate term to describe such reports may be that of **counter accounting**, which Gallhofer *et al.* consider to be:

> information and reporting systems employed by groups such as campaigners and activists with a view to promoting their causes or countering or challenging the prevailing official and hegemonic position.
>
> (Gallhofer *et al.*, 2006: 681–82)

Falling into this category would certainly be the myriad of reports produced by organisations campaigning on single issues (for example, Campaign against the Arms Trade) or focusing their attention on individual multinational corporations.[16] Additionally, many of the campaigning reports produced by more mainstream NGOs on issues such as corporate tax avoidance (see, for example, Christian Aid, 2005, 2008) could perhaps best be considered as examples of counter accounting.

Whilst accepting that information provided in counter accounts is generally partial and selective, Spence (2009) nevertheless argues that such external initiatives can claim the moral high ground over the (alleged) equally partisan corporate produced social reports. This is for two main reasons. First, in contrast to the myth of objectivity and completeness conveyed by the corporate version, counter accounts, it is suggested, lay no claim to such pretensions but are rather quite transparent over the political agenda underpinning their work. Second, counter accounts explicitly seek to open up dialogue by exposing contradictions and conflicts whereas corporate social reporting rather seeks to close down debate via the promotion of a ubiquitous 'business case' which serves to constrain what is 'thinkable' and 'doable'.

In somewhat similar vein, Gallhofer *et al.* (2006) argue that counter accounting is no more biased than the corporate propaganda appearing in company reports and can, moreover, serve to improve democratic functioning, engender progressive change and promote emancipatory action. They further suggest that the internet potentially provides a central resource in bringing about the realisation of such aims, with key features being the offering of relatively cheap and fast access in respect of website construction and usage, thereby allowing a wealth of unofficial, critical and alternative channels of information to compete with the official version

---

[15]Dey (2007b) also provides detailed comparative analysis of the respective shadow reporting initiatives of ASH (2002) and Friends of the Earth (2003).

[16]Nike, Nestlé, Apple, McDonald's, Coca-Cola and Gap are, for example, just some of the organisations that have found themselves the subject of a very uncomfortable public gaze produced by a variety of 'counter accounting' initiatives.

of events. The aim here is to create what Adams and Whelan (2009) term 'cognitive dissonance' which can then lead to an 'unfreezing' of the status quo and promotion of change. Adams and Whelan illustrate their argument by pointing to the success of the internet-based 'Kentucky Fried Cruelty' campaign conducted by the pressure group People for the Ethical Treatment of Animals (PETA) and more recent campaigns successfully run by university students against the treatment of staff employed by Nike subcontractors which employed social media in addition to the internet.[17]

Gallhofer *et al.* go on to offer an in-depth exploration of web-based counter accounting in practice, featuring a case study of Corporate Watch (a not for profit research and publishing organisation which, amongst other things, provides a growing number of company profiles on their website), an analysis of the websites of 20 campaign groups and a web-based questionnaire addressed to a further 100 such groups. The overall conclusion reached is that at least some of the positive potential of reporting online for counter accounting is being realised. It is noted, for example, that whilst the web is largely being used as a replacement for the traditional print medium, particularly in the case of Corporate Watch's company profiles which bear some affinity with the CIS *Anti-Reports* we considered earlier, it does enable a much larger audience to be reached. Additionally, information accessibility is further enhanced by the offering of search facilities and provision of links to websites of groups pursuing similar goals and objectives, hence contributing to the creation of a more global movement able to exert increased pressure on the corporate world. However, it is also noted that much of the information flow is in only one direction, from the campaign group to site visitors, whilst dominance of the English language on the net, together with lack of opportunity of access to many of the world's poor, further serves to limit the web's possibilities in facilitating democratic engagement. Most significantly though, lack of resources, a problem that has bedevilled many of the external social audit initiatives considered in this chapter, would appear to be the major constraint faced by the counter accounting groups studied.

10.6 The external social audits: where to now?

Lack of resources perhaps goes a long way towards explaining the patchy, *ad hoc* and largely unsystematic nature of much of the contemporary social audit practice outlined above. The contrast with the growing volume of detailed, apparently authoritative, and regularly produced corporate social and environmental reports is all too vivid. For Dey (2007b), one possible way forward lies in promoting collaboration between NGOs and campaigning groups and academic researchers. The point here is that the reporting initiatives of the former, whilst being all too often sketchy and partial in the extreme, do at least provide a valuable resource for the academic in collating more systematic and authoritative external social audit reports. Alternatively, it is suggested that academic involvement could extend to encouraging and offering guidance to NGOs to produce more consistent and complete reports themselves. This in turn may be instrumental in promoting constructive dialogue with corporations.[18] Whereas there may be some potential in going down this route, experiences with current stakeholder engagement and dialogue practices, which we discuss at

---

[17]Any time spent on the web will reveal the plethora of entities now engaging in various forms of counter accounting and the diversity of approaches taken. Examples such as 'Adbusters' and the 'Yes Men' sit at this borderline between what we are calling social audits and the use of information for direct confrontation and social activism (see Spence, 2009).

[18]One very rare example of a corporation and NGO cooperating in producing a form of social audit report itself is that of Oxfam and Unilever's joint exploration of the links between international business and poverty reduction centring on a case study of Unilever's activities in Indonesia (Oxfam and Unilever, 2005).

some length in Chapter 11, give little cause for optimism. Additionally, the resource issue remains unresolved with the external 'David' continuing to confront the corporate 'Goliath'.

The central issue to address in terms of enabling social audit and shadow reporting exercises to achieve their emancipatory potential is, in any event, as Dey *et al.* (2010) point out, not simply one of content but rather one of how the reports are used. It is perhaps significant to note here that whilst Social Audit Ltd succeeded in producing reports ranking amongst the most complete and objective of any, the practical impact achieved was negligible. By contrast, ASH, although avowedly partisan and adversarial in their reporting initiatives (Dey, 2007b), have undoubtedly been a highly effective force in promoting the restrictive (and some might say draconian in civil liberties terms) regulations surrounding the sale and consumption of tobacco in recent years. The crucial difference here is that the ASH shadow reports are but one part of an overall strategy designed to confront the activities of 'big tobacco', with lobbying, commission and dissemination of scientific research and participation in influential policy-making forums, amongst other interventions, forming additional parts of their considerable armoury.

In addition to the necessity for the production of external social audits to form part of a wider intervention strategy, Cooper *et al.* (2005) further suggest that their effectiveness is crucially dependent on their being articulated to 'social movements'. In particular, they argue that 'the production of something akin to early social audits aligned to contemporary social struggles and action groups (e.g. trade unions) would promote the potential to create a more equitable society' (p. 951). They go on to illustrate their argument by reporting on their own development and use of a novel and challenging form of social audit (which they term a social account) to confront and seek to influence the debate surrounding higher education funding in Scotland.

Briefly, the background to Cooper *at al.*'s intervention was that the UK government during the 1990s was moving steadily towards placing university education in a kind of pseudo-open market through the removal of grants for students followed by the slow introduction of tuition fees.[19] Such moves obviously affect low-income and poor students a great deal more than they affect middle-class and wealthy ones. Cooper *et al.* set about collecting data and compiling an 'account' of the impact this would have on poorer students by conducting a substantial questionnaire survey amongst all third year full-time students at the three Glasgow universities. Essentially, the purpose of the questionnaire was to obtain basic information concerning the financial state of students and the impact of financial imperatives upon their studies (in particular the adverse effects of having to work long hours in poorly paid jobs in order to finance them). The resultant account compiled from the questionnaire responses (featuring both quantitative data and qualitative information based on written comments made by respondents) succeeded in giving a very powerful voice to an otherwise unheard body of the population and caused considerable media and public policy interest.

Most notable in the latter context, the account comprised one of the first submissions to a commission of enquiry (the Cubie committee) set up to investigate student financing and to report back to the Scottish parliament, and its influence was clearly discernible in the commission's final report. Whilst the recommendations of the report were eventually considerably watered down by the Scottish Parliament, with material adverse effects on student finances, one significant outcome was that the principle of charging tuition fees was clearly rejected. The result of this outcome is that Scottish students are at least relatively better off in financial terms than their English counterparts, having, as Cooper *et al.* point out, benefited from the social movement in Scotland against tuition fees, a movement to which their social account was closely aligned.

---

[19]Students in all parts of the world are likely to recognise these issues!

A further lesson to be gleaned from the work of Cooper *et al.* is that social audits can be assembled around any coherent entity or issue and used in many ways to engage more widely with political economy. One notable example in this latter context is the shadow account produced by Collison *et al.* (2010) which sought to hold Anglo-American capitalism, with its uncompromising shareholder value focus to the exclusion of all else, to account for its social outcomes relative to alternative various forms of social market (or welfare) capitalism. Utilising a key social indicator, that of child mortality, Collison *et al.*'s detailed cross-sectional and longitudinal analysis of figures for the wealthiest OECD countries highlights the consistently poor, and worsening, performance of countries adopting the Anglo-American model which, significantly, are characterised by far greater levels of income inequality than the social market alternatives.

As far as the future of social audit is concerned, we would suggest that the only limits to its further development are those of imagination, time, effort, will and, it has to be said, political freedom. Certainly, as the examples in this chapter have shown, there is a plethora of data and information out there and a wide range of organisations who, broadly speaking, are determined to give voice to alternative views and to challenge the hegemony of dominant vested interests be they corporations, government agencies or even political and economic systems. Much of their work is made possible through the ever growing influence of the internet, and what the web has made possible seems unlikely to abate any time soon.

10.7 Conclusion

The central thrust of the 'external social audits' is that if organisations will not discharge their own social and environmental accountability, and if the state will not act to introduce such regulation enforcing that discharge (see Chapter 11), then civil society must act in its own interests. Such exercises are defined by Dey *et al.* (2010) as 'accounting for the other by the other', meaning the derivation and communication of accounts which offer different perspectives on organisational life by individuals and groups who are not dominant or powerful in themselves. Such ideas are essential to a complex and vibrant democracy and it is a source of some encouragement that worldwide such counter accounting seems to be growing, becoming more diverse and more sophisticated. However, on a somewhat less optimistic note, the lack of transparency and accountability exhibited by both private- and public-sector organisations which social audits are able to so effectively expose does raise a fundamental question as to the true extent of democracy prevailing in our supposedly 'democratic' society.

References

Adams, C. A. (2004) The ethical, social and environmental reporting-performance portrayal gap, *Accounting, Auditing and Accountability Journal*, **17**(5): 731–57.

Adams, C. A. and Whelan, G. (2009) Conceptualising future change in corporate sustainability reporting, *Accounting, Auditing and Accountability Journal*, **22**(1): 118–43.

ASH (2002) *British American Tobacco: The other report to society*. London: Action on Smoking and Health.

ASH (2005) *BAT in its own words*. London: Action on Smoking and Health.

Barnsley Metropolitan Borough Council (1984) *Coal Mining and Barnsley*. Barnsley: Barnsley Metropolitan Council.

CAITS (1984a) *Public Costs of Unemployment*. London: Centre for Alternative Industrial and Technological Systems.

CAITS (1984b) *Economic Audit of the Costs of Closure of the Foundry at Dagenham by Ford.* London: Centre for Alternative Industrial and Technological Systems.

Christian Aid (2003) *Behind the Mask: The real face of corporate social responsibility.* London: Christian Aid.

Christian Aid (2005) *The Shirts off Their Backs: How tax policies fleece the poor.* London: Christian Aid.

Christian Aid (2008) *Death and Taxes: The true toll of tax dodging.* London: Christian Aid.

CIS (1972 *et seq.*) *Anti-Reports.* London: Counter Information Services.

Collison, D., Dey, C., Hannah, G. and Stevenson, L. (2010) Anglo-American capitalism: the role and potential role of social accounting, *Accounting, Auditing and Accountability Journal,* 23(8): 956–81.

Cooper, C., Taylor, P., Smith, N. and Catchpowle, L. (2005) A discussion of the political potential of social accounting, *Critical Perspectives on Accounting,* 16(7): 951–74.

County of Cleveland (1983) *The Economic and Social Importance of the British Steel Corporation in Cleveland.* County of Cleveland.

Craig-Smith, N. (1990) *Morality and the Market: Consumer pressure for corporate accountability.* London: Routledge.

Dey, C. (2007a) Social accounting at Traidcraft Exchange plc: a struggle for the meaning of fair trade, *Accounting, Auditing and Accountability Journal,* 20(3): 423–45.

Dey, C. (2007b) Developing silent and shadow accounts, in Unerman, J, Bebbington, J. and O'Dwyer, B. (eds), *Sustainability Accounting and Accountability,* 307–26. Abingdon: Routledge.

Dey, C., Russell, S. and Thomson, I. (2010) Exploring the potential of shadow accounts in problematising institutional conduct, in Osborne, S. P. and Ball, A. (eds), *Social Accounting and Public Management: Accountability for the Common Good,* pp. 64–75. Abingdon: Routledge.

Friends of the Earth (2003) *Failing the Challenge: The other Shell report 2002.* London: Friends of the Earth.

Gallhofer, S., Haslam, J. and Monk, E. (2006) The emancipatory potential of online reporting: the case of counter accounting, *Accounting, Auditing and Accountability Journal,* 19(5): 681–718.

Geddes, M. (1992) The social audit movement, in Owen, D. (ed.), *Green Reporting: Accountancy and the challenge of the nineties,* pp. 215–41. London: Chapman and Hall.

Gibbon, J. (2012) Understandings of accountability: an autoethnographic account using metaphor, *Critical Perspectives on Accounting,* 23(3): 201–12.

Gibson, K., Gray, R., Laing, Y. and Dey, C. (2001a) *Tesco plc 1999–2000: The Silent Report and the Shadow Report* (www.st-andrews.ac.uk/csear/approaches/silent-and-shadow).

Gibson, K., Gray, R., Laing, Y. and Dey, C. (2001b) *HSBC Holdings plc 1999–2000: The Silent Report and the Shadow Report* (www.st-andrews.ac.uk/csear/approaches/silent-and-shadow).

Glyn, A. and Harrison, J. (1980) *The British Economic Disaster.* London: Pluto Press.

Gray, R. (1997) The practice of silent accounting, in Zadek, S., Pruzan, P. and Evans, R. (eds), *Building Corporate Accountability,* pp. 201–17. London: Earthscan.

Gray, R., Bebbington, J. and Walters, D. (1993) *Accounting for the Environment: The greening of accountancy Part II.* London: Paul Chapman.

Gray, R., Owen, D. and Adams, C. (1996) *Accounting and Accountability: Changes and challenges in corporate social and environmental reporting.* London: Prentice Hall.

Greenpeace (2005a) *Climate Crime File – Land Rover.* London: Greenpeace.

Greenpeace (2005b) *Climate Crime File – Esso.* London: Greenpeace.

Hamling, A., Kalolian, C., Lloyd, Z. and Yuill, Z. (2006) *RyanAir Ltd: The Silent and Shadow Report* (www.st-andrews.ac.uk/csear/approaches/silent-and-shadow).

Harte, G. and Owen, D. (1987) Fighting de-industrialisation: the role of local government social audits, *Accounting, Organizations and Society,* 12 (2): 123–41.

House of Lords (1982) *Report of the Select Committee on Unemployment.* London: House of Lords.

Humble, J. (1973) *Social Responsibility Audit – a management tool for survival.* London: The Foundation for Business Responsibilities.

IWC (1971) *UCS: A Social Audit.* London: Institute for Workers Control.

IWC (1975) *Why Imperial Typewriters Must Not Close.* London: Institute for Workers Control.

Medawar, C. (1976) The social audit – a political view, *Accounting, Organizations and Society,* 1(4): 389–94.

Newcastle City Council (1985) *Newcastle Upon Tyne Social Audit 1979–1984.* Newcastle upon Tyne: Newcastle City Council.

Owen, D., Swift. T., Humphrey. C. and Bowerman, M. (2000) The new social audits: accountability, managerial capture or the agenda of social champions?, *European Accounting Review,* 9(1): 81–98.

Oxfam and Unilever. (2005) *Exploring the Links between International Business and Poverty Reduction.* Oxford: Oxfam.

Pearce, J. and Kay, A. (2005) *Social Accounting and Audit – The Manual* (and CD). Edinburgh: Social Audit Network.

Puxty, A. G. (1991) Social accountability and universal pragmatics, *Advances in Public Interest Accounting,* **4**: 35–46.

Ridgers, B. (1979) The use of statistics in counter information, in Irvine, J. (ed.), *Demystifying Social Statistics.* London: Pluto Press.

Rowthorn, B. and Ward, J, (1979) How to run a company and run down an economy – the effects of closing down steel making in Corby, *Cambridge Journal of Economics,* **3**: 329–40.

Sheffield City Council (1985) *Sheffield Jobs Audit.* Sheffield: Sheffield City Council.

Spence, C. (2009) Social accounting's emancipatory potential: A Gramscian critique, *Critical Perspectives on Accounting,* **20**(2): 205–27.

SustainAbility (1999) *The Social Reporting Report.* London: SustainAbility.

Thomson, I., Dey, C. and Russell, S. (2010) *Social Accounting and the External Problematisation of Institutional Conduct: Exploring the Emancipatory Potential of Shadow Accounts,* paper presented at the APIRA conference, University of Sydney, 12–13 July.

Wells, D. (1985) Barclays is losing its fight with a shadow on points, *Accountancy Age,* 9 May: 9–10.

Zadek, S. and Evans, R. (1993) *Auditing the Market: A Practical Approach to Social Accounting.* London: New Economics Foundation.

CHAPTER **11**

Governance, attestation and institutional issues

11.1 Introduction

One of the most exciting things (or frustrating things, depending on your point of view) about social, environmental and sustainability accountability and accounting is the sheer diversity of practice and possibilities – not to mention the rate of change of those practices and possibilities. The field embraces, as we have seen, everything from costing through to stakeholder engagement; from risk assessment through to attempts to capture un-sustainability; from measuring liabilities through to externally produced counter-accounts. The positive side of this diversity is the vibrancy and experimentation. Social accounting, whatever else it might be, is not mired in convention and tradition. The frustrating side is that the field is impossible to define entirely coherently, the world-wide diversity is difficult to easily comprehend and, most importantly, most entities do not embrace social accounting in anything like a substantive manner. The reasons for this are relatively simple: the governance systems surrounding social accounting are themselves diverse, predominantly voluntary and largely superficial. And this arises, in large part it would seem, because the institutions of national and international governance and the powerful institutions of financial capitalism do not see either social responsibility or sustainability as sufficiently important to insist that it happens. And, for reasons which are slightly less obvious, these same institutional forces resist most attempts to develop any reliable mechanisms of social, environmental and sustainability accountability. Imagine what world-wide financial accounting might look like if there was no complex infrastructure, sanctions and institutions developing, implementing and monitoring its practice. It might look a lot like social accounting does now.

Governance very broadly might be thought of as '*processes of supervision and control (of "governing") intended to ensure that an entity's management acts in accordance with the interests of its constituents*'. This definition is adapted from Parkinson (1993: 159) – adapted to reflect the wider range of issues, organisations and constituents that exercise us in social accounting.[1] Governance then comes to include a very wide range of things – indeed, it can be taken to include culture, religion, ethics and all the factors that lead us to do (or not do) things. How governance comes to be, more formally, will inevitably start with the law and the state but, as we have already seen throughout the text, such law and/or regulation as it relates to corporate social responsibility (CSR), sustainability and social accounting is, at best, patchy and at worse superficial.

---

[1] More conventional governance definitions recognise that governance arose predominantly in the large corporate sector and was intended to protect the shareholders – not the wider stakeholders – and this tension continues throughout the debate as we shall see (see, for example, Solomon, 2007; Blowfield and Murray, 2011). Other definitions are used in different contexts – the World Bank is a stimulating place to start in this regard. We return briefly to definition later in the next section.

The principal argument (or political *motif*) underpinning and informing this text as a whole is that, in order to be effective in bringing about substantial social and environmental change, the development of social accounting must operate to empower stakeholders so that they can hold powerful institutions (notably large corporations) accountable for their societal impacts (see Chapter 1). For this to come about, there clearly needs to be some mechanism by which stakeholder views can feed directly into corporate decision-making. As Owen *et al.* (1997), drawing upon the work of Power (1994), put it, there is a need for **administrative** (in this case accounting) **reform** concerning the way things are done, and this needs to be accompanied by **institutional reform** designed to improve the structure and context within which the process occurs – see Figure 11.1.

Figure 11.1 Administrative versus institutional reform

- *Administrative (or technical) reform* – sometimes limited to accounting reform – relates to reforms designed to increase levels of procedural effectiveness. In the case of governance this may often be focused on increasing organisational transparency.
- *Institutional reform* – is broadly concerned with structures around the issue of concern and when related to governance will include reform of regulation and enforcement but also means of empowering stakeholders through greater participation.

A great deal of what we have covered so far might be thought of as focused principally upon administrative reform – the ways in which social and environmental accounting, reporting and accountability are practised, how the processes operate and how these can be developed to support improved responsiveness to social, environmental and sustainability issues.

A significant degree of administrative reform is clearly discernible over recent years in terms of a noticeable increase in the number of major companies across the globe voluntarily producing substantial paper, or web-based, environmental, social and sustainability reports. Perhaps not surprisingly in view of the fall-out from Enron and similar corporate scandals, reputation building appears to provide a primary motivating factor for companies going down the CSR path. Thus, for example, Business in the Community's 'business case' for CSR notes that it offers:

> . . . a means by which companies can manage and influence the attitudes and perceptions of their stakeholders, building their trust and enabling the benefits of positive relationships to deliver business advantage.
>
> (Business in the Community, 2003: 3)

Whether reporting change driven solely by concerns with enhancing competitive advantage is likely to bring about fundamental change in organisational attitudes and priorities is somewhat of a moot point. Furthermore, the question arises as to whether exclusive reliance on the business case to encourage CSR initiatives is capable of promoting institutional reform sufficient to empower organisational stakeholders so that any potential heightened accountability arising from the increased transparency inherent in reporting initiatives may be realised.

Consequently, it is the institutional context of reporting that particularly concerns us in this chapter. In the next section, we provide an overview of governance as it relates to CSR and sustainability issues. Then, in Section 11.3, we examine the degree of stakeholder empowerment manifested by both the considerable amount of corporate governance reform introduced in western economies in recent years and the apparent move on the part of many companies towards introducing myriad forms of stakeholder engagement in order to underpin their social and environmental reporting initiatives. We then move on to consider, in Section 11.4, the possibilities for an alternative, more indirect, form of potential stakeholder empowerment offered by civil regulation, whereby control is enforced chiefly by markets and which relies on public pressure to bring about socially responsible corporate behaviour. As Parkinson (2003) points out, effective civil regulation is highly dependent on an adequate disclosure regime being established. Clearly, adequacy (or completeness) of reporting can be called into question in a reporting environment largely characterised by voluntary, rather than legally mandated, disclosure driven predominantly by business case considerations. Section 11.5 presents what might be thought of as an object lesson/case study drawing from the largely disappointing British experience with (what is known as) the Operating and Financial Review (OFR). The final element in this brief governance jigsaw that we cover is that of **assurance** in Section 11.6. Independent verification, or assurance, of these social, environmental and sustainability (sic) reports, which an increasing number of companies have submitted themselves to in recent years, offers some potential for addressing the governance concerns. The final part of the chapter, Section 11.7, attempts to draw a few conclusions from this review.

11.2 CSR and corporate governance

As Solomon (2007) points out, there is no single accepted definition of corporate governance, with differences arising according to which country, and associated system of governance, is being considered.[2] In the latter context, broadly speaking there is generally considered to be two distinct systems of governance, although many national systems share at least some characteristics of both 'pure' forms. On the one hand, there are 'insider'-dominated forms of governance whereby companies are owned and controlled by a small number of major shareholders, which might include family members, other companies (most notably in the Japanese keiretsu system), banks and the government. A number of major Western European countries, notably France and Germany, can be considered to exhibit predominantly insider forms of corporate governance.[3] Alternatively, there is the 'outsider' model in which companies are controlled by their managers whilst being owned by a large number of individual shareholders and financial institutions, with the latter increasingly coming to dominate share ownership. The latter system largely prevails in Anglo–American capitalistic economies (upon which this text largely focuses), and is also how we understand the governance of most multi-national corporations (MNCs) regardless of where they are based. It is therefore this latter system with which we will be concerning ourselves in this chapter.

Although we introduced a definition of governance based on Parkinson's (1993) approach, a simpler working definition of corporate governance is offered by the Cadbury Report (Cadbury, 1992), this being 'the system by which companies are directed and controlled'.[4] Corporate governance codes and guidelines were issued in many countries following a spate of corporate scandals throughout the late 20th and into the 21st century.[5] One common feature of

---

[2]The differences between countries are a lot more than simple matters of regulation, and these differences have deep and profound influences on many aspects of these nations' lives. This is especially well illustrated in the work of David Collison (see, for example, Collison *et al.*, 2007, 2010).

[3]On the importance of these internal forms in non-profit sector organisations see Chapter 12.

[4]Further definitions can be found in, for example, Cadbury (1999) and OECD (1999) and a discussion of these can be found in Mallin, 2007.

[5]These scandals ranged from Maxwell Communications Corporation in the late 1980s and early 1990s through to the Enron scandal and on into the international banking crisis(es) (for more details see Mallin, 2007; Solomon, 2007).

these scandals was that those controlling the company, the directors, were perceived to have failed to act in the best interests of shareholders leading, in many cases, to the eventual collapse of the company. In simple terms, the principal (i.e. shareholder) and agent (i.e. director) relationship was seen to have broken down. The ensuing codes and guidelines produced were designed to repair this relationship and addressed various governance weaknesses pinpointed in the individual corporate failures. Particular attention was focused on the functioning of the board of directors. The aim of the codes seems to be to prevent too much control being exercised by particular individuals and to support this with both increased transparency (particularly in terms of risk disclosure), and a strengthening of the external audit function.[6] Additionally, emphasis was placed on the role of institutional investors, as a countervailing power to company management, who were encouraged to adopt a long-term activist approach to share ownership rather than regarding themselves as passive share traders.[7]

Although it is apparent that consideration of corporate governance issues on the part of researchers has started to broaden in coverage with '. . . a change in emphasis away from the traditional shareholder-centric approach towards a more stakeholder-orientated approach' (Brennan and Solomon, 2008: 890), practical reforms in the shape of the codes and guidelines referred to above do not appeared to have followed this path although, as Solomon (2007: 232) argues, it is probable that we should see the increasing concern with CSR in business – at least in part – as some sort of recognition of a widening governance agenda. There is also some suggestion of an interest in greater diversity in boardroom membership and, in particular, drawing non-executive board members from a broader group of constituencies than has traditionally been the case.[8] The suggestion would, we might think, lead to a greater understanding of stakeholder needs and thereby better relationships and, to a degree at least, to some convergence with the German model of two-tier boards. Whether this represents an acceptance of the need for stakeholder inclusion in corporate decision-making, as Solomon (2007), suggests is certainly contestable and, in any event, there is very little evidence of greater boardroom diversity world-wide (see, for example, McDougall and Cumming, 2007). Rather, it would appear that corporate governance reform, especially in countries like the UK and the USA (through the USA's highly publicised Sarbanes-Oxley Act), simply reflects an effort to preserve the status quo of shareholder primacy (Merino *et al.*, 2010).

This focus on shareholder primacy is probably to be expected in the increasingly neo-liberal atmosphere of the Anglo-American economies, and the occasional promises of a more stakeholder (or 'plural') approach (as, for example, was signalled by the UK company law review at the turn of the century – *Modernising Company Law*, cm 5553) tend to be just another false spring. The UK has adopted, what it is pleased to call, an inclusive, or enlightened, shareholder approach (Company Law Review Steering Group, 2000a, b) which entails a requirement that:

> . . . directors have regard to all the relationships on which the company depends and to the long, as well as short term, implications of their actions, **with a view to achieving company success for the benefit of shareholders as a whole**.
>
> (Company Law Review Steering Group, 2000a: viii, emphasis added)

---

[6]Compliance with corporate governance codes is a requirement for listing on most major stock exchanges rather than a legal obligation. Additionally, in the UK for example, the approach taken is not overly prescriptive in that a 'comply or explain' regime prevails whereby departure from recommended best practice is permitted if adequate explanation for such departure is offered.

[7]For a comprehensive and eminently readable guide to UK corporate governance reform see Solomon (2007).

[8]In the UK such an initiative was mooted in The Tyson Report on the Recruitment and Development of Non-Executive Directors (2003).

If one is optimistic, it is perhaps possible to discern the beginnings of an acknowledgement of some form of responsibility towards stakeholders within the ambit of directors' duties. This 'inclusive' concept of directors being responsible to stakeholders, whilst only owing accountability to the company (its shareholders), is one introduced by the South African King Report (2002) on corporate governance. Solomon (2007) considers the approach adopted in this report to be the most forward-looking and progressive of any code of practice and 'admirable in its attempt to address a genuine stakeholder approach to corporate governance' (p. 309). However, the report's drawing of a distinction between accountability to shareholders and responsibility to stakeholders may serve to lessen the impact it may have on CSR.

As we have seen throughout the text, though, there may be some grounds for optimism in the evidence concerning the extent to which CSR is being incorporated into business strategy and decision-making. McDougall and Cumming's (2007) survey of practices adopted by the UK FTSE 100 companies points to the establishment of board level or senior management CSR committees becoming the norm, with only 31% of the sample failing to provide evidence in their annual report of how CSR issues are addressed at board level. Similar evidence is provided by KPMG (2008, 2011, for example) which reports widespread disclosure of a corporate responsibility strategy and almost universal publication of a code of conduct or ethics amongst the largest companies.

The generally positive picture presented above is somewhat tempered by Kolk's (2008) observation that reporting on the incorporation of CSR issues in corporate governance mechanisms tends to be of a general nature rather than detailed information on actual structures and responsibilities being provided. In similar vein, KPMG's (2008) survey notes that only 59% of companies disclosing a code of conduct or ethics go on to report on any instances of non-compliance. Furthermore, it is pointed out that only a minority (43% of the G250 and 27% of the N100 samples, respectively) describe how good governance incorporates CSR, an observation which leads the survey's authors to conclude that '. . . although there seems to be an obvious link between corporate responsibility and good governance, in reality this is not widely recognised' (p. 44).[9]

Clearly, from the evidence of the above, corporate governance systems, whilst beginning to embrace the CSR dimension, will have to evolve much further for such issues to become central to corporate strategy as a whole. Whether further reform will achieve much in the way of empowering stakeholders is also somewhat questionable. Whilst one might expect external representation to increase over time, along with greater use being made of external advisory panels whose views on CSR strategy might be used to inform board level decision-making (see Cooper and Owen, 2007), their role in democratising the CSR process is in any event, arguably, rather limited. The point here is that external participants (as far as may be ascertained from scrutinising corporate publications) are appointed by corporate management, rather than being elected by those they purport to represent and are thereby only accountable to themselves.

Notwithstanding the lack of direct stakeholder representation in core corporate governance forums, the one aspect of CSR that we might infer – whether from surveys or from publication of standalone corporate reports – was becoming normal practice is the engagement in both informal and structured forms of dialogue with their stakeholders. Whether through the wide range of engagement approaches or the published CSR and sustainability reports, organisations are seeking to convey the impression, explicitly or implicitly, that the relationship established with stakeholders by such means is one of accountability of the

---

[9]The G250 sample represents the top 250 companies listed on the Fortune Global 500. Additionally, KPMG's survey includes an N100 sample, being the 100 largest companies by revenue in 22 of the world's major economies.

organisation to the latter (Cooper and Owen, 2007). The following section of the chapter examines the extent to which current engagement and dialogue practice can be considered to enhance accountability to, and empowerment of, non-capital providing stakeholders.

11.3 Stakeholder engagement and the issue of empowerment

Organisational engagement with stakeholders is an increasingly widespread activity and entails a diverse set of practices (see Chapters 5 and 10). The essence of the activity, as we have seen, involves some form of two-way communication between an organisation and each of its stakeholder groups through such approaches as focus groups, meetings, telephone interviews, web feedback forms and so on. Companies claim to utilise stakeholder engagement practices for a variety of purposes. It would appear that the majority of organisations engaging in stakeholder dialogue are simply seeking to obtain a better understanding of stakeholder expectations and identify risks, with somewhat fewer using dialogue to help define their corporate responsibility strategy or as an element in the elaboration of corporate responsibility reports (KPMG, 2008). On the whole, one might, however, expect stakeholder engagement in the reporting domain to continue to increase not least as the trend arising from the influence of the Global Reporting Initiative (GRI) seems unlikely to abate (see Chapter 9) – whether or not their mantle is taken over by other initiatives.

Stakeholder engagement remains a central theme of GRI Guidelines. For example, the 2006 Guidelines (G3) state that 'failure to identify and *engage* with stakeholders is likely to result in reports that are not suitable, and therefore not fully credible, to all stakeholders' (GRI, 2006: 10, emphasis added). Moreover, the Guidelines view reporting as an exercise in stakeholder accountability, a stance unequivocally adopted in their statement of the purpose of sustainability reporting:

> Sustainability reporting is the practice of measuring, disclosing, and being accountable to internal and external stakeholders for organizational performance towards the goal of sustainable development.
>
> (GRI, 2006: 3)

The stakeholder orientation of reporting exercises adopting the GRI approach is further apparent in the reporting principles laid down in the Guidelines. For example, 'inclusiveness' calls for the organisation to identify its stakeholders and explain how it has responded to their reasonable expectations and interests; 'materiality' of reported information is defined in terms of its influence on the assessments and decisions of stakeholders; and 'completeness' is defined in terms of information provided reflecting significant economic, social and environmental impacts sufficient to enable stakeholders to assess organisational performance.

The GRI approach well illustrates the point we opened with concerning the need for both institutional and administrative reform. Institutional reform is represented by the principle of stakeholder engagement and accountability as underpinning the whole reporting process whilst administrative reform is represented by the actual performance indicators themselves which form a central feature of the Guidelines. However, as we noted earlier, whilst many reporting organisations themselves apparently equally subscribe to the notion of reporting being an exercise in stakeholder accountability, economic considerations, as encapsulated in the 'business case', would nevertheless appear to provide the main driving force behind such initiatives. This somewhat seamless intertwining of the very different concepts of accountability and economic gain has attracted the interest of a growing number of academic researchers in recent years. Much of this research has questioned the degree of influence stakeholder engagement exercises do actually have on both the reporting function and, more fundamentally, corporate decision-making processes, with the over-riding conclusion being drawn that a form of 'managerial capture' permeates the whole process (Owen *et al.*, 2001; Baker, 2010).

An early study by Owen *et al.* (2000), drawing upon a series of interviews with leading practitioners and opinion formulators in the CSR reporting field, together with a critical examination of emerging standards and institutional processes, pointed to a discernible degree of managerial capture prevailing, with capture viewed as:

> . . . the concept that sees management take control of the whole process (including the degree of stakeholder inclusion) by strategically collecting and disseminating only the information it deems appropriate to advance the corporate image, rather than being truly transparent and accountable to the society it serves.
>
> (Owen *et al.*, 2000: 85)

Support for this contention is offered by O'Dwyer's (2003) interview based exploration of the perceptions of CSR held by 29 senior executives of Irish public limited companies who, despite instances of individual resistance, demonstrated a clear tendency to interpret the concept in a highly constricted fashion consistent with corporate goals of shareholder wealth maximisation. Later work by the same author (O'Dwyer, 2005) which examined the evolution of a social accounting process in an Irish overseas aid agency (the Agency for Personal Service Overseas) suggests that managerial capture can extend to the reporting process itself. This is evidenced by factors such as the denying of a voice to key stakeholders (notably local communities in developing countries); distrust of management, together with fear of the consequences of dissenting opinions being expressed, serving to inhibit dialogue; and the absence of board level commitment to acting on stakeholder concerns evident in a resistance to the inclusion of any critical comment in the published report.

Further evidence of managerial control and manipulation of stakeholder dialogue is provided by the work of Thomson and Bebbington (2005) and Unerman and Bennett (2004). The former draw on the experience of their own extensive interactions over a considerable time period with reporting organisations to argue that a perceived 'one way' managerial communication process, and associated lack of responsiveness to stakeholder concerns, seriously inhibits the potential for the reporting process to give rise to a change in organisational priorities. For their part, Unerman and Bennett's analysis of Shell's (since discontinued) internet-based stakeholder dialogue forum ('Tell Shell') highlights a fundamental flaw in the process in that it is simply not possible to tell the extent, if any, to which stakeholder views have actually affected corporate decisions:

> . . . internal decision making might be informed by stakeholder views expressed through the web forum [as Shell claims]. However, in the absence of transparency in decision-making processes, the web forum might just be a public relations exercise aimed at enhancing Shell's competitive advantage by attempting to convince economically powerful stakeholders that Shell behaves in a morally 'desirable' manner by taking account of all stakeholder views.
>
> (Unerman and Bennett, 2004: 703).

The lack of transparency in decision-making processes referred to in the Unerman and Bennett study is of fundamental importance in that, as Cooper and Owen (2007) point out, the crucial question concerning the whole engagement and dialogue process from a stakeholder accountability and empowerment perspective is whether participation can meaningfully influence specific aspects of corporate decision-making. In particular, is it possible for decisions to be reached that favour the interests of stakeholders other than shareholders? For the majority of reporting companies, this does not appear to be an issue as they happily subscribe to the business case 'win–win' scenario, whereby no conflict is seen between promoting shareholder interests whilst at the same time being responsive to the needs of other stakeholders. This is, of course, a highly contestable proposition. A somewhat different perspective is, for example, offered by Jones (1999) who most cogently argues that *stakeholder conflict*, rather than *harmony*, permeates much economic activity and, crucially, such conflict is invariably resolved in the favour of shareholders as a powerful combination of external financial hegemony (exerted by the capital markets) and internal bureaucratic control conspire to prevent organisations from being socially responsible in anything other than an instrumental sense.

A rare example of a non-financial stakeholder group appearing to have had some influence on corporate decision-making and, in this case, environmental reporting practice is provided by Deegan and Blomquist (2006). In the context of a detailed case study, which utilises a wealth of interview-based material drawing on the views of all concerned parties, it is illustrated how intervention by the environmental NGO WWF Australia led to revisions to both the Australian Minerals Industry Code for Environmental Management and, subsequently, the reporting practices of individual companies. However, the authors go on to question whether the observed change in reporting practice actually reflects any substantial change in business priorities. In particular, it is suggested that perhaps only fairly cosmetic change within a 'business as usual' framework occurred and that the support of a prominent 'moderate' NGO was simply utilised as a very useful legitimising device in order to deflect and downplay the concerns of more critical stakeholder groups.

The latter observation is a significant one in that in the prevailing climate of *voluntarism* underpinning stakeholder engagement exercises and any related reporting initiatives, corporate management are, of course, quite free to choose which stakeholders to engage with, and which voices to exclude.[10] For a number of commentators, this leads inevitably to a situation where companies confine their attention to the interests of powerful stakeholders, whose actions may have adverse consequences for the organisation, to the exclusion of those of the economically weak. This, it is suggested, is indicative of a focus being placed on issues of stakeholder management rather than stakeholder accountability (see, for example, Unerman and Bennett, 2004; O'Dwyer *et al.*, 2005a).

Despite the serious concerns over deficiencies in stakeholder engagement processes considered above, and associated weaknesses in reporting practice (notably its failure to address the needs of less economically powerful stakeholders), few studies have directly investigated the perceptions of non-capital provider stakeholder groups towards current practice. One notable attempt to address this deficiency is provided by the work of O'Dwyer *et al.* (2005a, b) which employs interview- and questionnaire-based approaches in order to explore the views of prominent Irish social and environmental NGOs concerning the current and potential adequacy of CSR reporting practice to meet their information needs and thereby help them in holding corporations to account. Significantly, respondents in both studies questioned the credibility of current reporting practice and, in particular, expressed the view that all too often there is a failure to provide information on any adverse social and environmental impacts. The over-riding consensus was that disclosure is primarily motivated by corporate self-interest and a concern with issues of stakeholder management rather than representing any genuine desire to discharge accountability to less economically

---

[10]Additionally, of course, certain stakeholder groups may choose to exclude themselves from the engagement process which can have repercussions on the credibility of the reporting exercise itself. Moerman and Van Der Laan (2005), for example, point to the decision of health groups to turn down an invitation to participate in dialogue with the tobacco company BAT. This, they suggest, led to a selective report being produced which ignored, amongst other issues, the harmful effects of smoking.

powerful stakeholders. Therefore, it was argued, the public's 'right to know' about the impacts of corporate activities on their lives is not being satisfied.[11]

The above views expressed by non-financial stakeholders provide some support for Swift's (2001) contention that current stakeholder dialogue and engagement activity can only, at best, deliver a 'soft' form of accountability. The key point here is that with no rights to information built into the process, power differentials between the organisation and its stakeholders remain unaltered. The only way in which reporting deficiencies, and *potential* power differentials, might be addressed in this scenario is via the introduction of mandatory, standardised and externally verifiable reporting. Certainly, this was an approach favoured by the respondents in the O'Dwyer *et al.* studies referred to above and finds favour more widely.[12] What is needed is full transparency of business operations so that public pressure may be applied in order to bring about more socially responsible corporate behaviour. Essentially, what is commonly termed civil regulation is viewed as providing a powerful means of introducing a greater measure of social control over business behaviour.

11.4 Civil regulation and institutional reform

Civil regulation, as Parkinson (2003) explains, relies on market forces rather than direct legal intervention in order to bring about socially responsible business behaviour. The practical advantages of going down this route are that:

> It does not depend on altruism, nor does it require problematic governance reform. Rather it works with the grain of the profit motive, by penalising companies for socially disapproved, and rewarding them for exemplary, conduct.
>
> (Parkinson 2003: 25)

The potential influence of civil regulation is perhaps best exemplified by the high-profile consumer campaigns waged against major companies such as Nestlé, GAP, Apple, Shell and Nike in recent years sparked by accusations of poor employment and marketing practices being operated in developing countries (see also Chapter 10). However, the potential for civil regulation to bring about systematic reform of corporate practice is limited in the absence of a rigorous and consistent social and environmental disclosure regime being established, without which reliance has to be placed on the somewhat *ad hoc* and partial contribution offered by investigative journalism and associated one-off, selective campaigns. Significantly, in common with social accounting researchers such as Gray (see, for example, Gray, 2001), Parkinson argues that such a regime is unlikely to evolve voluntarily and that for civil regulation to be truly effective, the adoption of a mandatory approach towards sustainability (sic) reporting and assurance is needed.

In light of the above, it would appear to be of some significance that survey evidence of current trends in mandatory and voluntary approaches towards social, environmental and (perhaps) sustainability reporting within 30 leading economies (KPMG *et al.*, 2010) does

---

[11]Other studies on this topic include: SustainAbility (2008); SustainAbility/UNEP (1997); GlobeScan/SustainAbility (2012); Heberd and Cobrda (2009); . . . and notably Pleon (2005) which offers a rather more optimistic view of developments.

[12]One illustration is provided by the UK-based Corporate Responsibility Coalition (CORE). CORE's central mission is to seek improvement in UK companies' impacts on people and the environment. CORE has over 130 members including representatives from ethical businesses, women's groups, religious groups, unions, academic and environment, development and human rights groups. Its work is led by a steering group of Amnesty International UK, Action Aid, Friends of the Earth, Traidcraft, War on Want and WWF (GB) (http://corporate-responsibility.org).

indeed discern a stronger regulatory role emerging for the state in ensuring at least a minimum level of disclosure. In particular, the survey is able to identify a total of 142 country standards and/or laws with some form of sustainability-related reporting requirement or guidance, of which two thirds can be classified as mandatory. However, caution is called for on two counts *if* one is tempted to interpret this finding as providing support for the notion of civil regulation providing a practical and effective means of establishing greater levels of social control over corporate conduct.

Firstly, it would appear that the phrase *minimum* level of disclosure referred to above is highly significant. This is particularly evidenced in the wording of the EU Modernisation Directive of 2003 which requires European companies to include non-financial information in their annual financial reports but only *if it is necessary* for an understanding of the company's development, performance or position. Furthermore, no detailed guidance is provided on what such reporting on environmental and employee matters might comprise. Secondly, in the few instances where individual country legislation ventures beyond the vague and generalised approach of the EU Directive, evidence suggests that disclosure practice has failed to meet the aspirations of the legislators. For example, France's New Economic Regulations (NRE) Act (2001) requiring all stock exchange listed companies to include social and environmental information in the management report section of their annual financial report, was followed a year later by detailed disclosure requirements based on a list of 40 indicators derived from the GRI and existing *bilan social* regulations. However, a review of the application of the NRE by the largest 40 companies affected in 2004 clearly suggested a reluctance, or perhaps inability, to provide quantified information, with only 20% of reporters having more than 20 of the 40 indicators specified and 10% none at all (see KPMG *et al.*, 2010: 80). In similar vein, a mandatory requirement introduced in 2007 for Swedish state owned companies to publish a sustainability report in accordance with the GRI G3 Guidelines has met with a somewhat muted response, with the majority complying at only the lowest[13] level (see KPMG *et al.*, 2010: 67).

Although one can expect to find at least some level of improvement in reporting quality as a learning process gradually evolves in response to regulation, Spence (2009) points out that, where social and environmental disclosure has been mandated in the past, compliance has been generally low and the quality of such information that is produced predominantly very poor. Indeed, this is an issue we have drawn attention to throughout the text – for example in Chapter 6 we report on the (non-)disclosure of statutorily required employee-related information by UK companies, a situation largely echoed in studies looking at US and Canadian companies' responses to various mandatory environmental disclosure provisions (see, for example, Freedman and Stagliano, 1995, 2002; Buhr, 2003).

Clearly, such a situation can only be remedied by the introduction of far more stringent enforcement regulations than governments have hitherto been prepared to contemplate. Spence (2009) suggests that this scenario is highly unlikely to be brought about in a world where governments are increasingly in thrall to corporate interests whose financial muscle and lobbying power dwarfs that of non-corporate civil society groups. As Bakan (2004) points out, when corporations lobby government their usual goal is to avoid regulation, either by preventing its introduction altogether or by seeking the repeal or watering down of existing regulations.[14] A neat example of corporate influence seemingly being

---

[13]This is the 'C' level of disclosure for which only a minimum of 10 out of a possible 84 indicators have to be provided.

[14]The subject of corporate influence over government actions has exercised a number of authors. The reader is particularly directed to the work of Hertz (2001), Korten (2001) and Monbiot (2000) whose painstaking analysis presents a damning picture of what amounts to a corporate takeover of the state and accompanying death of democracy.

brought to bear in the specific context of the introduction of social and environmental disclosure regulation is provided by the saga of the UK's OFR reporting requirement. Significantly, the sad fate of the OFR also tells us much about the true nature of corporate commitment to the notion of stakeholder accountability, notwithstanding the rhetoric underpinning many reporting initiatives.

11.5 A very British example: the case of the Operating and Financial Review[15]

The concept that a company should produce a narrative and analytical section as a component of its annual financial reporting package is not a new one. This might typically be known as the Management Discussion and Analysis (MD&A) in the USA whilst in the UK it has developed into the Operating and Financial Review (OFR). Essentially, the OFR is designed to provide more explanatory and interpretive material for the reader of the accounts, particularly concerning future business prospects and likely trends, than is conveyed by the bare figures contained in the financial statements themselves, with the overall aim being to:

> . . . give users of the annual report a more consistent foundation on which to make investment decisions regarding the company.

> (ASB, 2003: 2)

The focus on investors as the prime audience for the OFR suggests that social and environmental information might be unlikely to figure prominently, despite specific reference being made in the ASB statement to issues such as health and safety and environmental protection costs and potential liabilities providing examples of principal risks and uncertainties likely to affect the business and therefore worthy of specific identification. Indeed, Rutherford's (2003) analysis of reporting practice on the part of the Times UK 1000 companies confirmed that very little social and environmental information was included in the OFR. Additionally, Rutherford raised questions about the overall rigour of reporting, noting that many, especially smaller, companies were failing to comply with the ASB's guidance.

The UK government established the Company Law Review Steering Group in the late 1990s to review the law governing corporate activity and disclosure and, in particular, the prevailing voluntary OFR regime. Its final report (2001) called for mandatory publication of an OFR for large companies and recommended disclosure of the following information *to the extent that it is material* in the context of enabling an informed assessment of the business to be made:

- an account of the company's key relationships with employees, customers, suppliers and others on whom its success depends;
- policies and performance on environmental, community, social, ethical and reputational issues including compliance with relevant laws and regulations.

Additionally, whilst emphasising the need for companies to disclose information of sufficient quality to enable shareholders to exercise their powers to call directors to account, the report intriguingly goes on to suggest the need for disclosure to meet the needs of a much broader audience.

---

[15]The material in this section draws heavily on the work of Owen *et al.* (2005) and Cooper and Owen (2007) to which the reader is referred for more detailed analysis.

Others – whether employees, trading partners, or the wider community – also have a legitimate interest in the company's activities, particularly in the case of companies which exercise significant economic power. **Our proposals must also satisfy these wider concerns for accountability and transparency.**

(Company Law Review Steering Group 2001: 48, emphasis added)

Any initial optimism that the UK was moving towards a regime which embraced a wider and more inclusive sense of accountability was then slowly eroded. First, as the proposals made their way through Parliament, the notion of accountability and transparency was narrowed whilst the directors were only faced with a 'duty to consider' the disclosure of policies and performance concerning employment, environmental, social and community issues. Whether these items merited disclosure rested upon the directors' assessment of their 'materiality' in the context of enabling an informed assessment of the business to be made.

All was not yet lost. The Government appointed an independent group of experts which was charged with the task of providing guidance to directors on the disclosure issue. This group produced a consultation document in June 2003 which stated its belief in the importance of taking into account other potential user groups:

> . . . the primary audience for the OFR must be the members, albeit that the information provided in it **will clearly, and rightly, be of direct benefit to other users who may have an interest in the companies affairs**.
>
> (Operating and Financial Review Working Group on Materiality, 2003: paragraph 20, emphasis added)

and called for directors to take a 'broad view' of important stakeholder relationships together with the promotion of the strengthening of accountability for the way social and environmental issues are managed as a key objective of the OFR. Additionally, reference was made to the benefits of stakeholder consultation as a means of exploring stakeholder views on the materiality question along with the efficacy of consulting guidelines and standards used in the context of 'standalone reporting' (for example, the GRI and AA1000 amongst others).

The working group's consultation document was put out for comment and elicited a total of 79 responses from companies, business associations and civil society groups. In analysing these responses, two very conflicting views become apparent. The business response suggests a great deal of unease centred on a perception of a move towards some form of stakeholder reporting model being proposed. Particularly forthright were the Institute of Directors:

> We do not have, nor should we move towards, a stakeholder model. Directors' duties are and will continue to be owed to the members as a whole. The views of, and the impact of the activities of the company on, others ('stakeholders') may well be of relevance in certain circumstances, but the tenor of the Consultation Document is an insidious creep in the direction of the stakeholder model . . .

On the other hand, a very different response to the consultation document can be observed on the part of non-corporate interests. Representatives as diverse as the ACCA, the CORE Coalition and the fair trade organisation Traidcraft did not share the corporate fears of 'an insidious creep' in the direction of a stakeholder model of corporate governance being introduced in the consultation document of the Working Group on Materiality. Rather, they adopt the position that too much discretion is allowed to directors in making social and environmental disclosure decisions concerning OFR content and thereby draw attention to inherent weaknesses in terms of the efficacy of the OFR in promoting civil regulation.

Following the consultation period, the government produced draft regulations in May 2004 and subsequently a Draft Statutory Instrument laid before parliament in January 2005. The government took the opportunity to allay corporate fears concerning any drift towards stakeholder-centric, as opposed to shareholder-centric, reporting, and the draft regulations specified that the objective of the OFR is to 'allow shareholders to assess the company's strategies and their potential to succeed' (paragraph 3.5), expressed even more starkly in the statement that it is 'through shareholders exercising informed influence over companies that their expectations and those of the wider community will best be met' (paragraph 2.3). Furthermore, it is emphasised that the OFR 'must reflect the directors' view of the business' (paragraph 3.32). Finally, the government's commitment to the 'business case' for reporting was made apparent in the impression conveyed throughout the draft regulations (illustrated by reference to a number of potential reporting issues) that, to the extent that it is appropriate to consider social and environmental factors, this is apparently only necessary when financial loss may ensue from ignoring them.[16]

Notwithstanding the government's continuing insistence that the OFR proposals are designed to improve corporate governance via 'improved transparency and accountability' (paragraph 2.2) and that whilst prepared for shareholders it will also be relevant to 'other stakeholders' (paragraph 2.4) it would appear clear that corporate interests have increasingly prevailed as the legislation progressed. Certainly, the concerns of non-corporate interests, as conveyed in their responses to the initial consultation document of the Working Group on Materiality, would appear to have been completely disregarded. Indeed, the whole OFR saga, if nothing else, calls into question any possibility of civil regulation offering a meaningful route towards stakeholder empowerment. This latter observation is re-enforced by the fact that even the business-friendly OFR eventually introduced was regarded by then Chancellor of the Exchequer Gordon Brown as 'an extra administrative cost' on business which he took the opportunity to remove by abolishing the reporting requirement within a year of its introduction. Perhaps fittingly, this decision was announced in a speech made at the CBI's annual conference!

Following the demise of the OFR reporting requirement, the issue of non-financial disclosure within annual financial reports for UK companies is simply governed by the provisions of the EU Modernisation Directive of 2003 which, as we noted earlier, are far from onerous in nature. For the foreseeable future, the main vehicle for extensive corporate social and environmental disclosure will therefore clearly remain purely voluntarily despite the lack of credibility this has with key stakeholder groups (O'Dwyer et al., 2005a, b). Whether some encouragement may perhaps be gleaned from the apparently growing practice of companies including in these voluntary reports an externally prepared assurance statement is considered in the next section.

11.6 Strengthening civil regulation: a role for sustainability assurance?

One way in which social, environmental and sustainability accountability might be developed, if formal regulation is not to be used, is through developing both institutional and administrative reforms *around* the reporting. One important such set of reforms relates to the process of assessing and judging the reporting that takes place. A system of mechanisms

---

[16]For example, in considering environmental health and safety issues the concern is that a poor record 'could adversely affect a company's standing and business prospects' (paragraph 3.33) and that 'the financial loss to the company from poorly managing these issues could be direct . . . indirect . . . or from costs associated with missed opportunities' (paragraph 3.34).

through which the reporting was formally assessed (institutional reforms in effect) would *potentially* help expose the extent to which the reporting was fit-for-purpose. For this to be effective, the quality of that assessment process would need to be substantial (in effect requiring administrative reform as we shall see). This field of endeavour is now known as **assurance** – a broad term whose meaning is often (deliberately?) unclear but which probably does not carry the same rigour of attestation suggested by the term 'audit'. Why that might be is something we will touch upon here.

A comprehensive study of sustainability (sic) reporting practices worldwide undertaken by the ACCA and CorporateRegister.com drew particular attention to the importance of assurance, noting that it

> . . . represents the next stage of development in sustainability reporting as approaches become more developed and demands of report users more sophisticated. Organisations which fail to obtain assurance for their reports are likely to face issues of credibility.
>
> (ACCA/Corporate Register.com, 2004: 15)

In addition to assurance enhancing the credibility of 'sustainability'[17] reporting exercises, it should be pointed out that the reporting organisation itself may experience additional internal benefits, notably 'improved overall management of performance in relation to existing policies and commitments, improved risk management and better understanding of emerging issues' (Zadek *et al.*, 2004: 16). It is, however, the role of assurance in potentially strengthening civil regulation, and thereby contributing towards the empowerment of stakeholders, that is of particular concern to us here.

Key trends in sustainability assurance practice

The increasing prominence of sustainability assurance[18] provision over the years continues to be highlighted and commented upon in a number of reports. For example, CorporateRegister.com's (2008) survey of assurance practice utilising their comprehensive reports directory (profiling 90–95% of all published non-financial reports) points to an annual growth rate of approximately 20% between 1997 and 2007. KPMG's series of triennial surveys of corporate responsibility reporting practice indicates that, for the G250 sample at least (that is for the world's largest 250 companies), this trend continues with the number of reports including a formal assurance statement rising from 30% in 2005 to 46% in 2011 – although notably still less than 50%. However, as is the case with reporting itself, the level of assurance provision for the top 100 company sample[19] is somewhat lower (38% in 2011, for example). Whether this level of assurance is changing is not clear (see, for example, KPMG, 2008). Although caution should be exercised in interpreting the data from top 100 companies, as more (and different) countries form the basis of the 2011 sample as compared with that of 2008, what

---

[17]The relationship of 'sustainability' reporting with sustainability as understood in the Brundtland sense arguably remains elusive (see especially Chapters 1 and 9). The term 'sustainability' is employed very generally and loosely to mean social and environmental issues and, just possibly, triple bottom line (TBL) issues (see Chapters 3 and 9). The ubiquity of the term makes it difficult to avoid, but that does not mean the term should be used uncritically (Milne and Gray, 2012). For an introduction to governance driven by a direct consideration of sustainability see, for example, Kemp *et al.* (2005).

[18]This is a case in point of the use of the term 'sustainability' without either a direct reference to sustainability or some form of qualifying statement. A sustainability assurance might be expected to provide insight into the extent to which an organisation has or has not succeeded in reporting upon its contributions to un-sustainability. If one expected this, one would be (almost) entirely disappointed.

[19]That is, KPMG take a sample comprising the largest 100 companies in a wide range of countries. Inevitably these are, on average, somewhat smaller than the companies comprising the largest world 250 sample.

they do point to are major differences in the prevalence of assurance across individual countries. For example, whereas approximately two thirds of corporate responsibility reports from Denmark, Spain and Italy included an external assurance statement, fewer than a quarter of those from Japan, Canada and the USA did so. In addition to differences in the extent of assurance provision at individual country level, KPMG's (2011) survey also highlights differences at industrial sector level, with, for example, mining and utility companies being more likely to commission assurance than those from other sectors. A further observation to make concerning sustainability assurance provision is that, as Simnett *et al.* (2009) point out, it is very much a large company phenomenon. Indeed, at country level, when one ventures beyond the top 100 companies the incidence of assurance declines fairly dramatically.[20]

The major international accounting firms (the Big Four) dominate the market for the provision of assurance (KPMG, 2011) providing, for example, over 70% of assurance statements in KPMG's G250 sample. The majority of other assurance statements were provided by certification bodies, firms of technical experts and 'specialist' assurance providers. Interestingly, this pattern of assurance provision differs significantly from that found in an earlier study carried out for CPA Australia (2004) which drew on a comprehensive database of 170 assurance statements appearing between 2000 and 2003, predominantly from Australia, the UK, mainland Europe and Japan. In this latter study, whilst accountants provided the majority of assurance statements for the mainland Europe sample, they were very much in the minority for the Australian (15%), Japanese (37.5%) and UK (23%) samples.

There is, however, some evidence that provision patterns have changed markedly since publication of the CPA Australia study. For example, as far as the UK is concerned, later survey work (Owen *et al.*, 2009; Cooper and Owen, submitted) indicates that since 2005 accountants have taken an ever increasing share of the market to the extent that, at least for the FTSE 100 companies, they now provide around two thirds of assurance statements published. O'Dwyer (2011) similarly draws attention to the increasingly dominant position of the Big Four in the global sustainability assurance market. He puts this down to the fact that, as sustainability reports have become more significant and complex documents, reporters have increasingly gone for Big Four assurance providers on the grounds of their size (including geographical scope) together with their perceived reputation and assurance competencies. The ever-growing market dominance of accounting firms who tend to adopt a fairly standard and rigid approach towards sustainability assurance provision is not insignificant in terms of the potential for such assurance to strengthen civil regulation, as we shall shortly see.

The development of sustainability assurance standards

Early academic studies which examined the initial wave of assurance practice on environmental reports published in the 1990s (Ball *et al.*, 2000; Kamp-Roelands, 2002) raised serious concerns over its rigour and usefulness. Kamp-Roelands, for example, highlighted major inconsistencies in the subject matter addressed, scope of the exercise, objectives, assurance criteria and procedures applied, level of assurance provided and the wording of opinions offered. For their part, Ball *et al.* raised even more fundamental question marks over the key issues of assuror independence and thoroughness of the work carried out. Additionally, they drew attention to a perceived high degree of managerial control pervading the whole

---

[20]Looking at the UK, for example, Salterbaxter and Context's (2005) survey of reporting trends indicates that whilst 44 of the top 100 companies' reports contained an assurance statement, the figure only increases to 60 when the sample is extended to the top 250.

assurance process with, for example, an emphasis being placed on management systems as opposed to performance-based issues. This 'managerial turn', they argued, greatly limited the potential of assurance to act as a vehicle for enhancing corporate transparency and accountability to stakeholders.

A major problem facing early assurance providers lay (allegedly) in the absence of clear standards or guidelines that could be used to govern the approach adopted. This concern has been addressed in recent years by the issuing of sustainability assurance practice guidelines. Significantly, these fall into two distinct categories. Firstly, we have the somewhat cautious accountancy-based approach, particularly exemplified by the Fédération des Experts Comptables Européens (FEE, 2002) and the International Auditing and Assurance Standards Board (IAASB, 2004), which is particularly concerned with attesting the accuracy of published data and minimising assuror liability. Secondly, there are the AccountAbility series of standards (AccountAbility, 1999, 2003, 2008a, b, 2011) where, in stark contrast, the issue of stakeholder accountability is absolutely central to the assurance process.

The accountancy-based guidelines are largely informed by traditional financial auditing concepts and standards and very much concerned with formalising the structure of assurance statements so as to avoid creating any expectations gap 'whereby a user mistakenly assumes that there is more assurance than is actually present' (FEE, 2002: 17). To this end, amongst key suggested elements of an assurance statement are a description of the scope and objective of the exercise, the identification of criteria used to assess evidence and reach a conclusion, outlining of the respective responsibilities of reporter and assurance provider and specification of the assurance standards used.

The most detailed, and most influential in terms of practical impact, accountancy-based standard is the IAASB's (2004) ISAE 3000 which applies to all assurance engagements other than audits and reviews of historical financial information and whose application is mandatory for all assurance reports issued by professional accounting bodies. ISAE 3000 provides detailed guidance for conducting assurance work from initial acceptance of the engagement through to the issuance of the assurance statement itself. Of particular note in terms of the latter is the distinction drawn between 'reasonable assurance' engagements and 'limited review' engagements and the related nature of the conclusions that may be respectively drawn:

> The objective of a reasonable assurance engagement is a reduction in assurance engagement risk to an acceptably low level in the circumstances of the engagement as the basis for a **positive** form of expression of the practitioner's conclusion. The objective of a limited assurance engagement is a reduction in assurance engagement risk to a level that is acceptable in the circumstances of the engagement, but where that risk is greater than for a reasonable assurance engagement, as a basis for a **negative** form of expression of the practitioner's conclusion.
>
> (IAASB, 2004: paragraph 2, emphasis added)

An example of a negative form of conclusion would be 'nothing has come to our attention that causes us to believe that figures appearing in the report are not fairly stated'. The contrasting positive form would simply be 'figures in the report are fairly stated'. Clearly, the latter offers a far greater level of endorsement as, amongst other things, the significance of nothing coming to the assurance provider's attention depends on the depth, and rigour, of the exercise carried out, which is far from easy to ascertain simply by scrutinising the guarded statement generally offered.

A further feature of ISAE 3000 is that it is generic in nature in that it applies to a wide range of non-financial assurance exercises rather than being exclusively concerned with sustainability assurance. Therefore, although the standard notes, for example, that 'considering materiality requires the practitioner to understand and assess what factors might influence the decision of intended users' (paragraph 23), no specific attention is paid to issues such as stakeholder inclusion in the assurance process. Interestingly here, FEE (2004, 2006, 2009) has issued several, hitherto unheeded, calls for a specific international assurance standard for sustainability reports to be developed, whilst a small number of national standard setters have indeed issued such a standard, with the issue of stakeholder needs featuring prominently in those emanating from Sweden and the Netherlands (see FEE, 2006).

Notwithstanding the above-mentioned tentative moves on the part of the accounting profession towards the development of specific sustainability standards, the most developed stakeholder-centred approach to sustainability assurance appears within the AccountAbility series of standards referred to earlier. At the outset, it is made clear that the purpose of assurance is to provide:

> A comprehensive way of holding an organisation to account for its management, performance and reporting of sustainability issues by evaluating the adherence of an organisation to the AA1000 Accountability Principles and the quality of the disclosed information on sustainability performance.
>
> (AccountAbility, 2008a: 6)

The stakeholder-centred approach to reporting and assurance issues is particularly explicit in the definition applied to the Foundation Principle of 'Inclusivity'.

> For an organisation that accepts its accountability to those on whom it has an impact and those who have an impact on it, inclusivity is the participation of stakeholders in developing and achieving an accountable and strategic response to sustainability.
>
> (AccountAbility, 2008b: 7).

Indeed, the stakeholder focus is further emphasised in the other two Principles advanced – these being materiality, 'determining the relevance of an issue to an organisation and its stakeholders' (AccountAbility, 2008b: 9) and responsiveness, 'an organisation's response to stakeholder issues that affect its sustainability performance and is realised through decisions, actions and performance as well as communication with stakeholders' (AccountAbility, 2008b: 11).

Of further significance in the AA1000 approach to assurance is the suggestion that assurance providers engage directly with stakeholders in cases where high-level assurance is sought.[21] Additionally, assurance providers are encouraged to offer evaluative comment on the reporting organisation's systems and processes and to highlight perceived strengths and weaknesses in both the reporting and performance domains. In sum, a strategic and 'value added' perspective on assurance is offered which focuses on the usefulness of the report for stakeholders and is explicitly concerned with driving future performance. However, perhaps a note of caution should be sounded here in that such an approach carries the danger of combining what is essentially a consultancy function with a separate 'arms length' assurance exercise, thereby potentially compromising the integrity of the latter (O'Dwyer and Owen, 2005). Particularly problematical is a tendency within some assurance statements to include praise for the organisation's achievements which may well serve to undermine the perceived independence of the assurance provider.

---

[21]Two levels of assurance are specified in the standards, high and moderate, which are designed to be consistent with the reasonable and limited assurance levels of ISAE 300.

A critique of current sustainability assurance practice from a stakeholder empowerment perspective

Given the dominance of the sustainability assurance market by the accounting profession, it comes as no surprise to learn that ISAE 3000 is by far the most widely employed assurance standard (KPMG, 2011) and, perhaps equally, that only 'limited assurance' is provided in the vast majority of cases.[22] Of equal concern from a stakeholder empowerment perspective is the observation that, to the extent that assurance statements are addressed to any constituency at all, they are addressed to corporate management rather than external stakeholder groups, which implies that the former are the key perceived beneficiaries of the exercise.

Concern over the efficacy of social, environmental and (un)sustainability assurance in enhancing corporate transparency and accountability is heightened by the findings of a growing body of empirical research into the nature of assurance practice. Initially, such investigations were 'desk-based' and focused upon published assurance statements. The CPA Australia study referred to earlier is one such example which pointed to a number of problem areas including:

- variability in the title of assurance statements;

- a tendency not to identify an addressee;

- a wide range of objectives for, and scope of, the assignment (with the latter typically prescribed by management);

- variation in the amount of description of the nature, timing and extent of procedures employed, and variability in the wording of conclusions offered.

Somewhat damningly, the authors conclude that on the basis of their findings readers of assurance statements would 'often have great uncertainty in understanding how the assurance provider undertook the engagement, what they reviewed and what was the meaning of their conclusion' (p. 67).

A similar variability in practice is observed in O'Dwyer and Owen's (2005) study of 41 assurance statements which, somewhat significantly, appeared in the reports of 'leading edge' companies short-listed for the 2002 ACCA UK and European Sustainability Reporting Awards Scheme. O'Dwyer and Owen do acknowledge some improvements in assurance practice since the earlier Ball *et al.* (2000) study, notably in terms of the scope of work carried out and independence of the assurance provider, but also point to continuing managerial control over the exercise. In particular, it is noted that management appoints the assurance providers and can thereby place any restrictions they wish on the engagement, a minimal level of stakeholder involvement prevails and little in the way of evaluations of corporate responsiveness to stakeholder concerns is being offered.

Subsequent empirical work moved into fieldwork to explore the perceptions of key actors in the assurance process – corporate managers, stakeholder representatives and assurance providers . Owen *et al.*'s (2009) study featuring a series of interviews with senior corporate responsibility managers from ten FTSE 100 companies together with eight representatives of three key stakeholder groups (investors, NGOs and the trade union movement) raises a number of concerns relating to the role of assurance in promoting stakeholder

---

[22]That is, featuring negative form opinions – although in a small number of instances specific elements of the report may carry reasonable level assurance.

empowerment.[23] Evidence from the interviews with corporate managers suggests that the major driving force behind assurance is internal in nature, with major benefits being perceived as improvements in information and reporting systems together with increased confidence in the integrity and reliability of corporate data released into the public domain. Above all, the exercise must provide value for money. Significantly, the interviewees accepted that stakeholders were detached from the assurance process and didn't generally read assurance statements. The utilisation of stakeholder panels as an integral feature of the assurance process was suggested, providing a possible means of countering this problem, although even here it was recognised that there are major difficulties concerning how representative of (possibly widely diverging) stakeholder views such panels can be.[24]

Owen *et al.* (2009) also found that the different categories of stakeholder representatives themselves had markedly differing perceptions concerning the value of assurance. Those from the investment community, for example, viewed the exercise as having little relevance for their decision-making needs and were not particularly concerned with the issue of stakeholder inclusion in the process. Even more dismissive was the trade union official who expressed fundamental reservations concerning both the competency of assurance providers and the institutional legitimacy of the assurance industry itself, which he perceived as being dominated by corporate interests and the accounting 'industry'. Whilst the four NGO representatives were generally more supportive of assurance, and expressed a willingness to become involved in the process through inclusion on stakeholder panels, in common with the management respondents they foresaw problems in such panels being truly representative of stakeholder opinion. Additionally, concern was expressed that membership might compromise their independence whilst resource constraints could well give rise to 'stakeholder fatigue' should panels become widespread practice.

Most fundamentally, Owen *et al.* (2009) pointed to a stark difference in opinion between corporate managers and stakeholders concerning the issue of to whom assurance statements should be addressed. The former were clear that as commissioners of the exercise, assurance statements should be addressed to them, hence confirming the internally driven nature of the exercise from their perspective. However, for their part, stakeholders were strongly of the view that, as sustainability reporting itself is purportedly addressed to society at large, the accompanying assurance statement should be addressed to the same constituency.

Studies by Edgley *et al.* (2010) and O'Dwyer *et al.* (2011) focus on perceptions of assurance providers themselves and paint perhaps a slightly more positive picture concerning the efficacy of assurance in promoting stakeholder empowerment. Edgley *et al.*'s study, for example, based on a programme of 20 interviews with both accountant and consultant assurance providers' elicited an overall perception that stakeholders do benefit from assurance and that their views are incorporated into the process. Indeed, it is suggested that assurance represents 'a learning process for company management and stakeholders with consequent changes in managerial behaviour' (p. 553). Furthermore, a number of respondents appeared to consider themselves as acting as the 'voice' of the stakeholders and thereby being key agents in promoting their interests to corporate management. However, a note of caution should be sounded here in that, whereas the majority of respondents were of the view that stakeholder inclusion in the assurance process is likely to increase, the accountant assurors tended to be mostly concerned with the contribution stakeholders might offer to companies'

---

[23]Clearly the small-scale nature of this study suggests that any findings should be interpreted with caution. There is certainly scope for further, larger scale, studies focusing in particular on stakeholder views concerning the assurance process.

[24]Additionally, membership of such panels is highly likely to be decided by company management, a factor which (as we noted earlier in this chapter) calls into question the real degree of stakeholder accountability thereby established.

systems of internal control, materiality decisions and management processes. Additionally, accountant assurors, in contrast to their consultant counterparts, tended to favour incorporating stakeholder views into the assurance process through indirect mechanisms (such as perusing stakeholder feedback received by the client company) rather than by engaging directly with stakeholders. A final impediment to the desired stakeholder inclusion in the assurance process, noted in the study and worthy of particular mention, is simply the 'ignorance' or complete indifference of stakeholders themselves.

O'Dwyer *et al.*'s study, which features a series of interviews with sustainability assurance (sic) practitioners in one large professional services firm, whilst pointing to a similar commitment to opening up dialogue within the assurance process on the part of these practitioners, also identifies a number of obstacles towards achieving this aim. Foremost amongst these are firstly (in common with Edgley *et al.*'s findings) stakeholder indifference pervading the whole process. Indeed, practitioners are faced with having to construct 'a somewhat mythical audience' in order to convince corporate management of the continued necessity for assurance once any initial financial value derived from improvement in internal systems and enhancement of external credibility has been realised. Secondly, there is a 'strong resistance from [the professional services firm's] Risk Department [responsible for approving the wording of assurance statements] to the expansion of assurance statement content and, relatedly, moving towards higher levels of assurance' (p. 49). The attitude of the Risk Department here, with their concern over potential liability for misleading statements being issued, is perhaps understandable in view of the fact that fees for sustainability assurance are generally a small fraction (less than 10%) of those for the financial audit (see Park and Brorson, 2005; Cooper and Owen, submitted), a factor which arguably precludes the carrying out of the necessary amount of substantive testing to justify a positive form of conclusion.

The analysis presented above, notwithstanding the ambitions on the part of some assurance providers to move towards a more stakeholder inclusive form of assurance, suggests that, in its present guise, sustainability assurance offers little in the way of strengthening civil regulation and consequent stakeholder empowerment. With the market increasingly dominated by the Big Four accounting firms, bound by the dictates of ISAE 3000, limited form assurance, which conveys little in the way of useful information to stakeholders, generally prevails. Whilst, admittedly, there is a discernible trend for the AA1000 assurance standard to be used in conjunction with ISAE 3000 (see Cooper and Owen, submitted) stakeholder inclusion in the assurance process remains largely of an indirect nature and certainly falls far short of that envisaged in AA1000. Overall, sustainability assurance practice is driven by management who define the scope of the exercise and to whom (in cases where an audience is identified) the assurance statement is addressed. Small wonder therefore that empirical studies have pointed to a general level of stakeholder indifference to the whole process.

Most fundamentally, from a civil regulation perspective, the whole case for sustainability assurance would seem to be based on persuading corporate management as to its efficacy as a driver of improved financial value. Certainly, no transfer of power whereby stakeholders can hold the organisation to account and actively enforce responsiveness to their concerns is contemplated. Adams and Evans (2004), in a detailed analysis of the shortcomings of sustainability assurance as a vehicle for enhancing stakeholder accountability, have indeed suggested ways of transferring some degree of power over the process by, for example, enabling stakeholders to appoint assurance providers and to determine the scope of the exercise. Neither of these suggestions has, of course, found favour in practice. However, even in the unlikely event they should do so it still begs the question as to how stakeholders could use the assurance findings in order to influence organisational decision-making.

11.7 Summary and conclusions

What the review in this chapter brings sharply into focus is the almost entirely voluntary nature of social, environmental and sustainability reporting and assurance. Consequently, in the absence of substantive institutional reform – whether through regulation or other means – we should not be surprised to discover that reporting and assurance take place primarily at the whim of management as and when the management of organisations find such mechanisms useful to them. Equally, it is quite apparent that from a stakeholder accountability perspective the state of reporting and assurance is far from satisfactory.

To recap, there is very little substantive regulation that governs the accountability of organisations to their stakeholders for their social, environmental and (un)sustainability activities. There has been a range of substantial attempts by different elements of society to develop administrative reform (through, for example, developing the GRI guide to reporting) as well as institutional reform (through such initiatives as the AA1000 standards on stakeholder consultation and the various guides on assurance). If any of these various initiatives had succeeded in rigorously addressing *both* the detailed content of reporting and assurance *and* the widespread adoption and monitoring of reporting and assurance then the state of stakeholder social, environmental and sustainability accountability would be unrecognisable from its present state. But each component is either less complete than it needs to be and/or has been unable of itself to challenge the shareholder and management power that prevents accountability developing. Although there have been developments in governance, it is not yet apparent that governance of any substance is taking place.

The considerable strides made in both administrative and institutional reform over the last 20–30 years should not blind us to realisation that the discharge of a substantive social, environmental and (un)sustainability accountability is still a long way off (Owen, 2008). Furthermore, as this chapter has shown, the key to any developments of worth will be the recognition that learning *how* to do social, environmental and sustainability accounting and reporting (administrative reform) is empty without structures that ensure that it takes place in a sensible manner (institutional reform).

No amount of administrative reform can remedy . . . political errors, although . . . administrators . . . will be expected miraculously to produce good out of bad and success out of unavoidable failure. Administrative reform cannot substitute for political or economic or institutional reform. On the other hand, political, economic and institutional reforms can rarely succeed without administrative reform. Administration is not neutral or merely instrumental. It has a life of its own . . .

(Caiden, 1991: 11)

References

ACCA/Corporate Register.com (2004) *Towards Transparency: Progress on Global Sustainability Reporting.* London: Association of Chartered Certified Accountants.

AccountAbility (1999) *AA1000 Framework; Standard, Guidelines and Professional Qualification.* London: AccountAbility.

AccountAbility (2003) *AA 1000 Assurance Standard.* London: AccountAbility.

AccountAbility (2008a) *The AA1000 Accountability Principles Standard 2008.* London: AccountAbility.

AccountAbility (2008b) *AA 1000 Assurance Standard 2008.* London: AccountAbility.

AccountAbility (2011) *AA1000 Stakeholder Engagement Standard.* London: AccountAbility.

Adams, C. A. and Evans, R. (2004) Accountability, completeness, credibility and the audit expectations gap, *Journal of Corporate Citizenship*, **14**: 97–115.

ASB (2003) *Operating and Financial Review*. London: Accounting Standards Board.

Bakan, J. (2004) *The Corporation: The Pathological Pursuit of Profit and Power*. London: Constable.

Baker, M. (2010) Re-conceiving managerial capture, *Accounting, Auditing and Accountability Journal*, **23**(7): 847–67.

Ball, A., Owen, D. L and Gray, R. H. (2000) External transparency or internal capture? The role of third party statements in adding value to corporate environmental reports, *Business Strategy and the Environment*, **9**(1): 1–23.

BITC (2003) *Indicators that Count*. London: Business in the Community.

Blowfield, M. and Murray, A. (2011) *Corporate Responsibility*, 2nd Edition. Oxford: Oxford University Press.

Brennan, N. M. and Solomon, J. (2008) Corporate governance, accountability and mechanisms of accountability: an overview, *Accounting, Auditing and Accountability Journal*, **21**(7): 885–906.

Buhr, N. (2003) Mandatory environmental disclosure: the current practice in Canada with a comparison to the United States, *Accountability Quarterly*, No. 21, September.

Cadbury Sir A. (1999) *Corporate Governance Overview*. World Bank Report.

Cadbury, Sir A. (2002) *The Report of the Committee on the Financial Aspects of Corporate Governance: The Code of Practice*. London: Gee Professional Publishing.

Caiden, G. E. (1991) *Administrative Reform Comes of Age*. New York: de Gruyter.

Collison, D., Dey, C., Hannah, G. and Stevenson, L. (2007) Income inequality and child mortality in wealthy nations, *Journal of Public Health*, **29**(2): 114–17.

Collison, D., Dey, C., Hannah, G. and Stevenson, L. (2010) Anglo-American capitalism: the role and potential of social accounting, *Accounting, Auditing and Accountability Journal*, **23**(8): 956–81.

Company Law Review Steering Group (2000a) *Modern Company Law for a Competitive Economy: Developing the Framework*. London: DTI.

Company Law Review Steering Group (2000b) *Modern Company Law for a Competitive Economy: Completing the Structure*. London: DTI.

Company Law Review Steering Group (2001) *Modern Company Law for a Competitive Economy: Final Report*. London: DTI.

Cooper, S. M. and Owen, D. L. (2007) Corporate social reporting and stakeholder accountability: the missing link, *Accounting, Organizations and Society*, **32**: 649–67.

Cooper, S. M. and Owen, D. L. (submitted) Independent assurance practice of sustainability reports, in Unerman, J., Bebbington, J. and O'Dwyer, B. (eds), *Sustainability Accounting and Accountability*, 2nd Edition. Abingdon: Routledge (forthcoming).

CorporateRegister.com (2008) *Assure View: The CSR Assurance Statement Report*. London: CorporateRegister.com.

CPA Australia (2004) *Triple Bottom Line: A Study of Assurance Statements Worldwide*. Melbourne: CPA Australia.

Deegan, C. and Blomquist, C. (2006) Stakeholder influence on corporate reporting: An exploration of the interaction between WWF-Australia and the Australian minerals industry, *Accounting, Organizations and Society*, **31**: 343–72.

Edgeley, R. R., Jones, M. J. and Solomon, J. (2010) Stakeholder inclusivity in social and environmental report assurance, *Accounting, Auditing and Accountability Journal*, **23**(4): 532–57.

FEE (2002) *Providing Assurance on Sustainability Reports*. Brussels: Federation des Experts Comptables Europeens.

FEE (2004) *FEE Call for Action: Assurance for Sustainability*. Brussels: Federation des Experts Comptables Europeens.

FEE (2006) *Key Issues in Sustainability Assurance: An Overview*. Brussels: Federation des Experts Comptables Europeens.

FEE (2009) *Policy Statement: Towards a Sustainable Economy: The Contribution of Assurance*. Brussels: Federation des Experts Comptables Europeens.

Freedman, M. and Stagliano, A. J. (1995) Disclosure of environmental clean-up costs: the impact of the Superfund Act, *Advances in Public Interest Accounting*, **6**: 163–78.

Freedman, M. and Stagliano, A. L. (2002) Environmental disclosure by companies involved in initial public offerings, *Accounting, Auditing and Accountability Journal*, **15**(1): 94–105.

GlobeScan/SustainAbility (2012) *Rate the Raters: Polling the Experts.* Toronto and New York: GlobeScan and SustainAbility.

Gray, R. H. (2001) Thirty years of social accounting, auditing and reporting: what (if anything) have we learned?, *Business Ethics: A European Review*, **10**(1): 9–15.

GRI (2006) *Sustainability Reporting Guidelines.* Amsterdam: Global Reporting Initiative.

Hebard, A. J. and Cobrda, W. S. (2009) *The Corporate Reality of Consumer Perceptions: Bringing the consumer perspective to CSR reporting.* GreenBiz.com:Earthsense LLC (http://www.greenbiz.com/sites/default/files/document/GreenBizReports-ConsumerPerceptions.pdf).

Hertz, N. (2001) *The Silent Takeover: Global Capitalism and the Death of Democracy.* London: William Heinemann.

IAASB (2004) *International Standard on Assurance Engagements 3000: Assurance Engagements on Other than Audits or Reviews of Historical Information.* New York: International Federation of Accountants.

Jones, M.T, (1999) The institutional determinants of social responsibility, *Journal of Business Ethics*, **20**: 163–79.

Kamp-Roelands, N. (2002) *Towards a Framework for Auditing Environmental Reports*, unpublished PhD thesis. Tilburg University, The Netherlands.

Kemp, R., Parto, S. and Gibson, R. B. (2005) Governance for sustainable development: moving from theory to practice, *International Journal of Sustainable Development*, **8**(1/2): 12–30.

King Report, The (2002) *The King Report on Corporate Governance for South Africa.* Parktown, South Africa: Institute of Directors in Southern Africa.

Kolk, A. (2008) Accountability and corporate governance: exploring multinationals' reporting practice, *Business Strategy and the Environment*, **17**(1): 1–15.

Korten, D. C. (2001) *When Corporations Rule the World*, 2nd Edition. Bloomfield, CT: Kumarian Press.

KPMG (2008) *KPMG International Survey of Corporate Responsibility Reporting 2008.* Amsterdam: KPMG International.

KPMG (2011) *KPMG International Survey of Corporate Responsibility Reporting 2011* (kpmg.com).

KPMG, Unit for Corporate Governance in Africa, Global Reporting Initiative and United Nations Environment Programme (2010) *Carrots and Sticks - Promoting Transparency and Sustainability: An Up-date on Trends in Voluntary and Mandatory Approaches to Sustainability Reporting* (http://www.kpmg.com/ZA/en/IssuesAndInsights/ArticlesPublications/Advisory-Publications/Documents/Carrots_Sticks_2010.pdf).

Mallin, C. (2007) *Corporate Governance*, 2nd Edition. Oxford: Oxford University Press.

McDougall, L. and Cumming, J. F. (2007) *Incorporating CSR into Business Strategy and Decision-Making: Current FTSE 100 Board Level Practice.* London: Article 13.

Merino, B. D., Mayper, A. G. and Tolleson, T. D. (2010) Neoliberalism, deregulation and Sarbanes-Oxley: The legitimation of a failed corporate governance model, *Accounting, Auditing and Accountability Journal*, **23**(6): 774–92.

Milne, M. and Gray, R. (2012) W(h)ither ecology? The triple bottom line, the Global Reporting Initiative, and corporate sustainability reporting, *Journal of Business Ethics*, November SSN 0167-4544 DOI 10.1007/s10551-012-1543-8.

Moerman, L. and Van der Laan, S. (2005) Social reporting in the tobacco industry: all smoke and mirrors, *Accounting, Auditing and Accountability Journal*, **18**(3): 374–89.

Monbiot, G. (2000) *Captive State: The Corporate Takeover of Britain.* London: Macmillan.

O'Dwyer, B. (2003) Conceptions of social responsibility: the nature of managerial capture, *Accounting, Auditing and Accountability Journal*, **16**(4): 523–57.

O'Dwyer, B. (2005) The construction of a social account: a case study in an overseas aid agency, *Accounting, Organizations and Society*, **30**: 279–96.

O'Dwyer, B. (2011) The case of sustainability assurance: constructing a new assurance service, *Contemporary Accounting Research*, **28**(4): 1230–66.

O'Dwyer, B and Owen, D. L. (2005) Assurance statement practice in environmental, social and sustainability reporting: a critical evaluation, *British Accounting Review*, **37**(2): 205–29.

O'Dwyer, B., Unerman, J. and Bradley, J. (2005a) Perceptions on the emergence and future development of corporate social disclosure in Ireland: Engaging the voices of non-governmental organisations, *Accounting, Auditing and Accountability Journal*, **18**(1): 14–43.

O'Dwyer, B., Unerman, J. and Hession, E. (2005b) User needs in sustainability reporting: perspectives of stakeholders in Ireland, *European Accounting Review*, **14**(4): 759–87.

O'Dwyer, B., Owen, D. L. and Unerman, J. (2011) Seeking legitimacy for new assurance forms: the case of assurance on sustainability reporting, *Accounting, Organizations and Society*, **36**: 31–52.

OECD (1999) *Principles of Corporate Governance*. Paris: Organisation for Economic Cooperation and Development.

Operating and Financial Review Working Group on Materiality (2003) *A Consultation Document*. London: DTI.

Owen, D. L. (2008) Chronicles of wasted time? A personal reflection on the current state of, and future prospects for, social and environmental accounting research, *Accounting, Auditing and Accountability Journal*, **21**(2): 240–67.

Owen, D. L., Gray, R. H. and Bebbington, J. (1997) Green accounting: cosmetic irrelevance or radical agenda for change?, *Asia Pacific Journal of Accounting*, **4**(2): 175–98.

Owen, D., Swift, T., Humphrey, C. and Bowerman, M. (2000) The new social audits: accountability, managerial capture or the agenda of social champions?, *The European Accounting Review*, **9**(1): 81–98.

Owen, D., Swift, T. and Hunt, K. (2001) Questioning the role of stakeholder engagement in social and ethical accounting, auditing and reporting, *Accounting Forum*, **25**(3): 264–82.

Owen, D. L., Shaw, K, and Cooper, S. (2005) *The Operating and Financial Review: A Catalyst for Improved Corporate Social and Environmental Disclosure?* Research Report 89. London: ACCA.

Owen, D. L., Chapple, W. and Urzola, A. P. (2009) *Key Issues in Sustainability Assurance*, Research Report 115. London: ACCA.

Park, J. and Brorson, T. (2005) Experiences of and views on third party assurance of corporate environmental and sustainability reports, *Journal of Cleaner Production*, **13**: 1095–106.

Parkinson, J. E. (1993) *Corporate Power and Responsibility*. Oxford: Oxford University Press.

Parkinson, J. (2003) Disclosure and corporate social and environmental performance: competitiveness and enterprise in a broader social frame, *Journal of Corporate Law Studies*, 3(1): 3–39.

Pleon (2005) *Accounting for Good: The Global Stakeholder Report 2005*. Amsterdam: Pleon.

Power, M. (1994) Constructing the responsible organisation's accounting and environmental representation, in Feubner, G., Farmer, L. and Murphy, D. (eds), *Environmental Law and Ecological Responsibility: The Concept and Practice of Ecological Self Organisation*, pp. 370–92. Chichester: Wiley.

Rutherford, B. A. (2003) *Half the Story: Progress and Prospects for the Operating and Financial Review*. London: ACCA.

Salterbaxter and Context (2005) *Trends in CSR Reporting 2003-2004*. London: Salterbaxter and Context.

Simnett, R., Vanstraelen, A. and Chua, W. F. (2009) Assurance on sustainability reports: An international comparison, *The Accounting Review*, **84**: 937–67.

Solomon, J. (2007) *Corporate Governance and Accountability*, 2nd Edition. Chichester: Wiley.

Spence, C. (2009) Social accounting's emancipatory potential: a Gramscian critique, *Critical Perspectives on Accounting*, **20**(2): 205–27.

SustainAbility (2008) *Count me in: the reader's take on sustainability reporting* (pdf) (www.globalreporting.org/survey).

SustainAbility/UNEP (1997) *Engaging Stakeholders: The 1997 Benchmark Survey*. London: SustainAbility.

Swift, T. (2001) Trust, reputation and corporate accountability to stakeholders, *Business Ethics: A European Review*, **10**: 16–26.

Thomson, I. and Bebbington, J. (2005) Social and environmental reporting in the UK: A pedagogic evaluation, *Critical Perspectives on Accounting*, **16**(5): 507–33.

Tyson Report, The (2003) *The Tyson Report on the Recruitment and Development of Non-Executive Directors*. London: London Business School.

Unerman, J. and Bennett, M. (2004) Increased stakeholder dialogue and the internet: Towards greater accountability or reinforcing capitalist hegemony?, *Accounting, Organizations and Society*, **29**: 685–707.

Zadek, S., Raynard, P., Forstater, M. and Oelschaegel, J. (2004) *The Future of Sustainability Assurance*. London: ACCA.

CHAPTER 12

CSR and accountability in other organisations: the public and third sectors, not-for-profit organisations and social business

12.1 Introduction and background

The bulk of the literature and concern around corporate social responsibility (CSR) and social and environmental accounting (SEA) and accountability is focused on the issues as they arise from and affect larger (typically commercial) organisations (Ball and Grubnic, 2007). This text has been no different. Given the issues of concern, whether sustainability, social justice or accountability, the influence of the largest market-based organisations (typically the multi-national corporations, MNCs) and their supporting infrastructure (including international financial markets and multi-national financial institutions and accounting firms[1] for example) is colossal. Whilst such organisations are amongst the most visible and may, very properly, be those whose activities might concern us most, to concentrate solely upon them is to ignore the rest of the organisational and institutional world (Marcuccio and Steccolini, 2005; Ball and Osborne, 2011).

The field that comprises (what Broadbent and Guthrie, 2008, call) 'public services' is potentially enormous and certainly exceptionally diverse. Social and environmental accounting and accountability are at least as important here and to ignore them would produce a very partial picture of the practice and potential of SEA. It is not just that maybe as much as 50%[2] of a nation's economic activity passes through these organisations or that even maybe as many as 50% of a nation's employees are engaged in the sector but that, to varying degrees, the broad sector has the capacity to set the social, legal and physical infrastructure within which the rest of society operates.

It is often helpful to adopt a simplification of Gramsci's conception of human endeavour as comprising **state**, **market** and **civil society** (see Chapter 5). MNCs and their supporting infrastructure are the most obvious and influential elements of the market. Government – federal, state and local – comprises the most obvious organs of the state, whilst humanity as individuals and families, as well as NGOs, charities (probably churches) and a range of grass-roots organisations, are amongst the most typical manifestations of civil society.[3] But

---

[1] The big four accounting firms qualify as amongst the largest businesses on the planet and under some estimates would all qualify for the Fortune 100 companies with a collective revenue of over $100bn in 2012 (Economist, 2012).

[2] See Anheier (2005) for detail on the non-profit (*excluding* all government and public) sector in various countries which he demonstrates itself comprises typically over 10–20% of the country's activity by various measures.

[3] Although civil society may, more properly, be located 'somewhere between the state, the market and the family' (Chandhoke, 2002: 45) and defined as the place where 'people come together in projects of all kinds to make their collective histories' (Edwards, 2000: 7).

extensive though that might be, it still doesn't capture the full range of organisations and entities and institutions. There are those that are transnational – like the United Nations (UN) – which stretch across states. There are hybrid market state organisations – typically state-owned enterprises but also many universities, hospitals and health organisations. In addition, there are a range of hybrid market – civil society entities. These might include social enterprises and social businesses of various sorts and may even be taken to include not just 'values-driven businesses' but the many, many small and medium-sized enterprises (SMEs) that are as much a part of a community or society as they are of a market (Stubblefield Loucks *et al.*, 2010; Blowfield and Murray, 2011). So if for no other reason than their diversity, number and ubiquity it would be a mistake to ignore the CSR and accountability issues outside the large business sector (Ball and Osborne, 2011: 2).

As Ball and Grubnic (2007) state, '*[t]he agenda for research and practice in sustainability (sic) accounting and accountability has been played out in an almost exclusively for-profit, corporate setting . . .*' (p. 243), and this is a matter to which social and environmental accounting researchers are slowly awakening (see Ball and Bebbington, 2008; Owen, 2008; Guthrie *et al.*, 2010). Indeed Ball and Osborne (2011) suggest that the potential of a social accounting for the public and third sectors has barely been imagined as yet – let alone realised. Despite many instances over the years and across countries of non-profit organisations producing interesting and novel disclosures, the potential for a social accounting within a non-commercial organisation really has the potential to re-define and redirect not just the organisation but its stakeholders and their perception of those organisations. Ball and Osborne are encouraging us to seek out 'socially significant' accountings: that is accountings which 'are planned to render visible, challenge or confront changing patterns of social, political and economic control and those factors which are shaping the social roles which information is serving (Hopwood, 1978, p. 3)' (Ball and Osborne, 2011: 3). There remains a truly considerable agenda of important work to be done in social accounting in the public and third sectors – and, ironically perhaps, these sectors are likely to prove more fertile ground for experimentation and development than has the traditional focus of social and environmental accounting – the commercial sector.

This chapter is some attempt to offer more balance on this matter. However, given the range of entities, institutions and enterprises that we need to consider to offer even a flavour of the non-MNCs sector of human organisation, our coverage is inevitably sketchy. The chapter is structured as follows. The next section offers a brief overview of the non-profit sector. The sections that follow focus on a selection of the organisations that make up the sector and introduce a few of the key issues that we know about concerning SEA, accountability and disclosure. Section 12.3 is concerned with the transnational non-profit organisations whilst Section 12.4 explores the public or government sector. Section 12.5 looks at universities before we move onto explore non-governmental organisations (NGOs) and civil society organisations (CSOs) in Section 12.6. Section 12.7 briefly examines social business in its various forms before we draw the chapter to a conclusion in Section 12.8.

12.2 Structure + parameters of the public and third sectors (and beyond)

As Anthony and Young (1984) show, the distinction between profit-oriented and non-profit organisations is far from precise and the divisions and categories vary considerably from country to country (see also Anheier, 2005). So, in what follows, you might want to keep a cautious eye on the data and take the opportunity to consult sources more relevant to your own country or region. That said, the broad principles are going to be much the same wherever we are.

Essentially, the private or commercial sector (the market) refers to organisations which provide and trade goods and services for a profit, which profit is the property of the owners of the organisation. The government/public sector (the state) is funded predominantly by taxation revenues and uses these to finance the apparatus of the state including legislation and the national infrastructure as well as providing basic services such as health, education and social welfare. The private non-profit sector (civil society) locates itself between government and the market and comprises an enormous range of social, environmental and religious organisations operated for a variety of social ends and which are (generally) only loosely owned by the employees, contributors, members and beneficiaries.

The dominant component of the non-profit sector is, unsurprisingly, the **government/ public sector**. This sector consists of principally national (federal) government and local (sometimes confusing called state) government. National government comprises all the (typically elected) representatives of the people who form the government, the mechanisms that support them and the range of ministries (e.g. finance, law, education, health, environment, defence) that enact and manage the policies of the state. In addition, there is a range of institutions and organisations which, to varying degrees, have traditionally been a part of the public sector but whose status is often more fluid and, depending where you are, diverse. These include universities, hospitals, state-owned industries, central banks, national utilities and so on.

Local government, as you might expect, fulfils many of these functions either wholly or in part on behalf of national government. The local government may, for example, maintain the police and fire services, run the schools and maintain the roads whilst national government concentrates upon defence, national policies and the legislature. There is considerable variation in this.

The *private* non-profit sector is also known as the **third sector, civil society organisations** or the **social sector**. It will include religious groups, arts organisations, trades unions, charities, mutual societies, clubs, local associations of various sorts, housing associations, political action and pressure groups, NGOs, environmental associations as well as social enterprises and a variety of local enterprises for social and employment purposes. There is a very large number of such organisations – Roeger *et al.* (2011), for example, identify 1.5 million such entities in the USA alone.[4]

Third sector organisations will generally be funded privately through membership, fees and donations, although Salaman *et al.* (2004) show that across 34 countries, government is amongst the biggest financial supporter of the third sector, often accounting for a third or more of the sector's income – particularly in developed countries. This in turn (as we shall see below) can critically skew the nature of the relationships and, consequently, both the focus of the entity and the accountability processes involved (O'Dwyer and Unerman, 2007).

These distinctions are neither firm nor immutable (Anheier, 2005). It is not just, as we shall see, that organisations may have a partly private commercial sector element but (say) still be partly owned by the state, or that a private business may be more 'social' than 'commercial' (the cooperative movement is one example of this) but that policy and politics move organisations across the public/private and commercial/non-profit divide all the time. Indeed, the whole area of the responsibilities and appropriate domains of companies, the state and NGOs is highly contested terrain. Lehman (2007) is not alone in noting the

---

[4]By comparison, a British source suggests that there are around 200,000 such entities in the UK and, as you might expect, the vast majority of these are tiny. The third sector tends to be dominated, financially at least, by about 400 very large entities (http://knowhownonprofit.org/basics/what-is-non-profit).

way, for example, that NGOs have grown in response to spaces left by the state whilst simultaneously both contracting with and offering a buffer against commercial interests that have moved into social arenas. It continues to get very complicated.

There are a number of factors which require us to distinguish the non-profit sector from the profit sector, and these include such matters as ownership structures, legal governance and organisational purposes (but for more detail see, for example, Williams and Taylor, 2012). Amongst these factors, two will be particularly important to us, not least because they also involve accounting in its various forms. These two are **control** and **accountability**.

The control issue relates to one of the most fundamental differences between non-profit and for-profit organisations. All organisations need managing to some degree or other (this is actually a tautology as without management of some sort there is probably no organisation) and that, in turn, involves some degree of control. As Hofstede (1981) for example suggests, management control is easiest if four conditions obtain. These are that: (i) the objectives are unambiguous; (ii) the outputs are measurable; (iii) the effects of intervention are known; and (iv) the activity is repetitive. Whilst no organisation of any note will satisfy all four conditions in all its activities, for a great many non-profit organisations several of these key conditions will frequently be absent. What is the objective of an army during peace time? How does a church measure its overall impact on saving souls? How does Greenpeace measure its impact and how does it balance its political intentions with a need to reduce its own ecological impact? And so on. Furthermore, virtually all non-profit organisations have some social and/or environmental objectives – but many of them may well be either implicit or in conflict with each other. For example, a university would normally be expected to pursue academic excellence *and* increase its sustainability whilst responding to the needs of its community and simultaneously acting as the conscience of a society *and* maximising research grants from industry. All of these factors will conflict to some degree or other.

In a commercial organisation, the requirements of the owners and the requirements of the customers are likely to be different (the customer probably wants quality or satisfaction; the owners probably want income and capital stability, say), but they are linked through profit (no sales, no profit, no dividend). In a non-commercial organisation, the 'owners' (the state, the electorate, etc.) and the 'customers' (the citizens and society) can be presumed to have very similar objectives, but their goals are *not* linked through the financial system – there is, quite properly, no measurement of 'profit' (Anthony and Young, 1984). Moreover, this has a further, interesting implication. In one sense at least, for all its limitations and faults, the financial accounting system seeks to capture and represent a major category of the 'success' of a commercial organisation (with all the provisos and limitations that we know about). There is no equivalent for a non-commercial organisation as its goals cannot be expressed in any simple financial way and capturing the complexities of a charity's or a club's or a university's goal(s) is extremely testing and often nearly impossible in any simple way.

Consequently, a great deal of attention is given in non-profit organisations to trying to measure and report output – to find, essentially, some analogue for the figures of (typically) sales and profit/loss as they appear in the financial statements of a for-profit organisation (see especially Jones, 2010: 14). One way of measuring aspects of outputs is through the use of multiple social and environmental indicators: such things as (say, in a hospital setting) number of operations; number of successful operations; numbers seen in accident and emergency; waiting lists; proportion of bed-spaces filled; ratio of patients to nursing staff; average life-expectancy of the community; and so on (see, for example, Maddocks, 2011). The range and complexity is fairly obvious, the point being that a lot of what we might see as social and environmental measurement in a non-profit organisation may well be, crudely, attempts to capture the complex outputs and objectives of the entity. This, together with the increased complexity of the relationships that a non-profit organisation holds with its stakeholders, contributes to some of the additional challenges of organisational accountability (see, for example, Costa *et al.*, 2011).

It can be argued accountability might be best understood as a responsibility to provide accounts within a series of relationships (see Chapter 3). The more relationships an organisation has with its stakeholders then, potentially, the more accounts are necessary to discharge the accountability(ies). The more complex each of those relationships, the more complex the accounts may need to be to fully discharge the accountability(ies). Ebrahim (2003a) follows a similar line of reasoning and identifies a number of factors which make non-profit accountability a more elusive notion. Particularly, he identifies three themes.

First, Ebrahim argues that non-profit organisations can be both principals *and* agents in accountability processes (a point also made by Goetz and Jenkins, 2001). Thus, for example, the Environmental Protection Agency will be accountable to society whilst simultaneously holding (for example) corporations accountable for their pollution (see also Ebrahim, 2005; Weisband and Ebrahim, 2007).

The second point is that accountability operates as much through *internal* as *external* processes. That is, whilst external, visible forms of accountability discharge may not be apparent, this does not mean that systems of oversight, audit and professional governance are not operating successfully within the entity.

Third, and in a related fashion, Ebrahim suggests that a very important part of any non-profit accountability is the accountability that individuals, professionals and other committed people owe (and indeed offer) to their (what are known as) 'epistemic communities' – the people with whom they work and whose values they share, implement and ultimately support (Gray *et al.*, 2006). This, in turn, affects and is affected by the management control system that the organisation adopts (Chenhall *et al.*, 2010) . That is, control can also successfully operate through the levels (of what Chenhall *et al.* identify as) 'social capital' which create social ties between the actors and predispose actors to embrace and enact certain forms of accountability and behaviour.

Finally, it is very likely indeed that when looking at the non-profit sector one will see varied and different channels of accountability discharge as well as differing media (see, for example, Ebrahim, 2003b; Ball and Bebbington, 2008; Brandsen *et al.*, 2011). This is in part a result of the *closeness* and the role of *epistemic communities* that many (especially) smaller entities enjoy, but it also reflects the nature of the organisations and the relationships they hold with stakeholders. So, for example, a local government will probably have direct access to all its citizens through its various activities – charging local taxes, organising welfare benefits, schooling, waste collection and so on. And citizens may well be content with discharge of accountability through activity (the waste bins are emptied, the streets are safe, for example). In a broad sense, the non-profit organisation may have – in addition to social capital, a call upon epistemic community and closeness – a greater range of means of interaction within which the commercial and non-commercial are blurred and through which forms of accountability may, in theory at least, be discharged in part or in whole.

Having said all this, we think we would want to maintain that the broad principles of accountability should apply equally in both commercial and non-commercial organisations. How this plays out and what progress is being made in different sectors is something we want to try and examine in the following sections.

12.3 Multilateralism and intergovernmental – beyond the nation state

> As global governance expands, few can hold those who exercise power to account.
> The implications for democracy are profound. Within the boundaries of the state
> people enjoy at least the *potential* to hold their governments to account . . . Yet
> increasingly . . . governments cannot be held to account for a widening range of
> decisions.
>
> (Woods, 2007: 27)

Just as corporations and their apparatus have become multi-national and transnational, so
have aspects of the non-market sector: not least in an attempt to stay in touch with – and
control – the MNCs themselves (Albrow and Glasius, 2008). Banerjee (2007) argues that the
neo-liberal globalisation has changed the dynamics between civil society, the market and the
state, blurring the lines of demarcation and that new arenas and mechanism of accountability
have not been developed to meet this lacuna: 'What is needed is some kind of supranational
agency with enforcement powers working in partnership with a wide range of local advocacy
groups . . .' (p. 157). This lack of a world government is partly offset, Bennet and van der
Lugt (2004) show, by the emerging elements of a degree of global governance (p. 47).

MNCs are not the only multi-national entities. Some are strictly governmental (such as
the European Union and the G7) and subject, very broadly, to many of the same issues as
arise in the governmental sector (see Section 12.4). A very considerable number of these
multi-national entities are better thought of as international NGOs (INGOs) such as the
Global Reporting Initiative (GRI), Greenpeace, Oxfam, Amnesty and WWF, and therefore a part of discussion on NGOs below in Section 12.6, but there is also a range of (what
are perhaps best thought of as) inter-governmental organisations that have a crucial role in
global accountability and sustainability. The most obvious of these is the UN and the panoply of organs that work in and through the UN. As we have seen throughout the text and we
shall see again shortly, the UN figures very strongly in many global initiatives – not least
initiatives like the Earth Summit, the Global Compact (UNGC) and the Millennium
Development Goals. But the UN is by no means alone here and the impact of entities as
diverse as the World Bank, the International Monetary Fund, the International Labor
Organization and the World Trade Organization gives some brief flavour of the importance
of multilateralism in the global society.[5]

The complexities of how to account for these entities and the accountability(ies) involved
are exceptionally daunting. The relationships between these organisations, their component
states and private sector organisations such as companies and NGOs ensures that there can
be little that is essentially straightforward here (Kim, 2011). Furthermore, as Anheier and
Hawkes (2008) so persuasively argue, global society has not even begun to get abreast of
these issues or to recognise just how to overcome the increasingly apparent major difference
between domestic and global accountability. Into this mix we need to add Ebrahim's (2003a)
important reminder that these organisations crucially act as both principal and agent in
many relationships – they need to be held accountable, but they are critical in the holding of
other organisations to account.

As Woods (2007) shows, there has been a steady erosion of potential mechanisms of
accountability for these institutions and there is much to be done to re-institute new accountability mechanisms and processes. That is not to say that there are no accountability

---

[5]And a further range of entities whose character (and whose accountability) is even harder to locate. Such entities
might include the OECD, The World Economic Forum and NATO for example.

mechanisms, but they tend to be broadly internal and not visible to broader society. That is, for example, the UN's system of committees – each with oversight and responsibilities to each other, plus, especially, the Trusteeship Council, ensures some sort of surveillance and accountability. In other institutions there will normally be some equivalence through mechanisms such as inspections panels (in the case of the World Bank) and an ombudsman's office. This is not, however, the same as a robust and a truly substantial accountability.[6] Woods (2007), in fact, recommends increased participation to encourage 'better representativeness, more responsiveness and stronger accountability' (p. 42) in order to rebuild trust and return to an exploitation of the social capital that non-profit organisations should be able to exploit.

As far as we can assess, the multilateral organisations rarely produce accounts – social, environmental or otherwise, that might be a direct analogue for normal SEA or even financial accounting, but they are exceptionally important in the production of 'global accounts'. That is, our understanding of the state of the global commons depends predominantly on UN initiatives such as the UNEP *Global Environmental Outlook* (see, for example, UNEP, 2012), the United Nations Millennium Ecosystem Assessment (2005) and the UN's Development Programme Human Development Index (HDI) (which provides an aggregate measure of such matters as income, inequality, schooling and life expectancy). These data are of exceptional importance and form the basis for so many of our global narratives (see, for example, Randers, 2012; WWF, 2012). These are centrally important to global environmental and social accounts (Gray, 2006, 2010).[7] Moreover, important and useful though these data are, the accounting literature has yet to address (as far as we can tell) the accountability and social and environmental activities of these organisations. This is a striking lacuna.

There is rather more visible activity when it comes to the principal functions of these multilateral bodies. For example, UNEP has been very active (working especially with the consultancy and campaigning outfit, SustainAbility) in the development of environmental and sustainability reporting (see, for example, SustainAbility/UNEP, 2002) and the output from these projects represents an important snapshot of corporate and regulatory development in the field that complements the KPMG surveys (Milne and Gray, 2012). Equally active has been the UN Conference on Trade and Development which has been at the forefront of trying to develop environmental accounting for over 20 years (Moore, 2009) and, more recently, has also engaged in attempts to develop social responsibility indicators for use by transnational corporations (see, for example, UNCTAD, 2008).

Some of these initiatives find their way into the accounting literature, but perhaps the most telling research in this field is that of Rahman (1998) in which he tracks the UN Center for Transnational Corporations' attempts to impose accountability upon MNCs for their activities in host countries. It is a harrowing tale of a complete inability to hold capital to account at a global level and offers an insight into the really important events that (social and environmental) accounting research can potentially address. Equally, there is emergent work on both the World Bank (Woods, 2007) and the International Monetary Fund (IMF) (Woods and Narlikar, 2001; Kim, 2011), but this pioneering work simply illustrates how little is known from an accounting and accountability perspective and how much more needs to be addressed across the whole panoply of global institutions (see, for example, Bennet and van der Lugt, 2004).

---

[6] It has to be recognised that substantial accountability systems typically run into the standard problem that the price of increased democracy and accountability may be reduced efficacy and independence.

[7] It is worth drawing attention to the World Bank's datasets. Their website contains an astonishing range of detailed and reliable data about global economic issues, and any scholar interested in such matters should spend time on that site. In broad terms, these data are also a potential element in the organisation's accountability.

12.4 Government and the public sector

Ball and Grubnic (2007) remind us of what the public sector comprises: national government and its ministries; regional and local government; health care; emergency services; public corporations; educational and research institutions and so on (universities deserve separate consideration in Section 12.5). It is clear, as we have seen, that social, environmental and sustainability accounting and accountability matter here at least as much as they do in the commercial private sector. The sector, virtually world-wide, has been subject to pressure and change – not least under the neo-liberal assumptions that the public sector is necessarily less efficient and/or effective than the for-profit sector. This has, in turn, produced pressures for re-orientation of the traditional remit of the sector as well as increased calls for forms of accountability. That such a challenge is fundamentally mis-specified (Maddocks, 2011) has not prevented a steady encroachment of pseudo-market incursions into the traditionally professional and service-orientated entities. These incursions often occur under what is loosely called the 'new public management' and all that has come with that (Broadbent *et al.*, 1991; Ball and Grubnic, 2007).

Despite the focus of the public sector being upon service to the community and (traditionally at least) essential to the fabric of society (Ball and Grubnic, 2007: 246), there has, as Guthrie *et al.* (2010), show, been a relative dearth of research work on SEA and the public sector. We will touch upon a few key elements in what follows and, for simplicity, organise these brief comments into sections covering the government and the state; health provision; local authorities; and then, finally, state-owned enterprises (SOEs).

Government and the state

Much as we saw with multi-lateral entities, there is a general view that social, environmental and sustainability accounting and accountability are in their infancy in government.[8] In one sense, this can seem counter-intuitive in that, in addition to the holding to account of members of parliament and ministers through the diversity of democratic processes, there is an array of mechanisms through which some aspects of government action are visibly held accountable – commissions, policy impact assessments, national audit office, national statistics, such things as quality of life indicators[9] and, indeed, the public media itself. In addition, as we saw above, multi-lateral bodies have effectively been calling states to account for their social and environmental performances through the environmental accounts and mechanisms such as the HDI. In this regard, Bennet and van der Lugt (2004) report on the UN's attempts to encourage national and regional state-of-the-environment reports which, as Jones (2010) reports, have resulted in a range of different approaches across different countries.

For example, the UK government produces annual reports against 15 headline quality of life indicators which Porritt (2005: 64–5) argues are possibly the most advanced in the world (see Figure 12.1 and Jackson, 2009, for more detail). But governments are also crucial in helping set the social, environmental and economic agendas of their nations and, perhaps more importantly, how those agendas are framed. The continuing refusal (or inability?) of most governments to engage with the issues of economic growth, its questionable desirability and its impact on social and environmental well-being suggests that the key issues and the key debates remain excluded from consideration (Porritt, 2005; Jackson, 2009).

---

[8]Indeed, it is not that long ago that the financial accountability of governments and their ministries was only an emerging issue – and one which is a long way from solved even now (Broadbent and Laughlin, 2003).

[9]These suggest some of both the external and internal accountability mechanisms that we discussed earlier as key to public and third sector accountability.

Figure 12.1 Quality of life indicators

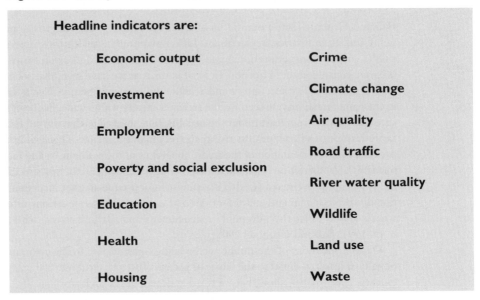

Headline indicators are:	
Economic output	Crime
Investment	Climate change
Employment	Air quality
	Road traffic
Poverty and social exclusion	River water quality
Education	Wildlife
Health	Land use
Housing	Waste

Source: Taken from *Part III: Guidelines for reporting on indicators for sustainable development at national level* (http://www.un.org/esa/agenda21/natlinfo/countr/uk/2003indicators_uk.pdf).

The development of things like the Index of Sustainable Economic Welfare (ISEW) as alternatives to gross domestic product (GDP) are asking for this next and fundamental step to be taken (Jackson, 2009). Until there is focused analysis of the major and growing gap between GDP and (such things as) ISEW, it seems unlikely that social, environmental and sustainability issues can be sensibly addressed. Government is responsible for these lacunae.

It is failure to undertake the discharge of such fundamental areas of social and environmental accountability that leads Wapner, for example, to argue that states (certainly many states) are broadly unaccountable to their people (Wapner, 2002). Bennet and van der Lugt's reporting that governments have been found to be reluctant to report upon the impact of their own operations lends support to this concern (Bennet and van der Lugt, 2004: 46).

Porritt (2005) reminds us that the state ultimately has the duty and the ability to demand and enforce regulation of accountability of other organisations – both from profit and non-profit sectors (Porritt, 2005; Hyndman and McMahon, 2011). It is the civic responsibility of government to provide for and enforce the infrastructure within which citizens operate. Little illustrates this better than Agenda 21 set at the Rio Summit in 1992 which lays out the policy commitments necessary to encourage nations along a path towards sustainable development, and it falls to the public sector to assume responsibility for moving the sustainable development agenda forward (Ball and Grubnic, 2007: 252).

There is no shortage of guidance on how public sectors might go about their reporting (Ball, 2004; Jones, 2010) but, as Dumay *et al.* (2010) argue, the dominant guidance in reporting – the GRI guidelines – are managerialist not ecological and 'do not contribute to sustainability'. They are still little used in the public and third sectors (see also Jones, 2010). Nevertheless, Ball and Grubnic (2007: 254–5) see potential for the development of broader accountability through GRI but, notably, supplemented by the mechanisms of stakeholder dialogue (see Chapter 5). It may well be that increasing the accountability mechanisms of government will clarify and make visible the intense conflicts that the public sector faces between its various pressures whilst pursuing agendas of social justice and sustainable development.

One of the major social and environmental accounting initiatives of the early years of the 21st century has been the Prince of Wales 'Accounting for Sustainability' (A4S) project (see, for example, Hopwood *et al.*, 2010). It is perhaps a sign of the changing times that two of the major case studies initiated as a result of the A4S project were related to non-profit organisations. The experiences at the English Environment Agency (EA) – an arm of government – offer novel and unique insights. Thomson and Georgakopoulos's (2010) investigation of the EA picks up on an entity that has pursued issues of social, environmental and sustainability accounting and accountability for many years and has consistently reported some of this in its annual report and accounts. As an organisation charged with ensuring the environmental performance of other entities, it has actively taken steps to suggest that it takes its own accounting and accountability seriously. What is clear from Thomson and Georgakopoulos's analysis is that there are few substitutes for experience and that years of experimentation, false starts and organisation-wide buy-in have allowed for a great deal more progress than could otherwise have been the case. Despite an exemplary endeavour in many areas, it is quite clear that even the EA is still struggling to identify, manage and report a full and coherent social and environmental response.

If there is little clarity about how government's social and environmental accountability is to be developed, this seems to be much more the case for health provision.

Health provision

Forms of health provision vary considerably nation by nation and typically combine elements of both the public non-commercial and private commercial sectors. The social impact of health provision is self-evident – health, longevity and child mortality being amongst the principle indicators of social welfare (Collison *et al.*, 2007). Health care entities are also ones with considerable social and environmental impacts in their own right. But one would be tempted to infer that, for most researchers, the question of interest is how social and environmental issues affect health – not how health care affects social and environmental issues.

Despite the fact that there is every reason to imagine that social, environmental and sustainability issues are just as important and that accountability and reporting are just as necessary as they are for the private sector commercial organisations, there appears to very little research attention given to the field – at least in the English language and at least outside Italy (see, for example, Monfardini *et al.*, 2013). This relative lack of research attention seems to be despite the innovations in reporting that health has shown over the years (see, for example, Gray, 1984; Gray *et al.*, 1987; Monfardini *et al.*, 2013) and the continuing complexity of the accountability relationships (see Kinder, 2011). The extract in Figure 12.2 (from an Ohio health care organisation) illustrates just one aspect of what reporting can entail.

Local government

Whilst many of the same things we have mentioned about other elements of the public sector apply equally to local government, there appears to be a far wider acknowledgement of the crucial importance of local authorities in (particularly) the sustainability agenda (Ball, 2002). As Ball argues, local governments the world over are tasked by Agenda 21 as well as by the exigencies of restructuring for sustainability. They have (if at times reluctantly) begun to embrace the new tools of social and environmental accounting as necessary steps towards understanding and managing change (Ball, 2005) as well as an antidote to the encroaching commercialisation of non-profit organisations.

Local government reporting in the UK and indeed elsewhere (see, for example, Joseph and Taplin, 2012) has often been particularly innovative and committed to the production of external

Figure 12.2 Extract from Cleveland Clinic's 2012 Global Compact Report

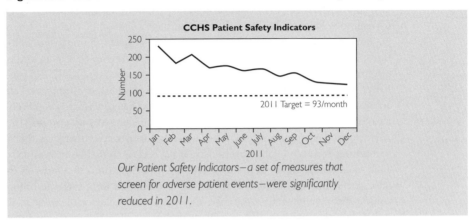

Our Patient Safety Indicators—a set of measures that
screen for adverse patient events—were significantly
reduced in 2011.

Source: Taken from http://my.clevelandclinic.org/Documents/About-Cleveland-Clinic/overview/
CC_UNreport_2012.pdf.

social and environmental documents (see Figure 12.3). Social and environmental-related
reporting by local government has a long history (see, for example, Butterworth *et al*., 1989),
and there is widespread recognition of the essential need for social and environmental disclosure by local government. As Jones (2010) shows in some detail, there is no shortage of
sources of guidance on reporting. However, as Ball and Grubnic (2007) have remarked,
there is (at the time of writing) still relatively little systematic analysis of these reports and
their contents. There are signs that this is changing (see especially Williams *et al*., 2011) and
in some countries – most notably Italy – research appears to be much further advanced (see
Marcuccio and Steccolini, 2005; and Mussari and Monfardini, 2010).

In terms of understanding what is happening within the organisation as it seeks to address
social and environmental issues, one of the most detailed analyses is offered by Grubnic and
Owen (2010). They report on an English county council's engagement with 'sustainability' as
the over-arching framework within which to embed social and environmental considerations
through the organisation. If only for the background material outlining the complex range of
activities with which even a relatively small local government entity is involved and the panoply
of national, regional and local initiatives and regulations and governance structures through
which ideas have to navigate, the experience reported by Grubnic and Owen is valuable. Coherent
embedding of social and environmental concerns is never going to be simple. Using a well-
guided, top-down, holistic team-based approach throughout the entity, the authority sought to
develop targets and plan their achievement across the whole organisation – a task that proved to
be difficult enough relating to environmental issues but exceptionally resistant on social issues
where consensus about objectives and how they might be measured was much less clear (p. 103).
The experiences echo so well those of other initiatives in the private sector in that some areas are
more amenable to measurement and control (energy, travel, buildings, for example) whilst
others are either elusive, embedded in conflict or simply cannot engage the appropriate
personnel or attract the apposite level of resource support. Nevertheless it is clear that the
council's efforts resulted in changes in behaviour, activity and information collection, and the
last of these directly influenced the council's reporting practices.

Finally, we must remind ourselves that, as with other public sector entities, local government has a dual role of both upwards accountability (to central government) and downwards
accountability (to ratepayers, citizens and suppliers) and acts, simultaneously, as an organisation with major social and environmental impacts and as an entity whose tasks include

Figure 12.3 Extract from *State of the Shire Report 2010/11*, Wyong Shire Council, NSW, Australia

Wyong Shire at a Glance

State of the Shire	2010/11
Population June 2010	150,338
Population projection	200,015
Families 2006 Census	38,035
Area	820,42 km² or 82,042 ha
Parks and Reserves	409 sites or 3,000 ha
Coastline	35 km
Lakes (Budgewoi, Munmorah, Tuggerah)	79.6 km²
Beach water quality testing (17 ocean beach sites)	100% compliance
Vulnerable, endangered or critically endangered species (NSW Threatened Species Conservation Act 1995)	129
Active Landcare groups	42
Active Landcare sites	56
Bushland	3,000 ha
Hospitals	1 public, 1 private
Number of neighbourhood/community centres	5 Neighbourhood and 29 Community Centres
Precinct committees	13
Library members	48,039
Obesity levels	60% obese or overweight
Constructed wetlands	25
Shire residents participation in the workforce (March 2010)	65,286 persons
Unemployment rate (June 2011)	6.4%
Median Weekly Individual Income (2006 Census)	$318/week

Source: Taken from http://www.wyong.nsw.gov.au/environment/state-of-the-shire/.

encouraging others whilst monitoring other behaviours – as both principal and agent (Goetz and Jenkins, 2001). Little illustrated this better than the local authority social audits so carefully analysed by Harte and Owen (1987; see also Chapter 10). Here the local government took steps to challenge the right of economic private self-interest to dominate the immediate needs of a community and its citizenry. At the heart of local authority activities – and their accountability – will always be an elusive and constantly debated notion of what is (or is not) the public good.

State-owned enterprises

State-owned enterprises (SOEs) vary considerably in the balance required of them regarding their social and economic goals and, consequently, in their orientation towards profit-seeking. Equally, SOEs can rarely be seen in isolation from the political dimension. That is, on the one hand for the neo-liberal, SOEs are little short of the spawn of the devil and should be released into the wilds of the market place as soon as possible (*The Economist*, 6th August 2012).[10] For others, SOEs represent a sensible marriage of state and community needs with the drivers of commercial efficiency.

---

[10]http://www.economist.com/node/21564274.

What is clear is that the hybrid nature of the organisations has – and always seems to have – caused difficulties for accountability mechanisms. This is because, precisely as we have seen elsewhere in the public sector, the ambiguity of goals makes control and the thrust of accountability much less clear. This can be (and is) solved in a variety of ways – and it seems to depend to a large degree upon the nation state involved. For example, the UK had a rich history of (what it called) nationalised industries, and its reporting and accountability was developed (to varying degrees it has to be said) principally through the use of suites of performance indicators (see Chapter 9, and see Figure 12.4 for an illustration). These indicators would seek to capture the social, the customer and the environmental elements of performance alongside the economic components. This still continues and so, for instance, you could look at the British Broadcasting Corporation's (BBC) report for 2011/12 and learn what proportion of programmes had subtitles, what proportion of audiences professed themselves satisfied with BBC programmes and what proportion of programmes were repeated at peak times – all set against their stated targets for the year. But as you might expect, perhaps, you will find relatively little of direct social, environmental or sustainability relevance in this report unless it forms part of these complex objectives. Such complexity has long been a challenge for these organisations (see, for example, Perks and Glendinning, 1981; Likierman and Creasey, 1985; Jones and Pendlebury, 2010).

For states of a neo-liberal inclination (as the UK now is, for example) the response to these difficulties is to *privatise* the entity and effectively remove many of these conflicts. Many of the UK's national utilities, its water, energy and transport systems, for example, have all been subject to this neo-liberal call. Certainly the ambiguity of the SOE and its objectives seems to make it more vulnerable to 'free market' arguments – although whether this improves the social, environmental and sustainability performance and accountability remains a moot question.

This all has major implications not just for the political economy of the so-called developed economies but also for the newer economies – especially China. China has famously sought to blur the lines between private and public sectors and the body of SOEs is a major component of both the Chinese economy and its push towards growth and 'development'. Yet the state retains considerable influence and control over these organisations. This, in turn, might be thought to have influenced a more responsible attitude to social and environmental disclosure, but this does not seem to be the case (Liu and Anbumozhi, 2009; Wei and Xiao, 2009; Moon and Shen, 2010; Rowe and Guthrie, 2010; Du and Gray, 2013). Broadly, despite the very different conditions obtaining, it is not at all apparent that Chinese accounting and disclosure is substantially different from that found elsewhere in the world.

This all seems to lead to an inference that whilst the accountability relationships and the social and environmental interactions as well as the putative influence over (un)sustainability

Figure 12.4 Example of data extracted from nationalised railway accounts

Indices of Staff Performance							
	1962	**1963**	**1964**	**1965**	**1966**	**1967**	**1968**
Traffic units* per employee	100	107	121	126	133	136	149
Loaded train miles per employee	100	104	109	111	115	117	123
Route miles per employee	100	104	109	112	113	113	116
*Traffic units are passenger miles plus net ton miles.							

Source: Taken from Table II, p. 163 in C. D. Jones (1970) 'The performance of British Railways 1962–1968', *Journal of Transport Economics and Policy* May (pp. 162–170).

are substantially different in the public sector from private sector commercial organisations, the disclosure, reporting and accounting practices need not necessarily be so. That is an important hypothesis that deserves further exploration.

12.5 Universities

Universities play a crucial role in most societies in knowledge creation and dissemination, in education and, potentially, in attitudes and skills (Tilbury, 2011; Adams, 2013). They also have significant social, environmental and economic impacts (Tilbury, 2011). As a consequence, one might have expected universities to be key innovators in, and widely committed to, SEA and related matters. Despite (as we shall see below) occasional engagements with the examination and control of social, environmental and sustainability issues, universities (and other educational institutions) have rarely provided best practice in either integration of these matters into their strategic planning or in their subsequent accountability practices.

Not that universities have been inert in the face of a changing social and environmental world. It seems to be, rather, that the world they face is so complex that developments in social and environmental matters happen more slowly in the light of more urgent, pressing and hostile forces. This is the point made clearly by Guthrie and Neumann (2007) in recognising the extent of financial, economic and social pressures to which universities world-wide are increasingly subject. Equally, Coy and Pratt (1998) articulate the complex political maelstrom in which universities sit. Both papers show that these factors directly influence the reporting and accountability decisions made by the institutions (see also Frost and Seamer, 2002).

The 21st century has seen the beginnings of what might be a potential change as the sustainability (sic) agenda gathers momentum. Certainly the turn of the century saw increasing attention given to what was thought to be the demands of sustainable development (Tilbury, 2011) and this, in turn, has been reflected in the published academic literature (see, for example, Lukman and Glavič, 2007).

The potential sources of change in universities are myriad, but experience suggests that where initiatives occur they are it is more likely to be led by students, the facilities management team or academics rather than as a strategic direction led by senior management. Unfortunately such initiatives do not lead to holistic change. That requires embedding sustainability in strategy (Adams, 2013). Senior management is frequently occupied by the issues of the day and faces limited external drivers to adopt a substantive focus on sustainability.

Despite (or perhaps because of) the turbulent external environment that universities face, there is an increasing level of support from associations and independent bodies dedicated to helping universities embrace social, environmental and sustainability strategy and accountability.[11] Initial experience suggests, however, that there seems to be some way to go if these initiatives are to fully embrace the global, cross-sectoral, multi-sector developments in accountability for sustainability performance: initiatives which have as yet not fully supported the university sector.

Whilst little is still known about the detailed adoption of social and environmental accounting practices in universities, rather more is known about the more obviously visible

---

[11]Examples include – all of which are well worth consulting – the Association for the Advancement of Sustainability in Higher Education (www.aashe.org); the Environmental Association for Universities and Colleges (www.eauc.org.uk) and their Green Gown Awards; People and Planet's Green League Table (http://peopleandplanet.org/green-league-2012/tables); the college sustainability report card (www.greenreportcard.com); and the Higher Education Academy (www.heacademy.ac.uk).

Figure 12.5 Extract from La Trobe University, *Creating Futures: Sustainability Report 2011*, p. 37, April 2012

Occupational health and safety

THREATENING BEHAVIOUR AND ASSAULTS

	2009	2010	2011
Threatening behaviour	14	6	5
Aggravated assaults	1	4	6

Includes staff, students, contractors and university visitors.

GRIEVANCES, BULLYING OR DISCRIMINATION

	2009	2010	2011
Initial contact /enquiries	NA	28	9
Formal cases /complaints	21	12	16
Continuing	0	3	1
Completed	21	9	15

Initial contact data not available in 2009.

OH&S INCIDENTS	2009	2010	2011
Total occupational health and safety incidents			
Hazard reports (no injury)	64	61	66
Incident reports (no injury)	166	261	284
Incident reports (injury)	223	262	258

Source: Taken from http://www.latrobe.edu.au/sustainability/documents/4906 Creating Futures v6b.pdf.

external reporting practices on social, environmental and sustainability performance. As we now know, non-profit sector reporting practices tend to reflect a hybrid and sometimes confused amalgam of financial, performance indicator and social/environmental reporting. The university sector is no different in this – although there is some suggestion that such reporting has consistently been more obviously responsive to external forces than we might have expected (see Gray and Haslam, 1990; Dixon *et al.*, 1991; Cameron and Guthrie, 1993; Guthrie and Neumann, 2007).

By the turn of the century, reporting was potentially influenced by two initiatives. The United Nations' publication in 2012 of *A Practical Guide to the United Nations Global Compact for Higher Education Institutions*[12] provided guidance on how universities might go about adopting CSR principles in line with the UNGC and, in line with company signatories to the Compact, universities were expected to report annually on progress.[13] Interesting initiative though this is, it still fails to directly address a whole raft of the really important issues in considering sustainability at university level. This is a drawback which also applies to using the GRI.

On the face of it, there seems no reason why universities might not embrace the GRI as a sound basis for reporting. However, this has happened only slowly. Not only did reporting itself develop slowly: in 2012, CorporateRegister.com included only 44 university sustainability reports for that year in the same year only 14 universities were shown on the GRI database as having produced GRI report (Adams, 2013). This broad picture is confirmed by Lozano (2011) who concludes that university reporting is still in its early stages. Equally, Tort (2010) concludes that the take up of GRI by public agencies generally is weak. This is confirmed by Fonseca *et al.* (2011) who found little engagement with reporting by Canadian universities and that those few which did adopted a diverse range of approaches which were very varied in adoption of GRI. (Figure 12.5 is taken from one award-winning report and suggests the sort of information one might discover in a university report.)

---

[12]http://www.unprme.org/resource-docs/APracticalGuidetotheUnitedNationsGlobalCompactforHigher EducationInstitutions.pdf.

[13]It is worth noting that the UN had already established (under the United Nations Principles of Responsible Management Education) the principle of encouraging business schools to adopt CSR principles *and* report upon them annually (Adams and Petrella, 2010).

This is not quite the whole story and whilst universities have a long way to go to fulfil their potential and/or catch up with other sectors on social, environmental and sustainability accounting and accountability discharge, interesting experiments are unfolding all the time and there will always be the occasional individual university that will innovate and buck the trend (Pineno, 2011). Figure 12.6 illustrates one approach to innovation in stakeholder engagement (see Chapter 5).[14]

There is clearly considerable scope for major changes in how universities address these issues – as well as considerable need for a great deal more research in the field.

12.6 Civil society organisations: NGOs and charities

When we turn to **civil society organisations (CSOs)** life becomes rather more complicated still. In the first place, to what extent do we seek to distinguish CSOs and NGOs from other grassroots organisations (see Bendell, 2000; Edwards, 2000; Teegen *et al.*, 2004)? Where do clubs and the local baby-sitting circle actually fit (but see Gibbon, 2012)? We will side step this problem and focus broadly on NGOs with a brief excursion to look at charities towards the end of this section.

NGOs

O'Dwyer (2007) argues that even defining NGOs is difficult as they include entities as diverse as trades unions, religious organisations and the National Rifle Association all the way across to international entities like WWF and Oxfam. The complexities of civil society raise new and challenging (frequently moral) questions regarding the accountability of NGOs (Kaldor, 2004; Unerman and O'Dwyer, 2010). The complexities start from the realisation that NGOs have upward accountability (to their donors and the state), downwards accountability (to their clients and communities) and horizontal accountability (to the NGO community and professional/care/social communities with which they interact – their 'epistemic communities') (see Ebrahim, 2003b). These complexities are increased by the realisation that NGOs are increasingly subject to co-option (by business and/or government) (Baur and Schmitz, 2011) and this must be contrasted with the crucial need for them to remain independent as a key element in a healthy democracy (Jepson, 2005). The essential need for most NGOs to constantly renew their pursuit of funding is clearly a source of potential loss of independence (Yang *et al.*, 2011). Furthermore, NGOs are stakeholders themselves whilst also key representative of stakeholder groups (O'Dwyer *et al.*, 2005a, b): they are key in holding others to account and are a principal source for **social audits** and **shadow accounts** (see Chapter 10). As such, they have become especially important as both representatives of civil society and as an increasingly key component in state–market negotiations and debates (Teegen *et al.*, 2004). As Unerman and O'Dwyer (2010) so eloquently express it, NGOs have an increasingly important role in both the delivery of social and environmental ends and the protection of those most affected by un-sustainability.

One might think that this was complex enough (Brown and Moore, 2001; Unerman and O'Dwyer, 2006)[15] but NGOs are under an increasing amount of pressure from

---

[14]Other examples would include the 2011 report by Leuphana Universität in Lüneberg and the 2011 University of Graz report. Other exemplars might include The University of North Carolina at Chapel Hill 2009 *Campus Sustainability Report* (http://sustainability.unc.edu) and University of Maryland *Campus Sustainability Report: 2010.* (http://www.sustainability.umd.edu/documents/2010_Campus_Sustainability_Report.pdf).

[15]These complexities escalate further still when one explores their role in emerging civil societies – especially China (Jia'ning and ChanHow, 2011) but also, for example, countries within the former Soviet Union (Hunŏvá, 2011).

Figure 12.6 An example of a stakeholder set identified in La Trobe University's 2011 sustainability report, *Creating Futures*

STAKE IN OUR SUSTAINABILITY PERFORMANCE

2011 ENGAGMENT

CURRENT AND FUTURE STUDENTS: INTERNATIONAL AND DOMESTIC

Mission Australia's National Survey of Young Australians (2011) asked young people to indicate three issues they thought were the most important in Australia today. The environment was by far the most frequently mentioned topic with 44.7% of respondents identifying it as a major issue. **missionaustralia.com.au/downloads/national-survey-of-young-australians/2011**

Our students are future leaders who want to be prepared for the workplace and attractive to employers. We are addressing this through our Design for Learning project (**latrobe.edu.au/ctlc/dfl**).

- Social media: La Trobe Generations facebook page, Twitter **@ltugenerations**, LinkedIn
- Videocasts, podcasts, iTunesU
- Website: **latrobe.edu.au/sustainability**
- Representation on the Sustainability Management Committee, Sustainability Forum, Fairtrade Steering Committee

- Blog entries from key international sustainability events
- Support to students in developing campaigns and addressing sustainability issues
- Fairtrade fortnight
- Market research on sustainability priorities and education needs
- *Responsible Futures*

STAFF

Our staff want to work for an organisation that is 'doing the right thing.' They recognise the importance of the University's reputation in managing sustainability. They want the University to improve its sustainability performance.

They also want to succeed in applying for sustainability-related research grants and have opportunities to make a contribution to the reduction of harmful environmental and social impacts.

- Presentations on *Responsible Futures*, approaches to sustainability, sustainability targets and plans to Faculties, Divisions and at Academic Board and Senior Staff Forums
- Social media: La Trobe Generations facebook page, Twitter **@ltugenerations**, LinkedIn, Yammer
- Videocasts, podcasts, iTunesU, UniNews
- Website: **latrobe.edu.au/sustainability**
- Generations email list
- Representation on the Sustainability Forum

- Blog entries from key international sustainability events
- Victorian Centre for Climate Change Adaptation Research – sponsored Think Tanks
- Business Forum and Academic Symposium on Leadership for Climate Change and Sustainability
- Lunchtime discussion series
- Assistance with sustainability research proposals
- Treadly Tuesdays, Ride to Work Day, Family Day and Greenies on the Green

GOVERNMENT AND GOVERNMENT AGENCIES

A number of government agencies are working to promote improvements in sustainability performance in the university sector, including: Australian Government's National Action Plan for Education for Sustainability (**environment.gov.au/education**) and Sustainability Victoria (**sustainability.vic.gov.au**). In addition we engage with a number of government agencies that promote and/or regulate aspects of sustainability practice across all sectors.

- Victorian Centre for Climate Change Adaptation Research – sponsored Think Tanks
- Business Forum on Leadership for Climate Change and Sustainability

- Conference keynote speeches
- *Responsible Futures*

OUR COMMUNITIES – URBAN, REGIONAL AND RURAL

We have signed Memoranda of Understanding with a number of local councils, which involve engaging on sustainability issues. Our regional and rural communities face social and environmental sustainability issues, and expect us to contribute to the development of solutions.

Public lectures and events

Social media: La Trobe Generations facebook page, Twitter **@ltugenerations**, LinkedIn

Videocasts, podcasts, iTunesU

Website: **latrobe.edu.au/sustainability**

Our partnerships with Melbourne Heart (**latrobe.edu.au/partnerships/partnerships/melbourne-heart**), the InterFaith/Intercultural Network (through our Centre for Dialogue **latrobe.edu.au/dialogue/projects/northern-interfaith/niin-hub/about-niin.html**) and out new partnership with Greening Australia are important to our communities

Responsible Futures

ALUMNI

The reputation of the University is important to our alumni in career enhancement. We are working to develop responsible leaders of the future and to find ways of involving our alumni in our sustainability initiatives.

Social media: La Trobe Generations facebook page, Twitter **@ltugeneration**, LinkedIn

Videocasts, podcasts, iTunesU

Website: **latrobe.edu.au/sustainability**

Articles in the *Bulletin*

Responsible Futures

EMPLOYERS

Public and private sector organisations are increasingly aware of the impact of climate change and sustainability issues on future operations, and are seeking graduates with the skills and knowledge to help them adapt and the ability to apply sustainability knowledge to business and policy issues.

Business Forum on Leadership for Climate Change and Sustainability

Market research on future sustainability skills needs

Conference keynote speeches

Social media: LinkedIn

Website: **latrobe.edu.au/sustainability**

Responsible Futures

FUTURE GENERATIONS AND SOCIETY AT LARGE

Social, environmental and economic sustainability requires action by all organisations, individuals and governments. Universities have an important role to play as educators of large numbers of future leaders and through their community connections.

Social media: La Trobe Generations facebook page, Twitter **@ltugenerations** LinkedIn, COP 16/17 blog

Videocasts, podcasts, iTunesU

Website: **latrobe.edu.au/sustainability**

Responsible Futures

STAFF AND STUDENT UNIONS

Unions represent current University staff and students. They are interested in efforts to embed sustainability in the University's operations, research and curriculum and can exert a positive influence on their members, encouraging behaviour change.

Collective Agreement

Social media: La Trobe Generations facebook page, Twitter **@ltugenerations**, LinkedIn

Committee representation

Videocasts, podcasts, iTunesU

Website: **latrobe.edu.au/sustainability**

Figure 12.6 (*continued*)

STAKE IN OUR SUSTAINABILITY PERFORMANCE	2011 ENGAGMENT	
SUPPLIERS		
Our suppliers provide goods and services that can have a direct impact on environmental and social sustainability. They are often willing to consider innovative approaches that lead to improved sustainability outcomes; we need to engage with our suppliers to embed sustainability within our (and their) supply chains.	Social media: La Trobe Generations facebook page, Twitter **@latugeneration**, LinkedIn Videocasts, podcasts, iTunesU	Website: **latrobe.edu.au/sustainability** Meetings with key suppliers
PARTNER INSTITUTIONS FOR POTENTIAL FUTURE STUDENTS		
NAVITAS and Northern Melbourne Institute of TAFE provide training opportunities for potential future students of the University. Although their interest in our sustainability performance may be low, our decisions can have an impact on these organisations.	Social media: La Trobe Generations facebook page, Twitter **@latugenerations** LinkedIn Videocasts, podcasts, iTunesU	Website: **latrobe.edu.au/sustainability** *Responsible Futures*
OUR GOVERNING BODY		
As the governing body of the University, the University Council's role is to 'identify the distinctive character of the institution, to plan strategically, and to ensure that the University responds to the wider social context in which it operates'. As such, it engages with sustainability across the University and is represented on the University's External Sustainability Advisory Board. University Council members are listed at: **latrobe.edu.au/about/management/governance**	Social media: La Trobe Generations facebook page, Twitter **@latugenerations** LinkedIn Videocasts, podcasts, iTunesU Website: **latrobe.edu.au/sustainability**	Presentations to Council by the Pro Vice-Chancellor (Sustainability) *Responsible Futures*

Source: Taken from http://www.latrobe.edu.au/sustainability/documents/4906_Creating_Futures_V6b.pdf.

growing and widespread discourse that seems designed to undermine the legitimacy of NGOs for their own failures of accountability (Jordan and Van Tuijl, 2006; Gray *et al.*, 2006). It is not disputed that NGOs have a wide range of accountabilities to a wide range of stakeholders – they have a shifting set of responsibilities for their social and environmental activities, for their fulfilment of objectives in addition to the direct accountability they owe to their funders and to service beneficiaries (Awio *et al.*, 2011). The problem is that NGOs use a range of different forms and mechanisms of accountability (Ebrahim, 2003b), some involving informal accounting processes (personal communication, and other mechanisms of *closeness* for example) but many of them relatively under-developed, and this can produce very serious dysfunctional behaviours on the part of the NGOs (Dixon *et al.*, 2006). That is, there is every danger that different emphases on (say) functional short-term accountability to solve immediate pressures can drive out the more strategic accountability required when addressing longer-term issues. This is nowhere better illustrated than in O'Dwyer and Unerman's (2008) convincing charting of the emergence of different accountability mechanisms in Amnesty Ireland and the contradictions and paradoxes this produces. In essence, they illustrate the point we have already made: a non-profit organisation may well have a long-term strategic goal (world peace, justice, eradication of starvation) but increasingly finds itself being held accountable for shorter-term, funding-related measurable goals (letters written, economic efficiency, money donated, etc.). This, it seems, is just the broader manifestation of the troublesome tendency amongst neo-liberal regimes to mischievously force (say) financial accountability onto a non-profit organisation in an entirely inappropriate way – they are not financial entities (Gray *et al.*, 2006).[16]

As these complexities increase, we come to realise (as O'Dwyer, 2007, and SustainAbility, 2003, argue) that much of the problem derives from the intersecting mesh of factors driving the NGOs' accountability agenda – factors which embrace a range of external forces as well as ambitious, moral and idealistic intentions. Coupled with the ambiguous nature of NGOs and their sheer diversity, it is little wonder that there is a confusion of duty and accountability. This seems to lead, in turn (just as it does with corporations), to the 'most influential stakeholders gaining prominence' (O'Dwyer, 2007: 289) and the resultant focus on immediate and short term rather than strategic aims which are so much more difficult to demonstrate – the classic non-profit problems writ large.

When O'Dwyer (2007) summarises Ebrahim (2003b), we can see some of the range of potential channels which NGOs can use to try to discharge their accountability. These channels include: formal reports and other mechanisms of formal disclosure; a range of other individuals and groups involved in performance assessments and evaluations; various degrees of participation and engagement; self-regulation (including the crucial roles of epistemic communities and social capital); and social audits of varying kinds. Despite this range, there seem to be virtually no surveys of the actual practices of NGOs regarding accountability discharge and their social and environmental accounting and reporting or even their disclosure practices. The area has attracted field work where the complexities and conflicts that we have seen above have been carefully explored (O'Dwyer, 2005; O'Dwyer and Unerman, 2007, 2008, 2010), but we still lack any detailed overview of what NGOs are up to, what they say about themselves and the extent to which (and how) they employ the different mechanisms that Ebrahim talks about. There is work to be done here.

---

[16]Figure 12.7 (below) is only an extract from a charity's annual report, but it illustrates the way in which a financial accountability might be seen as a crucial (rather than as a probably irrelevant) element of accountability.

Figure 12.7 Extract from Christian Aid Annual Report and Accounts 2011–2012, p. 36

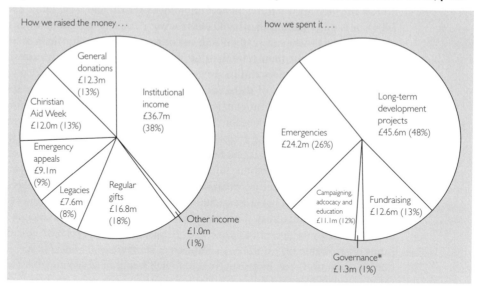

Source: Taken from http://www.christianaid.org.uk/images/2011-2012-annual-report.pdf.

Charities

Whilst it seems clear that we should consider charities as one element of the NGO sector, there does seem to be a great deal more clarity once we focus on charities specifically. Although they still experience exactly the same array of bewildering conflicts and ambiguities as the rest of the sector, it does seem that the separate legal definition that charities enjoy in many countries alongside the definitional strictures frequently offered by exemption from taxation makes the entities easier to identify, understand and isolate. An especially good review of this point of view is offered by Hyndman and Jones (2011) who offer the insight that charities will often have more singular, sometimes simpler, stated objectives that are sometimes amenable to better performance measurement.[17] This, in turn it seems, has been built upon in some jurisdictions (the UK is one such) where the requirements for financial accountability attract regulators and governments and provide a focus for the development of channels of accountability (Hyndman and McMahon, 2011).

Dhanani and Connolly (2012) offer some challenging and novel insights into reporting and disclosure by charities. They discover that the annual report has returned to being (or has always been?) the more important document for substantive disclosure – arguing that the standalone 'review' style documents are principally for advertising and advocacy (the Christian Aid Report illustrated in Figure 12.7 is suggestive of this). More interesting still, they argue that the forms of the disclosures – and, more importantly, the disclosures which are absent from the report – suggest that the primary reason that charities are reporting is for reasons of legitimacy. That, of itself, would not be especially striking were it not that we should expect charities (and perhaps NGOs more generally) to report in line with ethical or normative stakeholder theory (see Chapter 4) – and this evidence from Dhanani and Connolly (2012) suggests that they do not. We need to know more about this and, once again, we find ourselves needing, at a minimum, sound descriptive research on reporting practices in this sector as well as in others.

---

[17]Kreander *et al.*'s (2009) exploration of charities' investment policies and practices adds another, intriguing dimension to the ways in which accountability in a charity might (or might not!) be approached.

12.7 Cooperatives and social business

Our final port of call in this whistle stop tour of the public and third sector might, strictly, not be part of this sector at all. We are talking about those organisations which have some commercial or semi-commercial aims but whose activities, focus and *raison d'être* is social. As has been the case throughout this chapter, we are therefore talking about a wide diversity of entities with an equally diverse array of ownership arrangements, organisational structures and objectives. Indeed, we might (as do Crowther and Reis, 2011) see a continuous spectrum of organisation types from 'socially responsible' businesses through to explicitly social enterprises. In between, we find a whole raft of entities. There are the (what are often called) values-based businesses that one finds throughout this and any other texts concerned with social and environmental issues in organisations – Body Shop, Ben and Jerry's, Patagonia, Triodos Bank and Ecover, for example (Barter and Bebbington, 2010). These are companies with a clear profit-seeking orientation, but this is coupled with an explicit social aim. At the other extreme sit social enterprises which are established for explicit social functions (such as job creation, re-generating an impoverished community, extending fairtrade or other forms of self-help) and which use a corporate form for convenience. In these circumstances, profit may be elusive and their owners/shareholders are unlikely to expect much in the way of (or indeed any) dividends. Such entities are diverse, including such examples as Traidcraft (see Gray *et al.*, 1997; Dey, 2007), The Centre for Alternative Technology, Shared Earth and Shared Interest. As a phenomenon, they are particularly common in France (see, for example, Capron and Gray, 2000).

In addition, these hybrid social–commercial entities can also include credit unions, mutual societies, community interest companies (Nicholls, 2010), micro-financing initiatives (the Grameen Bank being the most famous) and, particularly, cooperatives (Webster *et al.*, 2011). Perhaps we might even include some small- and medium-sized enterprises (Stubblefield Loucks *et al.*, 2010). Increasingly, this broad and ill-defined category of organisations is known collectively as **social businesses** (see, for example, Baker, 2011, and Figure 12.8).

Each of these types of social business, in turn, presents varied and individual challenges in the whole matter of control, accountability and social and environmental accounting (Quarter *et al.*, 2003). Considerable efforts have been expended in supporting and encouraging the take up of (especially) social accounting practices in this sector, not least as a mechanism for social enterprises to demonstrate their social benefit in terms which predominantly financial stakeholders are likely to understand and accept (Pearce, 1996, 2003; Quarter *et al.*, 2003; Pearce and Kay, 2005). Third sector debates, conferences and texts give the impression of an active worldwide network of social accounting and audit activity[18] but the research literature – notably in accounting and business – seems to know relatively little about these aspects of these important organisations. What is clear is that each manifest, to varying degrees, the diversity of accountability(ies) and discharge mechanism that we have seen throughout the chapter.

In this regard, within the accounting literature, Dey's work on the fairtrade company and charity Traidcraft (Dey *et al.*, 1995; Dey, 2007) offers detailed glimpses into the unexpected struggles that a well-intentioned organisation faced in trying to come to terms with and discharge its formal accountability. Gibbon's work is a rare but detailed exploration of how a small CSO approached its own accountability and produced its own social accounts (Gibbon,

---

[18]See http://www.socialauditnetwork.org.uk/ for one particularly inspirational source of guidance and information, and see also the excellent http://socialeconomycentre.ca/. There is also an increasing range of support for cooperatives and www.eurisce.eu; www.cets.coop and www.uk.coop; are all valuable sources.

Figure 12.8 Social business?

> While the notion of 'social business' is central to all the world's great philosophies, its application and implementation has been overshadowed since the Industrial Revolution of the mid-Eighteenth century by theories of competition and what has come to be known as 'Anglo-Saxon' capitalism. It has become clear that, while this approach has brought many benefits to affluent western style economies, it is unsuited to two thirds of the world's population who survive at the bottom, or Tier 4, of the world economic pyramid. The need for an alternative business model became increasingly apparent during the second half of the 20th century . . .

Source: Taken from Michael Baker (2011) 'Aims and Scope', *Social Business* (http://www.westburn-publishers.com/social-business/aims-a-scope.html).

2012). By contrast, Nicholls (2010) offers a first insight into community interest companies (a relatively recent UK initiative) although this insight does not obviously encourage enthusiasm concerning their formal (annual report-based) social and environmental accountability discharge practices.

Developments in practice are more promising. We know from personal experience through the ACCA's Reporting Awards Schemes that social enterprises of all sorts engage in formal accounting and reporting of their activities – in 2001–2 a very small training organisation called 'The Cat's Pyjamas' (then part of the furniture cooperative FRC) won the reporting awards ahead of some major MNCs. Indeed, the third sector is especially active in this field[19] and the UK's Cooperative Society has been consistently one of the world's leading reporters (Harvey, 1984, 1995). Figure 12.9 is one brief illustration of the approach taken by a New Zealand third sector organisation to its social accounts.

12.8 Summary and conclusions

There is a wide range of issues raised by this chapter that more commercially-orientated analyses often fail to explore. Indeed, it is only when one undertakes a review of this sort that one becomes aware just how much diversity and possibility is missing from much of the mainstream SEA literature (Ball and Osborne, 2011). In addition to this diversity, variety and possibility – and our broad lack of knowledge of what happens in these sectors – a number of key factors emerged in this review.

First, our attention was drawn to the different forms of relationships that obtain between organisations and their many stakeholders in these sectors. The diversity of relationships was crucial in understanding what accountability was required and what channels of accountability might be appropriate. Issues of closeness, epistemic communities and social capital function in ways that purely economic relationships either cannot recognise or seek to expunge (Thielemann, 2000). This, in turn, was reflected in the existence of upward, downward and horizontal accountability(ies) and the challenges that each of these have for various accounting and accountability systems.

Then, and following from these points, we learned about the importance of both internal and external accountability. Although our focus is frequently upon the visible forms of accountability, there are many accountability mechanisms which may only be visible from within the organisation and from within certain communities. To a degree, this echoes the issues of control

---

[19]See the Social Audit Network materials and particularly consult both http://www.socialauditnetwork.org.uk/getting-started/social-reports/ and http://www.socialauditnetwork.org.uk/files/8713/2136/1687/updateddirectoryofsocial accounts-090508.pdf.

Figure 12.9 Extract from *Stop Violence NZ Social Accounts Summary 2004*

OUR KAUPAPA

VISION: "to enable all people in Aotearoa/New Zealand to live free of all forms of violence, abuse & oppression"

STATEMENT OF PARAMOUNTCY: "The safety of children and women is paramount"

OUR PRINCIPLES   Safety   Respect   Accountability
                     Responsibility   Gender partnership & equality   Excellence in practice

SCOPE OF THE SOCIAL ACCOUNTS:

Stakeholders:

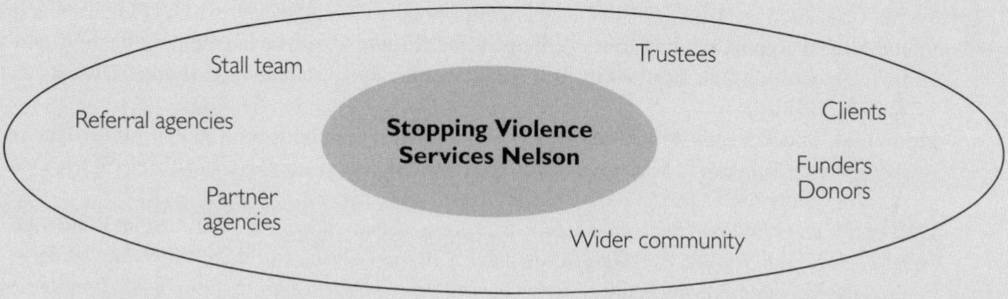

- Of the stakeholders identified above all were included with the exception of the 'wider community', this exclusion being for pragmatic reasons.

Social book-keeping:

- SVSN has many ongoing processes, practices and systems which enable ongoing feedback and monitoring of practice, hence much of the social book-keeping requirements were already in place. In addition, for the purpose of this first set of social accounts, a number of surveys were developed for stakeholders which were social accounts specific. All information and feedback used for these accounts is germaine to the audit year, i.e. July 1st 2003 – June 30th 2004.

- The intention was to use the New Economic Foundation model (Pearce et al) which means the accounts measure performance against the vision, mission, objectives and values of the organisations. This proved to be unachievable due to the scope and wording of our vision, objectives etc. – they were not 'social-audit' friendly! Therefore a key recommendation is that a review is held of our kaupapa, from which clear social objectives are identified, with strategies & actions & performance indicators

Source: Taken from http://www.socialauditnetwork.org.uk/files/7613/4633/2646/Stop Violence NZ Social Accounts Summary - 2004.pdf.

with which all organisations struggle except that, we might contend, a number of the non-profit sector's mechanisms are ultimately accessible to participants external to the organisation.

Finally, as commentators have noted, many organisations in this area are involved as both agent and principal; in being both held to account and holding others to account (as well as, on occasions, being held to account for the extent to which they have held others to account).

We met, even in this brief tour, many different accounts and forms of account – not least the crucial global and regional accounts that many of these organisations produce.

There are a number of overwhelming inferences we would draw from this chapter. First, we know too little currently about whether or not many of these organisations are, indeed, discharging their accountability and (to the extent that they do so) how they are doing this. Their influence – not least in establishing and implementing the frameworks within which other entities conduct themselves – is such that a much more critical eye needs to be brought to bear at international, national and local levels. It is also quite apparent that a greater knowledge of these organisations and their practices will greatly influence our understanding of social accounting and accountability. The opportunities for SEA innovation and experimentation are considerable – and the need for such innovation urgent.

References

Adams, C. A. (2013) Sustainability reporting and Performance Management in universities *Sustainability Accounting Management and Policy Journal*, **4**(3) (in press).

Adams, C. A. and Petrella, L. (2010) Collaboration, connections and change: the UN Global Compact, the Global Reporting Initiative, Principles for Responsible Management Education and the Globally Responsible Leadership Initiative, *Sustainability Accounting, Management and Policy Journal*, **1**(2): 292–6.

Albrow, M. and Glasius, M. (2008) Democracy and the possibility of a global public sphere, in Albrow, M., Anheir, H., Glasius, M., Price, M. and Kaldor, M. (eds), *Global Civil Society 2007/8*, pp. 1–18. London: Sage.

Anheier, H. K. (2005) *Nonprofit Organizations: Theory, management, policy*. Abingdon: Routledge.

Anheier, H. K and Hawkes, A. (2008) Accountability in a globalising world, in Albrow, M., Anheir, H., Glasius, M., Price, M. and Kaldor, M. (eds), *Global Civil Society 2007/8*, pp. 124–43. London: Sage.

Anthony, R. N. and Young, D. W. (1984) *Management Control in Nonprofit Organizations*. Homewood: Richard D Irwin.

Awio, G., Northcott, D. and Lawrence, S. (2011) Social capital and accountability in grass-roots NGOs: the case of the Ugandan community-led HIV/AIDS initiative, *Accounting, Auditing and Accountability Journal*, **24**(1): 63–92.

Baker, M. J. (2011) Editorial: why social business?, *Social Business*, **1**(1): 1–15.

Ball, A. (2002) *Sustainability Accounting in UK Local Government: An Agenda for Research*, ACCA Research Report 78. London: CAET.

Ball, A. (2004) *Advancing Sustainability Accounting and Reporting: An Agenda for Public Service Organisations*. London: CIPFA.

Ball, A. (2005) Environmental accounting and change in UK Local Government, *Accounting, Auditing and Accountability Journal*, **18**(3): 346–73.

Ball, A. and Bebbington, J. (2008) Editorial: Accounting and reporting for sustainable development in public sector organizations, *Public Money and Management*, **28**(6): 323–6.

Ball, A. and Grubnic, S. (2007) Sustainability accounting and accountability in the public sector, in Unerman J., Bebbington, J. and O'Dwyer, B. (eds), *Sustainability Accounting and Accountability*, pp. 243–65. London: Routledge.

Ball, A. and Osborne, S. P. (eds) (2011) *Social Accounting and Public Management: Accountability and the common good*. Abingdon: Routledge.

Banerjee, S. B. (2007) *Corporate Social Responsibility: The good, the bad and the ugly*. Cheltenham: Edward Elgar.

Barter, N. and Bebbington, J. (2010) *Pursuing Environmental Sustainability*, Research Report 116. London: ACCA.

Baur, D. and Schmitz, H. P. (2011) Corporations and NGOs: when accountability leads to co-optation, *Journal of Business Ethics*, **106**(1): 9–21.

Bendell, J. (ed.) (2000) *Terms for Endearment: Business, NGOs and Sustainable Development*. Sheffield: Greenleaf /New Academy of Business.

Bennet, N. and van der Lugt, C. (2004) Tracking global governance and sustainability: is the system working?, in Henriques, A. and Richardson, J. (eds), *The Triple Bottom Line: Does it add up?* pp. 45–58. London: Earthscan.

Blowfield, M. and Murray, A. (2011) *Corporate Responsibility* , 2nd Edition. Oxford: Oxford University Press.

Brandsen, T., OudeVrielink, M., Schillemans, T. and van Hout, E. (2011) Non-profit organisations, democratisation and new forms of accountability: a preliminary evaluation, in Ball, A. and Osborne, S. P. (eds), *Social Accounting and Public Management: Accountability and the common good*, pp. 90–102. Abingdon: Routledge.

Broadbent, J. and Guthrie, J. (2008) Public sector to public services: 20 years of 'contextual' accounting research, *Accounting, Auditing and Accountability Journal*, 21(2): 129–69.

Broadbent, J. and Laughlin, R. (2003) Control and legitimation in government accountability processes: the private finance initiative in the UK, *Critical Perspectives on Accounting*, 14(1/2): 23–48.

Broadbent J., Laughlin, R. and Read, S. (1991) Recent financial and administrative changes in the NHS: A Critical Theory analysis, *Critical Perspectives on Accounting*, 2(1): 1–30.

Brown, L. D. and Moore, M. H. (2001) Accountability, strategy, and international nongovernmental organizations, *Nonprofit and Voluntary Sector Quarterly*, 30(3): 569–87.

Butterworth P., Gray, R. H. and Haslam, J. (1989) The Local Authority Annual Report in the UK: an exploratory study of accounting communication and democracy, *Financial Accountability and Management*, 5(2): 73–87.

Cameron, J. and Guthrie, J. (1993) External annual reporting by an Australian university: changing patterns, *Financial Accountability & Management*, 9(1): 1–16.

Capron, M. and Gray, R. H. (2000) Experimenting with assessing corporate social responsibility in France: an exploratory note on an initiative by social economy firms, *European Accounting Review*, 9(1): 99–109.

Chandhoke, N. (2002) The limits of global civil society, in Glasius, M., Kaldor, M. and Anheie, H. (eds), *Global Civil Society 2002*, pp. 35–53. Oxford: Oxford University Press.

Chenhall, R. H., Hall, M. and Smith, D. (2010) Social capital and management control systems: a study of a non-government organization, *Accounting, Organizations and Society*, 35(8): 737–56.

Collison, D., Dey, C., Hannah, G. and Stevenson, L. (2007) Income inequality and child mortality in wealthy nations, *Journal of Public Health*, 29(2): 114–17.

Costa, E., Ramus, T. and Andreaus, M. (2011) Accountability as a managerial tool in non-profit organizations: evidence from Italian CSVs, *Voluntas: International Journal of Voluntary and Nonprofit Organizations*, 22(3): 470–93.

Coy, D. and Pratt, M. (1998) An insight into accountability and politics in universities: a case study, *Accounting, Auditing & Accountability Journal*, 11(5): 540–61.

Crowther, D. and Reis, C. (2011) Social responsibility or social business?, *Social Business*, 1(2): 129–48.

Dey, C. (2007) Social accounting at Traidcraft plc: a struggle for the meaning of fair trade, *Accounting, Auditing and Accountability Journal*, 20(3): 423–45.

Dey, C., Evans, R. and Gray, R. H. (1995) Towards social information systems and bookkeeping: a note on developing the mechanisms for social accounting and audit, *Journal of Applied Accounting Research*, 2(3): 36–69.

Dhanani, A. and Connolly, C. (2012) Discharging not-for-profit accountability: UK charities and public discourse, *Accounting, Auditing and Accountability Journal*, 25(7): 1140–69.

Dixon, K., Coy, D. and Towe, G. (1991) External reporting by New Zealand universities 1985–1989: improving accountability, *Financial Accountability and Management*, 7(3): 159–78.

Dixon, R., Ritchie, J. and Siwale, J. (2006) Microfinance: accountability from the grassroots, *Accounting, Auditing and Accountability Journal*, 19(3): 405–27.

Du, Y and Gray, R. (2013) The emergence of stand alone social and environmental reporting in Mainland China: an exploratory research note, *Social and Environmental Accountability Journal* 33(2) September (pp. 104–112).

Dumay, J., Guthrie, J. and Farnetia, F. (2010) Gri Sustainability Reporting Guidelines For Public And Third Sector Organizations: A critical review, *Public Management Review*, 12(4): 531–48.

Ebrahim, A. (2003a) Making sense of accountability: conceptual perspectives for northern and southern nonprofits, *Nonprofit Management and Leadership*, **14**(2): 191–212.

Ebrahim, A. (2003b) Accountability in practice: mechanisms for NGOs, *World Development*, **31**(5): 813–29.

Ebrahim, A. (2005) Accountability myopia: losing sight of organizational learning, *Nonprofit and Voluntary Sector Quarterly*, **35**(1): 56–87.

Economist (2012) The Big Four accounting firms: shape shifter, *The Economist*, September 29th.

Edwards, M. (2000) *NGO Rights and Responsibilities: A new deal for global governance*. London: The Foreign Policy Centre/NCVO.

Fonseca, L., Macdonald, A., Dandy, E. and Valenti, P. (2011) The state of sustainability reporting at Canadian universities, *International Journal of Sustainability in Higher Education*, **12**(1): 22–40.

Frost, G. and Seamer, M. (2002) Adoption of environmental reporting and management practices: an analysis of New South Wales public sector entities, *Financial Accountability and Management*, **18**: 103–27.

Gibbon, J. (2012) Understandings of accountability: an autoethnographic account using metaphor, *Critical Perspectives on Accounting*, **23**(3): 201–12.

Goetz, A. M. and Jenkins, R. (2001) Hybrid forms of accountability: citizen engagement in institutions of public-sector oversight in India, *Public Management Review*, **3**(3): 363–83.

Gray, R. H. (1984) The NHS treasurer and local accountability, *Public Finance and Accountancy*, April: 30–32.

Gray, R. (2006) Social, environmental, and sustainability reporting and organisational value creation? Whose value? Whose creation?, *Accounting, Auditing and Accountability Journal*, **19**(3): 319–48.

Gray, R. (2010) Is accounting for sustainability actually accounting for sustainability . . . and how would we know? An exploration of narratives of organisations and the planet, *Accounting, Organizations and Society*, **35**(1): 47–62.

Gray, R. and Haslam, J. (1990) External reporting by UK universities: an exploratory study of accounting change, *Financial Accountability and Management*, **6**(1): 51–72.

Gray, R. H., Owen, D. L. and Maunders, K. T. (1987) *Corporate Social Reporting: Accounting and accountability*. Hemel Hempstead: Prentice Hall.

Gray, R. H., Dey, C., Owen, D., Evans, R. and Zadek, S. (1997) Struggling with the praxis of social accounting: stakeholders, accountability, audits and procedures, *Accounting, Auditing and Accountability Journal*, **10**(3): 325–64.

Gray, R., Bebbington, J. and Collison, D. (2006) NGOs, civil society and accountability: making the people accountable to capital, *Accounting, Auditing and Accountability Journal*, **19**(3): 319–48.

Grubnic, S. and Owen, D. (2010) A golden thread for embedding sustainability in a Local Government context: The case of West Sussex County Council, in Hopwood A., Unerman, J. and Frie, J. (eds), *Accounting for Sustainability: Practical Insights*, pp. 95–128. London: Earthscan.

Guthrie, J. and Neumann, R. (2007) Economic and non-financial performance indicators in universities: the establishment of a performance-driven system for Australian higher education, *Public Management Review*, **9**(2): 231–52.

Guthrie, J., Ball, A. and Farneti, F. (2010) Advancing sustainable management of public and not for profit organizations, *Public Management Review*, **12**(4): 449–59.

Harte, G. and Owen, D. L. (1987) Fighting de-industrialisation: the role of local government social audits, *Accounting, Organizations and Society*, **12**(2): 123–42.

Harvey, B. (1984) Managing in the public interest: four British case studies, *Research in Corporate Social Performance and Policy*, **6**: 169–85.

Harvey, B. (1995) Ethical banking: the case of the Co-operative Bank, *Journal of Business Ethics*, **12**(14): 1005–13.

Hofstede, G (1981) Management control of public and not-for-profit activities, *Accounting, Organizations and Society*, **6**(3): 193–216.

Hopwood, A. G. (1978) Social accounting – the way ahead, in *Social Accounting*, pp. 53–64. London: CIPFA.

Hopwood, A., Unerman, J. and Fries, J. (eds) (2010) *Accounting for Sustainability: Practical Insights*. London: Earthscan.

Hunčová, M. (2011) Social and public value measurement and social audit: the Czech experience, in Ball, A. and Osborne, S. P. (eds), *Social Accounting and Public Management: Accountability and the common good*, pp. 254–7. Abingdon: Routledge.

Hyndman, N. and Jones, R. (2011) Good governance in charities – some key issues, *Public Money & Management*, 31(3): 151–5.

Hyndman, H. and McMahon, D. (2011) The hand of government in shaping accounting and reporting in the UK charity sector, *Public Money & Management*, 31(3): 167–74.

Jackson, T. (2009) *Prosperity Without Growth: transition to a sustainable economy*. London: Sustainable Development Commission.

Jepson, P. (2005) Governance and accountability of environmental NGOs, *Environmental Science & Policy*, 8(5): 515–24.

Jia'ning, Y. and ChanHow, K. (2011) NGOs advance corporate social responsibility in China, *The China Nonprofit Review*, 3(1): 99–113.

Jones, H. (2010) *Sustainability Reporting Matters: What are national governments doing about it?* London: ACCA.

Jones, R. and Pendlebury, M. (2010) *Public Sector Accounting*. Harlow: Pearson.

Jordan, L. and Van Tuijl, P. (2006) Rights and responsibilities in the political landscape of NGO accountability: introduction and overview, in Jordan, L. and Van Tuijl, P. (eds), *NGO Accountability, Politics, Principles and Innovations*, pp. 3–20. London: Earthscan.

Joseph, C. and Taplin, R. (2012) Local government website sustainability reporting: a mimicry perspective, *Social Responsibility Journal*, 8(3): 363–72.

Kaldor, M. (2004) Civil society and accountability, *Journal of Human Development*, 4(1): 5–27.

Kim, T. (2011) Contradictions of global accountability: the World Bank, development NGOs, and global social governance, *Journal of International and Area Studies*, 18(2): 23–48.

Kinder, T. (2011) Evolving accountabilities: experiences and prospects from Scottish public services, in Ball, A. and Osborne, S. P. (eds), *Social Accounting and Public Management: Accountability and the common good*, pp. 147–64. Abingdon: Routledge.

Kreander, N., Beattie, V. and McPhail, K. (2009) Putting our money where their mouth is: alignment of charitable aims with charity investments – tensions in policy and practice, *British Accounting Review*, 41(3): 154–68.

Lehman, G. (2007) The accountability of NGOs in civil society and its public spheres, *Critical Perspectives on Accounting*, 18(6): 645–69.

Likierman, A. and Creasey, P. (1985) Objectives and entitlements to rights in government financial information, *Financial Accountability and Management*, 1(1): 33–5.

Liu, X. and Anbumozhi, V. (2009) Determinant factors of corporate environmental information disclosure: an empirical study of Chinese listed companies, *Journal of Cleaner Production*, 17: 593–600.

Lozano, R. (2011) The state of sustainability reporting in universities, *International Journal of Sustainability in Higher Education*, 12(1): 67–78.

Lukman, R. and Glavič, P. (2007) What are the key elements of a sustainable university?, *Clean Technologies and Environmental Policy*, 9(2): 103–14.

Maddocks, J. (2011) Sustainability reporting: a missing piece of the charity-reporting jigsaw, *Public Money & Management*, 31(3): 157–8.

Marcuccio, M. and Steccolini, I. (2005) Social and environmental reporting in local authorities, *Public Management Review*, 7(2): 155–76.

Milne, M. and Gray, R. (2012) W(h)ither ecology? The triple bottom line, the global reporting initiative, and corporate sustainability reporting, *Journal of Business Ethics*, November.

Monfardini, P., Barretta, A. D. and Ruggiero, P. (2013) Seeking legitimacy: social reporting in the healthcare sector, *Accounting Forum*, 37(1): 54–66.

Moon, J. and Shen, X. (2010) CSR in China research: salience, focus and nature, *Journal of Business Ethics*, 94: 613–29.

Moore, D. (2009) Reporting on environmental performance, in UNCTAD (ed.), *Promoting Transparency in Corporate Reporting: A quarter century of ISAR*, pp. 21–52. Geneva: United Nations.

Mussari, R. and Monfardini, P. (2010) Practices of social reporting in public sector and non-profit organizations: an Italian perspective, *Public Management Review*, 12(4): 487–92.

Nicholls, A. (2010) Institutionalizing social entrepreneurship in regulatory space: reporting and disclosure by community interest companies, *Accounting, Organizations and Society*, 35(4): 394–415.

O'Dwyer, B. (2005) The construction of a social account: a case study in an overseas aid agency, *Accounting, Organizations and Society*, 30(3): 279–96.

O'Dwyer, B. (2007) The nature of NGO accountability: motives, mechanisms and practice, in Unerman J., Bebbington, J. and O'Dwyer, B. (eds), *Sustainability Accounting and Accountability*, pp. 285–306. London: Routledge.

O'Dwyer, B. and Unerman, J. (2007) From functional to social accountability: transforming the accountability relationship between funders and non-governmental development organisations, *Accounting, Auditing and Accountability Journal*, 20(3): 446–71.

O'Dwyer, B. and Unerman, J. (2008) The paradox of greater NGO accountability: a case study of Amnesty Ireland, *Accounting, Organizations and Society*, 33(7–8): 801–24.

O'Dwyer, B. and Unerman, J. (2010) Enhancing the role of accountability in promoting the rights of beneficiaries of development NGOs, *Accounting and Business Research*, 40(5): 451–71.

O'Dwyer, B., Unerman, J. and Bradley, J. (2005a) Perceptions on the emergence and future development of corporate social disclosure in Ireland: engaging the voices of non-governmental organisations, *Accounting, Auditing and Accountability Journal*, 18(1): 14–43.

O'Dwyer, B., Unerman, J. and Hession, E. (2005b) User needs in sustainability reporting: perspectives of stakeholders in Ireland, *European Accounting Review*, 14(4): 759–87.

Owen, D. (2008) Chronicles of wasted time? A personal reflection on the current state of, and future prospects for, social and environmental accounting research, *Accounting, Auditing & Accountability Journal*, 21(2): 240–67.

Pearce, J. (1996) *Measuring Social Wealth: A study of social audit practice for community and cooperative enterprises.* London: New Economics Foundation.

Pearce, J. (2003) *Social Enterprise In Anytown.* London: Calouste Gulbenkian Foundation.

Pearce, J. and Kay, A. (2005) *Social Accounting And Audit: The Manual* (and CD). Edinburgh: Social Audit Network.

Perks and Glendinning (1981) Little progress seen in published performance indicators, *Management Accounting*, December: 28–30.

Pineno, C. J. (2011) Sustainability reporting by universities: an integrated approach with the balanced scorecard, *American Society of Business and Behavioral Sciences eJournal*, 7(1): 90–104.

Porritt, J. (2005) *Capitalism: as if the world matters.* London: Earthscan.

Quarter, J., Mook, L. and Richmond, B. J. (2003) *What Counts: Social Accounting for Nonprofits and Cooperatives.* Englewood Cliffs, NJ: Prentice Hall.

Rahman, S. F. (1998) International accounting regulation by the United Nations: A power perspective, *Accounting, Auditing and Accountability Journal*, 11(5): 593–623.

Randers, J. (2012) *2052: A global forecast for the next forty years.* White River Junction, VT: Chelsea Green.

Roeger, K. L., Blackwood, A. and Pettijohn, S. L. (2011) *Public Charities, Giving, and Volunteering, 2011: The non profit sector in brief.* Washington, DC: Urban Institute.

Rowe, A. L. and Guthrie, J. (2010) The Chinese government's formal institutional influence on corporate environmental management, *Public Management Review*, 12(4): 511–29.

Salaman, L. M., Sokolowski, S. W. and Associates (2004) *Global Civil Society: Dimensions of the Nonprofit Sector, Volume Two.* Bloomfield, CT: Kumarian Press.

Stubblefield Loucks, E., Martens, M. L. and Cho, C. H. (2010) Engaging small- and medium-sized businesses in sustainability, *Sustainability Accounting, Management and Policy Journal*, 1(2): 178–200.

SustainAbility (2003) *The 21st century NGO: In the market for change.* London: SustainAbility/The Global Compact/UNEP.

SustainAbility/UNEP (2002) *Trust Us: The 2002 Global Reporters Survey of Corporate Sustainability Reporting.* London/Paris: SustainAbility/UNEP.

Teegen, H., Doh, J. P. and Vachani, S. (2004) The importance of nongovernmental organizations (NGOs) in global governance and value creation: an international business research agenda, *Journal of International Business Studies*, **35**: 463–83.

Thielemann, U. (2000) A brief theory of the market – ethically focused, *International Journal of Social Economics*, **27**(1): 6–31.

Thomson, I. and Georgakopoulas, G. (2010) Building from the bottom up, inspired from the top: accounting for sustainability and the Environment Agency, in Hopwood, A., Unerman, J. and Fries, J. (eds), *Accounting for Sustainability: Practical Insights*, pp. 129–48. London: Earthscan.

Tilbury, D. (2011) Higher education for sustainability: a global overview of commitment and progress, in Global University Network for Innovation (GUNI), *Higher Education's Commitment to Sustainability: From Understanding to Action*, pp. 1–21. Paris: Palgrave Macmillan.

Tort, I. E. (2010) *GRI Reporting in Government Agencies*. Amsterdam: GRI.

UNCTAD (2008) *2008 Review of the reporting status of corporate responsibility indicators*, TD/B/C.II/ISAR/CRP. 2 Intergovernmental Working Group of Experts on International Standards of Accounting and Reporting Twenty-fifth session Geneva, 4–6 November 2008 Item 3.

UNEP (2012) *Global Environmental Outlook (GEO - 5); Measuring Progress, Environmental Goals, and Gaps*. Nairobi: United Nations Environment Programme.

Unerman, J. and O'Dwyer, B. (2006) On James Bond and the importance of NGO accountability, *Accounting, Auditing and Accountability Journal*, **19**(3): 305–18.

Unerman, J. and O'Dwyer, B. (2010) NGO accountability and sustainability issues in the changing global environment, *Public Management Review*, **12**(4): 475–86.

United Nations Millennium Ecosystem Assessment (2005) *Living Beyond Our Means: Natural Assets and Human Well-Being: Statement from the board* (http://www.millenniumassessment.org/en/Products.BoardStatement).

Wapner, P. (2002) Defending accountability in NGOs, *Chicago Journal of International Law*, **3**(1): 197–206.

Webster, A., Shaw, L., Walton, J. K., Brown, A. and Stewart, D. (eds) (2011) *The Hidden Alternative: Co-operative values, past, present and future*. Manchester: Manchester University Press.

Wei, G. and Xiao, J. Z. (2009) Equity ownership segregation, shareholder preferences, and dividend policy in China, *The British Accounting Review*, **41**(3): 169–83.

Weisband, E. and Ebrahim, A. (2007) Introduction: forging global accountabilities, in Ebrahim. A. and Weisband, E. (eds), *Global Accountabilities: Participation, pluralism and public ethics*, pp. 1–24. Cambridge: Cambridge University Press.

Williams, A. P. and Taylor, J. A. (2012) Resolving accountability ambiguity in nonprofit organizations, *Voluntas: International Journal of Voluntary and Nonprofit Organizations* (online).

Williams, B., Wilmhurst, T. and Clift, R. (2011) Sustainability reporting by local government in Australia: current and future prospects, *Accounting Forum*, **35**(3): 176–86.

Woods, N. (2007) Multilateralism and building stronger international institutions, in Ebrahim, A. and Weisband, E. (eds), *Global Accountabilities: Participation, pluralism and public ethics*, pp. 27–44. Cambridge: Cambridge University Press.

Woods, N. and Narlikar, N. (2001) Governance and the limits of accountability: the WTO, the IMF, and the World Bank, *International Social Science Journal*, **53**(170): 569–83.

WWF (2012) *Living Planet Report 2012*. Gland: World Wide Fund for Nature.

Yang, A. H., Lee, P. and Chang, T. Y. (2011) Does self-financing matter? Recalibrating the rationale of NGOs' self-reliance in search of sustainability, *Journal of Asian Public Policy*, **4**(3): 263–78.

CHAPTER 13

Accounting and accountability for responsibility and sustainability: some possible ways forward?

13.1 Introduction and background

Throughout this text we have witnessed, explored and examined the astonishing array of issues, possibilities and practices that comprise social and environmental accounting (SEA) and accountability. There has been a remarkable increase in interest in the area and in the attention given to it in a relatively few years (Gray *et al.*, 1987). As we have seen, it is a rapidly changing area and, to a degree at least, its considerable importance and potential are (occasionally) being recognised. The basic principles do appear to be very widely accepted: namely, that certain forms of SEA can help organisations in their pursuit of their objectives, that organisations should (in some unspecified way) be responsible and that any such responsibility will include pursuit of sustainable development in some form or other. It also might look to be the case that it is finally accepted that organisations should be transparent in their discharge of their accountability.

However, whilst there is considerable progress to celebrate, it would be the height of foolishness to get carried away on a wave of optimism. One only needs to indulge in a little critical analysis to see that this is true. Consider any organisation you know – a multinational corporation (MNC) you buy your shoes from, the grocery store you frequent, the organisation you work for, etc. – and ask yourself a few simple questions. Ask: which responsibilities does this organisation accept and which does it fulfil? Which responsibilities doesn't it accept and/or fulfil? Why? Why not? To what extent is this organisation sustainable? How do I know? How does it resolve conflicts between social, environmental and sustainability criteria and economic criteria? How do you know?

The point is that if social and environmental and sustainability accountability and reporting by organisations was complete and reliable, you would be able to answer those questions thoroughly and with confidence. That you cannot – and we are fairly sure that you cannot except in very rare cases – tells you that whatever accounting and accountability exists is (at best) incomplete and unreliable. And that is a conclusion from which it is pretty difficult to escape.

The most basic point is, as Chapter 11 shows most directly, voluntary initiatives only work if everybody is willing to comply. They are not so willing. The proportions of organisations engaging in anything like substantive social, environmental and sustainability accountability is woefully small. The majority of the world's organisations – even larger organisations – produce pitiful levels of disclosure.

However, even if all organisations did undertake some voluntary systematic social, environmental and/or sustainability disclosure there would still be two issues needing attention. The first is that the standards established to guide disclosure and, to a degree, the management systems that underpin the accounting and disclosure, are still a long way short of what a full accountability requires. That is, ISO standards – notably the 14000 (environmental) and 26000 (social responsibility) series (see Chapters 7 and 5 respectively) – are predominantly management standards and not designed for accountability. Eco-Management and Audit Scheme (EMAS) (see Chapter 7) has both management and disclosure elements but is relatively limited. Both sets of standards talk more generally about improvement rather than absolute impacts. The Global Reporting Initiative (GRI) is a work in progress (see Chapter 9) and, although very successful in its own terms, will tell you nothing about the sustainability of an organisation (Dumay *et al.*, 2010). The Prince of Wales Accounting for Sustainability (A4S) project was concerned with corporate sustainability performance and the International Integrated Reporting Committee is aimed at the providers of capital addressing social, environmental and economic sustainability to the extent to which these are material to the organisation (see Chapter 9). The immensely powerful United Nations Global Compact (UNGC) is explicitly narrow in scope. The essence of the issue is that if an organisation complied fully with any one of these standards (or perhaps even with all of them), it is very uncertain whether one would have any reliable insight into either the organisation's (un-)sustainability or a full specification of the responsibilities that the organisation accepted and fulfilled (Moneva *et al.*, 2006; Milne and Gray, 2012).

This does not make any of these standards 'bad' standards in themselves, but what it does suggest (and this seems relatively un-contentious) is that the standards are designed to engage the support of the organisations whose accountability is being discussed. They are voluntary standards and how could they be otherwise? They are therefore standards designed to help management improve their social, environmental and sustainability behaviours to the extent that they consider apposite. These standards are not – and never have been – principally or primarily designed to enhance accountability to society. They are not designed to call organisations to account. For that, we need to approach the issue from society's point of view.

A further immediate problem then presents itself. In those relatively few instances when organisations do adopt social, environmental and sustainability standards, there is problem of the *extent* to which organisations are voluntarily complying with the standards. This is the problem of assurance (see Chapter 11). Most of the reports are not assured and, despite the best efforts of the AA1000 standards, not enough of those that are assured are assured to a sufficiently high standard. The parallel is with financial reporting – would any financial market believe the protestations of the directors about their financial performance without a substantial financial audit? Probably not (Gray, 2006). Assurance in all its forms is thus crucial to the reporting process (Cooper and Owen, submitted), and the importance of the standards applied in this regard probably cannot be overstated.

It is with this that this brief chapter is concerned. If one really wanted an organisation to address its social, environmental and sustainability management and accountability issues, how would it go about it? Now this must come with a *caveat*; namely, as we have tried to show throughout, a set of full and complete and entirely reliable and un-biased accounts is almost certainly an impossibility. What we try to show here is what represents the best steps (at the present point in time) that can be taken at this point on the journey (apologies for the metaphor, Milne *et al.*, 2006) away from un-sustainability and towards a slightly more nuanced sense of social responsibility.[1]

The chapter is organised as follows. The next section briefly synthesises a few of the approaches that an organisation might take with its internal processes before we move on, in Section 13.3, to look at the issues of stakeholders and 'closeness'. How far can entities and stakeholders cooperate successfully? Section 13.4 then offers an approach to attempting to

---

[1]These ideas are also briefly outlined in Spence and Gray (2007).

discharge the accountability of an entity for its social responsibilities, offering a model that includes a combination of voices. Section 13.5 briefly looks at accountability for sustainability (see Chapter 9) before we offer a brief conclusion in Section 13.6.

13.2 Management control for responsibility and un-sustainability

There is a considerable range of sources of guidance for managers on how their organisation might go about responding to social, environmental and sustainability issues. Whether it be the review of issues from a managerialist point of view (see, for example, Smith and Lennsen, 2009; Bansal and Hoffman, 2012), direct guidance on application (see, for example, Brady *et al.*, 2011; Moratis and Cochius, 2011) or direct support from business organisations and the bodies that produce the standards (see, for example, WBCSD, 2010), organisations do not lack for guidance and direction. The guidance, very broadly, tends to examine how to develop appropriate information systems (environmental management systems being the most obvious) having identified the relevance of the appropriate social, environmental and un-sustainability issues for the organisation's goals and strategy (although it is rarely straightforward – see Adams and Larrinaga-González, 2007; Adams and Whelan, 2009). Such guidance may typically be embedded in or directly orientated towards standards like ISO14000, ISO26000 or GRI for example. After all, as we asserted above (and hope to have shown throughout the text), these standards are principally designed to engage managers and to support them in their understanding of and response to social, environmental and sustainability issues.

In the same vein, any reporting that flows from this appropriately managerialist approach is bound to be managerially orientated. Such reporting will be primarily designed as an addition to the organisation's own strategic use of external information – whether (for example) as a legitimating, attention deflecting, impression managing or employee-informing device.

Accounting will play a role at this basic level. We have seen the capacity of management accounting to (for example) identify potential cost savings or refine and explore new areas of investment appraisal (see, for example, Chapter 7). In addition, we have seen that financial accounting has capacities to help the organisation manage its liabilities and respond appropriately to financial risk and the concerns of the financial community.

All of this is well provided for and well understood – the only uncertainties here are about the future and how new and emerging social, environmental or political issues may impinge on the organisation's functioning. For the organisation wishing to address the challenges in a more substantial manner (Adams and Whelan, 2009), there would seem to be three major routes.

The first route is that being blazed by the champions of environmental management accounting and costing in particular. These innovators have systematically explored how reaching the full potential of environmental management (including techniques such as life cycle assessment) and (what the Germans tend to call) **eco-controlling** offers the possibilities of taking organisations beyond the simple business case (Schaltegger and Burritt, 2000; Schaltegger *et al.*, 2008). The emphasis in these initiatives is upon exploring more imaginative ways in which economic and (typically) environmental considerations can be brought more into line. Carbon accounting is one such area illustrated in Figure 13.1.

If this first strategy might be thought of as an iterative seeking out of new forms of **eco-efficiency** (see Chapter 7) and developing new ways in which an organisation might develop its understanding of win–win, the second route is far more strategic and entrepreneurial (see also Adams and Whelan, 2009). What might be thought of as the 'social entrepreneur' blurs the lines between social business (see Chapter 12) and conventional business and determines

Figure 13.1 An illustration of how management accounting might push beyond the simple business case

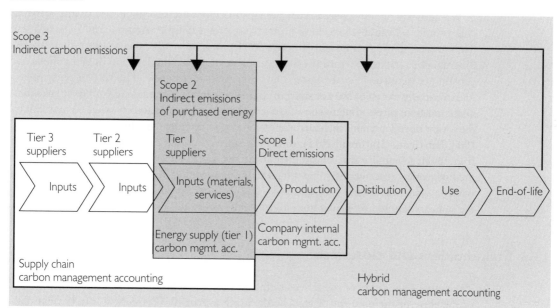

Note: The different scopes or levels of the 'GHG Protocol' challenge corporate managers to address the different aspects of their carbon impacts with the use of different carbon management accounting tools.

Source: Taken from Schaltegger and Csutora, (2012: 11).

to re-invent an economically sound business model that seeks to make positive social and environmental contributions (see, for example, Elkington and Hartigan, 2008). Whilst such optimism is to be applauded, it attracts a substantial level of scepticism as to whether or not such achievements are possible (see, for example, Young and Tilley, 2006; Barter and Bebbington, 2010). What does not yet seem clear is what accounting and SEA as currently understood have to offer here (yet another area needing development perhaps).

The third route is arguably the more difficult but, to our minds, potentially the most promising. This is the approach championed most vividly by Ernst von Weizsäcker in *Factor Four* (Weizsäcker *et al.*, 1997) and later in *Factor Five* (Weizsäcker *et al.*, 2009). It is an approach which initially steps outside the confines of the organisation and asks what would the organisation have to do if it were to be focused on responsibility and sustainability. (Such an approach accepts the possibility that the organisation might cease to exist – this is never an option in managerialist conceptions.) In so doing, the approach starts from a fundamental recognition of the potential for genuine and deep-rooted conflicts between (say) advertising encouraging increased consumption and disposal and a finite planet; increasing a firm's output whilst reducing its ecological footprint; and so on. Once these conflicts are recognised, then the manager has a series of difficult choices to make which, in essence, encourage the organisation to consider what it can potentially do; what it *cannot do under present circumstances* and, most innovatively, what needs to be done to allow a less un-sustainable activity (e.g. lobby for increased legislation, reduce the claims it makes, reduce the expectations of its shareholders and so on). One can find the essence of these approaches running through work such as Herbohn (2005), Bebbington (2007), McElroy and van Engelen (2012) as well as in the Forum for the Future developments that led up to and were incorporated in the initial plans of the A4S project.

We need say no more here except perhaps to emphasise two important points. First, as Adams and colleagues have long argued (see, for example, Adams, 2002; Adams and McNicholas, 2007; Adams and Whelan, 2009), there is insufficient research that engages directly – but critically – with the manager in an attempt to understand the limitations and constraints facing any organisation – however well-intentioned. The second point relates more centrally to one of the key *motifs* of this text: namely that a full and apposite accountability and transparency would allow organisations to explain to society what 'responsibility' they were simply unable (or perhaps unwilling in some cases) to adopt. As Bebbington and Gray (2001) and Gray and Bebbington (2000) show, there are many elements of a path away from un-sustainability that are simply outside the control of the individual organisation. To hold managers to account for that which they demonstrably have no power over is not productive: a full accountability would make this obvious. And it is in this regard that consulting stakeholders and reporting honestly and fully have such an important role to play.

13.3 Stakeholders and closeness[2]

Throughout our study of social accounting and accountability we have made extensive and explicit use of the concept of stakeholders (see especially Chapters 3 and 4). You may, indeed, have detected – either explicitly or implicitly – a number of different theoretical perspectives governing the roles that stakeholders might be assumed to take (see also Friedman and Miles, 2006). The most inclusive of these perspectives seeks to get away from the idea of an organisation having fixed boundaries and tries to imagine internal and external stakeholders in an evolving partnership through which the organisation is manifest and directed. Chapter 12, suggests some instances of this where the existence of professional and epistemic communities offered occasions when distinctions between external and internal participants and normative and positive stakeholder theory could begin to blur. It seems likely that social enterprises and other small organisations with high degrees of *closeness* can also manifest aspects of this view. Whether larger commercial organisations can operate like this is perhaps more contentious (but see, for example, Zadek *et al.*, 1997).

At the opposite end of the spectrum is a view that sees nothing but conflict between organisations and stakeholders (see Hudson and Harris, 2013). In essence, an organisation is a battleground between, in the first place, the directors and the shareholders, and then between managers and employees and then between the organisation and its external stakeholders. The key stakeholders are those that have salience – that is, those who can influence the organisation for their own ends (Mitchell *et al.*, 1997). Understanding the organisation is then a question of power and influence: who has that power and influence? Shareholders will nearly always be fairly dominant, as will directors, but communities will only occasionally manage to create salience. Consequently, the organisation will be run for those who have the

---

[2]This would be a good place to make a small confession. The notion of 'closeness' relates to the notion that the closer individuals and groups are physically, intellectually, professionally and in terms of their values the less formal need be the mechanisms of accountability. The discharge of accountabilities can arise casually or even non-verbally between peoples. We have used the notion extensively in our work and have always attributed the notion to John Rawls (Rawls, 1972). A recent re-reading of Rawls reveals that Rawls' notion is one of 'close knitness' not 'closeness'. They are not the same idea and, whilst our notion of 'closeness' may well be Rawlsian in intent, it is not taken from Rawls directly. It looks as though we have introduced a notion and coined a phrase which, whilst it works, has no academic pedigree outside our own work. Apologies for any confusion caused as a result.

power and as those most exercised by social, environmental and sustainability issues tend to be relatively un-powerful, the assumption must be that an organisation (at least a commercial organisation) cares considerably more about economic issues than about social and environmental ones. Such issues are typically in conflict and, typically, economic issues dominate.

Which view one subscribes to will influence the role one envisages for stakeholders in supporting organisations in a more responsible and less un-sustainable mode of operation (Murray *et al.*, 2010), and it is broadly in the space between these two extremes that important initiatives such as the AA1000 series of standards sit (see, for example, Forstater *et al.*, 2007). At its ideal, stakeholder engagement encourages a degree of convergence between the needs and wants of the organisation and its various constituencies, and it is this that documents like *AA1000 Stakeholder Engagement Standard 2011* are designed to encourage. Whatever else, there seems to be no question that the development of any substantial accountability and/or sustainability must encompass stakeholder consultation as a *sine qua non* (Cooper and Owen, 2007).

That stakeholder engagement and consultation are essential elements of social accountability does not necessarily mean, though, that such consultation is always effective or that it is all that is required (Owen *et al.*, 2001). Questioning the impetus for and the outcomes of stakeholder engagement – especially when it is recognised that the power, knowledge and time differentials may endanger any idea of a balanced and mutually reinforcing relationship (Forstater *et al.*, 2007) – remains essential to understanding accountability, not least because there is always that nagging doubt that stakeholders may not indeed know what responsibility the organisations should follow or what responsibility is due to them. It also looks very probable that, even if an organisation was managed entirely in harmony with its principal stakeholders, it would be no more likely to be environmentally sustainable (although it might be expected to be rather more socially sustainable).

The role for and of stakeholders is therefore not clear cut. That they have an essential role is obvious, quite how that role might be best understood and made manifest is less so. But in one regard above all, the accountability requires that the voice of the stakeholder is understood clearly as an essential element of the accountability relationship. It is this to which we now turn.

13.4 Social accountability to society

It is apposite to remind ourselves at this point about two important distinctions that influence how we proceed. First, we continue to try and distinguish between 'social responsibility' and 'sustainability'. Whilst the two notions overlap, they are not the same.[3] Social responsibility, as Friedman so eloquently demonstrates, is a social construction, a matter of opinion and political judgement. It is therefore a relative idea and it is perfectly possible to argue that a corporation should see its primary social responsibility as making money for shareholders, maintaining decent employment standards and increasing/satisfying the wants of its customers. This may well be a version of social responsibility, but it is probably not sustainable in the social and environmental sense as we have used it throughout the book. Equally, whilst sustainability is obviously socially constructed to a degree (see, for

---

[3]Although, as we have noted, it is difficult to think of a social responsibility that did not include a duty to at least attempt to pursue sustainability, one would expect any discharge of social accountability to include some reference to a sustainability and the social justice component of sustainability might well be approximated (as we do here) by corporate social responsibility (CSR).

example, Joseph, 2012), it also has a major empirical element – child starvation, species extinction and so on are clearly empirical phenomena as well as being clearly not sustainable. But by the same token, it is as well to note that to act sustainably in a manner consonant with deep ecology would probably be seen as socially *irresponsible* in some quarters. For this reason we will address social responsibility in this section and briefly revisit sustainability accounting in the next section.

The second matter we should briefly remind ourselves about relates to the notion of the environment (nature) as a stakeholder. We have just seen that the voices of stakeholders are important, but how does nature get a voice? And if nature does not shout as loud as the community and the shareholders will it be drowned out? These are not simple questions. In essence, some explicit steps must be taken to ensure that nature's 'voice' is recognised: for some this might be achieved through the environmental NGOs, for others some means of representing local ecologies (for example) might be developed.[4] For Dillard (2007) each human being, to have a legitimate claim on a voice, must explicitly have an ethic of accountability – an awareness of their place in society and nature. This ideal would ensure that we spoke not as economic or short-term self-interested creatures but as mature and civilised human beings. Dillard's exhortations are well made and need to be borne in mind in what follows.

There are many ways in which accounts of social responsibility might be constructed as we have seen throughout the text: indeed the GRI and UNGC approaches have something of this in them. Other approaches have been suggested.[5] One such is the **compliance-with-standard** approach which takes law and quasi law as the minimum of the terms of accountability and demands that organisations report on the extent to which they have – or have not – complied with those statutes and standards. This basic idea has merit, but it is partial and so we have incorporated it into the idea of the stakeholder map which we develop here.[6]

Figure 13.2 represents one of the few approaches that seem capable of addressing the crucial question of how a social account can be presumed to be 'complete' – or as near complete as we can make it. This simplified representation is a **stakeholder map** – a depiction of the different groups with which the entity interacts. (The solid shapes behind each stakeholder group are intended to indicate that stakeholder groups are not themselves homogeneous either.) A sensible accountability will recognise that, for example, 'employees' is not a homogeneous category and senior management at plush offices in a western city may not have the same relationship with the entity as temporary manufacturing workers employed in a developing country.

If, as we have argued, accountability derives from relationships, then we need to identify the full array of relationships to which an entity must pay attention. The stakeholder map does this (or should do this).[7] The next stage is identifying what information is needed in

---

[4]One approach to this is ensuring that at all meetings concerned with stakeholders there is a physical representation of threatened species and oppressed people. One academic, Tom Gladwin, would bring cardboard cut outs of fish and Somalian children to a meeting to ensure they were not forgotten.

[5]Chapter 10 looks, briefly, at the idea of the 'silent account' which is the collation of an organisation's existing social and environmental disclosure (typically spread around a variety of sources) reporting into a single place to produce a first pass at a CSR report. This approach is probably better than nothing and is almost costless, but it lacks any measure of *completeness*.

[6]This next section is drawn extensively from the Traidcraft experience which still represents in our view one of the more serious attempts at discharging a substantial accountability (Dey *et al.*, 1995; Gray *et al.*, 1997; Dey, 2007).

[7]A number of organisations have employed the stakeholder map as a basis for their reporting, Cooperative Financial Services in the UK being one of the most prominent in this regard – that experience having been transferred from Traidcraft itself.

Figure 13.2 A simplified stakeholder map

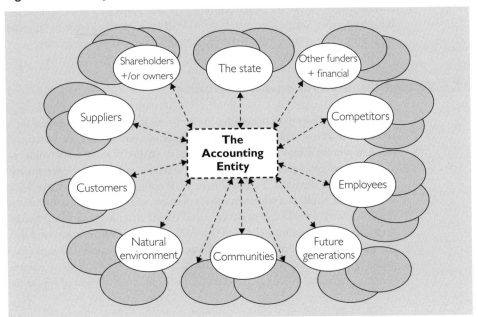

each channel – each relationship – if the discharge of accountability is to take place. This is less easy than it might sound – what are the terms of each relationship and who defines them? Gray *et al.* (1997) suggested that to approach a full accountability each relationship would need four categories of information relating to that relationship. These are shown in Figure 13.3.

A full accountability report would contain all of this data on each of the four categories for each relationship. Now it is quite apparent that this could easily become a ridiculous burden on any entity – regardless of its size or intentions. Consequently, we might expect a report to focus on certain of these relationships and on certain aspects of these relationships. That is, one can easily see that the discharge of accountability is very *unlikely* to be *complete . . . but* each reader will be able to assess the degree of incompleteness.

Figure 13.3 Types of information needed in a full CSR report

- *Descriptive data*: raw basic data which describes the parameters of the relationship, e.g. numbers and categories of employees.
- *Entity-defined/preferred data*: this is the entity's own voice and should be heard – data which the company believes is important – related to its mission perhaps.
- *Society-defined statute and standards information*: these are the basic rules of the relationship and, consequently, the data to which the parties to the relationship have a right (e.g. law and quasi-law – comprising such things as industry codes, voluntary and UN standards – plus conformance with the entity's own mission statement and standards).
- *The stakeholder voices*: what information do the parties to the relationship believe they have a right to?

Few organisations have managed to produce CSR reports to anything like this level of completeness[8] and so the opportunity for serious leadership in the matter of social responsibility reporting is still very much alive. Will (commercial) entities voluntarily adopt these standards of reporting, disclosure and accountability? It seems unlikely.

13.5 Accountability and un-sustainability

It should be apparent that nothing in the foregoing section examining social responsibility reporting addresses any fundamental aspects of sustainability or sustainable development (see Chapter 9). That is the entity's accountability for its social responsibility and its accountability for its un-sustainability, although clearly related and perhaps even overlapping notions, are not identical. To develop the CSR report that is suggested in Section 13.4 in order that it might begin to embrace un-sustainability needs an account of un-sustainability. For this, the entity would need to develop and report accounts which, alongside the stakeholder map and the information related to those relationships that are expressed by such a map, also encompassed, something like an eco-balance and an ecological footprint (see Chapter 7) plus some attempt to account for social justice (see Chapter 9). Such accounts are scarce and likely to remain so as long as these matters are voluntary. We need repeat no more here.

13.6 Ways forward? Conclusions? And the need for shadow accounts?

As virtually no organisations, of which we are aware, have delivered either of the sorts of accounts that would be necessary to envisage the extent of their un-sustainability (and our suggestions are only part of the available possibilities) and as few have managed to approach the levels required for a proper substantive CSR account, it is difficult not to conclude that organisational accountability is a long way from reaching fruition. That no entity (that we know of) has done both rather emphasises the point. As Owen (2008) concludes, accountability is a truly radical notion and the way forward, in our view, is for entities to begin to address these matters intelligently and to avoid the extensive and pathological claims to being responsible and sustainable with no evidence to support them.

After over 50 years of interest in social responsibility and over two decades of explicit engagement with standalone reporting, two decades of apparent concern for both stakeholders and the issues of environmental accountability, there is a sense in which the worldwide progress is unimpressive. It is precisely this lack of progress, despite all the bluster and pomp surrounding 'sustainability' issues, that the external social audits and the shadow accounts are designed to address (see Chapter 10). In this connection, they become so very important indeed. If organisations cannot willingly be accountable, and if governments will not hold organisations to account, then civil society will need to undertake this task itself.

In this regard, it is valuable to note that the range of external social audits (reviewed in Chapter 10) is really only the tip of the possible iceberg of new and challenging accounts. Figure 13.4 lists a few of the newer ideas that begin to suggest how a civil society might begin to offer different, new and more challenging accounts of the world we all inhabit.

---

[8]Although note the Traidcraft Exchange Report on the CSEAR website under 'Approaches to Practice' and some of the exemplars offered by the Social Accounting Network (see Chapter 12).

Figure 13.4 Some of the newer forms of accounts emerging from civil society

- Accounts of capitalism
  - e.g. Collison *et al.* (2007, 2010)
- Accounts of un-sustainability
  - e.g. Gray (2006)
- Accounts of the oppressed/silenced
  - e.g. Cooper *et al.* (2005)
- Accounts of the profession/corruption
  - e.g. Sikka (2010a, b)
- Silent + shadow accounts
  - e.g. Gibson *et al.* (2001); Dey *et al.* (2011)
- External social audits
  - e.g. Harte and Owen (1987)
- Counter accounts
  - e.g. Gallhofer *et al.* (2006); Steiner (2010)
- Performance-portrayal gaps
  - e.g. Adams (2004)
- Accounts for sustainability
  - e.g. Bebbington (2007)
- Accounts as imagining
  - e.g. Davison and Warren (2009)
- Regional accounts of water, air, land, etc.
  - e.g. Lewis and Russell (2011)

References

Adams, C. A. (2002) Internal organisational factors influencing corporate social and ethical reporting, *Accounting, Auditing and Accountability Journal,* **15**(2): 223–50.

Adams, C.A. (2004) The ethical, social and environmental reporting-performance portrayal gap, *Accounting, Auditing and Accountability Journal,* **17**(5): 731–57.

Adams, C.A. and Larrinaga-González, C. (2007) Engaging with organisations in pursuit of improved sustainability accountability and performance, *Accounting, Auditing and Accountability Journal,* **20**(3): 333–55.

Adams, C.A. and McNicholas, P. (2007) Making a difference: sustainability reporting, accountability and organisational change, *Accounting, Auditing and Accountability Journal,* **20**(3): 382–402.

Adams, C.A. and Whelan, G. (2009) Conceptualising future changes in corporate sustainability reporting, *Accounting, Auditing and Accountability Journal,* **22**(1): 118–43.

Bansal, P. and Hoffman, A. J. (eds) (2012) *The Oxford Handbook of Business and The Natural Environment.* Oxford: Oxford University Press.

Barter, N. and Bebbington, J. (2010) *Pursuing Environmental Sustainability,* Research Report 116. London: ACCA.

Bebbington, J. (2007) *Accounting for Sustainable Development Performance.* London: CIMA.

Bebbington, K.J. and Gray, R. H. (2001) An account of sustainability: failure, success and a reconception, *Critical Perspectives on Accounting,* **12**(5): 557–87.

Brady, J., Ebbage, A. and Lunn, R. (eds) (2011) *Environmental Management in Organizations: The IEMA Handbook,* 2nd Edition. London: Earthscan.

Collison, D., Dey, C., Hannah, G. and Stevenson, L. (2007) Income inequality and child mortality in wealthy nations, *Journal of Public Health*, **29**(2): 114–17.

Collison, D., Dey, C., Hannah, G. and Stevenson, L. (2010) Anglo-American capitalism: the role and potential of social accounting, *Accounting, Auditing and Accountability Journal*, **23**(8): 956–81.

Cooper, C., Taylor, P., Smith, N. and Catchpowle, L. (2005) A discussion of the political potential of social accounting, *Critical Perspectives on Accounting*, **16**(7): 951–74.

Cooper, S. M. and Owen, D. L. (2007) Corporate social reporting and stakeholder accountability: the missing link, *Accounting, Organizations and Society*, **32**(7/8): 649–67.

Cooper, S. M. and Owen D. L. (submitted) Independent assurance practice of sustainability reports, in Unerman, J,. Bebbington, J. and O'Dwyer, B. (eds), *Sustainability Accounting and Accountability*, 2nd Edition. Abingdon: Routledge (forthcoming).

Davison, J. and Warren, S. (2009) Imag(in)ing accounting and accountability, *Accounting, Auditing and Accountability Journal*, **22**(6): 845–57.

Dey, C. (2007) Social accounting at Traidcraft plc: a struggle for the meaning of fair trade, *Accounting, Auditing and Accountability Journal*, **20**(3): 423–45.

Dey, C., Evans, R. and Gray, R. H. (1995) Towards social information systems and bookkeeping: A note on developing the mechanisms for social accounting and audit, *Journal of Applied Accounting Research*, **2**(3): 36–69.

Dey, C., Russell, S. and Thomson, I. (2011) Exploring the potential of shadow accounts in problematising institutional conduct, in Osborne, S. and Ball, A. (eds), *Social Accounting and Public Management: Accountability for the common good*, pp. 64–75. Abingdon: Routledge.

Dillard, J. (2007) Legitimating the social accounting project: an ethic of accountability, in Unerman J., Bebbington, J. and O'Dwyer, B. (eds), *Sustainability Accounting and Accountability*, pp. 37–54. London: Routledge.

Dumay, J., Guthrie, J. and Farneti, F. (2010) GRI sustainability reporting guidelines for public and third sector organisations, *Public Management Review*, **12**(4): 531–48.

Elkington, J. and Hartigan, P. (2008) *The Power of Unreasonable People: How social entrepreneurs create markets that change the world*. Boston, MA: Harvard Business Press.

Forstater, M., Dupre, S., Oelschlaegel, J., Tabakian, P. and de Robillard, V. (2007) *Critical Friends: The Emerging Role of Stakeholder Panels in Corporate Governance, Reporting and Assurance*. London: AccountAbility.

Friedman, A. L. and Miles, S. (2006) *Stakeholder: Theory and practice*. Oxford: Oxford University Press.

Gallhofer S., Haslam, J., Monk, E. and Roberts, C. (2006) The emancipatory potential of online reporting: the case of counter accounting, *Accounting, Auditing and Accountability Journal*, **19**(5): 681–718.

Gibson K., Gray, R., Laing, Y. and Dey, C. (2001) *The Silent Accounts project: Draft Silent and Shadow Accounts Tesco plc 1999–2000*. Glasgow: CSEAR (www.st-andrews.ac.uk/management/csear).

Gray, R. (2006), Social, environmental, and sustainability reporting and organisational value creation? Whose value? Whose creation?, *Accounting, Auditing and Accountability Journal*, **19**(3): 319–48.

Gray, R. H. and Bebbington, K. J. (2000) Environmental accounting, managerialism and sustainability: is the planet safe in the hands of business and accounting?, *Advances in Environmental Accounting and Management*, **1**: 1–44.

Gray, R. H., Owen, D. L. and Maunders, K. T. (1987) *Corporate Social Reporting: Accounting and accountability*. Hemel Hempstead: Prentice Hall.

Gray, R. H., Dey, C., Owen, D., Evans, R. and Zadek, S. (1997) Struggling with the praxis of social accounting: stakeholders, accountability, audits and procedures, *Accounting, Auditing and Accountability Journal*, **10**(3): 325–64.

Harte, G. and Owen, D. L. (1987) Fighting de-industrialisation: the role of local government social audits, *Accounting, Organizations and Society*, **12**(2): 123–42.

Herbohn, K. (2005) A full cost accounting experiment, *Accounting, Organizations and Society*, **30**(6): 519–36.

Hudson, L. J. and Harris, M. (2013) CSR and collaboration, in Haynes, K., Murray, A. and Dillard, J. (eds), *Corporate Social Responsibility: A research handbook*, pp. 41–7. London: Routledge.

Joseph, G. (2012) Ambiguous but tethered: an accounting basis for sustainability reporting, *Critical Perspectives on Accounting*, **23**: 93–106.

Lewis, L. and Russell, S. (2011) Permeating boundaries: accountability at the nexus of water and climate change, *Social and Environmental Accountability Journal*, **31**(2): 117–23.

McElroy, M. W. and van Engelen, J. M. L. (2012) *Corporate Sustainability Management: The art and science of managing non-financial performance*. London: Earthscan.

Milne, M. and Gray, R. (2012) W(h)ither ecology? The triple bottom line, the Global Reporting Initiative, and corporate sustainability reporting, *Journal of Business Ethics*, November

Milne M. J., Kearins, K. N. and Walton, S. (2006) Creating adventures in wonderland? The journey metaphor and environmental sustainability, *Organization*, **13**(6): 801–39.

Mitchell, R., Agle, B. R. and Wood, D. J. (1997) Toward a theory of stakeholder identification and salience: defining the principle of who and what really counts, *Academy of Management Review*, **22**(4): 853–86.

Moneva, J. M., Archel, P. and Correa, C. (2006) GRI and the camouflaging of corporate unsustainability, *Accounting Forum*, **30**(2): 121–37.

Moratis, L. and Cochius, T. (2011) *ISO 26000: The business guide to the new standard on social responsibility*. Sheffield: Greenleaf.

Murray, A., Haynes, K. and Hudson, L. (2010) Collaborating to achieve corporate social responsibility and sustainability? Possibilities and problems, *Sustainability, Accounting, Management and Policy Journal*, **1**(2): 161–77.

Owen, D. (2008) Chronicles of wasted time? A personal reflection on the current state of, and future prospects for, social and environmental accounting research, *Accounting, Auditing and Accountability Journal*, **21**(2): 240–67.

Owen, D. L., Swift, T. and Hunt, K. (2001) Questioning the role of stakeholder engagement in social and ethical accounting, auditing and reporting, *Accounting Forum*, **25**(3): 264–82.

Rawls, J. (1972) *A Theory of Justice*. Oxford: Oxford University Press.

Schaltegger, S. and Burritt, R. (2000) *Contemporary Environmental Accounting. Issues, Concepts and Practice*. Sheffield: Greenleaf.

Schaltegger, S. and Csutora, M. (2012) Carbon accounting for sustainability and management: Status quo and challenges, *Journal of Cleaner Production*, **36**(1): 1–16.

Schaltegger, S., Bennett, M., Burritt, R. and Jasch, C. (eds) (2008) *Environmental Management Accounting for Cleaner Production*. Dordrecht: Springer.

Sikka, P. (2010a) Smoke and mirrors: corporate social responsibility and tax avoidance, *Accounting Forum*, **34** (3–4): 153–68.

Sikka, P. (2010b) Using the media to hold accountants to account: some observations, *Qualitative Research in Accounting and Management*, **7** (3): 270–80.

Smith, N. C. and Lenssen, G. (eds) (2009) *Mainstreaming Corporate Responsibility*. Chichester: Wiley.

Spence, C. and Gray, R. (2007) *Social and Environmental Reporting and the Business Case*. London: ACCA.

Steiner, R. (2010) *Double standard: Shell practices in Nigeria compared with international standards to prevent and control pipeline oil spills and the Deepwater Horizon oil spill*. Amsterdam: Friends of the Earth.

WBCSD (2010) *Translating ESG into sustainable business value: key insights for companies and investors*. Geneva: World Business Council for Sustainable Development and UNEP Finance Initiative.

Weizsäcker, E. Von, Lovins, A. B. and Lovin, L. H. (1997) *Factor Four: Doubling Wealth, Halving Resource Use*. London: Earthscan.

Weizsäcker, E. Von, Hargroves, K. C., Smith, M. H., Desha, C. and Stasinopoulos, P. (2009) *Factor Five: Transforming the Global Economy through 80% improvements in resource productivity*. London: Earthscan.

Young, W. and Tilley, F. (2006) Can business move beyond efficiency? The shift toward effectiveness and equity in the corporate sustainability debate, *Business Strategy and the Environment*, **15**: 402–15.

Zadek, S., Pruzan, P. and Evans, R. (1997) *Building Corporate Accountability: Emerging Practices in Social and Ethical Accounting, Auditing and Reporting*. London: Earthscan.

CHAPTER 14

What next? A few final thoughts

14.1 Introduction

At the very heart of the theory, practice and study of social (and environmental, ethical and sustainability) accounting lie two very serious conundrums.

The first of these is to do with the nature of (conventional) accounting itself and, by association, with notions of business and management as well. At a micro-level (at the level of the individual or the individual entity), the efficacy of accounting is probably judged by its ability to help us manage the finances and cash flow of the organisation and help us make decisions that will, broadly, maximise our wealth and perhaps that of shareholders. Accounting and the accounting profession occupies a very special place in all developed economies, often enjoying a monopoly position and typically supported by a wide range of legislation. It enjoys this situation because at the macro- (nation-state) level it is thought that a 'well-run economy' is an essential part of a civilised state and, further, that accounting and its complex paraphernalia are essential components of such an economy. This may be correct – to a degree at least. Accounting's efficacy is judged at this level by the extent to which it is thought to be in the public interest and, if conventional accounting is satisfying the public interest, then we might well be able to conclude (as do most businesses, accountants and governments) that there is no actual need for social accounting. If, on the other hand, one comes to a conclusion that species extinction, vast inequality, climate change, over-consumption and a major democratic deficit are not in the public interest – and consequently that social accounting is an essential element of any complex and civilised society[1] – then one may well be led to conclude that conventional accounting is probably fatally flawed. Pushed to its extreme, the only ground for opposing social accounting is that it offers nothing of value that conventional systems of accounting do not already offer. This is clearly not so, and the resistance for nearly half a century to social accounting at business, government and professional levels is altogether more troublesome. If business and government and the accounting professions had genuinely backed substantive social accounting, the present situation would likely be very different indeed. As we have shown throughout the text, it is perfectly feasible and practicable.

The second conundrum sort of follows from the first and relates to the fact that social, environmental and sustainability accounting have to be seen simultaneously as trivial *and* exceptionally important. Many of you will have recognised this by now. Although some aspects of social accounting – most obviously those to do with either the calculation and reporting of liabilities in financial accounting and the exploitation of win–win situations in

---

[1]There is an important but much simpler argument that in smaller, less complex and much 'closer' societies, there would be very little use for accounting generally and any need for social accounting would probably be discharged informally.

environmental management accounting (see Chapter 7) – are clearly substantive and have real impacts, the majority of opportunities in social accounting and reporting remain at the level of experiment, suggestions or occasional exhibitions. The levels of substantive social, environmental and sustainability accounting and reporting, as we have seen, are very low and the quality of that accounting and reporting is itself often relatively unimpressive. Clearly and incontrovertibly, there is no danger of wider social accountability being discharged any time soon. Thus, in this sense, we could see social accounting as fairly trivial. However, two other issues offer a completely different view. The first is that expressed so directly by Owen (2008): namely that accountability – whether for social responsibility or sustainability – is amongst the most radical notions of which modern society can conceive. Its potential to clarify so much of what we see as the increasing parade of un-sustainable and irresponsible actions by large organisations, the opportunities to add transparency to the claims made by entities and the possible insights into the range of environmental and social issues the planet and its peoples face on a daily basis would have considerable potential to transform social and economic relationships – nationally and globally. Relatedly, then, the enormous efforts expended by business, government and the professions to ensure that such substantive accountability cannot come to fruition looks all the more telling. Powerful people the world over appear to know just how transformative a full accountability and transparency would be – and work very hard to prevent it coming to fruition. It seems difficult to avoid this conclusion – however unsettling it might be.

These conundrums bring us back to the key issue with which we started. That is, by examining individual and organisational social, environmental and sustainability relationships we are confronted with the most basic of questions about such things as: who are we?, what sort of society do we live in?, and just how many of our taken-for-granted assumptions about business, economics, growth, democracy and wealth are actually supportable or defensible? It is this that primarily makes the study of social accounting so very engaging and potentially so important. It is also this which can, despite the vast array of examples of good news, potentially depress one about the broad lack of progress.

So what can we do about it?

This final and very brief chapter offers a few of the ways forward that we can see and, in so doing, suggests that counsels of despair are of no use to us here. This chapter then, in effect, passes the baton over to you the reader – we are getting old and tired – now it is your turn to do what you can (if anything). We suggest, again very briefly, three areas of action: education (which is touched upon in the next section), engagement, experimentation and challenge (which is touched upon in Section 14.3) and finally a brief note on activism and counter-narratives in Section 14.4.

14.2 Education

Teachers and researchers involved in social accounting have frequently also had a strong interest in education itself. To a degree at least, this seems to be so because any growing disquiet about the conventional 'wisdom' of accounting and business inevitably leads to a questioning about the foundations of that 'conventional wisdom' and, from there, to asking 'how did it come about?' Once one begins to ask awkward questions, then one begins to see some of the absurdities that are taught as 'common sense' in business and accounting. A dissatisfaction with accounting will often lead directly to social accounting. If accounting and business is the problem, then social accounting may well be the answer. This then directs us towards trying to deconstruct what we were/are taught and how might it all be taught differently.

For many years now, accounting has been subjected to a barrage of critique, and the way it is taught has attracted a lot of scholarly and analytical attention (Mayper *et al.*, 2005). This has, in turn, led to considerable attention being given to what is taught and why (Owen *et al.*, 1994; Spence, 2007); how it is taught (Thomson and Bebbington, 2004); and how more exciting and engaging approaches to education might be developed (Coulson and Thomson, 2006).

For teachers, this should, you might think, offer fantastic opportunities to break free of the shackles of conventional accounting and business teaching (Springett, 2005; Stubbs and Cocklin, 2008). After all, what is education for? Really? Taken across the board, education is inevitably some combination of both the inculcation of skills and knowledge *and* the development of the thinking muscle – the development of analysis and critical thinking. Whilst different educations will emphasise more of one or the other, in most circumstances connected with applied and professional disciplines such as management, business, accounting and finance, we might well expect a balance between skills and critical thinking. Indeed, it is probably our moral duty as educators to reflect upon how we balance the 'received wisdom' and the analytical and critical – to reflect upon what it is that we are doing with the minds of young students (McPhail, 2004). Social accounting offers remarkable potential to explore both the practicable and the theoretical in an articulate and critical framework and in a personally challenging manner.

If, however, social accounting is to be seen for the liberating and emancipatory and even essential innovation that it is, it requires a much wider consideration amongst accounting and business teachers and a recognition that current dogma does not represent the best of all possible worlds. A commitment to critical analysis of existing practices and dogma will, inevitably, lead to a questioning and that allows the *possibilities* of social accounting to at least be considered. Great progress has been made in recent decades, but there is clearly a lot further yet to go.

The most important challenge of the present is clearly that of sustainable development, and, as we have seen, social accounting has a great deal to offer in this regard. The promulgation of sustainable development education by, for example, the United Nations Educational, Scientific and Cultural Organization (UNESCO) 'aims to help people to develop the attitudes, skills, perspectives and knowledge to make informed decisions and act upon them for the benefit of themselves and others, now and in the future'[1]. Such development of attitudes and understanding through education must recognise the central roles played by economics, business and accounting in un-sustainability and the possibilities of a sustainable future. Understanding and initiating social accounting is potentially a major part of that process. One important element of this arises, we would suggest, from the potential disjunction between the global and development initiatives and the actions and policies of international finance and business in (what is loosely called) globalisation.

> Globalisation, as defined by rich people like us, is a very nice thing . . . you are talking about the Internet, you are talking about cell phones, you are talking about computers. This doesn't affect two-thirds of the people of the world.
>
> (Ex-US President Jimmy Carter on the UNESCO website)

Education at all levels clearly needs to address these issues and, without a substantive attention to these issues in formal education, it is difficult to see how matters are likely to change much. In this regard, both the universities and the professional bodies have influential roles to play and teachers who are often connected to both are key to emancipation in this regard. Universities and the professional bodies (especially through the professional syllabi) have been both liberating and restrictive. Such restriction needs very careful analysis. But there is also the less formal matter of education in practice.

---

[1](UNESCO website http://www.unesco.org/education/tlsf/mods/theme_c/mod18.html).

14.3 Engagement, experimentation and challenge

One of the things we have sought to do is to chronicle the considerable and often startling developments that have occurred in practice. It is quite apparent that it is through the efforts of individuals and groups, understanding the issues and looking for ways to improve matters that many of the arresting initiatives, in communities, environmental management and finance for example (see Chapters 5, 7, and 8, respectively), have pushed the boundaries of the possible. The first task, then, for all of us is to seek out ways in which we can make a contribution in our workplaces, in our professions and in our classrooms.

This is not only a matter of teasing out win–win situations for organisations. It is at least as much about imagination and experimentation. It seems highly likely that the really effective and influential financial initiatives and the really serious accounts of sustainability have yet to be imagined, let alone developed. But developed they must be, and active experimentation needs to be much more widely embraced than it has been for the last 30 years if serious innovation is to be explored and initiated.

But, a word of warning. This cannot be achieved without conflict. Personal experience tells us, somewhat ruefully, that many people simply do not want to hear why they need to change or why current ways of doing things are un-sustainable or that a more just orientation is needed. It is actually fairly obvious that there will continue to be major resistance to the ideas that we have sought to represent. Many of the ideas and suggestions have been around for nearly 50 years and humanity has still barely got this social accountability off the ground.

Little will attract aggression in this field more directly than calls for law (as we know to our cost). It is quite clear that governance needs to be legally and/or regulation driven and that such legal backing must be imaginative and substantial. The opposition to such a call for law – especially in some countries and regions of the world – verges on the fanatical.

If you are going to become someone who seeks to educate themselves, their workplace, their classroom and/or their profession, then we are sorry to have to say that you should expect unpleasantness in the process. Sometimes, however, the resistance and unpleasantness are just so over-whelming that it might then be at least as productive to meet the issue head on and engage in explicit activism.

14.4 Activism and counter-narratives

Ultimately, the principal objective of social accounting is to inform and enhance democracy through encouraging the substantive discharge of wider social, environmental and sustainability accountabilities. The means of achieving this is through the production of new accounts that tell, if not the 'truth' then, either better truths or more balanced truths. If organisations were to be required to regularly produce such accounts (as opposed to the frequently vacuous and selective occasional output we are currently blessed with), it seems highly likely that our societies would be transformed. Such transformation would be, we would hope, for the better. Encouraging organisations to produce these accounts seems to us to be a noble cause and can achieve much. We might never know for certain what brought standalone reporting to the fore, or how socially responsible investment (SRI) became the force it now is or why carbon accounting was (in principle at least) so widely accepted, but there were many dedicated and thoughtful individuals pushing away at these initiatives well before they became manifest and more widely accepted. The ability to achieve social and environmental aims – however distant they might seem – is always with us as a possibility (albeit a frequently frustrating and frustrated possibility).

If one cannot achieve the aims of social, environmental and sustainability accountability through direct involvement, active participation in committees and professional initiatives or personal contact with government and politicians, then there is a need to embark upon (at least) two somewhat more oppositional paths. As Wright argues, change comes and will come from the niches inside capitalism (Wright, 2010). The first of these paths involves the active seeking out of those niches in capitalism. The public sector, non-governmental organisations (NGOs) and trades unions, churches and other groups are all committed to a view of humanity which is not exclusively about economic efficiency and consumption, a view which has space for justice and ecology. These 'niches' in our experience often have relatively little understanding of finance and accounting and are often fairly naïve about business. Whether exploring oppositional or cooperative strategies, one can bring one's social accounting to the party in order to add to the efforts of these groups and institutions.

The second of these paths comprises the development and dissemination of the external social audits. As we have seen, these represent a very diverse and challenging approach to social accounting and, they can be undertaken by any individual or group with the time and motivation (see Chapter 10, and the CSEAR website). These accounts challenge claims, bring pressure to bear and offer new and alternative narratives about the world in which we live. They give voice to the oppressed and silenced whilst challenging those with the loudest voices. At their best, they encourage cooperation and new initiatives such as that between Oxfam and Unilever. (For more detail see, for example, Medawar, 1976.)

If the state and the market won't supply the essential social, environmental and sustainability accountability, civil society will need to do so. Social accounting will be an important part of that process.

References

Coulson, A. and Thomson, I. (2006) Accounting and sustainability, encouraging a dialogical approach; integrating learning activities, delivery mechanisms and assessment strategies, *Accounting Education: an International Journal*, **5**(3): 261–73.

Mayper, A. G., Pavur, R. J., Merino, B. and Hoops, W. (2005) The impact of accounting education on ethical values: an institutional perspective, *Accounting in the Public Interest*, **5**: 32–55.

McPhail, K. (2004) An emotional response to the state of accounting education: developing accounting students' emotional intelligence, *Critical Perspectives on Accounting*, **15**: 629–48.

Medawar, C. (1976) The social audit: a political view, *Accounting, Organizations and Society*, **1**(4): 389–94.

Owen, D. (2008) Chronicles of wasted time? A personal reflection on the current state of, and future prospects for, social and environmental accounting research, *Accounting, Auditing and Accountability Journal*, **21**(2): 240–67.

Owen, D., Humphrey, C. and Lewis, L. (1994) *Social and Environmental Accounting Education in British Universities*, Certified Research Report, 3. London: ACCA.

Spence, C. (2007) Social and environmental reporting and hegemonic discourse, *Accounting, Auditing and Accountability Journal*, **20**(6): 855–81.

Springett, D. (2005) 'Education for sustainability' in the business studies curriculum: a call for a critical agenda, *Business Strategy & the Environment*, **14**(3): 146–59.

Stubbs, W. and Cocklin, C. (2008) Teaching sustainability to business students: shifting mindsets, *International Journal of Sustainability in Higher Education*, **9**(3): 208–21.

Thomson, I. and Bebbington, J. (2004) It doesn't matter what you teach?, *Critical Perspectives on Accounting*, **15**(4–5): 609–28.

Wright, E. O. (2010) *Envisioning Real Utopias*. London: Verso.

Glossary

A4S	Accounting for Sustainability (Prince of Wales)	**HDI**	Human Development Index
AAA	American Accounting Association	**HRA**	human resource accounting
ACCA	Association of Chartered Certified Accountants	**HRM**	Human Resource Management
AICPA	American Institute of Certified Public Accountants	**ICA**	intellectual capital accounting
		ICAEW	Institute of Chartered Accountants in England and Wales
ANT	actor network theory		
BSC	balanced scorecard	**ICAS**	Institute of Chartered Accountants in Scotland
CBI	Confederation of British Industry	**ICC**	International Chamber of Commerce
CCI	corporate community investment	**IDEW**	Index of Sustainable Economic Welfare
CERES	Coalition for Environmentally Responsible Economies	**IFC**	International Finance Corporation
		IIRC	International Initiative to Reassure Capitalism
CICA	Canadian Institute of Chartered Accountants	**IMF**	International Monetary Fund
CRE	Commission for Racial Equality	**Inc**	Incorporated
CSEAR	The Centre for Social and Environmental Accounting Research	**INGO**	international NGO
		IR	indescribable rubbish
CSO	civil society organisation	**ISO**	International Standards Organisation
CSR	corporate social responsibility (sometimes reporting)	**LDC**	lesser developed countries
		Ltd	limited liability
DCF	discounted cash flow	**MDG**	Millennium Development Goals
EA	Environment Agency	**MNC**	multi-national company/corporation
ECMH	efficient capital market hypothesis		
EMAN	Environmental and Sustainability Management Accountign Network	**NATO**	North Atlantic Treaty Organisation
		NGO	non-governmental organisation
EMAS	Eco-Management and Audit Scheme	**NIC**	newly industrialised country
EOC	Equal Opportunities Commission	**OECD**	Organisation for Economic Cooperation and Development
ESG	environmental, social and governance		
ESOP	employee share ownership scheme	**OFR**	Operating and Financial Review
EU	European Union	**plc**	public listed company (UK)
FDI	foreign direct investment	**PLED**	'pristine' liberal economic democratic
GC	Global Compact (UN)	**RDT**	resource dependency theory
GDP	gross domestic product	**SEA**	social and environmental accounting
GmbH	Gesellschaft mit beschränkter Haftung	**SEAR**	Social and Environmental Accounting and Reporting
GNP	gross national product		
GRI	Global Reporting Initiative	**SEER**	Social, Environmental and Ethical Reporting
GST	General Systems Theory		
HCM	human capital management	**SER**	Social and Environmental Reporting

SME	small and medium-sized enterprise
SOE	state-owned enterprise
SRI	socially responsible investment
TBL	triple bottom line
TNC	Trans-national Corporation
TUC	Trades Union Congress
UN	United Nations
UNCED	United Nations Conference on Environment and Development
UNCTAD	United Nations Conference on Trade and Development

UNCTC	United Nations Center for Transnational Corporations
UNEP	United Nations Environment Programme
UNESCO	United Nations Educational, Scientific and Culturial Organisation
UNGC	United Nations Global Compact
UNPRI	United Nations Principles of Responsible Investment
WBCD	World Council for Sustainable Development
WEF	World Economic Forum
WWF	World Wide Fund for Nature

Index